From Molecules to Living Organisms:
An Interplay Between Biology and Physics

École de Physique des Houches
Session CII, 7 July–1 August 2014

From Molecules to Living Organisms: An Interplay Between Biology and Physics

Edited by

Eva Pebay-Peyroula, Hugues Nury, François Parcy,
Rob W. H. Ruigrok, Christine Ziegler,
Leticia F. Cugliandolo

OXFORD
UNIVERSITY PRESS

Great Clarendon Street, Oxford, OX2 6DP,
United Kingdom

Oxford University Press is a department of the University of Oxford.
It furthers the University's objective of excellence in research, scholarship,
and education by publishing worldwide. Oxford is a registered trade mark of
Oxford University Press in the UK and in certain other countries

© Oxford University Press 2016

The moral rights of the authors have been asserted

First Edition published in 2016

All rights reserved. No part of this publication may be reproduced, stored in
a retrieval system, or transmitted, in any form or by any means, without the
prior permission in writing of Oxford University Press, or as expressly permitted
by law, by licence or under terms agreed with the appropriate reprographics
rights organization. Enquiries concerning reproduction outside the scope of the
above should be sent to the Rights Department, Oxford University Press, at the
address above

You must not circulate this work in any other form
and you must impose this same condition on any acquirer

Published in the United States of America by Oxford University Press
198 Madison Avenue, New York, NY 10016, United States of America

British Library Cataloguing in Publication Data

Data available

Library of Congress Control Number: 2015944582

ISBN 978–0–19–875295–0

Links to third party websites are provided by Oxford in good faith and
for information only. Oxford disclaims any responsibility for the materials
contained in any third party website referenced in this work.

École de Physique des Houches

Service inter-universitaire commun
à l'Université Joseph Fourier de Grenoble
et à l'Institut National Polytechnique de Grenoble

Subventionné par l'Université Joseph Fourier de Grenoble,
le Centre National de la Recherche Scientifique,
le Commissariat à l'Énergie Atomique

Directeur:
Leticia F. Cugliandolo, Sorbonne Universités, Université Pierre et Marie Curie
Laboratoire de Physique Théorique et Hautes Energies, Paris, France

Directeurs scientifiques de la session:
Eva Pebay-Peyroula, Institut de Biologie Structurale, CEA-CNRS Université Joseph
  Fourier, France
Hugues Nury, Institut de Biologie Structurale, CEA-CNRS Université Joseph
  Fourier, France
François Parcy, Laboratoire de Physiologie Cellulaire et Végétale, CNRS-CEA-
  INRA-Université Joseph Fourier, France
Rob W. H. Ruigrok, Unit for Viral Host Cell Interactions, Université Joseph
  Fourier -EMBL-CNRS, France
Christine Ziegler, Institute of Biophysics and Physical Biochemistry, University of
  Regensburg, Germany
Leticia F. Cugliandolo, Sorbonne Universités, Université Pierre et Marie Curie
  Laboratoire de Physique Théorique et Hautes Energies, Paris, France

Previous Sessions

I	1951	Quantum mechanics. Quantum field theory
II	1952	Quantum mechanics. Statistical mechanics. Nuclear physics
III	1953	Quantum mechanics. Solid state physics. Statistical mechanics. Elementary particle physics
IV	1954	Quantum mechanics. Collision theory. Nucleon–nucleon interaction. Quantum electrodynamics
V	1955	Quantum mechanics. Non equilibrium phenomena. Nuclear reactions. Interaction of a nucleus with atomic and molecular fields
VI	1956	Quantum perturbation theory. Low temperature physics. Quantum theory of solids. Ferromagnetism
VII	1957	Scattering theory. Recent developments in field theory. Nuclear and strong interactions. Experiments in high energy physics
VIII	1958	The many body problem
IX	1959	The theory of neutral and ionized gases
X	1960	Elementary particles and dispersion relations
XI	1961	Low temperature physics
XII	1962	Geophysics; the earth's environment
XIII	1963	Relativity groups and topology
XIV	1964	Quantum optics and electronics
XV	1965	High energy physics
XVI	1966	High energy astrophysics
XVII	1967	Many body physics
XVIII	1968	Nuclear physics
XIX	1969	Physical problems in biological systems
XX	1970	Statistical mechanics and quantum field theory
XXI	1971	Particle physics
XXII	1972	Plasma physics
XXIII	1972	Black holes
XXIV	1973	Fluids dynamics
XXV	1973	Molecular fluids
XXVI	1974	Atomic and molecular physics and the interstellar matter
XXVII	1975	Frontiers in laser spectroscopy
XXVIII	1975	Methods in field theory
XXIX	1976	Weak and electromagnetic interactions at high energy
XXX	1977	Nuclear physics with heavy ions and mesons

XXXI	1978	Ill condensed matter
XXXII	1979	Membranes and intercellular communication
XXXIII	1979	Physical cosmology
XXXIV	1980	Laser plasma interaction
XXXV	1980	Physics of defects
XXXVI	1981	Chaotic behavior of deterministic systems
XXXVII	1981	Gauge theories in high energy physics
XXXVIII	1982	New trends in atomic physics
XXXIX	1982	Recent advances in field theory and statistical mechanics
XL	1983	Relativity, groups and topology
XLI	1983	Birth and infancy of stars
XLII	1984	Cellular and molecular aspects of developmental biology
XLIII	1984	Critical phenomena, random systems, gauge theories
XLIV	1985	Architecture of fundamental interactions at short distances
XLV	1985	Signal processing
XLVI	1986	Chance and matter
XLVII	1986	Astrophysical fluid dynamics
XLVIII	1988	Liquids at interfaces
XLIX	1988	Fields, strings and critical phenomena
L	1988	Oceanographic and geophysical tomography
LI	1989	Liquids, freezing and glass transition
LII	1989	Chaos and quantum physics
LIII	1990	Fundamental systems in quantum optics
LIV	1990	Supernovae
LV	1991	Particles in the nineties
LVI	1991	Strongly interacting fermions and high Tc superconductivity
LVII	1992	Gravitation and quantizations
LVIII	1992	Progress in picture processing
LIX	1993	Computational fluid dynamics
LX	1993	Cosmology and large scale structure
LXI	1994	Mesoscopic quantum physics
LXII	1994	Fluctuating geometries in statistical mechanics and quantum field theory
LXIII	1995	Quantum fluctuations
LXIV	1995	Quantum symmetries
LXV	1996	From cell to brain
LXVI	1996	Trends in nuclear physics, 100 years later
LXVII	1997	Modeling the earth's climate and its variability
LXVIII	1997	Probing the Standard Model of particle interactions
LXIX	1998	Topological aspects of low dimensional systems
LXX	1998	Infrared space astronomy, today and tomorrow
LXXI	1999	The primordial universe
LXXII	1999	Coherent atomic matter waves

LXXIII	2000	Atomic clusters and nanoparticles
LXXIV	2000	New trends in turbulence
LXXV	2001	Physics of bio-molecules and cells
LXXVI	2001	Unity from duality: Gravity, gauge theory and strings
LXXVII	2002	Slow relaxations and nonequilibrium dynamics in condensed matter
LXXVIII	2002	Accretion discs, jets and high energy phenomena in astrophysics
LXXIX	2003	Quantum entanglement and information processing
LXXX	2003	Methods and models in neurophysics
LXXXI	2004	Nanophysics: Coherence and transport
LXXXII	2004	Multiple aspects of DNA and RNA
LXXXIII	2005	Mathematical statistical physics
LXXXIV	2005	Particle physics beyond the Standard Model
LXXXV	2006	Complex systems
LXXXVI	2006	Particle physics and cosmology: The fabric of spacetime
LXXXVII	2007	String theory and the real world: From particle physics to astrophysics
LXXXVIII	2007	Dynamos
LXXXIX	2008	Exact methods in low-dimensional statistical physics and quantum computing
XC	2008	Long-range interacting systems
XCI	2009	Ultracold gases and quantum information
XCII	2009	New trends in the physics and mechanics of biological systems
XCIII	2009	Modern perspectives in lattice QCD: quantum field theory and high performance computing
XCIV	2010	Many-body physics with ultra-cold gases
XCV	2010	Quantum theory from small to large scales
XCVI	2011	Quantum machines: measurement control of engineered quantum systems
XCVII	2011	Theoretical physics to face the challenge of LHC
Special Issue	2012	Advanced data assimilation for geosciences
XCVIII	2012	Soft interfaces
XCIX	2012	Strongly interacting quantum systems out of equilibrium
C	2013	Post-Planck cosmology
CI: Special Issue:	2013:	Quantum optics and nanophotonics Statistical physics, optimization, inference and message-passing algorithms
CII:	2014:	From molecules to living organisms: An interplay between biology and physics

Publishers

Session VIII: Dunod, Wiley, Methuen
Sessions IX and X: Herman, Wiley
Session XI: Gordon and Breach, Presses Universitaires
Sessions XII–XXV: Gordon and Breach
Sessions XXVI–LXVIII: North Holland
Session LXIX–LXXVIII: EDP Sciences, Springer
Session LXXIX–LXXXVIII: Elsevier
Session LXXXIX–: Oxford University Press

Preface

The 4-week school held in July 2014 addressed current approaches and concepts for understanding the relation between the regulation of gene expression, synthesis and structural assembly of proteins, the forces that dictate the structural, dynamic and functional properties of protein complexes and the properties of cells and their interactions in the formation of tissues and organisms. Students and early stage researchers from different fields of biology, physics and chemistry attended these multidisciplinary lectures. Several practical courses and seminars were also organized.

From molecules to cells and organisms

With the generalized application of genomics and proteomics the molecular level has become an important aspect of biology. At the core of molecular biology is the structure–function hypothesis. In recent years, new imaging approaches allowing the visualization of single molecular complexes within cells have paved the way for further understanding of biological processes at the molecular level. Integrating the detailed molecular description of individual proteins or complexes into the cellular environment and understanding the structures, architectures and interactions that guide fundamental cellular pathways and cellular responses to external stimuli are major challenges for the near future. At higher levels of organization, organisms that adopt precise shapes are built up from cells. Although several signaling pathways that guide development have been identified, many aspects cannot be explained just by the chemical nature of the molecules involved. Indeed, in several examples, physical forces have been shown to be responsible for the shapes of cells or cellular compartments and cellular the assemblies forming organisms. The generation of these forces is often related to structural assemblies of macromolecules and their dynamical rearrangement.

The context of this school is the emerging field of integrated biology (biomolecule↔cell↔organism) in the light of recent advances in cellular biophysics and modeling approaches. In particular, bridging data from different types of approaches and the provision of information on various scales (space and time) are far from well established. The aim of the school was to teach these new topics. The audience comprised structural biologists looking towards cell and organismal biology, biologists interested in the molecular view of biological pathways and physicists interested in biological processes for both the biology and the physics underlying the biology. Understanding the principles behind each method, and also their limitations and the complementarity between methods, was an important aim of the lectures. A large number of the interdisciplinary lectures were on subjects at the frontier between biology and physics.

A new way of thinking and teaching biology: a combination of interdisciplinarity and cutting-edge methods

Important new developments are expected in the coming years that may well introduce paradigm shifts in biological science. This school aimed to prepare participants to become major actors in these breakthroughs. It looked at opening a new way of teaching (and thinking) biology, bridging physics and biology beyond current biophysics. This book contains the proceedings of the main lectures given at the school. After an introduction to cell biology in Chapter 1 (Franz Bruckert), the power of integrated approaches from molecules to cells and organisms, including imaging, biophysics and structural biology, is illustrated through two examples. In Chapter 2 Hans-Georg Kräusslich and colleagues, demonstrate how the interactions between HIV and host cells can be deciphered, while in Chapter 3 François Parcy and his team highlight how floral development can be understood from the gene to the flower. Concepts in physics such as thermodynamics, important for understanding the behavior of biological macromolecules in solution, are recalled by Giuseppe Zaccai in Chapter 4, and in Chapter 6 Albert Guskov and Dirk Jan Slotboom show how some aspects of these behaviors can be experimentally characterized. Emerging and novel approaches such as in-cell NMR are described by Enrico Luchinat in Chapter 5. The next part of the book is dedicated to plant development, from innovative biological approaches, described by George Coupland in Chapter 7, to experimental evidence of the role of forces in plant development by Olivier Hamant in Chapter 8, and mathematical modeling based on this experimental knowledge by Christophe Godin and colleagues in Chapter 9. Forces also drive the shapes of membranes and their remodeling, as described in Chapter 10 by Michael Kozlov, Winfried Weissenhorn and Patricia Bassereau. These authors nicely illustrate the complementarity between experiments exploring physical parameters of proteins embedded in membranes, theoretical modeling based on physical principles and applications to a biological question, namely the budding of viruses out of host-cell membranes. The most predominant molecules in cells are proteins, and their shapes but also their conformational changes are responsible for their functional properties, as analyzed in Chapter 11 by Yves Gaudin. Membrane proteins are naturally embedded in a lipid bilayer with mesoscopic properties. Their handling necessitates special treatments, as shown by Christine Ziegler in Chapter 12. François Dehez, in Chapter 13, illustrates how such studies have benefited from an ensemble of tools based on molecular simulations in the light of experimental work. Chapters are grouped into six parts as indicated in contents list to facilitate structured reading. Altogether, the chapters show how the examination of a biological system from different viewpoints in a multidisciplinary fashion often brings new ideas to controversial arguments. Please note, that the online version of this book provides color figures that will be helpful to the reader.

Science and art

Observation was a key element in the development of biology, and more specifically botany. By drawing what they observed in the field, botanists could deduce important features in plants and define the various classifications. Nowadays, structural biologists spend a substantial amount of their time examining three-dimensional protein

structures on a computer screen in order to relate structural features to function by comparison with structures that are already known. Serge Aubert guided the participants through the alpine garden of the Col des Montets, illustrating the recent findings on molecular-level adaptation of plants to harsh mountain conditions that he presented during the conference. Anja Kieboom organized a few afternoon sessions during which she demonstrated some basic techniques for drawing flowers. These practical sessions were very successful and contributed to social team building. Graphic printouts of protein structures as well as plant drawings are not only informative, they are also intended to be attractive in themselves. Art in science contributes to the message that scientists aim to deliver.

While we were finishing this book, our colleague Serge Aubert died. Serge was a very talented and passionate scientist and he liked to share this passion with students. We would like to dedicate this book to him.

Acknowledgments

This Les Houches summer school was made possible by substantial financial support from:

- French-German University (UFA)
- Centre National de la Recherche Scientifique (CNRS)
- GRAL, a Labex program from the Investissements d'Avenir (IA-ANR)
- FRISBI, an Infrastructure program from the Investissements d'Avenir (IA-ANR)
- Instruct, a European network of infrastructures in integrated structural biology (ESFRI)
- Institut National de Recherche Agronomique (INRA)

This financial support made a large contribution to the summer school, in particular to the funding of several students, and permitted a broad international participation.

The organizers wish to express their gratitude to the scientific committee. In particular, discussions with Yves Gaudin, Giuseppe Zaccai, Lucia Banci and Gideon Schreiber helped to refine the program and find appropriate speakers. The organizers are very grateful to the speakers who dedicated their time to the school and gave excellent and highly appreciated courses. The participants also deserve thanks: by their questions during the lectures, coffee breaks and at other times and discussion sessions they organized in the evening they were the major contributors to the excellent ambience during the month, despite some very unpleasant weather. Finally, the organizers would also like to thank the administrative staff of the "Les Houches" physics school, Murielle Gardette, Isabelle Lelièvre and Flora Gheno, for their support before, during and after the session, as well as the restaurant staff; technical assistance from Jeff Aubrun also contributed to the success of the school.

E. Pebay-Peyroula
H. Nury
F. Parcy
R. Ruigrok
C. Ziegler
L. F. Cugliandolo
Grenoble, Paris, Regensburg, February 2015

Contents

List of participants	xxi
Part 1 Concepts in cell biology and examples of multiscale studies in biology	1

1 Introduction to cell biology
Franz BRUCKERT — 3
- 1.1 Levels of organization in cells — 5
- 1.2 Protein localization within cells — 13
- 1.3 Protein activation in cells — 16
- 1.4 Practical conclusions — 22
- References — 24

2 A small leak will sink a great ship: HIV–host interactions
Nikolas HEROLD, Hans-Georg KRÄUSSLICH, and Barbara MÜLLER — 25
- 2.1 Introduction — 27
- 2.2 HIV assembly and release — 29
- 2.3 HIV maturation — 31
- 2.4 HIV entry — 33
- 2.5 Future of integrative HIV research — 38
- Acknowledgement — 39
- References — 39

3 Floral development: an integrated view
Hicham CHAHTANE, Grégoire DENAY, Julia ENGELHORN, Marie MONNIAUX, Edwige MOYROUD, Fanny MOREAU, Cristel CARLES, Gabrielle TICHTINSKY, Chloe ZUBIETA, and François PARCY — 43
- 3.1 Description of floral development — 47
- 3.2 Genetic control of floral development in Arabidopsis — 54
- 3.3 Transcription and chromatin dynamics in the "seq" era — 61
- 3.4 Modeling TF–DNA binding to predict transcriptional regulation — 82
- 3.5 Layers of chromatin regulation and their actors — 86
- 3.6 Floral transcription factors: a structural view — 94
- References — 100

Part 2 Concepts in physics and emerging methods   117

4   Thermodynamics (a reminder)
Giuseppe (Joe) ZACCAI   119
- 4.1   Brief historical review   121
- 4.2   Why structural biologists need to be reminded about thermodynamics   121
- 4.3   Calorimetry   122
- 4.4   Current methods for the study of biological macromolecules in solution   123
- 4.5   Thermodynamics of aqueous solutions   124
- 4.6   Macromolecular solutions   127
- 4.7   A thermodynamics approach to scattering methods: macromolecule–solvent interactions (hydration and solvation)   129
- 4.8   Relating the thermodynamics and particle approaches   130
- 4.9   NMR and neutrons give molecular meaning to thermodynamics parameters   131
- References   132

5   NMR spectroscopy: from basic concepts to advanced methods
Enrico LUCHINAT   134
- 5.1   Basic concepts of NMR spectroscopy   136
- 5.2   Advances in protein NMR   148
- 5.3   In-cell NMR: towards integrated structural cell biology   157
- References   165

6   Size exclusion chromatography with multi-angle laser light scattering (SEC-MALLS) to determine protein oligomeric states
Albert GUSKOV and Dirk Jan SLOTBOOM   169
- 6.1   Introduction   171
- 6.2   Multi-angle light scattering   174
- 6.3   Applications   178
- 6.4   Outlook   180
- References   180

Part 3 Plant development: from genes to growth   185

7   Mechanisms controlling time measurement in plants and their significance in natural populations
George COUPLAND   187
- 7.1   Introduction   189
- 7.2   The plant circadian clock   189
- 7.3   Seasonal timing   194
- 7.4   Timing in natural populations   198
- 7.5   Conclusion   201
- References   202

8	**Forces in plant development** **Olivier HAMANT**	**209**
	8.1 Overview	211
	8.2 Understanding the role of forces at the tissue scale	221
	8.3 Understanding the role of forces at the cell scale	236
	8.4 Conclusion	242
	References	243
9	**An introduction to modeling the initiation of the floral primordium** **Christophe GODIN, Eugenio AZPEITIA, and Etienne FARCOT**	**247**
	9.1 Introduction	250
	9.2 Specifying growth points on meristem domes	251
	9.3 Modeling the regulation of floral initiation	265
	9.4 Conclusion	277
	References	278

	Part 4 Forces in biology: reshaping membranes	**285**
10	**Membrane remodeling: theoretical principles, structures of protein scaffolds and forces involved** **Michael M. KOZLOV, Winfried WEISSENHORN, and Patricia BASSEREAU**	**287**
	10.1 Introduction	290
	10.2 Theoretical principles of membrane remodeling	291
	10.3 Structural basis for membrane remodeling by a protein scaffold	304
	10.4 Forces involved in remodeling biological membranes	314
	Acknowledgments	331
	References	331

	Part 5 Conformational changes and their implications in diseases	**351**
11	**Protein conformational changes** **Yves GAUDIN**	**353**
	11.1 Protein conformational changes	355
	11.2 Conformational changes in viral fusion proteins	364
	11.3 From conformational diseases to cell memory	377
	References	388

	Part 6 Membrane transporters: from structure to function	**397**
12	**The dos and don'ts of handling membrane proteins for structural studies** **Christine ZIEGLER**	**399**
	12.1 Introduction	401
	12.2 Membrane proteins—Fragile! Handle with care!	401

    12.3 Determining the structure of a membrane protein: difficult
         but not impossible   408
    References   411

13 **Molecular simulation: a virtual microscope in the toolbox
    of integrated structural biology**
    François DEHEZ   413
    13.1 Introduction   415
    13.2 Modeling forces in molecular simulation: the force field   417
    13.3 Modeling the time evolution of biological systems using
         molecular dynamics   421
    13.4 Integrating molecular simulations with structural biology   426
    13.5 Conclusion   432
    References   433

List of participants

Organizers

PEBAY-PEYROULA Eva
Institut de Biologie Structurale, Université Joseph Fourier, CNRS, CEA, Grenoble, France

ZIEGLER Christine
Department of Membrane Protein Crystallography, Faculty of Biology and Preclinical Studies, University of Regensburg, Germany

NURY Hugues
Institut de Biologie Structurale, Université Joseph Fourier, CNRS, CEA, Grenoble, France

PARCY François
Laboratoire de Physiologie Cellulaire et Végétale, iRTSV, CNRS, Université Joseph Fourier, CEA, INRA, Grenoble, France

RUIGROK Rob
Unit of Virus Host Cell Interactions, Université Joseph Fourier, CNRS, EMBL, Grenoble, France

Lecturers

BASSEREAU Patricia
Institut Curie, Centre de Recherche, Paris, France

BRUCKERT Franz
Grenoble-INP Phelma, France

COUPLAND George
Max Planck Institute for Plant Breeding Research, Cologne, Germany

DEHEZ François
Laboratoire International Associé CNRS, France; University of Illinois Urbana-Champain, IL, USA; SRSMC, Université Lorraine, CNRS, Nancy France

GAUDIN Yves
Institute for Integrative Biology of the Cell, Université Paris-Saclay, CEA, CNRS; Université Paris-Sud, Gif-sur-Yvette, France

GODIN Christophe
INRIA, Virtual Plants Inria-Cirad-Inra Team, Montpellier, France

HAMANT OLIVIER
Plant Reproduction and Development lab. ENS Lyon France

KOZLOV MICHAEL
Department of Physiology and Pharmacology, Sackler Faculty of Medicine, Tel Aviv University, Israel

KRÄUSSLICH, HANS-GEORG
Department of Infectious Diseases, Virology, University Hospital Heidelberg, Germany

LUCHINAT ENRICO
Magnetic Resonance Center, CERM and Department of Biomedical, Clinical and Experimental Sciences, University of Florence, Italy

PARCY FRANÇOIS
Laboratoire de Physiologie Cellulaire et Végétale, iRTSV, CNRS, Université Grenoble Alpes, CEA, INRA, Grenoble, France

ROYER CATHERINE
Rennsselaer Polytechnic Institute, Troy, NY, USA

SCHERTLER GEBHARD
Laboratory of Biomolecular Research, Paul Scherrer Institute, Switzerland

SLOTBOOM DIRK JAN
University of Groningen, The Netherlands

WEISSENHORN WINFRIED
Unit of Virus Host Cell Interactions, Université Grenoble Alpes, CNRS, EMBL, Grenoble, France

ZACCAI GIUSEPPE
Institut de Biologie Structurale, Université Grenoble Alpes, CNRS, CEA, Grenoble, France; Institut Laue Langevin, Grenoble, France

ZIEGLER CHRISTINE
Department of Membrane Protein Crystallography, Faculty of Biology and Preclinical Studies, University of Regensburg, Germany

PRACTICALS

BELRHALI HASSAN
EMBL, Grenoble, France

FARIAS ESTROZI LEANDRO
IBS, Grenoble France

NURY HUGUES
IBS, Grenoble, France

KIEBOOM ANJA
Grenoble, France

Public lecture

AUBERT SERGE
Station Alpine Joseph Fourier, Université Joseph Fourier, CNRS, Grenoble, France

Students and auditors

AFANZAR OSHRI
Weizmann Institute, Rehovot, Israel

ALI OLIVIER
ENS, Lyon, France

ARNAUD CHARLES-ADRIEN
IBS, Grenoble, France

ARRANZ GIBERT POL
IRB/UB, Barcelona, Spain

AUBAILLY SIMON
Centre de Biophysique Moléculaire, Orléans, France

AZPEITIA EUGENIO
INRIA, Montpellier, France

BARRAGAN ANGELA
University of Illinois, Urbana-Champaign, IL, USA

BASSUNI MONA
Yale University, New Haven, CT, USA

BASU MAHASHWETA
Saha Institute of Nuclear Physics, Kolkata, India

BRUCHLEN DAVID
IGBMC, Illkirch, France

CHERVY PIERRE
CEA Saclay, Gif-sur-Yvette, France

CLAVEL DAMIEN
IBS Grenoble, France; Université Paris-Sud, France

COLLANI SILVIO
Max Planck Institute, Tübingen, Germany

CORTINI RUGGERO
LPTL, Paris, France

DA SILVEIRA TOME CATARINA
IBS, Grenoble, France

DE BRUIJN SUZANNE
University of Wageningen, The Netherlands

DENAY GRÉGOIRE
LPCV CEA-CNRS-UJF, Grenoble, France

EL KHATIB MARIAM
IBS, Grenoble, France

GALANTI MARTA
University of Firenze and INFN, Italy; Université Orléans, France

GAUTAM LOVELY
India Institute of Medical Sciences, New Delhi, India

GORETTI DANIELA
University of Milan, Italy

HAKENJOS JANA
University of Heidelberg, Germany

KRAJNC MATEJ
Jozef Stefan Institute, Ljubljana, Slovenia

KUMAR MUKESH
India Institute of Medical Sciences, New Delhi, India

LARRIVA-HORMIGOS MARIA
Univerity of St Andrews, Scotland, UK

LE TREUT GUILLAUME
CEA/CNRS, Gif-sur-Yvette, France

LUKARSKA MARIYA
EMBL, Grenoble, France

METOLA ANE
Unidad de Biofisica, Leioa, Spain

NIELSEN GLENN
University of Southern Denmark, Denmark

PORTIER FRANÇOIS
LCMCP, Paris, France

POSSNER DOMINIK
Karolinska Institute, Stockholm, Sweden

PULWICKI JULIA
University of Calgary, Canada

SHILOVA ANASTASYA
ESRF, Grenoble, France

STUBBE HANS CHRISTIAN
University of Hamburg, Eppendorf, Germany

TANG QIAN-YUAN
University of Nanjing, Jiangsu, China

UZDAVINYS POVILAS
University of Stockholm, Sweden

VARMA SIDDHARTHA
LIPhy UJF, Grenoble, France
VERDIER TIMOTHÉE
ENS, Lyon, France
VERHAGE LEONIE
University of Wageningen, The Netherlands
VEYRON SIMON
LEBS/CNRS, Gif-sur-Yvette, France
WANG SHOUWEN
CSRC and University of Tsing, Beijing, China
WOODHOUSE JOYCE
Université Pierre et Marie Curie, Paris, France; IBS, Grenoble, France
WOZNICKA ALEKSANDRA
IBS, Grenoble, France

Part 1

Concepts in cell biology and examples of multiscale studies in biology

1
Introduction to cell biology

Franz BRUCKERT

Grenoble INP Phelma, France

Abstract

Living cells are complex: they are made of a myriad of different molecules and their structure results from the dynamics of the interactions between these molecules. In this short introductory chapter some levels of cellular organization are first briefly described: membranes, cytoskeletons, adhesion structures and signaling pathways. Then, some mechanisms that specifically localize proteins in the cell are reviewed: signal and targeting sequences, vesicular transport. A third section deals with protein activation, emphasizing how energy is consumed to drive cycles of assembly and disassembly of protein complexes. A key problem is how the different parts and processes of the cell are coordinated. Some general mechanisms can help with that: changes in the transmembrane potential that spread rapidly along large distances and the bistable behavior of biochemical reactions combining non-linear activation and positive feedback. A remarkable example is the well-ordered pattern of gene expression that appears during cell differentiation.

Keywords

Cell structure, cytoskeletons, cell adhesion, cell differentiation, membrane proteins, protein targeting, protein activation, vesicular transport, synchronization of cell activity, control of gene expression

Chapter Contents

1 Introduction to cell biology — **3**
Franz BRUCKERT

- 1.1 Levels of organization in cells — 5
  - 1.1.1 Membrane structure and cell compartments — 5
  - 1.1.2 The cytoskeleton — 6
  - 1.1.3 Cell adhesion — 7
  - 1.1.4 Cell–cell communication — 9
  - 1.1.5 Cell culture — 10
  - 1.1.6 Purification of intracellular compartments — 13
- 1.2 Protein localization within cells — 13
  - 1.2.1 Cell targeting signals — 13
  - 1.2.2 Vesicular transport mechanisms — 15
- 1.3 Protein activation in cells — 16
  - 1.3.1 Phosphorylation as an example of a protein activation mechanism — 16
  - 1.3.2 Formation of protein complexes — 16
  - 1.3.3 Synchronization of cell activity — 18
  - 1.3.4 Control of gene expression — 21
- 1.4 Practical conclusions — 22

References — 24

1.1 Levels of organization in cells

The organization of living cells is usually described at two complementary levels, structural and functional. There are multiple levels of cellular structure: molecular complexes made of proteins and small molecules (ribosomes, proteasomes, etc.), larger polymeric protein structures (cytoskeletons, cilia) and membrane-based structures (vesicles, organelles, nucleus, plasma membrane). Membranes delineate cell compartments in which different reactions take place. Cytoskeletal filaments are often used for directed movements within the cell. These different structural elements collaborate to carry out the many functions of the cell: energy production, macromolecule synthesis and degradation, intracellular transport, uptake and secretion of molecules, cell movement, cell replication and division. A single structural element may therefore fulfill several functions—for example, both lipid synthesis and the first steps of protein secretion take place in the endoplasmic reticulum.

1.1.1 Membrane structure and cell compartments

Biological membranes are composed of proteins inserted into a lipid bilayer. The bilayer is only 5 nm thick, but vesicle diameters range from 50 nm to 5 µm. The protein content of biological membranes varies between 25 and 75%. It should be noted that although 70% of all proteins interact with membranes they constitute only 20% of total proteins by weight—this shows that membrane proteins are less abundant than other proteins but that they perform important roles. The main classes of lipids are phospholipids, sphingolipids, glycolipids and cholesterol. The composition of membrane lipids is highly complex, for several reasons:

- the diversity of alcohol groups found in the hydrophilic head;
- the diversity of fatty acids in the hydrophobic part;
- the lipid composition of biological membranes is asymmetric, meaning that the two layers do not have the same composition;
- the two-dimensional (2D) composition of membranes is also not uniform, with lipid rafts having a different composition from the surrounding bilayer.

One of the main reasons for this diversity is to prevent the solidification of lipid membranes. Pure lipids freeze at a defined melting temperature. A lipid mixture does not have a clear phase transition, which is beneficial because it allows conformational changes of the proteins embedded in the bilayer. Proteins such as ion channels indeed change their shape when they catalyze chemical reactions or transport. The fluctuations of lipid atoms in the layers create holes that can accommodate these changes. In other words, below the melting temperature the lipid structure is rigid and protein activity is hindered. These fluctuations also explain why water molecules diffuse rather easily through lipid bilayers.

Two types of proteins are associated with membranes: integral and peripheral.

Integral membrane proteins are embedded in the lipid bilayer. They necessarily span the entire hydrophobic interior of the bilayer and protrude from both sides of the membrane. Many of them contain one or several transmembrane alpha-helices containing mostly hydrophobic amino acids. This structure is stabilized by internal H-bonds and exposes the hydrophobic side chains to the hydrophobic interior of the membrane. Twenty-one amino acids correspond to six helical turns, or about 4 nm, which is the thickness of the hydrophobic part of the bilayer. The presence of these hydrophobic alpha-helices is easily predicted by sequence analysis software, such as TM Pred (http://www.ch.embnet.org/software/TMPRED_form.html). Some transmembrane proteins, for instance some transmembrane carriers, adopt another structure; they have a cylindrical shape formed from adjacent beta-sheets (beta-barrel). This structure also exhibits hydrophobic amino acids on the external side of the beta-barrel that faces the interior of the membrane.

Other important examples of transmembrane proteins involving beta-sheets are pore-forming toxins. These molecules are secreted by cells in soluble form in an alpha-helical conformation. Upon binding to the membrane, the proteins self-associate and experience a cooperative conformational change that results in the formation of intermolecular beta-sheets that form a transmembrane pore. The formation of this pore induces the release of ions and small molecules, which kills the cell. These pore-forming toxins are secreted by bacterial pathogens and by certain cells of the immune system that kill other cells (cytotoxic cells).

Peripheral membrane proteins are attached to one side of the lipid bilayer only. Different mechanisms of interaction exist. Positively charged protein domains can interact with negatively charged phospholipids. Proteins can also be post-translationally modified by lipids, which themselves are inserted in a membrane layer. Finally, some proteins interact with membrane proteins, and this interaction anchors them to the membrane. Experimentally, it is often possible to separate peripheral membrane proteins from the membranes themselves without destroying the membrane, for instance by varying the salt concentration or changing the pH (electrostatic interactions and H-bonds). Integral membrane proteins, however, cannot be separated from the membrane without destroying the membrane, because the separation would expose hydrophobic parts of the molecules to water.

In organelles, membranes delineate several compartments, the interior and exterior of the organelle, and each side of the membrane. For interacting proteins, these are regions that have different biochemical compositions.

1.1.2 The cytoskeleton

Animal cells possess two or three cytoskeletal networks made of polymerized proteins: the actin cytoskeleton (microfilaments), the tubulin cytoskeleton (microtubules) and intermediate filaments. The cycle of polymerization and de-polymerization of microfilaments or microtubules is coupled to hydrolysis of ATP or GTP, respectively. Intermediate filaments, for their part, are controlled by protein phosphorylation and de-phosphorylation. Microfilament and microtubule polymers have a "plus end", where the association and dissociation of monomers is fast, and a "minus end", where it is slow. The complexity arises from the fact that, upon integration in the filament, actin or tubulin hydrolyzes ATP or GTP, respectively. The linear composition of the filament is therefore not uniform. Furthermore, about 100 proteins control the growth and dissociation rates or are associated with microfilaments and microtubules:

actin-binding proteins, microtubule-associated proteins, bundling proteins, severing proteins, molecular motors and so on. Some of these influence the nucleation of new polymers, the nucleotide state of monomers within the filament.

Microfilaments are oriented polymers of actin, an ATP-binding and ATP-hydrolyzing protein. They constitute the cortex of the membrane, which allows the plasma membrane to deform. Stress fibers are actin microfilaments linking adhesion focal points. Podosomes are structures involved in cell motility. Actin microfilaments are reversibly linked to the plasma membrane by specific proteins (e.g., talin, catenin, ezrin). Mechanical forces are exerted at the plasma membrane by the polymerization of actin and between actin microfilaments by molecular motors.

Microtubules are oriented polymers of tubulin, a protein that binds and hydrolyzes GTP. They are organized radially around the centrosome, also called the microtubule organizing center (MTOC), which is usually located near the nucleus. The minus ends of microtubules are locked at the centrosome. Tubulin is incorporated in protofilaments as $T\alpha_{GTP}T\beta_{GTP}$, then the tubulin beta subunit rapidly hydrolyzes GTP into GDP. A ring of $T\alpha_{GTP}T\beta_{GTP}$, called a GTP cap, is thus present at the extremity of growing microtubules and can be revealed by proteins that specifically bind the GTP form of tubulin, such as EB1 (Mimori-Kiyosue et al. 2000). Since the $T\alpha_{GTP}T\beta_{GDP}$ protofilaments are unstable, the plus end of the microtubule depolymerizes about 100 times faster when it contains GDP tubulin than when it contains GTP tubulin. A GTP cap therefore favors growth and when it is lost rapid depolymerization occurs. Individual microtubules therefore alternate between a period of slow growth and a period of rapid disassembly, a phenomenon called dynamic instability. Microtubules are rigid, and their polymerization exerts mechanical forces that allow the centrosome to reach a position along the cell cortex determined by the microtubule depolymerization activity. They allow centripetal or centrifugal transport of organelles within the cell. During cell division, the centrosome replicates and organizes the separation of chromosomes by forming the mitotic spindle.

1.1.3 Cell adhesion

Tissues consist of differentiated cells and of extracellular matrix (ECM) bathed in interstitial fluid. Cells are attached to other cells (via homotypic or heterotypic cell–cell interactions) or to the ECM by means of specific adhesion receptors.

The ECM is made of macromolecules (proteins and polysaccharides) synthesized and secreted by cells. It provides a specific mechanical and chemical environment for the cells. Cell adhesion molecules are proteins expressed at the surface of cells that mediate cell–cell binding (e.g., cadherins) or binding to the ECM (e.g., integrins). They trigger intracellular signals and therefore act as receptors. These adhesion receptors interact with the cytoskeleton. Cadherins and integrins are indeed able to influence the growth of microfilaments and microtubules.

The interstitial fluid is a solution that surrounds cells and the ECM. Its composition depends on exchanges between the cells, permeability barriers and the blood, and is similar to that of blood plasma. The interstitial fluid also contains growth factors and

8 Introduction to cell biology

Table 1.1 The main receptor families involved in signaling in eukaryotic cells

RECEPTOR FAMILY	MAIN FUNCTION OF THE RECEPTORS	MAIN CELLS	APPROXIMATE NUMBER OF RECEPTORS
G-protein-coupled receptors	Differential detection, fine tuning, adaptation	Sensory systems (vision, hearing, taste, etc.)	907
Ion channel receptors	Fast responses, "digital" communication, frequency-dependent responses	Neuron and muscle action potentials and synaptic communication	400
Receptors with tyrosine kinase activity	Threshold detection, coincidence detection, multiple inputs, cell fate decisions	All cells	58
Receptors with associated enzyme activity (including kinases)	Threshold detection, coincidence detection, multiple inputs, cell fate decisions	All cells	115
Intracellular receptors	Response to hydrophobic molecules	All cells	48

hormones that influence cell behavior by the means of specific receptors present at the plasma membrane. (These receptors are described in Section 1.1.4 and Table 1.1.) Note that many growth factors strongly and specifically interact with the ECM, which therefore acts as a reservoir of bioactive molecules.

Connective tissue (mesenchyme) contains fibroblasts and is rich in fibrous ECM. Muscle tissues contain muscle cells and have strong contractile activity. Nervous tissues are made of neurons and glial cells and conduct action potentials. Epithelial tissues consist of monolayers of cells that provide a selective barrier; for instance, endothelial cells line the inner surface of blood vessels and separate them from the blood. Similarly, a complex epithelium delimits the border of the intestine (see Section 1.1.5 and Fig. 1.1). Organs are complex structures formed by several tissues carrying out specific functions. Different tissues are separated by specific basement membranes, also called basal lamina; these in fact are not membranes as already defined, but a very thin layer of ECM.

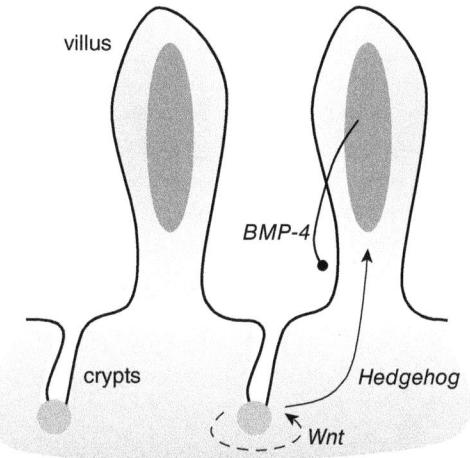

Fig. 1.1 Self-organization of the crypt–villus axis. The epithelium (black line) consists of a layer of cells covering the villi and the crypts. Stem cells at the bottom of the crypts (gray balls) are exposed to Wnt, locally secreted by Paneth cells and underlying mesenchymal cells. Wnt stimulates cell proliferation while preventing cell differentiation. The differentiation of stem cells into Paneth cells provides a positive feedback mechanism for their own proliferation. Mesenchymal cells (dark gray ovals) in the villi secrete BMP-4, that diffuses and stimulates the differentiation of epithelial cells in the villi. As a result, cell proliferation in the villi stops. Crypt cells secrete Hedgehog, a molecule that diffuses toward the mesenchymal cells of the villi and induces secretion of BMP-4.

1.1.4 Cell–cell communication

Cells are highly sensitive to their extracellular environment. Molecules dissolved in the interstitial fluid, in the blood, or eventually in air, bind to cell surface receptors where they elicit specific responses or enter the cell through channels or transporters. Hydrophobic molecules can even enter cells directly because they cross membranes spontaneously. All cells are highly sensitive to temperature and oxygen concentration. In tissues cells are less than 200 μm from a capillary to ensure a sufficient supply of oxygen. Conditions of chronic hypoxia cause the activation of hypoxia-inducible factors (HIF) that result in the secretion of vascular endothelial growth factor (VEGF), a growth factor that attracts endothelial cells and stimulates the growth of new capillaries (angiogenesis). The geometry of the ECM adhesion zone (because of the adhesion receptors such as integrins) influences the organization of the cellular cytoskeleton, and therefore the positioning of the nucleus and organelles (Théry et al. 2006). In addition, cells respond to mechanical forces, for instance transmural pressure and the flow shear stress in blood vessels and nephrons. They are also sensitive to the stiffness of the ECM. Discher's group (Discher et al. 2005) were the first to demonstrate that cell differentiation is influenced by this parameter.

Cells also sense and respond to neighboring cells. Gap junctions are cellular structures that allow the direct exchange of small molecules and ions between adjacent cells. Cells secrete diffusible molecules either in the interstitial fluid or in the blood; these bind to receptors on the surface of another cell or enter other cells through channels or transporters. Many cells express specific molecules that can be recognized by other cells as ligands to their own receptors. Well-known examples are the Fas/FasL pair that control apoptosis, the notch receptors and Delta-like and Jagged ligands, implicated in organ development. As with cell–ECM adhesion, cell–cell adhesion also induces cytoskeletal remodeling. Cells are therefore sensitive to mechanical forces (stress) exerted within living tissues (Desprat et al. 2008, Martin et al. 2010).

Cell responses can be classified as early or delayed depending on the involvement of *de novo* protein expression. Early responses involve none or limited new protein expression and include:

- cytoskeletal reorganization and changes in cell morphology,
- reorganization of cell–cell or cell–ECM contacts,
- metabolic changes (a shift in energy sources),
- cell migration (directional motility or chemotaxis),
- secretion of specific molecules (exocytosis)
- uptake of specific molecules (endocytosis).

These changes are induced by protein phosphorylation or other post-translational modifications, degradation signals or changes in the concentration of ions and second messengers. (Second messengers are small molecules produced in the cell in response to an external stimulus, e.g., cAMP.) These modifications can appear rapidly, in less than a second.

Delayed responses involve major changes in protein expression. These include cell division, the arrest of cell proliferation, cell differentiation and apoptosis. In addition to the mechanisms found in early responses, these changes also involve the activation of transcription factors and *de novo* protein synthesis. They are usually rather slow (20 min to 1 hour). Cell division is a complex mechanism consisting, first, of the replication of all cell components, including the genomic DNA. A set of several specific kinases (cyclin-dependent kinases) control the different stages of cell division. Cell differentiation involves the activation of "master genes" that control many other genes. Finally, during apoptosis, caspase genes are expressed. Caspases degrade specific intracellular cell components.

Proteins involved in cell signaling activity are usually grouped together in "signaling pathways". Many cell signaling proteins exist, because there are an enormous number of cell surface receptors. Table 1.1 gives an overview of the main categories of receptors, their main tasks and cell types and how many of then there are in the human genome.

1.1.5 Cell culture

Most biological experiments are performed with isolated cells or with wild-type or genetically modified organisms.

Primary cell cultures consist of cells that have been extracted directly from tissues, generally using trypsin and ethylenediaminetetraacetic acid (EDTA) to dissociate them. They divide a limited number of times *in vitro*. Note that some differentiated cells do not divide (neurons, myofibers). The tissues are often obtained from genetically modified animals; this allows the protein of interest to be labeled with green fluorescent protein (GFP), the knocking-down some gene of interest or the expression of mutations.

Secondary cell cultures consist of cancer cells extracted from tumors or "immortalized" cells modified by the expression of an oncogene (a gene whose permanent activity induces tumor growth and proliferation). These days many cell types are available as secondary cultures; their properties can nevertheless be different from those of primary cell cultures.

Stem cells are increasingly being used to obtain differentiated cell types. More precisely, in adult organisms, different classes of stem cells (in plants, meristem cells) exist that allow the renewal of different types of tissues. Pluripotent stem cells are able to divide and differentiate into any cell type. Multipotent progenitor cells are able to divide and differentiate into a limited number of cell types once an irreversible signal has been received (commitment). Note that, quite often, cells obtained from a primary cell culture will have originated from the stem cells present in the starting culture. A feature common to all stem cells, which differentiates them from ordinary cells, is that they divide asymmetrically. This means that, after division, one cell remains a stem cell whereas the other undergoes a series of symmetrical divisions as a multipotent progenitor cell. Stem cells reside in a "niche", with a very specific molecular and cellular environment that defines a polarity axis orienting cell division in such a way that two different cells are produced. One is a progenitor cell that will differentiate and proliferate, the other remains in the niche as a stem cell.

Intestinal crypts are a good example of an epithelial stem cell niche. The intestine contains four main epithelial cell types: enterocytes, or absorptive cells, goblet and Paneth cells that secrete mucus and anti-microbial molecules, respectively, in the intestinal lumen, and enteroendocrine cells that secrete hormones into the blood and neurotransmitters. Enterocytes, goblet cells and enteroendocrine cells are found in protruding structures called villi, consisting of an epithelium covering mesenchymal cells. Intestinal crypts are small recesses located between the villi where new cells are produced. These crypts are surrounded by connective tissue, itself supported by layers of smooth muscle cells whose waves of contraction ensure the movement of food boluses within the lumen of the intestine. Paneth cells are specifically located at the bottom of the intestinal crypts. Intestinal stem cells also reside in the crypts, in contact with Paneth cells, and divide asymmetrically to produce transit amplifying cells that divide rapidly (a 10-h cell cycle). Most of these transit amplifying cells move out the crypt and differentiate into enterocytes, goblet cells and enteroendocrine cells, while a few of them move down the crypt and differentiate as Paneth cells. Cell proliferation and differentiation are controlled by two signaling pathways, Wnt and BMP-4, respectively. Wnt is secreted by Paneth cells and underlying mesenchymal cells, and acts on stem cells and transit amplifying cells to stimulate their division. This provides a positive feedback mechanism that maintains cell proliferation. BMP-4 is secreted by mesenchymal cells in the villi and blocks the proliferation of epithelial cells while stimulating their differentiation. Intestinal epithelial cells indeed express cell surface receptors for both Wnt and BMP-4 and are thus sensitive to the relative concentration of these molecules. BMP-4 is called a morphogen because it controls the location of differentiated cell types in a tissue. Crypt cells also secrete Hedgehog, a signaling molecule that activates the secretion of BMP-4 by mesenchymal cells in the villi. The self-organization of crypts and villi results from the interplay between these three signaling pathways and these different cell types (Fig. 1.1). The concept of "stem cell niche" in this case is therefore relatively complex.

Eukaryotic cells are often grown on solid surfaces covered by ECM macromolecules. Surface stiffness is quite important for cell growth and differentiation (Discher et al. 2005). ECM molecules are often provided exogenously but they can also be secreted by the cells themselves. The exact nature of the ECM molecules is therefore often not known with precision. Cells require a specific medium (e.g., Dulbecco's modified Eagle's medium, DMEM) for growth, often supplemented with growth factors, vitamins and antibiotics. Initial ECM molecules and growth factors are often supplied by a certain proportion of fetal calf serum (FCS), the exact composition of which is usually not defined and probably varies from batch to batch. Later cells can secrete their own growth factors and their own ECM. Eukaryotic cells are usually grown at 37°C, in the presence of saturating H_2O, air and 5% CO_2 to maintain the pH. Chemical buffers can also be used to stabilize the pH. When desired, cells are dissociated from the ECM and other cells using trypsin and EDTA, a Ca^{2+}-chelating molecule. When the cell density reaches saturation (meaning that cells enter into contact), they should be dissociated, diluted and seeded onto a new plastic surface at a lower concentration. The current trend is to switch from 2D cultures to 3D, and possibly stem cells, using micro- and nanotechnology to engineer their specific environment. Note that the same cell type can switch from 2D to 3D culture simply by changing the nature of the ECM molecules provided. On fibronectin-coated surfaces, MDCK cells for instance form a monolayer. On Matrigel®, a complex set of macromolecules secreted by tumor cells, the same cell type forms spherical or cylindrical structures called cysts that resemble glands (acini) or vessels (Kleinman and Martin 2005).

Two *in vitro* cell culture systems have been described for intestinal cells that allow the growth and differentiation of stem cells. Intestinal *tissue fragments* can be grown on a collagen gel (Ootani et al. 2009). This gives rise to a 3D culture consisting of a hollow spherical epithelium growing on a layer of mesenchymal cells embedded in ECM. The spherical epithelium contains all cell types, and the culture does not need externally added growth factors because they are supplied by the mesenchymal cells. Alternatively, *intestinal crypts* can be grown on a laminin-rich gel (Matrigel®) (Sato et al. 2009). In this 3D culture, cysts form that look like hollow spheres with extending crypt-like protrusions. The lumen of the cysts therefore resembles an intestinal lumen while the newly formed crypts are similar to the initial ones and can even be grafted into a host tissue. Three growth factors are necessary for this: EGF (a broad-spectrum growth factor), Noggin (to block BMP-4 signaling) and R-spondin (to enhance Wnt activity). These cell cultures recapitulate the differentiation pattern of epithelial cells in the intestine and are thus very useful for studying their physiology and pathologies.

1.1.6 Purification of intracellular compartments

It is difficult to purify intracellular compartments. First, one needs to prepare a cell suspension. Since most cells are adherent, this implies using trypsin, a protease, and EDTA to cleave adhesion proteins and the base structure. Then, the plasma membrane needs to be gently disrupted, to free internal components. Mechanical or chemical techniques can be employed (Goldberg 2008). Mechanical disruption often relies on the shear stress created by passing the cell suspension through a narrow space between two surfaces (e.g., a Dounce homogenizer or Balch homogenizer, high-pressure devices, etc.). Chemical techniques use mild detergent to solubilize the plasma membrane or osmotic shock. Once the plasma membrane has been disrupted, intact cells and nuclei are separated from the cytoplasm by low-speed centrifugation. The resulting suspension is called a "post-nuclear supernatant". It can be further fractionated by a combination of high-speed centrifugation techniques: velocity centrifugation that separates organelles, vesicles and particles according to their sedimentation coefficient S and equilibrium centrifugation that separates them according to their buoyant density ρ. The sedimentation coefficient S depends on the size and the shape of the object as well as the difference between the density of the particle and that of the surrounding fluid. Note that the buoyant density of an organelle may vary because of osmotic effects.

All these purification methods are rather time-consuming and cumbersome. Furthermore, it is difficult to work with fewer than 10^8 cells. It is anticipated that microfluidic techniques will allow more rapid preparation of intracellular compartments with less starting material.

1.2 Protein localization within cells

At a given time, there are at least 10^4 different proteins expressed in a given cell. Since protein interaction only occurs over short distances (less than a few nanometers), the proper localization of a protein within the cell is essential. Many proteins are therefore bound to specific membrane surfaces or at specific places along cytoskeletal filaments. In this section we review the basic features that define the localization of a given protein as well as some common mechanisms used in eukaryotic cells for protein localization, focusing on the main cellular organelles.

1.2.1 Cell targeting signals

Targeting or sorting signals are stretches of amino acids encoded in the primary sequence that define the journey of a given protein in the cell and its final localization. A single protein may contain several targeting and sorting signals. Specific to a given compartment and well conserved in eukaryotes, they allow reversible interaction with the proteins that organize intracellular protein transport. They are generic and allow targeting of a protein of interest, for instance a fluorescent protein. In the absence of any signal, a protein is targeted to a default localization: the cytosol

14 Introduction to cell biology

Table 1.2 The main targeting signals in eukaryotic cells

TARGET COMPARTMENT	TYPICAL SEQUENCE	TRANSPORT MECHANISM
Import into the ER (signal sequence)	+MMSFVSLLLVGILF WATEAEQLTKCEVFN	Translocation through the membrane
Import into mitochondria (signal sequence)	+MLELRNSIRFFKPATRTLCSSRYLL	Translocation through the membrane(s)
Import into the nucleus	PPKKKRKV	Gated transport
Retention in the ER	KDEL-	Vesicular transport
Sorting to endosomes or lysosomes	YxxP, ExxLL	Vesicular transport
Plasma membrane	+GSSKSKPK, CxxL-, CCxx- and acylation (modification by a lipid group)	Direct binding of a peripheral membrane protein
Plasma membrane	Glycosylation at NxS/T sequence of membrane proteins and secreted proteins	Vesicular transport

ER, endoplasmic reticulum.

for a soluble protein and the plasma membrane for an integral membrane protein. Table 1.2 summarizes the main targeting signals used to address proteins in eukaryotic animal cells. Targeting signals are either constitutive (always active) or may be activated by phosphorylation and/or conformational changes. Their efficiency will depend on the accessibility of such sequences to the proteins of the targeting and sorting machinery.

A signal sequence consists of about 20 amino acids at the N-terminal end of the primary sequence of a protein. It allows insertion of the protein in the membrane of an organelle (e.g., endoplasmic reticulum, mitochondrion) or translocation of the protein through one or several organellar membranes. Once the protein has been imported into the lumen of the organelle, the signal sequence is often cleaved by a specific protease and degraded.

Retention signals maintain some proteins in a given compartment, usually by interacting with specific membrane receptors. The steady-state localization of these receptors results from the balance of anterograde and retrograde traffic.

1.2.2 Vesicular transport mechanisms

Vesicular transport is a complex intracellular mechanism that carries membrane proteins or soluble proteins contained within the lumen of an organelle to other locations in the cell. The basic mechanisms of vesicular transport are reviewed in Fig. 1.2. First, the proteins to be transported (the cargo) are sorted in budding vesicles, with the help of cargo receptors in the case of soluble proteins or by the direct interaction of transmembrane cargo proteins with coat proteins (1 and 2 in Fig. 1.2). Coat proteins (clathrin, coatomer) assemble around the nascent vesicle in large multiprotein structures that deform the membrane. Activation of a small G protein such as ARF is necessary for the assembly of the protein complex. Specific proteins such as dynamin pinch off the vesicle from the donor membranes (3). Dissociation of the coat is coupled to nucleotide hydrolysis by ARF (4). The vesicle is then carried along microtubules by specific molecular motors, possibly over several tens of micrometers. The vesicle then interacts with target membranes, involving another set of small G proteins called Rab, which ensures the specific docking of the vesicle with the proper target compartment (5 and 6). Membrane fusion between the incoming vesicle and the target membrane is performed by the action of SNARE proteins (7). These proteins are present on transport vesicles (v-SNAREs) and on target membranes (t-SNARE). The specificity of membrane fusion in ensured by specific v–t SNARE complexes that form a coiled-coil structure consisting of four alpha-helices. v-SNAREs and t-SNAREs are also called R- and Q-SNAREs, referring to the identity of the amino acid at the center of this coiled-coil structure. The positioning of v- and t-SNAREs in the vesicle and in the target membrane is strict: exchanging protein positions abolishes the membrane fusion activity. After membrane fusion, SNARE disassembly is catalyzed

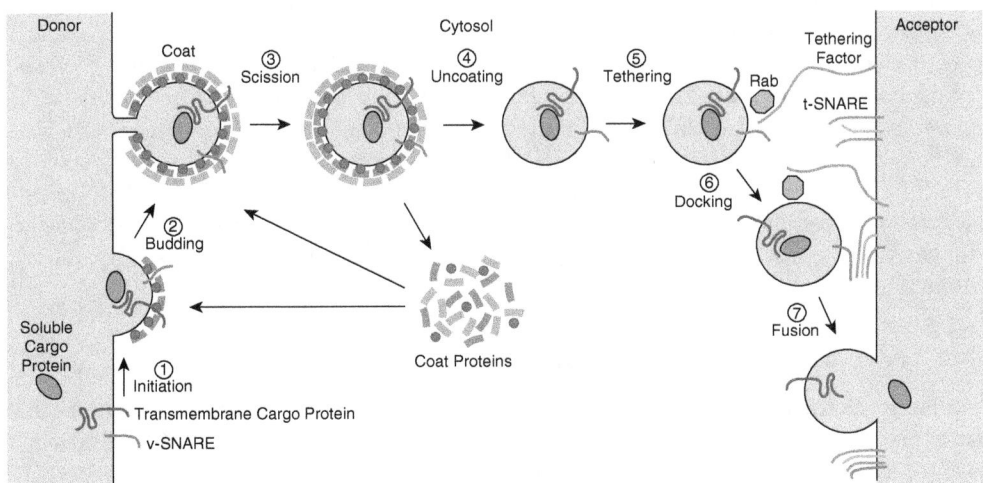

Fig. 1.2 Basic mechanisms of vesicular transport (Bonifacino and Glick 2004 *Cell* **116**, 153–166). See text for details.

by NSF and SNAP proteins. The cycle of SNARE complex assembly and assembly is driven by ATP hydrolysis.

1.3 Protein activation in cells

Protein activity is tightly regulated, either by post-translational modifications (Table 1.3) or by interaction with other proteins, resulting in conformational changes. This regulation provides another level of complexity, because for two proteins to interact they must not only be at the same place at the same time (control of protein expression and localization signals) but must also both be active at the same place at the same time! Protein activity is difficult to monitor *in situ*: mass spectrometry techniques allow monitoring of post-translational modifications and fluorescent reporter techniques have been developed, but they only monitor a single protein species at a time.

1.3.1 Phosphorylation as an example of a protein activation mechanism

Phosphorylation of the amino acids serine, threonine, tyrosine or, less frequently, histidine is a widespread mechanism for regulating protein activity. This biochemical reaction is catalyzed by specific kinases that bind both ATP and the protein (polypeptide) to be modified. Kinases catalyze the transfer of the γ-phosphate of ATP to the hydroxyl group of the amino acid. The phosphorylated protein is then released. Conversely, phosphatases catalyze the release of the phosphate group from the amino acids. Kinases are specific and recognize "consensus sequences". The accessibility of these consensus sequences on a potential target protein may be regulated by conformational changes. Kinase activity is often regulated, and kinases themselves are often regulated by phosphorylation. The main targets of regulation by kinase activity are the substrate-binding site and the catalytic site. The binding of the substrate can be controlled by the position of a "regulation loop". The catalytic transfer of the γ-phosphate of ATP may also depend on the position of some critical amino acids. Kinases are therefore regulated by many different molecules: small molecules such as cAMP in protein kinase A (PKA; cAMP-dependent kinase) or other proteins such as cyclin in cyclin-dependent kinase (CDK). Kinase domains are also present in proteins and are often regulated by conformational changes (e.g., receptors with tyrosine kinase activity). A total of 518 kinases and 73 phosphatases have been described in the human genome.

1.3.2 Formation of protein complexes

Cellular mechanisms often involve the formation of protein complexes. A remarkable feature is that several components can reversibly associate, have an effect and then dissociate. This cycle of complex formation and dissociation is linked to energy consumption. For instance, a small G protein containing a non-covalently bound GTP molecule can be incorporated during the formation of the complex. Completion of

Table 1.3 The main post-translational modifications that affect protein activity

POST-TRANSLATIONAL MODIFICATION	BIOCHEMICAL MODIFICATION	EFFECT
Proteolytic cleavage	Polypeptide bond cleavage	Introduces two new N-terminus and C-terminus ends. Possibly releases one of the polypeptides
Co-factor addition	Covalent or non-covalent bonding of a small molecule	Expands the span of catalytic activity of proteins
Lipid addition	Covalent binding of one or several fatty acids to specific amino acids	Addition of hydrophobic fatty acids anchors the protein or a portion of the protein into the membrane where the modification takes place
Acetylation	Covalent addition of an acetyl group, either at the N-terminus or at lysine residues	Eliminates the positive charge and enhances the hydrophobic character of the amino acid; in histones, this regulates DNA accessibility
Phosphorylation	Covalent addition of a phosphate group to specific amino acids (serine, threonine, tyrosine, histidine)	Change in H-bonding capacity and in the charge of the amino acid. Modifies protein conformation and interaction with partners
Ubiquitination	Covalent linkage of ubiquitin (a small protein) to specific lysine residue	Target the protein for degradation. Modifies protein conformation and interaction with partners
SUMOylation	Covalent linkage of the SUMO protein (small ubiquitin-related modifier) to a lysine in the consensus motif Ψ-K-x-D/E	Modifies protein conformation and interaction with partners—protein localization may change

complex formation triggers GTP hydrolysis by the small G protein and the release of phosphate. The energy released is used to disassemble the complex. This mechanism is at work in vesicular budding and docking (Arf and Rab proteins), delivery of tRNA to the ribosome and translocation of mRNA (EF-Tu, EF-G).

Another common mechanism involves the disassembly of protein complexes by AAA ATPases (AAA stands for ATPases associated with diverse cellular activities). These proteins often assemble in hexamers that form a toroidal structure. Good examples of this protein family are NSF, which disassembles SNARE complexes in collaboration with SNAP proteins, chaperones and proteasome AAA ATPases that unfold or refold proteins.

The provision of energy for the formation of protein complexes may also enhance the specificity of molecular recognition, in a manner similar to proofreading mechanisms (Hopfield 1974, Ninio 1975). Another way to achieve a high level of specificity is the presence of "scaffold proteins" that bind the components of a complex but do not play an active role in the process. They restrict the number of interacting molecules and avoid losing time in non-productive dissociation–reassociation steps. Scaffold proteins were initially described in cell signaling pathways, where they play an essential role in the response kinetics (Harris and Lim 2001).

Once the core proteins involved in cellular processes have been identified, and the basic biochemical mechanisms deciphered, it is interesting to reconstitute the process or part of the process in a simplified system, with purified components (lipids, proteins, small molecules). Nevertheless, the kinetics of the reconstituted process is often much reduced from that of intracellular processes. This can be due to a lower concentration of active proteins and difficulties in reconstituting the proper orientation of molecules in membrane structures.

1.3.3 Synchronization of cell activity

Time is an essential factor in self-organization. Although at the level of individual molecules, movement is governed by diffusion, things can be different at the level of cells, and information can propagate much faster than individual molecules. This is similar to the case of a metallic conductor. The velocity of individual charges is much slower (100 μm/s) than the propagation of the electrical field across the wire (the speed of light). In the following, we will give two examples showing how information can spread rapidly throughout a cell.

The first example corresponds to the well-known case of contraction of skeletal muscle. Muscle fibers (myofibers) are centimeter-long cells created by the fusion of smaller myoblasts. They contain myofibrils, an intracellular contractile structure made of repeated units called sarcomeres. The active part of the sarcomere consists of myosin bundles that move along actin microfilaments when their ATPase activity is unleashed in the presence of intracellular Ca^{2+}. The movement of myosin molecular motors along actin filaments reduces the length of the sarcomere, exerting force at the ends of the myofibrils. Ca^{2+} is mainly contained in intracellular compartments, called sarcoplasmic reticulum, that extend throughout the cell. How can a centimeter-long cell control the release and uptake of Ca^{2+} in a synchronized manner? This is the result of three fast biochemical mechanisms: (1) propagation of an action potential along the plasma membrane, (2) local release of Ca^{2+} at the T-tubules and (3) Ca^{2+}-induced release of Ca^{2+} at the surface of the sarcoplasmic reticulum.

Action potentials originate from a motor plate, a specialized synapse between a motor neuron and a muscle, and propagate along the myofiber plasma membrane, at a very high speed (around 1 m/s). This electrical signal is due to the fast oscillation of the transmembrane potential induced by the rapid transient opening of Na^+ ion channels, followed by the transient opening of K^+ ion channels. In this process, adjacent ion channels interact through the variation in transmembrane potential they generate, which propagates rapidly and passively along membrane structures.

The plasma membrane of myofibers is invaginated and forms T-tubules that come into close contact with the sarcoplasmic reticulum inside the myofiber (Fig. 1.3). Voltage-sensitive Ca^{2+} channels open upon membrane depolarization induced by the action potential. The initial rise in Ca^{2+} in the cytoplasm activates specific calcium channels in the sarcoplasmic reticulum (RyR) that release more Ca^{2+} from the calcium stores (calcium-induced calcium release). These calcium channels therefore open each other, and very rapidly the entire sarcoplasmic reticulum releases Ca^{2+}. The activity of these channels is transient since they rapidly inactivate, and the cytoplasmic calcium is pumped back into the sarcoplasmic reticulum by a specific ATPase (SERCA).

Using this example we can recall that propagation can be speeded up by two simple mechanisms: (1) the activation of one protein activates others nearby (amplification or positive feedback) and (2) some proteins spontaneously deactivate, preventing the full simultaneous activity of all molecules (delayed negative feedback). Instead, a traveling wave propagates rapidly.

A second example comes from signal transduction pathways. The mitogen-activated protein (MAP) kinase (MAPK) cascade consists of a set of three kinases that regulate each other and relay signals from the plasma membrane to the nucleus: kinase 1 phosphorylates and activates kinase 2, kinase 2 phosphorylates and activates kinase 3. At one end, kinase 1 is activated by receptor tyrosine kinase (RTK) receptors, and at the other kinase 3 activates transcription factors that determine cell differentiation, among other targets. *Xenopus* oocyte development involves such a cascade that mediates the response of oocytes to progesterone. *Xenopus* oocytes are millimeter-sized cells: Ferrell (1998) demonstrated that for an intermediate level of progesterone, at the level of a single cell, either all kinase molecules were active or none were. This remarkable result shows that these big cells somehow synchronize the activation state of the MAPK cascade molecules. This bistable behavior is explained by two features of the *Xenopus* MAPK pathway: a non-linear response and a positive feedback loop that provides an amplification mechanism. The non-linear response, also called ultra-sensitivity, results from the successive activation of the three kinases in the presence of constant antagonist phosphatase activity (Huang and Ferrell 1996). At a low signal concentration phosphatase activity overcomes kinase activity, whereas above a certain threshold the opposite occurs. The positive feedback loop is provided by a specific serine–threonine protein kinase (Mos), which activates MAPK 2 (Matten et al. 1996). As a consequence, activation of the full cascade requires a minimum activation signal (threshold) which can even be transient (hysteresis effect). This is explained by a simplified biochemical model which is presented in Fig. 1.4. Cooperative effects in protein interactions therefore provide non-linear responses that help cells take decisions. In this manner, all proteins in the same cell may have the same activation state at the same time.

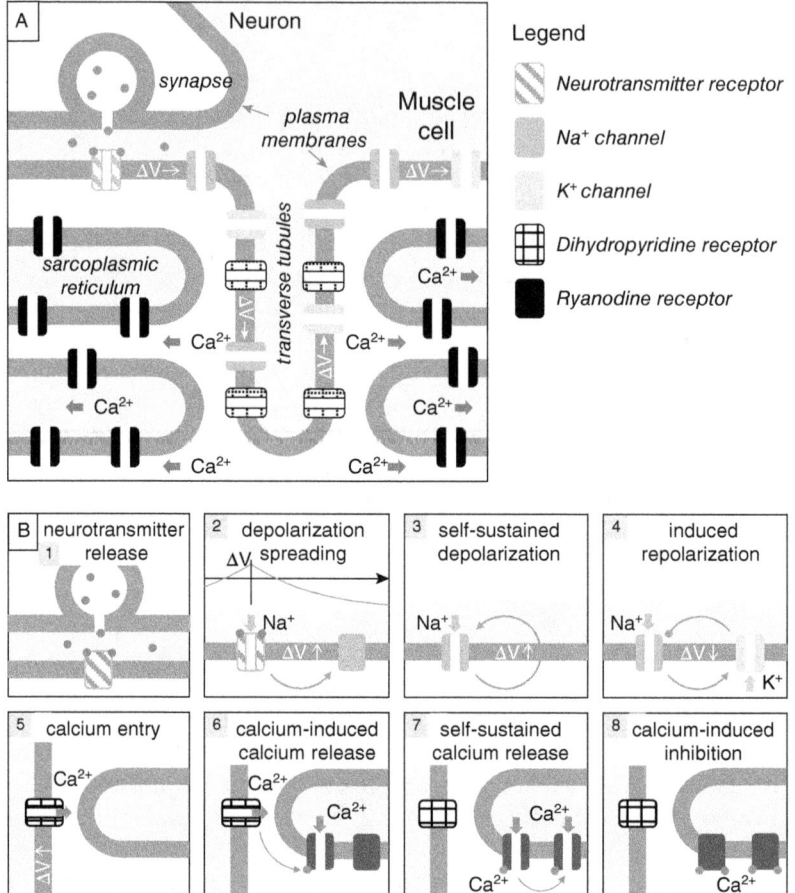

Fig. 1.3 Mechanisms of muscle cell activation. A. Structure of muscle cells: a motor neuron forms a synapse at the surface of the plasma membrane. The muscle plasma membrane is folded in T tubule. A specialized compartment, the sarcoplasmic reticulum, contains Ca^{2+} ions necessary for myosin-actin interaction and muscle contraction. B. The molecular mechanisms of the action potential and the wave of Ca^{2+} release illustrate the role of **positive feedback** in the fast propagation of activation signals. The synchronous Ca^{2+} release from sarcoplasmic reticulum ensures proper muscle cell contraction. The motor neuron synapse induces an action potential that propagates along the plasma membrane and the T tubules, inducing Ca^{2+} release from the sarcoplasmic reticulum by Ca^{2+}-induced Ca^{2+} calcium release mechanisms. The initial release of neurotransmitter at the synapse (1) induces a depolarization of the plasma membrane (2) that opens nearby Na^+ channels (grey). Na^+ entry further depolarizes the membrane that allows further signal spreading (3, **positive feedback**). After a delay, membrane depolarization induces K^+-channel (light grey) opening, that brings back the transmembrane potential to resting value (4). In T tubules, the plasma membrane depolarization due to the incoming action potential opens dihydropyridine receptors (5, square filling) that mediate Ca^{2+} entry. Ca^{2+} entry induces the activation of ryanodine receptors (6, dark grey). Ryanodine receptors are Ca^{2+} channels activated by cytosolic Ca^{2+}. The self activation of ryanodine receptors spreads rapidly in the sarcoplasmic reticulum (7, **positive feedback**). After a delay, ryanodine receptors get inactivated (8).

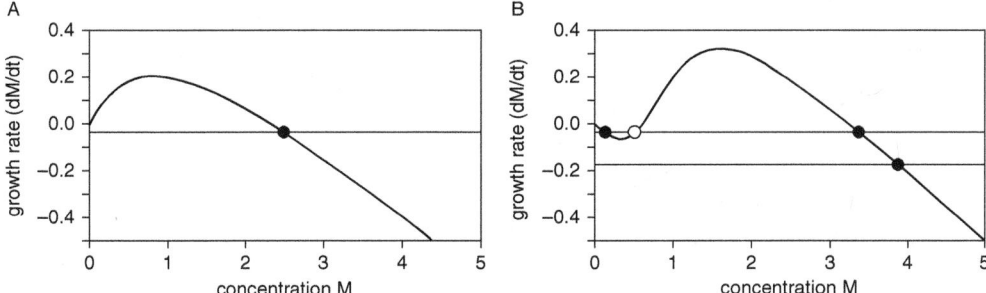

Fig. 1.4 A simple (minimal?) model of "molecular memory." A molecule M is considered, whose concentration evolves along time due to three phenomena: (1) external production (a control parameter), (2) internal production by a self-activation mechanism (a positive feedback loop) and (3) spontaneous degradation. M may represent a MAPK. The differential equation that M obeys is thus $dM/dt = C + P(M) - k_{deg}M$, where C is the control parameter, $P(M)$ is a function relating the rate of production to the concentration of M and k_{deg} is the rate of degradation. The relationship between the internal rate of production of M and the concentration of M is described by a Hill equation, with $N_{Hill} = 1$ (panel A) and $N_{Hill} = 3$ (panel B). Panels A and B, therefore, correspond to a linear or non-linear situation ($k_{deg} = 0.3$). The thick curve represents the rate of internal production and degradation of M, $P(M) - k_{deg}M$, as a function of the concentration of M. The thin horizontal lines represent the control parameter $(-C)$. The intercepts between the curves and the lines define fixed points (black circles), which can be stable or unstable (closed and open circles) depending on the position of the curves relative to the thin horizontal lines. When the curve has a negative slope at the fixed point, the fixed point is stable. For concentrations of M larger than the fixed point, the internal production and degradation rates are lower than the control parameter, so the amount of M decreases until it reaches the fixed point. Conversely, for concentrations of M lower than the fixed point, the internal production and degradation rates are higher than the control parameter, so the amount of M increases until it reaches the fixed point. By the same reasoning, it can be shown that when the curve has a positive slope at the fixed point, the fixed point is unstable. In the case of a linear response ($N_{Hill} = 1$; panel A) a single fixed point exists, which is attained whatever the value of the control parameter. In the case of a non-linear response ($N_{Hill} = 3$; panel B), two stable fixed points exist separated by an unstable one. Below the threshold (top thin line, panel B), the lower fixed point is stable and above the threshold (bottom thin line, panel B), the higher fixed point is reached and remains stable whatever the value of the control parameter afterward (hysteresis). Resetting the system requires an increase in the degradation rate, for example.

1.3.4 Control of gene expression

A remarkable case of controlled protein activity is gene transcription, the process of RNA synthesis catalyzed by RNA polymerase. In eukaryotic cells, the transcription rate can be modulated from 1- to 10^6-fold by two complementary mechanisms.

First, the binding of the TATA box protein and a set of transcription factors to specific regions of DNA (core promoter) upstream of the gene to be transcribed allows the assembly and activation of RNA polymerase (transcription initiation complex). Second, the accessibility of these DNA sequences to these proteins is regulated by their interaction with histones. About 200 DNA base pairs are indeed wrapped around eight histone molecules by electrostatic interactions, H-bonds and salt bridges, forming a structure called a nucleosome. Post-translational modifications of histones, especially lysine acetylation, arginine methylation and serine phosphorylation, modulate DNA–histone interaction. One particularly important example of such a modification is trimethylation of the fourth lysine in histone H3 (H3K4me3), whose presence in the vicinity of a promoter is a good indicator of gene transcription. In addition, histone ubiquitination allows specific histone turnover in the nucleosome.

Activation of a specific gene therefore requires both the freeing of parts of the DNA and the binding of specific transcription factors. The example of interferon transcriptional control is a good illustration of how transcription factors and histones cooperate (Goodsell 2010). Interferon transcription requires the binding of four transcription factors, the ATF-2/c-Jun complex, the interferon response factors (IRF) and nuclear factor kB (NF-kB). Their assembly on DNA allows the binding of the large CREB-binding protein p300 that contains a histone acetyl transferase domain. This enzymatic activity disassembles the nucleosome and frees the core promoter of the gene, allowing transcription initiation.

Applications of next-generation sequencing to the study of DNA and RNA have recently revolutionized cell biology by providing information about the identity and regulation of gene transcription at the level of the whole genome with an unrivalled precision. The potential use of these techniques is shown in Fig. 1.5, from the ENCODE (Encyclopedia of DNA Elements) Project, a consortium that studies gene expression and transcription factor activity in many cell types in order to build a complete database of gene regulation in eukaryotes. Several techniques are of great interest:

- RNAseq is a technique in which all RNA produced by a certain cell type under certain circumstances is sequenced and quantified.
- ChIPseq is a technique in which proteins that interact with DNA as histones or transcription factors are immunoprecipitated with specific antibody (IP), and the DNA sequence they bind to is identified by sequencing (seq).
- DNAse seq is a technique that provides the pattern of accessible DNA. DNAse I is an enzyme that can cleave DNA unless it is protected by histones or transcription factors. After cleavage by DNAse I the remaining DNA is identified by sequencing.

1.4 Practical conclusions

This brief description of cellular organization shows that cells are extremely complex objects to study. The enormous number of components is a memory challenge. Even a simple description of all these components in an organized and hierarchical manner

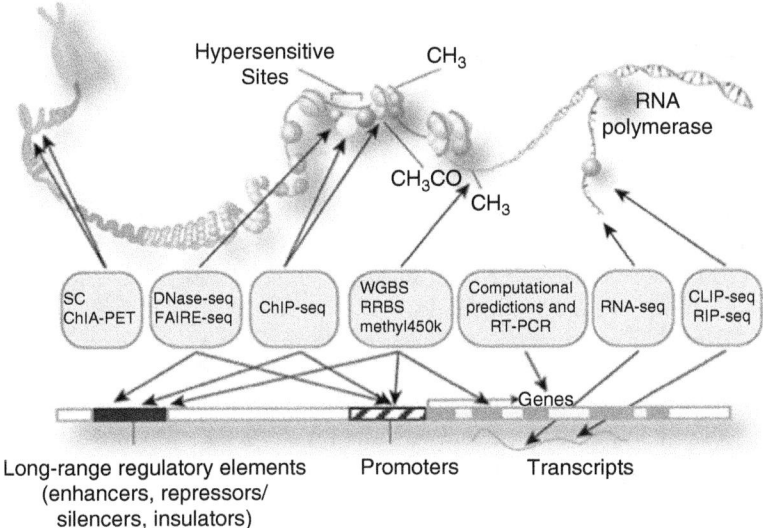

Fig. 1.5 Structure of DNA and the regulation of gene transcription. The top diagram represents the structure of DNA, from the chromosomes to the double strand, with the nucleosome structure in the middle. The bottom image represents the different elements in DNA that define genes: the coding and intervening sequences (dark and light gray to the right of the promoter region), the core promoter (hatched region) and the long-range regulating elements that bind transcription factors (black). The boxes in the middle give the different techniques used to study genomic and transcriptomic information at the level of the whole cell. Image credits: Darryl Leja (NHGRI), Ian Dunham (EBI), Michael Pazin (NHGRI) (https://www.encodeproject.org)

is difficult to construct. Nevertheless, the modular architecture of cells is apparent: different functions and different structures make use of different sets of proteins. For a defined function one also needs to delineate the proteins involved in the core of the function, the proteins involved in its regulation (e.g., kinases) and the proteins that speed up or localize the function (e.g., scaffold proteins).

When working with cells, one should be aware of the many parameters they are sensitive to. Differentiated or non-differentiated cells live in specific niches but *in vitro* studies usually provide them with a very different environment. Microfabrication technologies can help provide cells with a better-controlled environment. The presence of capillaries and blood flow, however, is difficult to mimic. Because of hysteresis phenomena in cell signaling pathways, cells can also retain the memory of a previous exposure to different stimuli. For instance, fibroblasts originating from different locations in the body (primary cultures) differentially express certain genes, even after five passages (Rinn et al. 2006). Finally, one should also bear in mind that growing cell populations are not usually synchronized, meaning that different cells are in different phases of the cell cycle.

References

Desprat N et al. (2008) Tissue deformation modulates twist expression to determine anterior midgut differentiation in Drosophila embryos. *Developmental Cell* 15: 470–477.

Discher DE et al. (2005) Tissue cells feel and respond to the stiffness of their substrate. *Science* 310: 1140–1143.

Ferrell JE Jr et al. (1998) The biochemical basis of an all-or-none cell fate switch in Xenopus oocytes. *Science* 280: 895–898.

Goldberg S (2008) Mechanical/physical methods of cell disruption and tissue homogenization. *Methods in Molecular Biology* 424: 3–22.

Goodsell D (2010) The enhanceosome. Protein Data Bank Molecule of the Month. doi: 10.2210/rcsb_pdb/mom_2010_2

Harris BZ and Lim WA (2001) Mechanism and role of PDZ domains in signaling complex assembly. *Journal of Cell Science* 114: 3219–3231.

Hopfield JJ (1974) Kinetic proofreading: a new mechanism for reducing errors in biosynthetic processes requiring high specificity. *Proceedings of the National Academy of Sciences of the United States of America* 71: 4135–4139.

Huang CY and Ferrell JE Jr (1996) Ultrasensitivity in the mitogen-activated protein kinase cascade. *Proceedings of the National Academy of Sciences of the United States of America* 93: 10078–10083.

Kleinman HK and Martin GR (2005) Matrigel: Basement membrane matrix with biological activity. *Seminars in Cancer Biology* 15: 378–386.

Martin AC et al. (2010) Integration of contractile forces during tissue invagination. *Journal of Cell Biology* 188: 735–749.

Matten WT et al. (1996) Positive feedback between MAP kinase and Mos during Xenopus oocyte maturation. *Developmental Biology* 179: 485–492.

Mimori-Kiyosue Y et al. (2000) The dynamic behavior of the APC-binding protein EB1 on the distal ends of microtubules. *Current Biology* 10: 865–868.

Ninio J (1975) Kinetic amplification of enzyme discrimination. *Biochimie* 57: 587–595.

Ootani A et al. (2009) Sustained in vitro intestinal epithelial culture within a Wnt-dependent stem cell niche. *Nature Medicine* 15: 701–706.

Rinn JL et al. (2006) Anatomic demarcation by positional variation in fibroblast gene expression programs. *PLoS Genetics* 2: 1084–1096.

Sato T et al. (2009) Single Lgr5 stem cells build crypt-villus structures in vitro without a mesenchymal niche. *Nature* 459: 262–265.

Théry M et al. (2006) The extracellular matrix guides the orientation of the cell division axis. *Nature Cell Biology* 7: 947–953.

2
A small leak will sink a great ship: HIV–host interactions

Nikolas HEROLD*, Hans-Georg KRÄUSSLICH, and Barbara MÜLLER

Department of Infectious Diseases, Virology, University Hospital Heidelberg, Germany
*Present address: Childhood Cancer Research Unit, Karolinska Institutet, Astrid Lindgren Children's Hospital, Karolinska University Hospital, Stockholm, Sweden

Abstract

As well as often being pathogenic, viruses are expert cell biologists that exploit and modulate many functions of the host cell. In this chapter the course of human immunodeficiency virus is traced from formation of the cell-free particle through extracellular maturation until entry into a new target cell and its nucleus. Full understanding of this elaborate pathway and the identification of new options for interfering with individual steps requires an integrative approach tracing the interaction of the virus with its host at multiple scales and in different experimental systems. Recent technological advances have provided us with powerful tools allowing quantitative analysis at high temporal and spatial resolution and for individual stochastic events.

Keywords

Virus, assembly, uncoating, maturation, cell entry, HIV

From Molecules to Living Organisms: An Interplay Between Biology and Physics. First Edition. Eva Pebay-Peyroula et al. © Oxford University Press 2016.
Published in 2016 by Oxford University Press.

Chapter Contents

2 A small leak will sink a great ship: HIV–host interactions   **25**
Nikolas HEROLD, Hans-Georg KRÄUSSLICH, and Barbara MÜLLER

2.1	Introduction	27
2.2	HIV assembly and release	29
2.3	HIV maturation	31
2.4	HIV entry	33
	2.4.1 Attachment	33
	2.4.2 Receptor binding	35
	2.4.3 Fusion at the plasma membrane or *via* endocytosis	35
	2.4.4 Uncoating	36
	2.4.5 Nuclear import	37
	2.4.6 Integration	37
2.5	Future of integrative HIV research	38
	Acknowledgement	39
	References	39

2.1 Introduction

Viruses are infectious agents that differ from all other organisms in that they occupy the twilight area between matter and life. Outside of a host cell, viruses share properties of inanimate matter: they neither move nor replicate and they have no metabolism. Simple regular viruses, such as poliovirus, can even be completely described by a chemical formula [1]. However, upon entry into a suitable host cell, these tiny pathogens acquire properties of living organisms: they alter and shape their cellular environment, usurp cellular resources and organize synthesis, transport and assembly of all building blocks to generate progeny virions.

All it takes to build an infectious virus can be encrypted in the nucleotide sequence of the viral genome. Although this is a heuristic simplification, it has been formally proven in the case of poliovirus. In the laboratory of Eckard Wimmer, the complete poliovirus genome sequence was assembled from chemically synthesized building blocks and used to transcribe viral RNA genomes *in vitro*. This artificially generated genome was sufficient to direct the formation of fully infectious poliovirus particles when added to cell lysates in test tubes [2].

Not all viruses are as simple as poliovirus. The traditional concept of viruses—they are smaller than bacteria, have smaller genomes and carry only very limited sets of their own proteins—has been challenged in recent years by the discovery of fascinating large and complex viruses belonging to the class of nucleocytoplasmic large DNA viruses [3]. A number of these so-called giant viruses have been discovered since the description of the first example, mimivirus, in 2003 [4]. Their genomes, like the virus particles themselves, are larger than those of small bacteria and encode numerous functional proteins and enzymes. Some giant viruses even host their own small viruses [5].

Nevertheless, even the most complex giant viruses share three features that are characteristic of viruses: (1) they are obligatory parasites that rely on a suitable host cell for their replication; (2) they multiply through a cycle of disintegration of the infectious particle ("eclipse") followed by synthesis of new building blocks and assembly of progeny virus; (3) they are characterized by a protein capsid that encases and protects their genome. Thus, viruses can simply be defined as an infectious genome encased by a protective proteinaceous shell, or in the words of Sir Peter Medawar, "Viruses are bad news, wrapped in protein".

Pathogenic viruses are thus, in essence, a simple stretch of packaged nucleic acid that has a substantial impact on a complex host organism. This makes them a prime example of the value of integrative multiscale research. True understanding of the viral replication cycle requires insights spanning large spatial and temporal scales: from detailed molecular analyses of viral genetic information and virion structure, to the dynamics of virus–host cell interactions, to cell–cell transmission and spread in an infected organism. Beyond that, spread within a population results in viral evolution, the development of resistance and the mutual adaptation of virus and host species over extended periods of time. Analysis of detailed biochemical and biophysical features and their integration into a broader picture require a wide range of expertise, from classical virology, molecular and cell biology, biochemistry, structural biology, imaging, chemistry, physics, material sciences, bioinformatics and epidemiology. The knowledge gained not only helps us to understand the molecular events involved in viral replication, but also contributes to the unraveling of the complex cellular pathways and machineries employed by these pathogens. It also provides us with defined targets for antiviral intervention.

Such an ideal, comprehensive understanding has not yet begun to emerge for any viral system. However, numerous technological developments in recent years have equipped virologists with novel tools to address questions on the integrative scale, from "very small, very rapid" events, to the larger cellular and eventually organismal context.

X-ray crystallography and cryoelectron tomography allow the detailed analysis of viral structures, while focused ion beam milling/scanning electron microscopy and correlative light and electron microscopy can visualize viruses within their cellular environment [6,7]. Advanced live cell imaging, together with a constantly increasing toolbox of fluorescent labeling techniques and image analysis algorithms, allow us to investigate rapid, dynamic interactions of individual virions with a host cell [8]. In combination with three-dimensional (3D) tissue culture or defined 3D environments this may be expanded to follow viruses on their path from one infected cell to the next [9]. Finally, super-resolution fluorescence microscopy, originally developed by Eric Betzig, Stefan W. Hell and William E. Moerner in work for which they received the Nobel Prize in Chemistry in 2014, provides a bridge between light and electron microscopy [10]. Applying these methods, we can follow molecules and events in live cells at a resolution that allows the visualization of subviral details.

A combination of these novel techniques has greatly advanced our understanding of the formation of human immunodeficiency virus (HIV) and its entry into cells, which

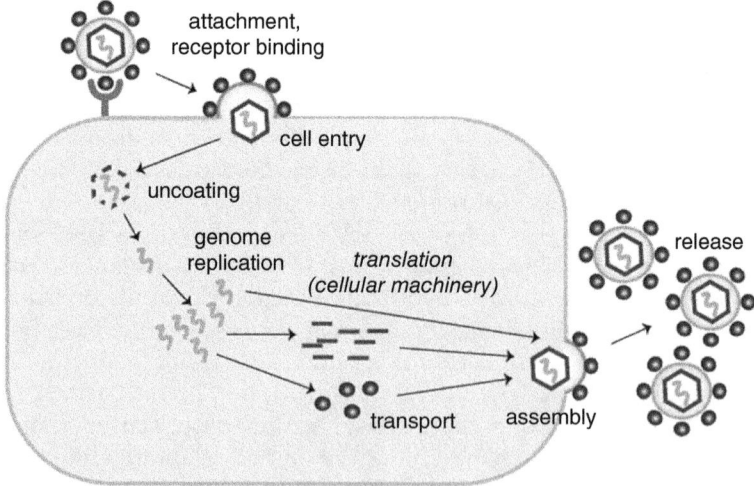

Fig. 2.1 Generic viral replication cycle. The replication of all viruses follows this basic sequence of events. Details of the individual replication steps vary from virus to virus.

is the focus of this chapter. Like all viruses, HIV undergoes a series of interactions with its host cell, from attachment to specific receptor molecules on the cell surface to cell entry, transport of the genome to the replication site, production of viral components, transport and assembly of these components into new virions and finally the release of these newly generated particles from the cell (Fig. 2.1). The molecular and mechanistic details of all of these generic steps vary greatly between different viruses and have evolved in response to the specific cellular environment. Research from ours and many other laboratories has shed light on the sequence of events in the case of HIV-1.

2.2 HIV assembly and release

As already outlined, viral particles can be considered as a biological shuttle for delivering the viral genome to the cellular environment that provides the biomolecules, enzymatic machinery and energy for multiplication of the viral genome and production of progeny virus. Thus, slightly shifting our point of view of the viral replication cycle, one may consider the formation of the infectious virion as the initial step that determines the outcome of subsequent interactions with the next host cell and leads to delivery of the packaged genome into the target cell (Fig. 2.2).

The assembly and release of HIV-1 are organized by the main structural protein of the virus, Gag (group specific antigen; Fig. 2.3). Gag is a 55-kDa polyprotein comprising all parts of the inner proteinaceous shell of the particle and accounts for around 50% of the mass of the virus [11]. It is transported *via* a pathway that is not yet

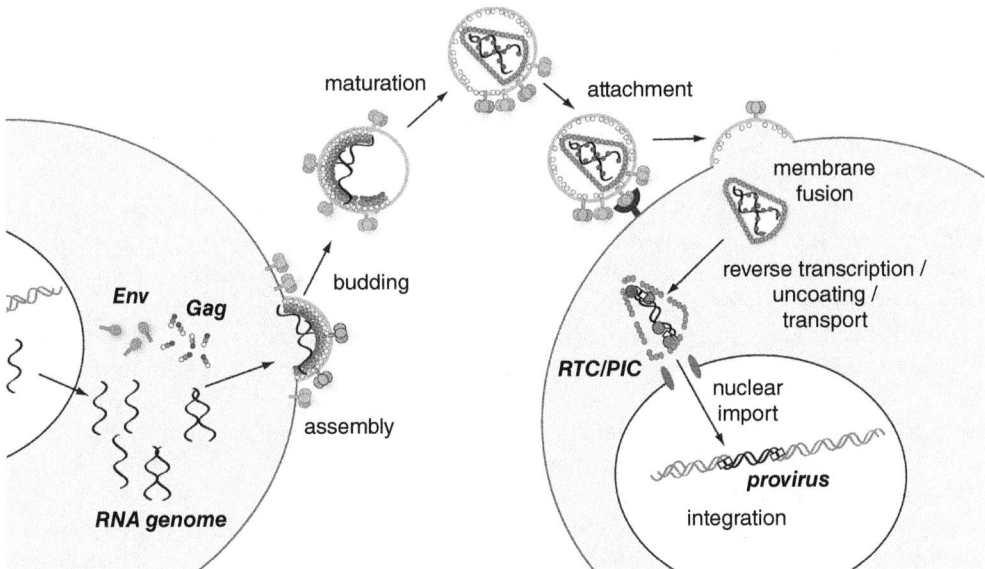

Fig. 2.2 Formation of infectious HIV particles and entry into the next host cell (RTC, reverse transcription complex; PIC, pre-integration complex).

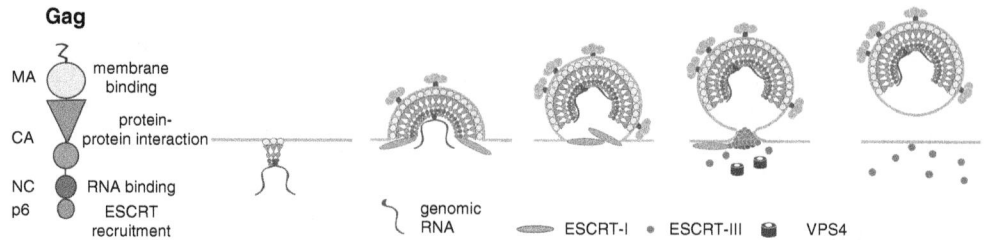

Fig. 2.3 HIV assembly is orchestrated by the Gag polyprotein.

understood to the plasma membrane, where it self-assembles into semi-spherical shells comprising about 2500 Gag molecules [12]. Gag also recruits other crucial virion components, such as the viral RNA genome, the viral replication enzymes encoded as a Gag-Pro-Pol polyprotein and the viral membrane proteins (Env), to the viral budding site. Furthermore, Gag interacts with components of the cellular endosomal sorting complex required for transport (ESCRT), which mediate fission of the viral envelope from the membrane of the host cell to release the nascent virion [13].

Gag is a polyfunctional protein, comprising subunits required for cell membrane attachment (matrix protein, MA), protein–protein interactions (capsid protein, CA), virus genome binding (nucleocapsid protein, NC) and recruitment of host cell factors that mediate particle release (p6) (Fig. 2.3). These four subdomains are linked in the Gag polyprotein, and functional regions are separated by flexible linker regions and two small spacer peptides. The intrinsic flexibility of the polyprotein, as well as the possibility of altering protein structure by proteolytic processing (see Sections 2.3 and 2.5), render Gag a highly regulatable assembly module. Conformational changes within the Gag monomer, interaction with RNA and lipid membranes, oligomerization of Gag as well as exposure of a myristoyl moiety at the N-terminus of the MA domain appear to be intertwined in a complex manner to mediate Gag trafficking and assembly [14].

Advanced imaging techniques have provided novel insights into the process of assembly of HIV-1. Using fluorescently labeled Gag derivatives, it has been possible to analyze the time course of assembly through live-cell fluorescence microscopy. These experiments revealed that HIV-1 bud formation is completed within an average time of 10 min, and that a further delay of several minutes occurs before particle release can be detected [15]. Components of the cellular ESCRT machinery (including the ATPase Vps4A that catalyzes the final step in the ESCRT membrane abscission cycle) were observed to be recruited either during or at the end of the Gag assembly period, but before particle release could be detected, indicating an active role for all these cellular proteins in release of the virus [16].

The lipid envelope that surrounds the Gag shell of HIV-1 is derived from the plasma membrane of the host cell. Lipid mass spectroscopic analyses and analysis of membrane protein composition indicate that HIV-1 buds from distinct membrane subdomains, which share characteristics of so-called lipid rafts [17]. Gag is the major determinant of viral membrane selection, and it has been proposed that Gag reorganizes the cell membrane, leading to coalescence of lipid-raft-like and tetraspanin-enriched microdomains [18]. Super-resolution fluorescence microscopy has provided a more detailed insight into the architecture of assembly sites. The presence of Gag assembly sites at the plasma membrane results in a distinct reorganization and recruitment of plasma membrane-bound HIV-1 Env molecules. Interestingly, Env molecules cluster around, rather than at, nascent Gag assemblies, indicating that Gag is able to reshape its membrane surroundings in order to create a distinct lipid–protein environment [19]. Further studies are needed in order to characterize the composition and nature of this environment and the mechanism that generates it.

2.3 HIV maturation

The underlying principle of all viral replication cycles is the disintegration of the infectious particle in the infected host cell, followed by synthesis of new virion components by the host machinery and their assembly into new particles, which are again released from the cell. This cycle presents a problem referred to as the assembly–disassembly paradox: a virus particle needs to be stable in order to protect the encased genome against harsh environmental conditions. At the same time, it has to be unstable in order to release its genome upon infection of a new host cell. This problem is often solved by a process called virus maturation: defined morphological changes convert the stable virus particle into a metastable state. In the case of HIV, this is accomplished by proteolytic cleavage of the main structural proteins Gag and Gag-Pro-Pol by a virus-encoded protease (PR). Around the time of budding, PR cleaves these proteins at nine distinct sites [20]. The mature subunits undergo a dramatic reorganization within the virion, converting the semi-spherical immature protein shell into the mature conical capsid characteristic of HIV (Fig. 2.4). Complete proteolytic cleavage at all sites as well as the ensuing morphological transitions are essential for the infectivity of HIV, and PR inhibitors that block HIV maturation are a central part of antiretroviral therapy [21].

Whereas advanced cryoelectron tomography methods and 2D crystallography have recently provided us with sub-nanometer resolution models of the immature Gag shell [22] and the mature capsid [23], the process of morphological maturation itself is less well understood. Whereas hexamers of the CA subunit form the main building block of both the immature Gag lattice and the mature capsid, the conformation of these hexameric subunits differs significantly, suggesting that morphological maturation entails disassembly of the immature lattice, followed by re-formation of the mature lattice. It is clear that temporal control of proteolytic maturation and correlation with the process of Gag assembly is crucial—not only impairment but also enhancement of PR activity abolishes particle formation. Furthermore, analyses of site-directed mutants, as well as of HIV variants emerging under PR inhibitor selection pressure, have shown that processing at the different cleavage sites in Gag and Gag-Pro-Pol occurs with different kinetics, and that maintenance of an ordered sequence of cleavage events is crucial [12,24]. Cleavages of the spacer peptides from the C-termini of NC and CA, respectively, appear to represent temporal switches in coordinating condensation of the genome with formation of the mature capsid. However, due to a lack of suitable assay systems, it is currently unknown how long morphological maturation of HIV-1 takes,

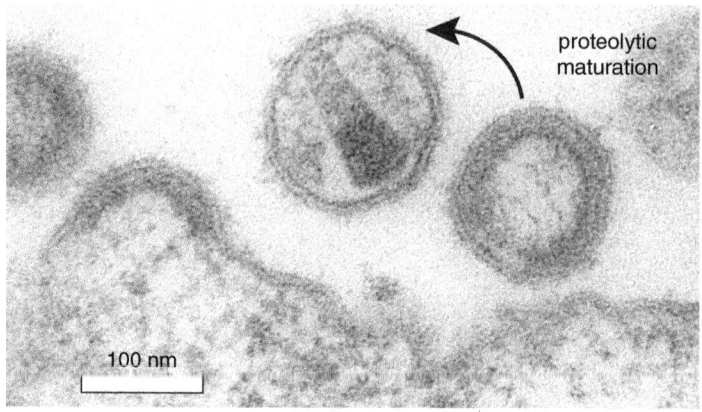

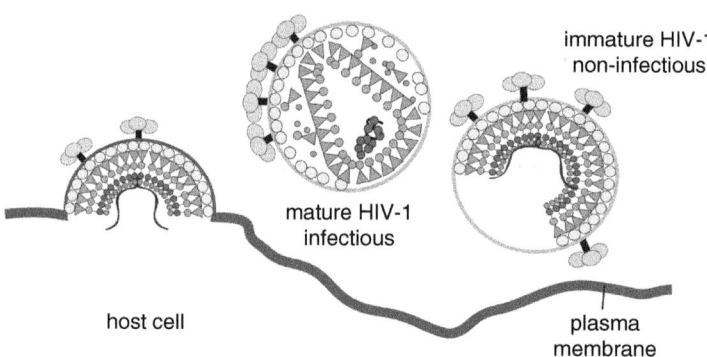

Fig. 2.4 Maturation of HIV-1.

how precisely the process is triggered and how the close correlation with particle assembly and release is maintained.

HIV maturation has three main consequences that are crucial for full virus infectivity [11]. First, the viral replication enzymes reverse transcriptase (RT) and integrase (IN), as well as PR itself, are produced as parts of the Gag-Pro-Pol polyprotein and require processing to their active forms. Second, as already outlined, conversion of the stable Gag shell into a metastable conical capsid is essential for release of the viral genome into the host cell. Third, it has been shown that immature HIV particles are deficient in cell entry, i.e., in fusion of the viral lipid envelope with the host cell membrane. At first glance this is surprising, since membrane fusion is mediated by the viral Env protein which is not a substrate of PR. Two reasons for the increased fusogenicity of mature HIV have been proposed. Atomic force measurements indicated that immature particles with the stable Gag shell lining the viral envelope are physically more rigid than mature particles [25]. This change in mechanical properties would generate a tendency of the HIV membrane to unbend and could increase the propensity of the membrane to fuse with other membranes. Another maturation-induced change in the virion has been discovered recently by super-resolution fluorescence microscopy. In contrast to other enveloped viruses, whose surface is tightly packed with viral fusion proteins, HIV carries only about 15 trimeric Env molecules on its lipid envelope. Stimulated emission depletion (STED) super-resolution microscopy of HIV particles revealed that Env molecules are distributed on the surface of immature particles, while they form a single cluster on the surface of the mature virus [26]. Efficient clustering of Env apparently depends on full cleavage of the Gag polyprotein and is correlated with enhanced single-round infectivity; clustering fusion proteins at virus–cell contact sites may stabilize attachment and promote the formation of membrane fusion pores. By coupling inner structural protein processing to Env clustering on the viral surface, HIV accomplishes an inside-out signaling mechanism ensuring that fusion with the host cell occurs when the inner structure of the virus is ready for infection.

2.4 HIV entry

2.4.1 Attachment

A virus that is not within a cell is only capable of non-active movement, i.e., passive diffusion, hydrodynamic convection (e.g., in the blood stream) and direct transfer from one infected cell to a new target cell. Hence, the stronger the affinity of the virus envelope for the cell surface, the higher the propensity for successful infection. Virologists distinguish two different modes of virus–host interaction on the target cell surface. Virus binding in the narrow sense is defined as binding of the virus to a specific viral receptor expressed on the cell surface of the target cell. Binding of HIV-1 is mediated by interaction of the viral Env glycoproteins with the receptor CD4 and a so-called co-receptor (one of the two chemokine receptors CXCR4 or CCR5) on the plasma membrane (Fig. 2.5) [27]. The events that precede specific binding of Env and CD4 are referred to as attachment. Attachment is usually less strong than Env–CD4 binding. Different membrane-associated components are involved in this process, and both the presence and the density of these components depend on the producer and target cell. Among the best known and most abundantly expressed attachment

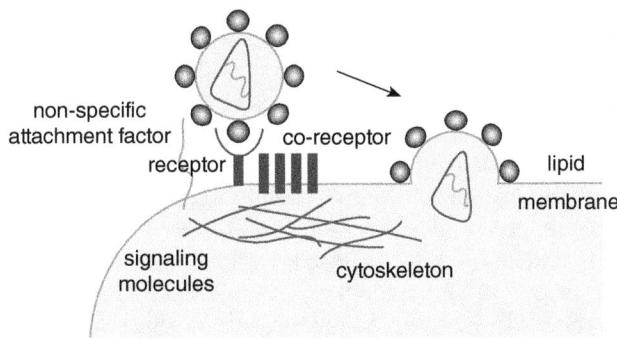

Fig. 2.5 Important cellular components involved in entry of HIV-1 into the cell.

factors for HIV and other viruses are heparan sulfate proteoglycans (HSPGs), or more generally glycosaminoglycans (GAGs). They do not require the presence of the viral glycoprotein on the particle surface. Although the interactions between the viral envelope and GAGs are relatively weak, the abundance of these sugar chains in sum makes an important contribution to viral attachment. Attachment in this case is mediated by summation of small-scale electrostatic effects within a large surface area rather than high-affinity binding of a small number of molecules [28]. Another attachment factor is the lectin Siglec-1, expressed by a subset of myeloid cells that binds to sialyllactose residues of the viral glycocalyx [29]. Cellular membrane-associated proteins that bind to Env, but do not mediate fusion, can also constitute attachment factors. As they interact directly with the viral glycoprotein Env, these factors are specific for HIV. The two best characterized members of this group are the C-type lectin DC-SIGN [30], expressed by dendritic cells, and the integrin $\alpha_4\beta_7$ [31], present on lymphocytes. Interestingly, the glycoprotein Env interacts with membrane proteins that play important roles in the recognition and uptake of pathogens (DC-SIGN) and directing blood cells to organs and tissues where they are needed to defend the host against pathogenic intruders (integrin $\alpha_4\beta_7$). This illustrates an evolutionary strategy for subverting host defense mechanisms.

The attachment of a free virus to possible target cells is rather inefficient. However, there is a set of mechanisms that catalyze virus–cell attachment, enhancing it by several orders of magnitude. They all share the ability to locally concentrate virus at the site of viral entry. Human protein fragments found in seminal fluid were shown to enhance infection by free virus by up to about 300-fold. Hence, this peptide was named semen-derived enhancer of viral infection (SEVI) [32]. Attachment to a prospective target cell can also be enhanced by cell-to-cell spread, establishing a so-called virological synapse between donor and target cells. A virological synapse is a cell–cell contact reminiscent of an immunological synapse, which plays an important role in the maturation and differentiation of immune cells. An immunological synapse is a tight cell–cell contact conferring different functions, most importantly the activation of immune cells, the presentation of antigens to lymphocytes and the release of signaling and defense molecules [33]. In addition to the cellular proteins that mediate synapse formation, Env and CD4 are part of the virological synapse. Again, HIV usurps the cellular machinery designed for immunological defense to ensure efficient spread. The donor cell can be productively infected: synapses between infected and uninfected T-cells are thought to be the major mode for the spread of the virus in lymphatic tissue [34]. The production of HIV particles polarizes to the synapse and viral particles bud into the synaptic cleft and can enter the target cell. In a process termed *trans*-infection, myeloid cells capture and store virus in a DC-SIGN- or Siglec-1-dependent manner in specialized compartments and transfer the virus to a target T-cell. Establishment of a synapse between infected donor T-cells or virus-carrying donor dendritic cells and uninfected target T-cells is, at least partially, mediated by binding of the integrin LFA-1 and the adhesion molecule ICAM-1 [34], molecules that are well known for their role in establishing cell–cell contacts.

Low-affinity attachment of viral particles can be followed by lateral movement along the cell surface. This process, termed "virus surfing" [35], is prominent on cellular extensions like microvilli or filopodia and is achieved by a cell-driven mechanism of actin–myosin-dependent membrane flow towards the cell body. Alternatively, viral particles may continuously dissociate from and reattach to low-affinity attachment factors on the cell surface until high-affinity binding with the viral receptor can eventually be established.

2.4.2 Receptor binding

The first, but not exclusive, determinant of whether a cell can be infected by HIV-1 is the surface expression of the primary viral receptor CD4. CD4 binding can induce CD4-mediated signaling, which may play a role in priming the cell for subsequent infection, thus mimicking the physiological ligands IL-16 and MHC-II, but the contribution of such signaling events is not well established. To mediate HIV entry, CD4 binding induces conformational changes within the gp120 subunit of HIV Env. This leads to exposure of the variable loop 3 (V3), which in turn can now bind the respective co-receptor. This results in dramatic rearrangements of the ternary Env–CD4–co-receptor complex causing insertion of the viral fusion peptide into the target membrane and formation of a six-helix bundle within the viral *trans*-membrane protein gp41, which leads to membrane approximation and distortion followed by fusion of the outer leaflet (hemifusion) and subsequent formation of a full fusion pore [36].

2.4.3 Fusion at the plasma membrane or *via* endocytosis

Many enveloped viruses depend on a low-pH trigger to mediate membrane fusion. At a cellular scale a low pH is confined to endo-lysosomal compartments, and these viruses therefore enter cells via the endosomal route. Making use of a low-pH trigger may serve to prevent premature (and irreversible) triggering of the viral machinery prior to reaching the susceptible target, thus explaining why many enveloped viruses have adopted this mechanism. HIV-1 fusion does not require a low pH, however, and fusion could thus occur at the plasma membrane or via an endosome. The fact that cells expressing Env on their surface can fuse with cells that express receptor and co-receptor, the pH independence and the observation that the CD4 endocytosis signal is not required for HIV-1 infection all suggest that HIV entry can occur at the plasma membrane. On the other hand, HI virions have been observed within endosomes in many studies. Accordingly, the pathway of HIV entry has been controversial, and evidence for exclusive fusion at the plasma membrane, exclusive fusion *via* an endosome or a combination of both has been reported [37]. Most of these studies were performed in transformed epithelial reporter cell lines that are hypersusceptible for HIV infection, but do not represent a natural target cell. CD4-mediated endocytosis also occurs in T-cells, the primary target cells of HIV, but this endocytosis does not appear to contribute to productive infection in this cell type, at least when infection by cell-free virus is analyzed [38]. Direct fusion at the plasma membrane creates the problem of transferring the incoming viral nucleocapsid through the dense meshwork of the cortical actin cytoskeleton and bringing it into close proximity to the nuclear membrane for subsequent nuclear import. The latter problem may be less relevant in the case of HIV infection of T-cells as these cells have a relatively small cytoplasm and incoming viral structures have been shown to associate with microtubules [39]. Bringing the incoming capsid through the dense actin meshwork, which would be easily overcome by endocytosis, would appear to be a significant obstacle to HIV infection, on the other hand, and further studies will be needed to understand this part of the entry pathway.

2.4.4 Uncoating

The events that directly follow viral fusion are the most enigmatic in the retroviral life cycle. Reverse transcription of the plus-stranded viral RNA genome into a linear double-stranded cDNA by the virus-encoded RT occurs in the cytoplasm of the infected cell, and the resulting pre-integration complex (PIC) is subsequently imported into the nucleus, where integration of the viral genome into a host cell chromosome occurs. In contrast to simple retroviruses, which require breakdown of the nuclear envelope during mitosis for nuclear entry, HIV can import its PIC through the intact nuclear pore. While the enzymatic processes involved in reverse transcription and integration have been studied in detail and inhibitors targeting both processes are widely used in HIV therapy, the composition, structure and evolution of the viral replication complexes as well as the role of host cell factors promoting or inhibiting these events are poorly understood. Following membrane fusion, the viral capsid encasing the genome and replication machinery is released into the cytoplasm. At some stage, this viral capsid must uncoat to release the genetic information, but the precise stage of uncoating and the role of the intact or remodeled capsid in the early replication processes is currently not well understood.

Initially it was presumed that the viral capsid would disintegrate immediately upon membrane fusion, releasing a subviral complex in which the genome is complexed with viral replication enzymes. Several lines of evidence indicate that this model is incorrect. First, mutational analyses revealed that sequence alterations that either stabilize or destabilize the mature capsid are detrimental for infection, indicating that temporal control of uncoating is essential [40]. Second, reverse transcription and capsid disassembly appear to be functionally linked to each other, again indicating the requirement for spatio-temporal control [41]. Finally, host cell factors identified to play a role in nuclear import of the HIV genome in different cell types (e.g., RanBP2, Nup153, CPSF6) have been shown to target the viral CA protein or even CA hexamers, indicating that the capsid, or a capsid lattice, remains an integral part of the viral replication complex up to the nuclear pore [42].

Carrying out reverse transcription within the shielded environment of the viral capsid or a capsid-like structure could represent a strategy for avoiding intrinsic host cell immunity. DNA is not normally present in the cytoplasm of healthy mammalian cells, but is a hallmark of viral infection by retroviruses and DNA viruses. In response to this, cells have developed DNA sensor proteins able to induce intrinsic antiviral defense pathways. Recently, the c-GAS/STING pathway has been implicated in innate sensing of HIV-1 genomes [43]. Possibly, delayed capsid uncoating coupled to the progress of reverse transcription and intracellular transport contributes to shielding the viral replication complex from these nucleic acid-sensing pathways, and formation of the viral capsid in the maturation process may thus have an important role in escape of the virus from intrinsic immunity.

2.4.5 Nuclear import

The nuclear pore complex allows free diffusion of small molecules (up to about 40 kDa) and permits active transport of larger structures provided that transport is facilitated by nuclear import (or export) factors. The HIV PIC (with or without a capsid) is far bigger than the nuclear exclusion limit, and nuclear import must therefore involve an active import machinery. Various viral and cellular proteins have been implicated in nuclear import of HIV, but the actual mechanism is still unclear. The HIV CA protein interacts with several nucleoporins, which suggest they have a role in nuclear import, but other studies have implicated the accessory protein Vpr in nuclear entry as well [40]. In addition, the central DNA flap (a structural element produced during reverse transcription) has also been suggested to facilitate nuclear entry [44]. The size of the viral capsid has led to the suggestion that uncoating and loss of the capsid must occur at the cytoplasmic face of the nuclear membrane at the latest, but CA has also been reported to interact with Nup153, which resides at the inner side of the nuclear pore [41], potentially challenging this hypothesis. The smallest unit that harbors all components that are essential for integration is the intasome, comprising the viral cDNA and an IN tetramer, but the actual PIC is clearly much larger, and probably contains additional cellular and viral proteins, including the cellular integration cofactor LEDGF (see Section 2.4.6). Studies tracking viral replication using live cell, correlative and super-resolution microscopy and identifying both the constituents of the various complexes and the kinetics of their formation and alteration are urgently needed to shed light on this largely enigmatic stage of viral replication. It is likely that these studies will also produce new insight into fundamental cellular mechanisms and unravel novel principles of antiviral defense.

2.4.6 Integration

HIV integrase (IN) binds to each end of the viral cDNA, catalyzing the removal of two nucleotides from each 3′ strand. This results in a reactive 3′ hydroxyl group that will later mediate the nucleophilic attack (DNA strand transfer) on the chromosomal DNA. HIV-1 does not select for one or several specific integration sites, but also does not randomly integrate into the host genome. Favored integration sites are genes within gene-dense areas in euchromatin regions with active transcription. More specifically, *quasi*-palindromic sequences and histone deacetylation are common features of these integration sites. The host factor p75/LEDGF interacts directly with IN and has been identified as the major determinant for HIV-1 integration site specificity [45]. This selection is achieved by tethering the PIC to specific regions of the host DNA. Once integrated into the chromosomal DNA, the HIV genome may be considered as an additional cellular gene, which will remain for the life of that cell and all its daughter cells. Genomic integration may lead to active gene expression and productive infection (in most cases) or the viral genome may remain silent, yielding a latently infected cell. Since such cells cannot be distinguished by human defense systems from their uninfected counterparts, the elimination of latently infected cells with silent HIV genomes remains the ultimate challenge on the way to a potential cure for HIV.

2.5 Future of integrative HIV research

The results outlined in this chapter show that detailed information on individual steps in the process of transferring HIV-1 from one cell to the next has been acquired in recent years, and that newly developed technologies will allow us to zoom in further and obtain even more detailed, quantitative insights. Modeling of individual processes remains a current challenge due to the fact that important quantitative or even qualitative information is still missing and that the intricate interplay between viral replication and the host cell's machinery results in a high level of complexity and cell type specificity, even for individual steps like virus–cell attachment. For example, the step of HIV-1 proteolytic maturation presumably involves only the viral PR and its substrates Gag and Gag-Pro-Pol in the relatively defined environment of the virus particle. Nevertheless, it entails at least 66 distinct substrates, intermediates or products and numerous competing intermolecular interactions occurring simultaneously [20]. Steps that involve not only the conversion of viral components but also various transient, and potentially also competing, interactions with cellular factors (e.g., trafficking of the viral post-entry complex to the nucleus) are significantly more complex.

On the other hand, zooming in to nanoscale details will never yield the complete picture: a zoomed out view of multicellular, complex systems or whole organisms is indispensable for putting nanoscale details into a functional context. At the simplest level, this involves understanding the transfer of infectious HIV from the virus-producing cell to the next host cell. In the case of HIV, it is believed that direct cell–cell transfer through virological synapses is the most important transmission mode *in vivo* [46]. Studying transmission through these contacts requires the visualization of cellular details over a comparatively large 3D volume. Serial electron tomography and, in particular, focused ion beam milling/scanning electron microscopy, have provided the first glimpses into the contact zones. Different types of virological synapses have been described in the case of HIV-1, representing different strategies for viral spread. As we have outlined, HIV-1 infected T-cells can engage in contacts with uninfected T-cells in a manner analogous to the immunological synapse, but mediated by specific Env–CD4 interaction. Viruses are apparently released in a targeted manner into the confined and partly protected intracellular space, from where they can directly enter their next host cell [47]. Macrophages, which are the other important HIV-1-producing cells *in vivo*, have been shown to release newly formed particles into a large, seemingly intracellular compartment, which is formed by a large invagination of the plasma membrane [48]. They can be stored within this compartment for prolonged periods of time and be released in a targeted manner upon contact with an infectable T-cell. Dendritic cells are specialized in just another variation of the theme. Mature dendritic cells are not infectable, but they can bind and sequester virus in a large compartment connected to the plasma membrane [49], the precise nature and mode of generation of which is currently unknown.

Zooming out one more step, cell–cell transmission *in vitro* still represents an artificially reduced and simplified model. Integrating data obtained from such models into a more *in vivo*-like system makes it necessary to move from 2D tissue culture to cells in 3D environments, then further to complex multicellular systems or organotypic cultures and finally to animal models. These models will hopefully not only help us to elucidate viral spread and pathogenesis within an infected individual, but also to understand why HIV transmission *via* sexual intercourse is rather inefficient. Understanding the determinants for successful transmission might help to develop new prevention strategies. Complex experimental systems are difficult to establish, but ongoing developments in material sciences, labeling strategies and intravital microscopy and nanoscopy will provide virologists in the coming years with new tools and possibilities for tackling this challenging task, and this is the central topic of our collaborative research program.

Acknowledgement

Work in the authors' laboratory is supported by grants from the Deutsche Forschungsgemeinschaft within SFB1129 (Integrative analysis of pathogen replication and spread) and by DFG grant MU885/5-1.

References

1. Molla A, Paul AV, Wimmer E. 1991. Cell-free, de novo synthesis of poliovirus. *Science* 254: 1647–1651.
2. Cello J, Paul AV, Wimmer E. 2002. Chemical synthesis of poliovirus cDNA: generation of infectious virus in the absence of natural template. *Science* 297: 1016–1018.
3. Filee J. 2013. Route of NCLDV evolution: the genomic accordion. *Current Opinion in Virology* 3: 595–599.
4. La Scola B, Audic S, Robert C, Jungang L, de Lamballerie X, Drancourt M, Birtles R, Claverie JM, Raoult D. 2003. A giant virus in amoebae. *Science* 299: 2033.
5. Desnues C, Boyer M, Raoult D. 2012. Sputnik, a virophage infecting the viral domain of life. *Advances in Virus Research* 82: 63–89.
6. Earl LA, Lifson JD, Subramaniam S. 2013. Catching HIV "in the act" with 3D electron microscopy. *Trends in Microbiology* 21: 397–404.
7. Schorb M, Briggs JA. 2014. Correlated cryo-fluorescence and cryo-electron microscopy with high spatial precision and improved sensitivity. *Ultramicroscopy* 143: 24–32.
8. Chojnacki J, Muller B. 2013. Investigation of HIV-1 assembly and release using modern fluorescence imaging techniques. *Traffic* 14: 15–24.

9. Fackler OT, Murooka TT, Imle A, Mempel TR. 2014. Adding new dimensions: towards an integrative understanding of HIV-1 spread. *Nature Reviews Microbiology* 12: 563–574.
10. Muller B, Heilemann M. 2013. Shedding new light on viruses: super-resolution microscopy for studying human immunodeficiency virus. *Trends in Microbiology* 21: 522–533.
11. Sundquist WI, Krausslich HG. 2012. HIV-1 assembly, budding, and maturation. *Cold Spring Harbor Perspectives in Medicine* 2: a006924.
12. Briggs JA, Krausslich HG. 2011. The molecular architecture of HIV. *Journal of Molecular Biology* 410: 491–500.
13. Votteler J, Sundquist WI. 2013. Virus budding and the ESCRT pathway. *Cell Host and Microbe* 14: 232 241.
14. Balasubramaniam M, Freed EO. 2011. New insights into HIV assembly and trafficking. *Physiology* 26: 236–251.
15. Jouvenet N, Simon SM, Bieniasz PD. 2011. Visualizing HIV-1 assembly. *Journal of Molecular Biology* 410: 501–511.
16. Baumgartel V, Muller B, Lamb DC. 2012. Quantitative live-cell imaging of human immunodeficiency virus (HIV-1) assembly. *Viruses* 4: 777–799.
17. Lorizate M, Krausslich HG. 2011. Role of lipids in virus replication. *Cold Spring Harbor Perspectives in Biology* 3: a004820.
18. Hogue IB, Grover JR, Soheilian F, Nagashima K, Ono A. 2011. Gag induces the coalescence of clustered lipid rafts and tetraspanin-enriched microdomains at HIV-1 assembly sites on the plasma membrane. *Journal of Virology* 85: 9749–9766.
19. Muranyi W, Malkusch S, Muller B, Heilemann M, Krausslich HG. 2013. Super-resolution microscopy reveals specific recruitment of HIV-1 envelope proteins to viral assembly sites dependent on the envelope C-terminal tail. *PLoS Pathogens* 9: e1003198.
20. Konnyu B, Sadiq SK, Turanyi T, Hirmondo R, Muller B, Krausslich HG, Coveney PV, Muller V. 2013. Gag-Pol processing during HIV-1 virion maturation: a systems biology approach. *PLoS Computational Biology* 9: e1003103.
21. Wensing AM, van Maarseveen NM, Nijhuis M. 2010. Fifteen years of HIV protease inhibitors: raising the barrier to resistance. *Antiviral Research* 85: 59–74.
22. Schur FK, Hagen WJ, Rumlova M, Ruml T, Muller B, Krausslich HG, Briggs JA. 2015. Structure of the immature HIV-1 capsid in intact virus particles at 8.8 Å resolution. *Nature* 517: 505–508.
23. Pornillos O, Ganser-Pornillos BK, Yeager M. 2011. Atomic-level modelling of the HIV capsid. *Nature* 469: 424–427.
24. Ganser-Pornillos BK, Yeager M, Pornillos O. 2012. Assembly and architecture of HIV. *Advances in Experimental Medicine and Biology* 726: 441–465.
25. Pang HB, Hevroni L, Kol N, Eckert DM, Tsvitov M, Kay MS, Rousso I. 2013. Virion stiffness regulates immature HIV-1 entry. *Retrovirology* 10: 4.
26. Chojnacki J, Staudt T, Glass B, Bingen P, Engelhardt J, Anders M, Schneider J, Muller B, Hell SW, Krausslich HG. 2012. Maturation-dependent HIV-1 surface protein redistribution revealed by fluorescence nanoscopy. *Science* 338: 524–528.

27. Doms RW. 2004. Unwelcome guests with master keys: how HIV enters cells and how it can be stopped. *Topics in HIV Medicine* 12: 100–103.
28. Marsh M, Helenius A. 2006. Virus entry: open sesame. *Cell* 124: 729–740.
29. Izquierdo-Useros N, Lorizate M, Puertas MC, Rodriguez-Plata MT, Zangger N, Erikson E, Pino M, Erkizia I, Glass B, Clotet B, Keppler OT, Telenti A, Krausslich HG, Martinez-Picado J. 2012. Siglec-1 is a novel dendritic cell receptor that mediates HIV-1 trans-infection through recognition of viral membrane gangliosides. *PLoS Biology* 10: e1001448.
30. Geijtenbeek TB, Kwon DS, Torensma R, van Vliet SJ, van Duijnhoven GC, Middel J, Cornelissen IL, Nottet HS, KewalRamani VN, Littman DR, Figdor CG, van Kooyk Y. 2000. DC-SIGN, a dendritic cell-specific HIV-1-binding protein that enhances trans-infection of T cells. *Cell* 100: 587–597.
31. Arthos J, Cicala C, Martinelli E, Macleod K, Van Ryk D, Wei D, Xiao Z, Veenstra TD, Conrad TP, Lempicki RA, McLaughlin S, Pascuccio M, Gopaul R, McNally J, Cruz CC, Censoplano N, Chung E, Reitano KN, Kottilil S, Goode DJ, Fauci AS. 2008. HIV-1 envelope protein binds to and signals through integrin alpha4beta7, the gut mucosal homing receptor for peripheral T cells. *Nature Immunology* 9: 301–309.
32. Munch J, Rucker E, Standker L, Adermann K, Goffinet C, Schindler M, Wildum S, Chinnadurai R, Rajan D, Specht A, Gimenez-Gallego G, Sanchez PC, Fowler DM, Koulov A, Kelly JW, Mothes W, Grivel JC, Margolis L, Keppler OT, Forssmann WG, Kirchhoff F. 2007. Semen-derived amyloid fibrils drastically enhance HIV infection. *Cell* 131: 1059–1071.
33. Dustin ML. 2014. The immunological synapse. *Cancer Immunology Research* 2: 1023–1033.
34. Kulpa DA, Brehm JH, Fromentin R, Cooper A, Cooper C, Ahlers J, Chomont N, Sekaly RP. 2013. The immunological synapse: the gateway to the HIV reservoir. *Immunological Reviews* 254: 305–325.
35. Zhong P, Agosto LM, Munro JB, Mothes W. 2013. Cell-to-cell transmission of viruses. *Current Opinion in Virology* 3: 44–50.
36. Blumenthal R, Durell S, Viard M. 2012. HIV entry and envelope glycoprotein-mediated fusion. *Journal of Biological Chemistry* 287: 40841–40849.
37. Permanyer M, Ballana E, Este JA. 2010. Endocytosis of HIV: anything goes. *Trends in Microbiology* 18: 543–551.
38. Herold N, Anders-Osswein M, Glass B, Eckhardt M, Muller B, Krausslich HG. 2014. HIV-1 entry in SupT1-R5, CEM-ss and primary CD4+ T-cells occurs at the plasma membrane and does not require endocytosis. *Journal of Virology* 88: 13956–13970.
39. Gaudin R, de Alencar BC, Arhel N, Benaroch P. 2013. HIV trafficking in host cells: motors wanted! *Trends in Cell Biology* 23: 652–662.
40. Ambrose Z, Aiken C. 2014. HIV-1 uncoating: connection to nuclear entry and regulation by host proteins. *Virology* 454–455: 371–379.
41. Hilditch L, Towers GJ. 2014. A model for cofactor use during HIV-1 reverse transcription and nuclear entry. *Current Opinion in Virology* 4: 32–36.
42. Arhel N. 2010. Revisiting HIV-1 uncoating. *Retrovirology* 7: 96.

43. Gao D, Wu J, Wu YT, Du F, Aroh C, Yan N, Sun L, Chen ZJ. 2013. Cyclic GMP-AMP synthase is an innate immune sensor of HIV and other retroviruses. *Science* 341: 903–906.
44. Arhel NJ, Souquere-Besse S, Charneau P. 2006. Wild-type and central DNA flap defective HIV-1 lentiviral vector genomes: intracellular visualization at ultrastructural resolution levels. *Retrovirology* 3: 38.
45. Krishnan L, Engelman A. 2012. Retroviral integrase proteins and HIV-1 DNA integration. *Journal of Biological Chemistry* 287: 40858–40866.
46. Martin N, Sattentau Q. 2009. Cell-to-cell HIV-1 spread and its implications for immune evasion. *Current Opinion in HIV and AIDS* 4: 143–149.
47. Felts RL, Narayan K, Estes JD, Shi D, Trubey CM, Fu J, Hartnell LM, Ruthel GT, Schneider DK, Nagashima K, Bess JW, Jr., Bavari S, Lowekamp BC, Bliss D, Lifson JD, Subramaniam S. 2010. 3D visualization of HIV transfer at the virological synapse between dendritic cells and T cells. *Proceedings of the National Academy of Sciences of the United States of America* 107: 13336–13341.
48. Welsch S, Groot F, Krausslich HG, Keppler OT, Sattentau QJ. 2011. Architecture and regulation of the HIV-1 assembly and holding compartment in macrophages. *Journal of Virology* 85: 7922–7927.
49. Izquierdo-Useros N, Lorizate M, McLaren PJ, Telenti A, Krausslich HG, Martinez-Picado J. 2014. HIV-1 capture and transmission by dendritic cells: the role of viral glycolipids and the cellular receptor Siglec-1. *PLoS Pathogens* 10: e1004146.

3
Floral development: an integrated view

Hicham CHAHTANE,[*,1,2,3,4] Grégoire DENAY,[*,1,2,3,4] Julia ENGELHORN,[*,1,2,3,4] Marie MONNIAUX,[*,5] Edwige MOYROUD,[*,6] Fanny MOREAU,[1,2,3,4] Cristel CARLES,[1,2,3,4] Gabrielle TICHTINSKY,[1,2,3,4] Chloe ZUBIETA,[1,2,3,4] and François PARCY[1,2,3,4]

[1] CNRS, Laboratoire de Physiologie Cellulaire et Végétale, Grenoble, France
[2] Université Grenoble Alpes, Grenoble, France
[3] CEA, DSV, iRTSV, LPCV, Grenoble, France
[4] INRA, LPCV, Grenoble, France
[5] Department of Comparative Development and Genetics, Max Planck Institute for Plant Breeding Research, Köln, Germany
[6] Department of Plant Sciences, University of Cambridge, UK
*These authors made equal contributions.

Abstract

This chapter reviews current knowledge about the development of flowers in the model plant *Arabidopsis thaliana*. It describes how a flower develops at the morphological level and the nature of the gene regulatory network that controls the different steps of this process. The chapter focuses on regulations occurring at the transcriptional and chromatin levels and lists the most prevalent proteins that act as key regulators. The latest methods available for modeling how these regulators work and studying

From Molecules to Living Organisms: An Interplay Between Biology and Physics. First Edition. Eva Pebay-Peyroula et al. © Oxford University Press 2016.
Published in 2016 by Oxford University Press.

their role at the genome-wide level are presented. Finally, the structural specificities of transcription factors acting specifically to define the flower are described.

Keywords

Floral development, transcription factor, chromatin regulators, genomics, modeling, structural biology, *Arabidopsis thaliana*

Chapter Contents

3 Floral development: an integrated view   43
Hicham CHAHTANE, Grégoire DENAY, Julia ENGELHORN, Marie MONNIAUX, Edwige MOYROUD, Fanny MOREAU, Cristel CARLES, Gabrielle TICHTINSKY, Chloe ZUBIETA, and François PARCY

3.1	Description of floral development		47
	3.1.1	Flowers in a nutshell	47
	3.1.2	Types of flowers	49
	3.1.3	The role of meristems in flower formation	50
	3.1.4	Types of inflorescence	51
	3.1.5	A model for understanding floral development: *Arabidopsis thaliana*	51
3.2	Genetic control of floral development in Arabidopsis		54
	3.2.1	Introduction	54
	3.2.2	Localization of the emerging primordium: auxin transport in the SAM	54
	3.2.3	Acquisition of meristematic identity: initiation of the floral primordium	55
	3.2.4	The determination of floral identity	56
	3.2.5	The ABC model of floral organ development	57
	3.2.6	Transcriptional regulation during floral development	60
3.3	Transcription and chromatin dynamics in the "seq" era		61
	3.3.1	Classical approaches	61
	3.3.2	NGS: what it is, different types	62
	3.3.3	RNA-seq	66
	3.3.4	ChIP-seq	71

	3.3.5	Chromatin accessibility and nucleosome mapping	75
	3.3.6	Physical interaction between distant chromatin regions	77
	3.3.7	Approaches and technologies for space and time resolution	79
	3.3.8	Summary	81
3.4		Modeling TF–DNA binding to predict transcriptional regulation	82
	3.4.1	Introduction	82
	3.4.2	Generating TF–DNA binding models	83
	3.4.3	Using TFBS models	85
	3.4.4	Prospects	86
3.5		Layers of chromatin regulation and their actors	86
	3.5.1	Nucleosome density and remodeling	86
	3.5.2	Histone variants	87
	3.5.3	Post-translational modifications (PTMs) of histone tails	88
	3.5.4	Evidence for the interplay between TFs and chromatin features in the floral development framework	93
3.6		Floral transcription factors: a structural view	94
	3.6.1	Introduction	94
	3.6.2	Basis for TF–DNA interaction at the structural level	94
	3.6.3	Examples of plant-specific floral TFs	95
	References		100

3.1 Description of floral development

Flowers are the emblematic reproductive structure of angiosperms, also known as "flowering plants". Flower production requires a profound change in the developmental program of the plant: from making leaves and shoots (vegetative development) to making flowers that form an inflorescence, i.e., a cluster of flowers arranged along the stem (reproductive development).

3.1.1 Flowers in a nutshell

3.1.1.1 *The appearance of flowering plants during evolution*

There are more than 350,000 living species of flowering plants, making the angiosperms an exceptionally species-rich group, only outnumbered by the arthropods (insects, spiders and crustaceans) in the animal kingdom. A striking feature of angiosperm evolution is the extremely rapid (40 Myr) diversification and dominance of this group over the land (Soltis et al., 2008) (Fig. 3.1). Many features of the angiosperms (e.g., true wood vessels, pollen grains with a supportive structure, two integuments surrounding the ovules) could account for the establishment of their dominance. The flower is one of those features and certainly played a prominent part in the success of the angiosperms. Despite the omnipresence and evolutionary importance of flowers, their origin remains enigmatic (Frohlich and Chase, 2007).

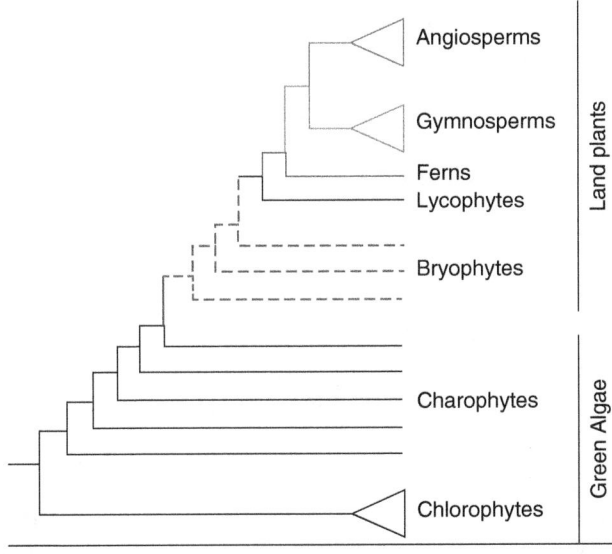

Fig. 3.1 Simplified plant phylogeny. Chlorophytes and charophytes together constitute the green algae while the other phyla (bryophytes, lycophytes, ferns, gymnosperms and angiosperms) are all land plants. The phylogenetic position and relationships of the three lineages of bryophytes (hornworts, liverworts and moss) are still debated and are thus indicated with a dashed line.

The earliest known fossils of flowers date back to 130 million years ago (Early Cretaceous) but molecular data and new fossil records suggest that the angiosperms, and thus the flower, could have emerged much earlier (Soltis et al., 2008). In addition, the identity and morphology of the last common ancestor of the angiosperms (i.e., the "first" flower) is still debated: gymnosperms (conifers, gingko, gnetales and cycads) are the closest living relatives to the angiosperms but their reproductive organs are very different from flowers, making comparative studies difficult (Frohlich and Chase, 2007). Consequently, we still do not know how the first flower was assembled.

3.1.1.2 Function of the flower

The flower is the reproductive structure of angiosperms, and ensures the fertilization of ovules to give seeds which are contained within a fruit. It represents a key innovation in the evolution of land plants: the flower gathers for the first time both male and female reproductive organs, together with protective and attractive organs, in a single compact structure (for review see Frohlich and Chase, 2007). The corolla provides visual, olfactory and tactile cues used by animal pollinators to detect flowers efficiently, thus increasing the rate of pollination. Also, unlike those of gymnosperms, angiosperm ovules are not naked but are enclosed within an ovary divided into carpels. Ovaries restrict access of the male gametes to the ovules and, together with self-incompatibility mechanisms, participate in controlling the specificity of fertilization, favoring outcrossing and consequently increasing genetic diversity. Fertilization is a complex mechanism in flowering plants: one of the two nuclei contained in the pollen will fertilize the egg cell (which will produce the zygote) while the other will combine with another set of maternal cells (which will develop into the endosperm, a nutritive tissue). This process of double fertilization is specific to the angiosperms, and constitutes an important evolutionary novelty since it triggers simultaneously the development of the zygote and its nutritive tissues (which will be crucial for seed maturation and germination) (Lord and Russell, 2003). After fertilization, the ovaries develop into fruits that contain the seeds and have a crucial role in their dispersion: fruits of the maple tree (*Acer*) develop thin membranes that allow wind carriage, fleshy fruits are generally eaten by animals which can thus carry the seeds on a long distance and some fruits (like the ones of *Impatiens balsamina*) even propel their seeds away from the mother plant.

By increasing the odds of pollination, making fertilization a highly efficient and specific process and providing the embryo with an optimum growth environment and means of dissemination, the flower is indeed one of evolution's greatest outcomes.

3.1.1.3 Morphological description of the flower

As everyone can observe in nature, the architecture of flowers is so diverse that giving a consensual definition of a flower would appear to be difficult. Yet, there is a single organizational plan shared by all flowers, which attests their common origin and the presence of a single common ancestor for all flowers. Flowers typically display four types of organs arranged on four concentric whorls (Fig. 3.2A). Sepals form the outermost whorl that develops first and encloses the floral bud to protect other organs until they are mature. Petals, just internal to sepals, have a primary role in flower

Fig. 3.2 Organization and diversity of flowers. (A) Typical body plan of a flower showing the four archetypal whorls of organs. (B) Sexual organs arranged on a spiral in *Magnolia*, note the traces of the stamen insertion point on the receptacle. (C) Actinomorphic flower of *Hibiscus trionum*. (D) Zygomorphic flower of *Lobelia erinus*. (E) Missing petal in a ginger (*Asarum canadense*) flower. (F) Male and (G) female flowers of holly (*Ilex verticillata*): here the stamens fail to develop in female flowers while carpels abort in male flowers. (H) Tepals of *Tulipa linifolia*. (I) Modified sepals forming the pappus in dandelions (*Taraxacum officinale*). (J) Nectar spur in *Linaria triornithophora*. (K) Modified flowers of grasses: grasses are generally wind pollinated and, as such, their flowers tend to be extremely reduced and devoid of conspicuous petals.

opening. Indeed, in some species such as the perennial violet the flower remains closed when petals fail to develop (Winn and Moriuchi, 2009). This is of ecological importance since closed flowers typically reproduce by self-pollination (cleistogamy) whereas open flowers are more prone to cross-pollination (chasmogamy). The second most evident function of the petal is the attraction of pollinators, which is mediated not only by the shape and the patterns (color and structures) of these organs but also by the scent they often produce. Stamens arise in the third whorl, and produce pollen grains, the male gametophytes involved in reproduction. Finally, carpels emerge at the center of the flower, organized in a structure termed the pistil, and form the ovules. These ovules contain the two female reproductive cells that will be involved in the double fertilization process later on (Lord and Russell, 2003). Although this classical body plan is common to a large number of angiosperm species, it can also show dramatic variations, which generate the amazing diversity of floral forms that we can observe.

3.1.2 Types of flowers

Flowers have experienced phenomenal diversification in the past 135 Myr. Variations in the arrangement of the floral parts combined with the morphological plasticity of individual organs account for the extraordinary diversity we can observe today.

First, the disposition of the four types of floral organs on the receptacle (phyllotaxis) varies: organs of the same type can emerge forming a ring-like structure (whorl) or adopt a spiral phyllotaxis. In this case, successive organs emerge following a Fibonacci pattern so that an organ arises at an angle of 137.5° from the previous organ (Fig. 3.2B). Monocots and most core eudicots display a whorled phyllotaxis, while in early diverging angiosperms and magnoliids, spiral and whorled organizations are both present (Endress and Doyle, 2007). Remarkably, the order of the successive organ types is extremely stable as only one species departs from the typical sequence "sepals, petals, stamens and carpels" (from the periphery to the center of the flower): in *Lacandonia schismatica* the position of the sexual organs is inverted so that stamens are found in the center of the flower, surrounded by a whorl of carpels (Vergara Silva et al., 2003).

Variations in floral symmetry also contribute to species diversity (Sargent, 2004). As the corolla (petals) is often the most conspicuous part of a flower, the symmetry of the flower generally depends on the symmetrical arrangement of the petals. For instance, flowers can display a radial symmetry (actinomorphic flower; Fig. 3.2C) or a single symmetry plane (zygomorphic flower; Fig. 3.2D). In this case, petals are often very different from each other so that "dorsal" petals are clearly distinct from "ventral" and/or lateral petals. In many species, one type of organ can also be absent (Fig. 3.2E–G), either because the corresponding organ primordia aborted prematurely during floral development or because they are converted into another organ type (e.g., many modern rose varieties are almost devoid of stamens because decades of intense breeding to produce flowers with many petals have selected for floral mutants in which stamens are converted into petals).

The number, shape and visual appearance of floral organs fluctuate enormously between species, contributing to the morphological diversity of flowers. Individual organs can also be dramatically modified: in lilies, sepals are very colorful and almost indistinguishable from petals, so that sepals and petals are commonly referred to as "tepals" (Fig. 3.2H). In many daisies, sepals are devoid of chlorophyll and are extremely reduced to resemble feather-like bristles called pappus (Fig. 3.2I). Pappus is functionally important as it enables seeds to be carried by the wind. In certain families, part of an organ can also be reshaped to create a novel structure. Nectar spurs found in many species of the snapdragon family (Antirrhineae) are a good example. Here, part of the petal extends to form an elongated tube that contains the nectar. Only pollinators with a tongue long enough to reach the end of the tube can pollinate the flower, and consequently structures like nectar spurs participate directly in genetic isolation and the creation of new species.

3.1.3 The role of meristems in flower formation

All plant organs arise from particular structures called meristems that contain undifferentiated stem cells. These meristems are precisely localized in the plant and influence plant architecture (Fig. 3.3A) (Grbić, 2005; Holt et al., 2014). Two main pools of stem cells are established during embryogenesis: the root apical meristem (RAM) and the shoot apical meristem (SAM). These two meristems produce different organs. The SAM, found at the tip of the primary stem, gives rise to leaves and flowers (Fig. 3.3B) whereas the RAM corresponds to the quiescent zone of the root tip and

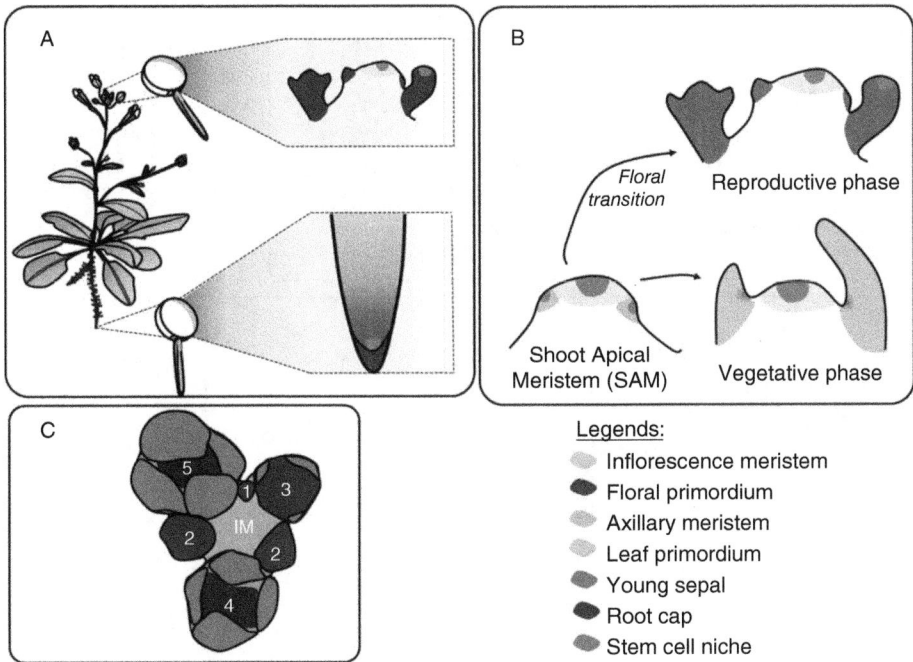

Fig. 3.3 Meristems as a source of undifferentiated cells for the formation of new organs. (A) Morphological description of the model plant *Arabidopsis thaliana* and positioning of the shoot apical meristem (SAM) and root apical meristem (RAM). (B) Focus on a SAM, producing a vegetative meristem during the vegetative phase and an inflorescence meristem (IM) and floral meristem (FM) during the reproductive phase. (C) Close-up of an IM with several FM.

contributes to root elongation (Sparks et al., 2013). The SAM produces individual meristems at their periphery, called axillary meristems (AM), that can generate axillary organs such as stems or flowers (see Section 3.2). Once the plant reaches the reproductive stage, the SAM becomes an inflorescence meristem (IM), which generates the floral meristems (FM) to give rise to flowers (Pidkowich et al., 1999) with a specific phyllotaxis (Fig. 3.3C; for more details see Section 3.2).

3.1.4 Types of inflorescence

An inflorescence represents a cluster of flowers precisely arranged along a reproductive shoot. Various inflorescence types exist in nature and are briefly illustrated in Fig. 3.4.

3.1.5 A model for understanding floral development: *Arabidopsis thaliana*

In order to overcome the enormous diversity observed in floral development, a few models have been chosen to study this complex process in detail. *Arabidopsis thaliana*,

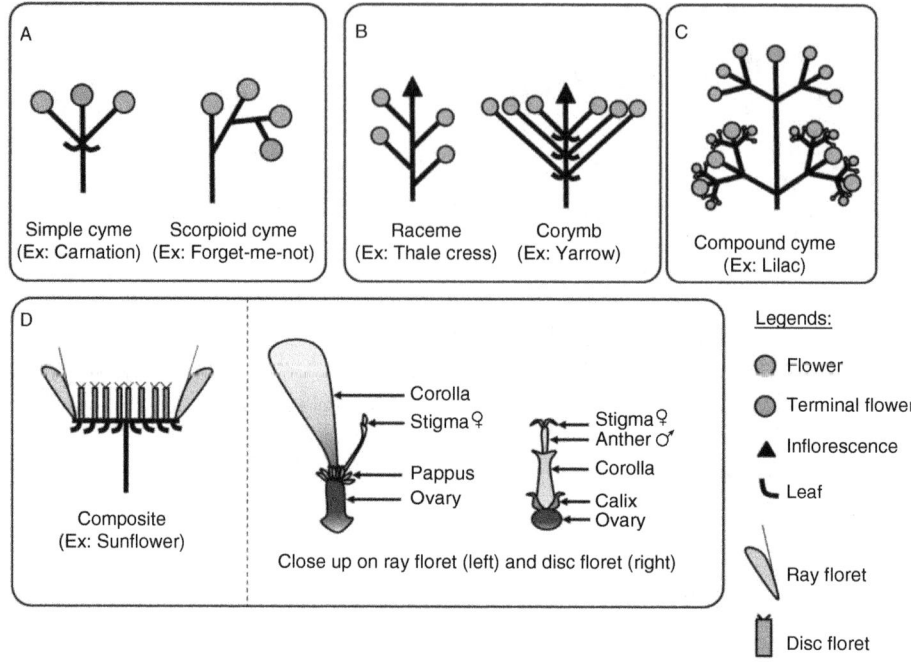

Fig. 3.4 The diversity of inflorescence architectures. (A) Examples of determinate inflorescences. (B) Examples of indeterminate inflorescences. (C) Example of a compound inflorescence (lilac). (D) Example of a flower-like inflorescence (daisy/sunflower) with details of the highly modified individual flowers called ray florets or disc florets, depending on their morphology and position within the inflorescence.

which is the most studied plant species in other domains of plant biology, is also one of the most common models used for understanding floral development. However, one has to bear in mind that a single model can never account for the large variability observed in floral development, which is why other genera such as *Antirrhinum* (snapdragon), *Aquilegia*, orchids and many more are also investigated—a crucial step in order to understand this process in all its complexity.

Arabidopsis thaliana belongs to the family Brassicaceae, as do cabbage and rapeseed, and it has features that make it an excellent model plant, such as a short life cycle, the capacity to self-pollinate and a high seed yield (for a review see Koornneef and Meinke, 2010). Its genome was fully sequenced in 2000 and transgenic plants are easily obtained by transformation with the bacterium *Agrobacterium tumefaciens*. Its inflorescence is a simple indeterminate raceme that produces numerous small flowers (Fig. 3.5A). In addition to the primary stem, secondary stems displaying the same inflorescence structure can initiate from the rosette. Flowers are not subtended by bracts (leaf-like organs) and produce sequentially four sepals, four white petals, six stamens (four long and two short, as characteristic of most Brassicaceae) and two fused carpels (Fig. 3.5B).

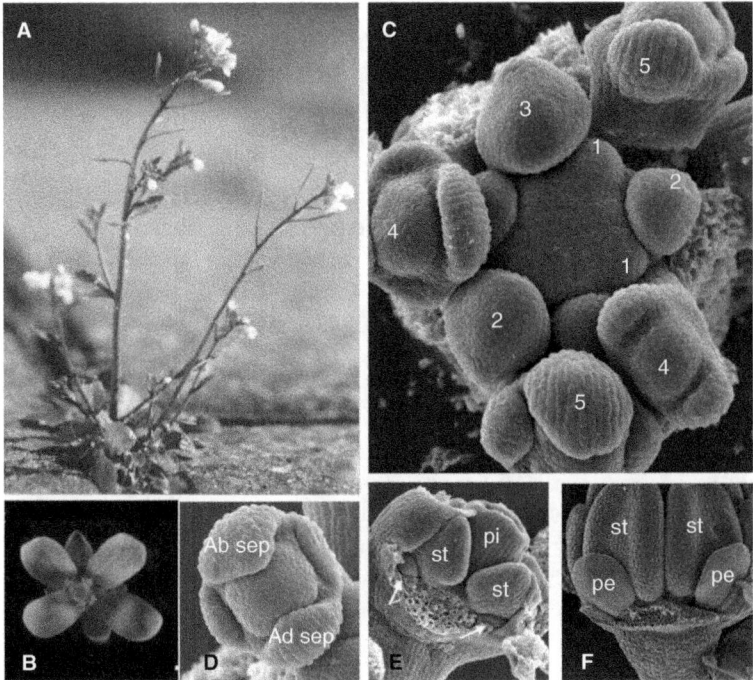

Fig. 3.5 *Arabidopsis thaliana*, a model for studying floral development. (A) An entire *A. thaliana* plant showing rosette leaves, primary stem and two secondary stems, with a simple raceme structure. (B) *Arabidopsis thaliana* flower seen from the top. (C)–(F) Scanning electron microscope images of inflorescence meristem and flowers at various stages of development: inflorescence meristem (center) producing flowers from stages 1 to 5 (C); detail of a stage 4 flower with sepals emerging (D); stage 7 flower where petals (indicated with an arrow) are just becoming visible at the base of already developed stamens (E); stage 8 flower (F). For pictures (E) and (F), one sepal was removed in order to see the internal organs. Ad sep, adaxial sepal; Ab sep, abaxial sepal; pe, petal; st, stamen; pi, pistil. Stages are defined in Smyth et al. (1990). Pictures (B)–(F) were provided courtesy of Angela Hay.

The precise spacing and timing of floral organ development has been assessed in *A. thaliana*, leading to the definition of 20 stages of reproductive development, from the barely noticeable flower meristem to the mature fruit (Smyth et al., 1990) (Fig. 3.5C). Floral development *per se* has been divided into 12 stages that together last for approximately 13 days. Nearly two flower buds emerge every day, making this process very rapid. The internal organs are only visible during the first stages (up to stage 6); later they are fully covered by sepals. Although organs initiate in a whorled manner, not all organs within the same whorl initiate at exactly the same moment: this is particularly evident when looking at sepal development, since the abaxial sepal clearly arises before the adaxial and the two lateral ones (Fig. 3.5D). Petals arise slightly earlier than stamen but grow more slowly, so that at stage 8 stamens are much bigger than petals (Fig. 3.5E,F). Later on, petal size increases to the size of the long stamens, and the pistil elongates further to give the final morphology of the mature flower. Precise description of the stages of floral development in a model species opens the door to a deep understanding of the genetic processes underlying each of these steps.

3.2 Genetic control of floral development in Arabidopsis

3.2.1 Introduction

In Section 3.1.5 we described the ontogeny of flowers. Here we will present the main genes involved in controlling floral development. We will discuss the molecular events that take place sequentially from the earliest emergence of the floral bud to the development of floral whorls.

The steps in floral development are described in Fig. 3.6. The position of an emerging primordium is first localized on the flank of the SAM (stage 0). This emerging primordium then acquires a meristematic identity, i.e., the capacity to produce new primordia on its sides (stage 2). It must also acquire either a floral or an inflorescence identity, depending on the architecture of the plant. In the case of floral meristems, the last developmental step is the production of floral organs (stage 3 onwards).

3.2.2 Localization of the emerging primordium: auxin transport in the SAM

Plant architecture is determined by the arrangement and nature of organs emerging on the flanks of the SAM. The position of emergence of new organs (called phyllotaxy) follows strict rules that are partly genetically regulated. In Arabidopsis, leaves or flowers are initiated in a spiral pattern following an angel of 137.5°.

The main regulator of this phyllotactic pattern is the plant hormone auxin. Mutants in the auxin efflux transporter PINFORMED1 (PIN1) as well as plants treated with the polar auxin transport inhibitor 1-N-naphthylphthalamic acid do not form

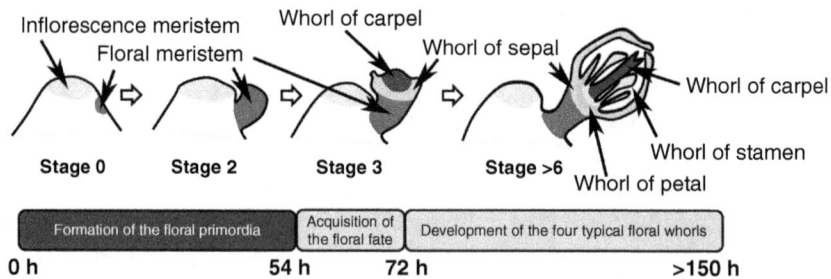

Fig. 3.6 Development of the floral meristem at the flank of the SAM. Three successive stages of floral development are shown: formation of the floral primordial, acquisition of the floral fate and the development of each floral organ in a stereotyped arrangement. Indicative hours of each step and stage are from Smyth et al. (1990).

lateral organs (Okada et al., 1991; Reinhardt et al., 2000). *In vivo* imaging of PIN1 and the localization pattern of the auxin response reporter DR5 in the SAM showed that PIN1 directs auxin transport towards the site of future primordia initiation through the epidermal cell layer, creating a local auxin maximum. PIN1 localization is then switched toward the inner tissues, depleting auxin in the forming primordium (Benková et al., 2003; Heisler et al., 2005). The auxin maximum triggers the expression and the activity of the AUXIN RESPONSE FACTOR 5 (ARF5)/MONOPTEROS (MP) transcription factor (TF) at the site of initiation of the floral primordium. Mutants in *MP* display a naked inflorescence resembling that of *pin1* mutants (Cole et al., 2009), pointing to MP as the integrator of auxin signaling in the SAM (Fig. 3.7).

3.2.3 Acquisition of meristematic identity: initiation of the floral primordium

Auxin maxima localize the zone of the SAM where a new primordium develops. Once activated in this zone, MP triggers the formation of a proper meristem.

MP first activates the floral marker genes *AINTEGUMENTA* (*ANT*) and *LEAFY* (*LFY*) (Blázquez et al., 1997; Long and Barton, 2000; Yamaguchi et al., 2013). These two TFs are crucial for defining the initiation of the meristem, although only a subset of their targets are known to act in this pathway. Among the targets of

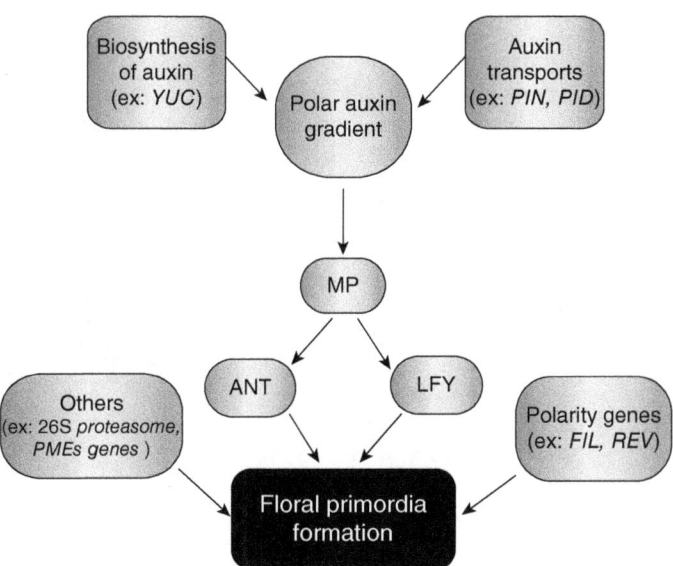

Fig. 3.7 A simplified genetic cascade leading to the formation of floral primordia. The auxin signaling pathway (the top four boxes) is the first part of this process, leading to the activation of precocious floral markers (ANT, LFY) through MP activity. ANT and LFY, as well as other important pathways (left and right bottom boxes), control downstream events leading to completed initiation of primordia.

56 *Floral development: an integrated view*

LFY, *REGULATOR OF AXILLARY MERISTEMS1* (*RAX1*) and *CUC-SHAPED-COTYLEDONS2* (*CUC2*) are both involved in the first events leading to the formation of axillary meristems (Chahtane et al., 2013; Yamaguchi et al., 2014). Mutations of either *rax1* or *cuc2* alone do not lead to a defect in initiation of floral primordia, probably due to strong genetic redundancy downstream of LFY and ANT (Yang et al., 2012). Nevertheless, all these genes are involved in the formation of axillary meristems in plants (leading preferentially to a side shoot), a type of meristem that is very similar to a floral meristem (Grbić and Bleecker, 2000). Thus, study of the direct targets of LFY and ANT involved in this process is required to fully understand how they work together to induce the formation of floral primordia.

Many other genes are also involved in the formation of floral primordia, including *FILAMENTOUS FLOWER* (*FIL*) and *REVOLUTA* (*REV*), which define the polarity of the emerging meristem towards the SAM (Chen et al., 1999; Otsuga et al., 2001) or the 26S proteasome (a complex responsible for protein degradation).

3.2.4 The determination of floral identity

During reproductive development, lateral meristems can develop into new inflorescences or into flowers. Inflorescence meristems have the same genetic identity as the SAM from which they are derived. However, to develop a flower, the newly produced meristem needs to acquire a floral identity.

Three main factors determine the balance between inflorescence meristem and floral meristem identity (Fig. 3.8): *LEAFY* (*LFY*), *APETALA1* (*AP1*) and *TERMINAL FLOWER 1* (*TFL1*) (Shannon and Meeks-Wagner, 1993). The *tfl1* mutant in *A. thaliana* shows a precocious flowering phenotype with a terminal flower, revealing that *TFL1* normally represses floral production (Shannon and Meeks-Wagner, 1991).

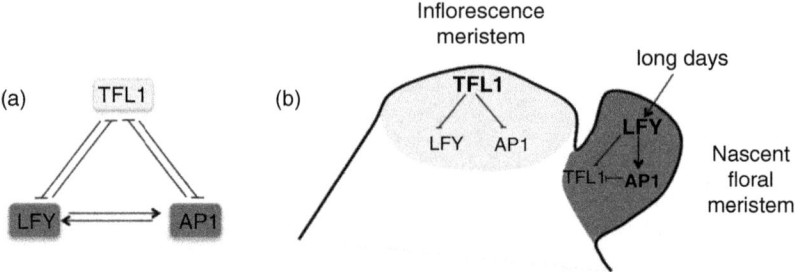

Fig. 3.8 Genetic control of floral meristem identity. Transcriptional relationships between *TFL1*, *LFY* and *AP1* are indicated in (a). The action of *TFL1* is antagonistic to the action of *LFY* and *AP1* by mutual transcriptional repression, whereas *AP1* and *LFY* mutually activate each other (Liljegren et al., 1999; Shannon and Meeks-Wagner, 1993). In the inflorescence meristem (b) *TFL1* is predominantly expressed, preventing the expression of *LFY* and *AP1*. In the nascent floral primordium, *LFY* is induced (e.g., by long days or the GA pathway). This leads to an increase in *AP1* expression and a concomitant decrease in *TFL1* (Ratcliffe et al., 1999).

In contrast, the *lfy* mutant develops indeterminate leafy branches in place of flowers, hence *LFY* is necessary for the transition between the inflorescence and the floral meristem (Schultz and Haughn, 1991; Weigel et al., 1992). In addition, *AP1*, together with it paralog *CAULIFLOWER* (*CAL*), is required for acquisition of floral identity. In the SAM, *TFL1* is predominantly expressed, conferring the inflorescence identity. The balance between *TFL1* and *LFY* will determine the identity of the emerging primordium.

Both LFY and AP1 are TFs: *LFY* is a unique gene, in a category of its own (Weigel et al., 1992; Blázquez et al., 1997), whereas *AP1* is a MIKC-type MADS-box gene whose members are particularly involved in reproductive development. *TFL1* belongs to the CETS family of transcriptional regulators (together with the flowering time gene *FT*; see Chapter 7). TFL1 was proposed to act as a co-regulator of FD, a flowering time TF (see Chapter 7): TFL1 would promote transcriptional repression of FD targets (Hanano and Goto, 2011), especially *LFY* and *AP1*. A small gene network therefore controls the identity of the emerging floral meristem (Fig 3.8). This network results in a developmental switch from vegetative to reproductive development (Parcy et al., 1998; Ratcliffe et al., 1999).

It has been recently demonstrated that AP1 acts with four other MADS-box TFs, namely SHORT VEGETATIVE PHASE (SVP), SUPPRESSOR OF OVEREXPRESSION OF CO 1 (SOC1), AGAMOUS-LIKE 24 (AGL24) and SEPALLATA 4 (SEP4), to downregulate *TFL1* expression (Liu C. et al., 2013). SOC1, SVP and AGL24 are flowering time genes, participating in inflorescence meristem identity. They are repressed by AP1 (Liu et al., 2007), and therefore act very transiently in the nascent floral primordium to allow the transition between the inflorescence and the floral meristem identity.

Other secondary players have been identified in this reproductive switch, such as *CAL*, a Brassicaceae-specific paralog of *AP1* (Kempin et al., 1995), *LATE MERISTEM IDENTITY 1* (*LMI1*), a leucine zipper homeodomain TF (Saddic et al., 2006; Grandi et al., 2012), *LMI2* (Saddic et al., 2006; Pastore et al., 2011), *FRUITFUL* (*FUL*) (Ferrándiz et al., 2000), *BOP1* and *-2* (Xu et al., 2010).

3.2.5 The ABC model of floral organ development

Around stage 2 of floral development (Fig. 3.6), the floral primordium is a small bulge of cells that acquires both its floral and meristematic identity. From stage 3 onwards, this bulge develops and forms several organs, distributed in Arabidopsis in successive whorls: sepals, petals, stamens and carpels. A small group of genes participate in the determination of floral organ identity. They were initially discovered through the study of homeotic floral mutants of *A. thaliana* and *Antirrhinum majus*, in which floral organs developed with the wrong identity (Bowman et al., 1989, 1991; Coen and Meyerowitz, 1991). These studies led to the so-called "ABC model", developed in the 1990s (Fig. 3.9; for a review see Meyerowitz et al., 1989). According to this model, the combined expression of class A, class B and class C genes defines the regions where sepals, petals, stamens and carpels will develop in the floral meristem.

58 *Floral development: an integrated view*

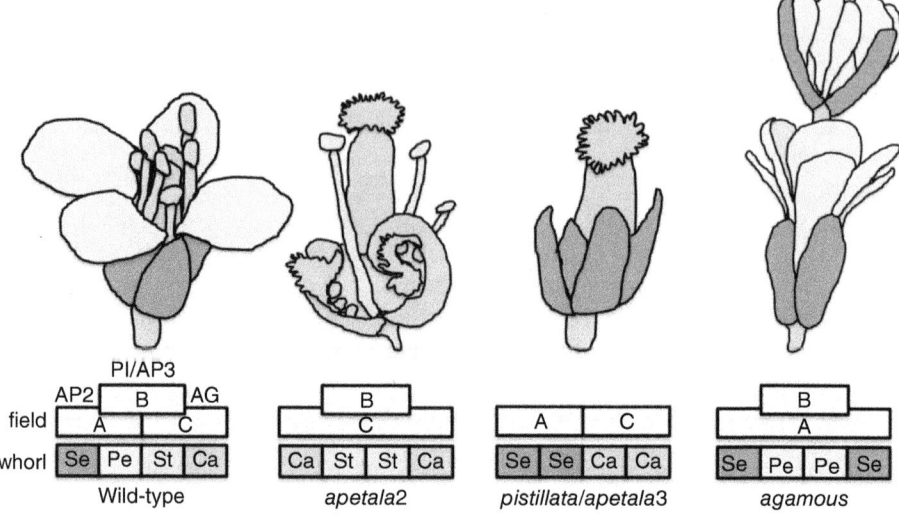

Fig. 3.9 The ABC model of floral development in Arabidopsis. Schematic representation of the floral organ identity determination model according to Bowman et al. (1991), showing how *APETALA2 (AP2), APETALA3 (AP3), AGAMOUS (AG)* and *PISTILLATA (PI)* genes can control organ identity. In the wild-type plant, *AP1* is expressed in the external field (A), *PI* and *AP3* in the medial field (B) and *AG* in the inner field (C; i.e., in the center of the flower). The combination of fields determines organ identity: the A function differentiate sepals, combination of A and B forms petals, combination of B and C leads to stamen formation and C function alone would give carpels. Sketches of the *ap2*, *ap3/pi* and *ag* mutants are shown above the representation of the genes' areas of action in each of these mutants. Whorls are represented in consecutive order and organ identity is shown for each whorl. Se, sepals; Pe, petals; St, stamens; Ca, carpels. The *agamous* mutants lack meristem termination as illustrated by the production of a flower in the center of the flower.

The ABC model has been further extended as the ABCDE model, where the D function determines the ovule and fruit identities and is carried by *SEEDSTICK (STK)* and *SHATTERPROOF (SHP)* (Pinyopich et al., 2003), while the E function is necessary for all organ identities to be complete, and is fulfilled by the *SEPALLATA (SEP)* genes (Pelaz et al., 2000).

LFY is the main activator of the ABCDE genes, since it directly activates the expression of *AP1, AG, AP3* and *SEP3* (Weigel and Nilsson, 1995; Parcy et al., 1998; Moyroud et al., 2011; Winter et al., 2011). LFY is expressed in the whole floral meristem and the whorl-specific activation of these ABCE genes is due to the interaction of LFY with different partners: for instance, UNUSUAL FLORAL ORGANS (UFO) is expressed in the second and third whorls of the floral meristem, and interacts with LFY to activate the expression of B genes (Lee et al., 1997). LFY-specific interaction with SEP3, present in whorls 2, 3 and 4, is necessary for the proper expression of B and C genes (Liu et al., 2009), while the LFY–WUSCHEL complex activates *AG* expression (Lohmann et al., 2001). In addition to this whorl-specific expression, the precise patterning of ABC gene expression in the floral meristem is also maintained by their cross-regulatory actions: *AP1* and *AP2* inhibit *AG* expression in whorls 1 and 2, while *AG* inhibits *AP1* expression in whorls 3 and 4 (Gustafson-Brown et al., 1994; Sridhar et al., 2006).

Strikingly, almost all of the ABCDE proteins, except AP2, are part of the MADS-domain family. These proteins are organized in four domains, named M, I, K and C. Whereas the M domain is the DNA-binding domain (DBD), the other domains mediate different degrees of protein interaction: indeed, the whorl-specific and combinatorial action of the ABC proteins for floral organ identity is mediated by complex patterns of homo- and heterodimerization and tetramerization, as proposed by the "floral quartet model" (Fig. 3.10). According to this model, a quartet of AP1 and SEP proteins would specify sepal identity, while a quartet of AP1, AP3, PI and SEP would specify petal identity (Theissen and Saedler, 2001). Interestingly, the model relies on DNA looping

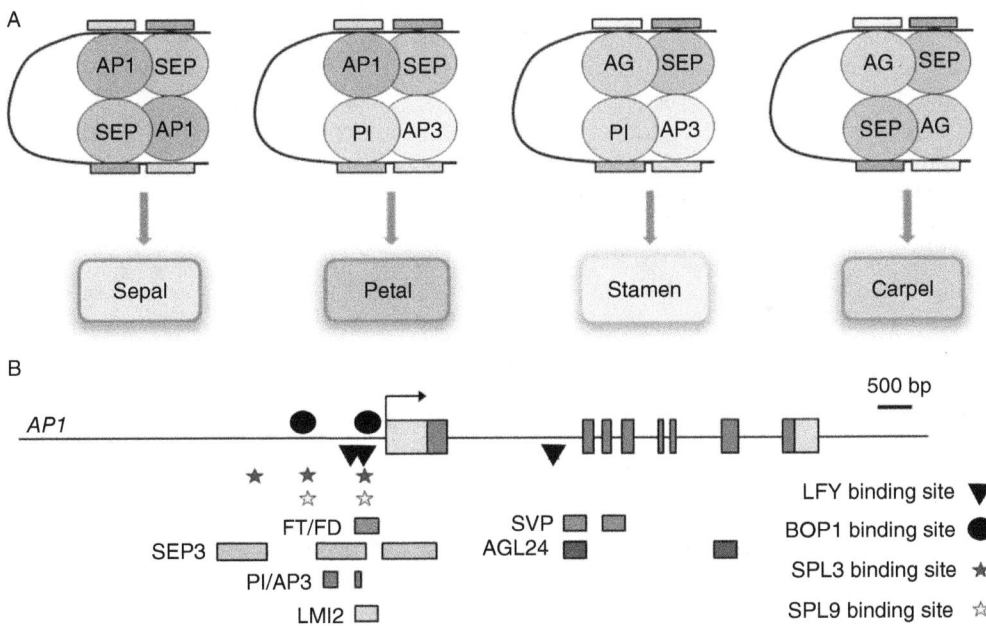

Fig. 3.10 The activity of MADS-box transcription factors depends on proper gene regulation, interaction with the correct partners and binding to specific *cis*-elements. (A) Regulation of MADS-box transcription factor activity through interaction with different partners: the quartet model. For instance, sepal identity would be controlled by an AP1–SEP quartet, where each protein binds to a specific *cis*-element, and the quartet is formed through DNA looping. (B) Gene model of *AP1*, showing binding sites or regions for all direct regulators characterized. Stars indicate precisely characterized binding sites, while rectangles show broader regions of interaction as determined by chromatin immunoprecipitation.

in order to explain the binding specificity of these complexes, a physical property that has been shown for several TFs. Physical evidence of all these different quartets has not been fully obtained yet, but a complex of AP3, PI and SEP3 proteins can form *in vitro* (Melzer and Theissen, 2009), and the STK–SEP3 complex as well as SEP3 homotetramers are able to bend DNA (Melzer et al., 2009; Mendes et al., 2013).

3.2.6 Transcriptional regulation during floral development

3.2.6.1 cis-regulation

All actors in floral development display a complex pattern of genetic and physical interactions. Whereas relatively little is known about their physical interactions, the regulatory interactions between these genes have been abundantly described. The precise spatio-temporal expression pattern of each floral regulator, a key feature in its function, is controlled by the combined action of several transcription factors. The case study of the regulation of *AP1* expression illustrates general principles of *cis*-regulation during floral development.

As previously mentioned, *AP1* is involved in floral meristem identity as well as in sepal and petal identity (it is a class A gene) (Irish and Sussex, 1990). *AP1* is expressed in the floral meristem from the early stages, and its expression later becomes restricted to whorls 1 and 2. Several regulators of *AP1* expression have been identified, and direct binding to *AP1* regulatory regions has been demonstrated for AGL24, SVP (Grandi et al., 2012), BOP1 (Xu et al., 2010), the FT/FD complex (Wigge et al., 2005), LMI2 (Pastore et al., 2011), SEP3 (Kaufmann et al., 2009), LFY (Parcy et al., 1998; Moyroud et al., 2011), the AP3/PI complex (Sundström et al., 2006) and a few SPL (SQUAMOSA PROMOTER BINDING PROTEIN-LIKE) factors (SPL3 and SPL9) (Wang et al., 2009; Yamaguchi et al., 2009) (Fig. 3.10B). The binding regions of these factors have been characterized with various degrees of precision, by chromatin immunoprecipitation (ChIP) on a large genomic fragment or by gel-shift assays allowing the precise location of the *cis*-element. Strikingly, most factors bind in the same region of the *AP1* promoter, i.e., the proximal promoter approximately 1 kb upstream of the start codon, while a few factors bind in the first intron (Fig. 3.10B). Hence, many transcription factors can bind very close to each other (although they can act in different tissues or at different stages of development) in *AP1* regulatory regions, suggesting that physical interactions between these transcription factors could occur and participate in the full integration of *cis*-regulatory signals for proper *AP1* expression.

Several of the above-mentioned regulators of *AP1* expression are MADS-box proteins, namely SVP, AGL24, AP3, PI and SEP3. For a long time, MADS-box protein-binding sites have been described as a CArG box, i.e., a stretch of A/T nucleotides surrounded by CC and GG. Obviously, this global definition of all MADS-box *cis*-elements could not explain the specificity of target gene activation of class A, B and C genes for instance, which specify the identity of different organs. Some of these factors (SEP3, SVP and AP1 itself) have recently been more thoroughly characterized, and subtle differences in their binding sites could account for their target gene specificities. Additionally, interaction with other partners increases the combinatorial specificity of these transcription factors, as proposed in the floral quartet model

(Fig. 3.10A). Thus, precise characterization of closely related transcription factors is of fundamental importance in order to understand the complexity of *cis*-regulations during floral development.

3.2.6.2 Transcriptional regulation at the genome-wide scale

As described before, the main genes controlling floral development encode transcription factors. Classical genetic and molecular studies allowed the characterization of transcriptional networks involving most of these regulators. In the last 5 years, many genome-wide scale studies, including ChIP-seq and RNA-seq (see Section 3.3), have been performed to determine the targets of many of these regulators: work has been done on LFY, SVP, SOC1, SEP3, AP1, AP2, AP3, PI, etc. (Kaufmann et al., 2009a, 2010; Yant et al., 2010; Moyroud et al., 2011; Winter et al., 2011; Tao et al., 2012; Wuest et al., 2012; for review see Pajoro et al., 2014b). These works have revealed that cross-regulation between these transcription factors is the rule, probably allowing complex retrocontrols or feed-forward loops. Besides this, the targets of the floral regulators are much more diverse than previously described, including genes controlling hormonal signaling, cell division or cell expansion. All these processes may account for their role in organ initiation and growth. Such studies have also confirmed proposed interactions between transcription factors. This is particularly the case for AP1 and SEP3. The hypothesis that they work in heterodimers is strongly reinforced by the fact that they share most of their regulating genes (Kaufmann et al., 2010a). The combined outcome of all these genome-wide studies is likely to lead to a better understanding of the interconnection between floral transcription factors. The spatio-temporal integration of these gene data remains a challenge for a full understanding of floral development.

3.3 Transcription and chromatin dynamics in the "seq" era

Many of the genes identified during the genetic analyses presented in Section 3.2 are transcription factors (TFs) or chromatin regulators. To further understand the functions of these factors, it is of special interest to identify the target genes/regions to which they bind. In correlation with this information, it is of great relevance to observe the read-out of their binding, that is, the transcriptional and chromatin status of the corresponding genes. In this section we will review the current methods employed to answer these questions at a genome-wide scale.

3.3.1 Classical approaches

Techniques for biochemically isolating mRNAs from a tissue or DNA for TF binding or chromatin features have existed for many years (see Section 3.3.4 for a description of the process for isolating DNA). However, analysis of isolated pieces of RNA or DNA remained limited to single target sequences that were already known and thus could be analyzed by polymerase chain reaction (PCR). Development of the quantitative PCR technique became a powerful tool for precisely studying the effects of TFs and chromatin factors on certain genes.

The first technique that was developed to analyze the abundance of nucleic acid fragments at a genome-wide scale was the DNA microarray (MA) technique. A DNA MA is a small slide onto which fragments of DNA (probes) are stably attached. These probes can be cDNAs or oligonucleotides. Each fragment is present multiple times in a so-called spot and the identity of the fragments present at each spot position is known. The DNA fragments of the sample to be analyzed are labeled with a fluorescent dye and incubated with the MA. These fragments will only hybridize to their complementary sequence on the array. Thus, the fluorescence intensity of a spot will correspond to the abundance of the complementary sequence in the original sample. Light intensities are measured by a reader and subsequently analyzed.

Two kinds of oligonucleotide arrays can be distinguished: expression MAs and tiling arrays. Expression microarrays usually contain probes representing the annotated genes of an organism and can be employed for expression analysis when no novel transcripts need to be identified. Tiling arrays, on the other hand, span the whole genome of an organism and can thus be used for discovery of novel transcripts or of target sequences for TF binding or chromatin features.

Thus, microarrays allow genome-wide analysis of transcriptional and chromatin states. However, the technique remains expensive and the data analysis can be very complex, especially regarding normalization between two arrays. For these reasons, MA technology is progressively being replaced by the so-called next-generation sequencing (NGS) techniques.

3.3.2 NGS: what it is, different types

NGS, or high-throughput sequencing, refers to the simultaneous sequencing of millions of DNA fragments. Outputs of the technique are sequences of a certain length, representing the boundaries of the fragments submitted for sequencing. NGS platforms differ in the sequencing technique employed, the number of sequences produced per run, the length of the reads and the amount of initial material required. Detailed reviews of the historical development of the NGS technique, the various platforms available and their mode of function can be found in Metzker (2010), Pareek et al. (2011) and Zhang et al. (2011).

The choice of platform depends on the planned application. The three leading platforms today are ABI/SOLID, Roche/454 and Illumina. While ABI/SOLID and Illumina produce relatively short reads of 50–150 bp, Roche/454 produces longer reads of 400–500 bp. Longer reads are useful for genome sequencing and the assembly of repetitive regions, while shorter reads are usually sufficient for the analytical methods described here. We will further focus our descriptions on the Illumina technology, which currently is the most widely used NGS platform (Fig. 3.11).

3.3.2.1 *Organization of sequencing units and latest developments*

Usually, sequencing machines are not operated by one laboratory alone but as service unit. These can be external companies, internal laboratories or non-profit organizations performing NGS services for external customers. Very often, the service units offer support for the planning of the NGS experiments and also perform some basic

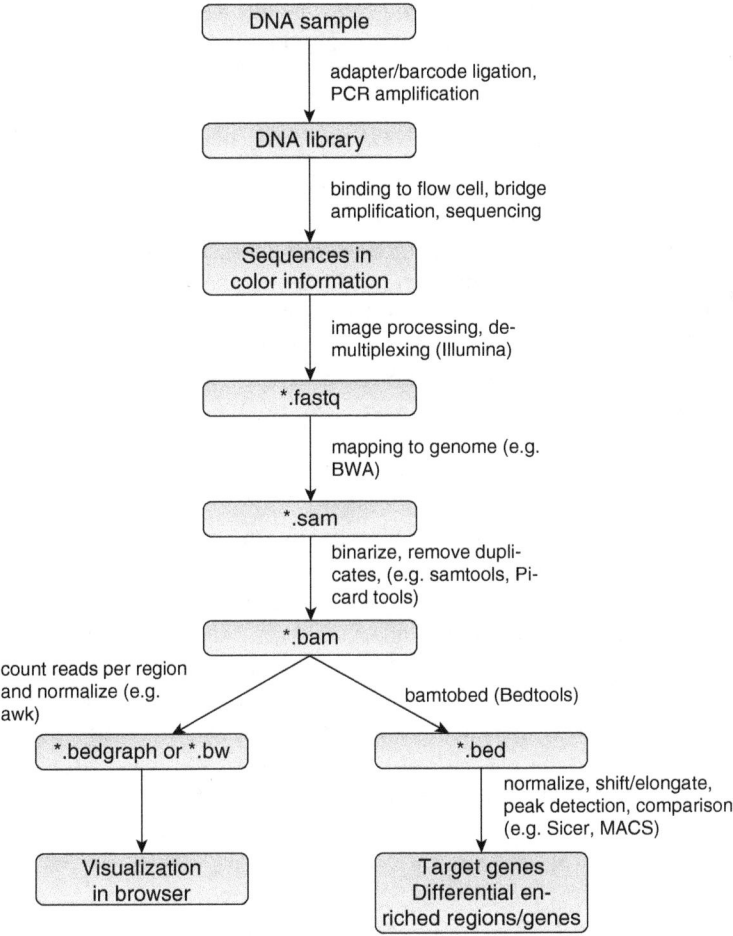

Fig. 3.11 Typical workflow of a NGS experiment using an Illumina sequencer. The top three boxes show library preparation and sequencing steps; the next five data analysis (with examples of programs to be used for the respective steps) and the bottom two boxes show the results. See the text for detailed explanation of the individual steps.

analysis. Some units also offer full service data analysis. It is therefore important to contact the service units before the experiments are performed, to coordinate the experiments and select the best sequencing modalities. One drawback of the constant growing demand for the performance of sequencing experiments is that many sequencing units have long waiting lists and the time from sample submission to receiving the sequences can be several months.

Recently, some companies have developed desktop sequencers, which are affordable for single laboratories and allow the generation of sequencing data within hours or days. Comparisons of different desktop sequencers can be found in Loman et al. (2012).

64 *Floral development: an integrated view*

To overcome the potential bias created by the PCR amplifications during sequencing, new third-generation NGS platforms for single-molecule sequencing are entering the market, as reviewed in Pareek et al. (2011).

3.3.2.2 Typical workflow for an NGS run in an Illumina machine

First, the so-called libraries are prepared for each sample (Fig. 3.11). For this, the DNA fragments of the sample are ligated to two different adapters and amplified by PCR. Often size selection is performed at this step to ensure the right size distribution of the fragments for sequencing. Then, a so-called flow cell is loaded with the library. The flow cell contains immobilized DNA adapters that serve as primers for bridge amplification. This results in clusters (spots on the flow cell) containing up to a thousand copies of the initial molecule (http://res.illumina.com/documents/products/techspotlights/techspotlight_sequencing.pdf). After the amplification, the sequencing reaction is started using one adapter as a primer. Through the use of labeled nucleotides it is possible to detect each base that is incorporated during the sequencing reaction. The resulting color information is then converted into sequence information along with a quality score (Metzker, 2010).

A typical Illumina flow cell contains eight lanes, which are read out separately. Since one lane usually produces more reads than needed for a single sample, several samples can be sequenced in one lane, using the bar-code technique for multiplexing. For this, a short, unique nucleotide sequence is ligated to the fragments of each sample prior to adapter ligation. This bar code is sequenced along with the sample DNA, thus allowing each fragment to be assigned to the sample it belonged to.

The output of the sequencing is delivered as a .fastq file containing the sequences of each read and quality information. This file can then be submitted to the desired analysis.

3.3.2.3 Data analysis

Although some of the tools mentioned in this section might function on a Windows operating system, there are various tools which only function on Unix-based operating systems such as Linux and MacOS. Some of the tools only work on Linux. Therefore, the operating system for the analysis should be carefully chosen. For most if not all of the functions we mention below, several tools are available. The ones we mention are examples that we have tried out ourselves (Fig. 3.11). A detailed list of available tools can be found in Zhang et al. (2011).

Quality control A graphically appealing tool for quality control of NGS read data is FastQC (http://www.bioinformatics.babraham.ac.uk/projects/fastqc). Among other parameters, it allows one to assess the following:

1. The overall quality per base (to see if the quality drops towards the end of the sequences).
2. The G/C content distribution over all sequences (which should be a bell-shaped curve in the absence of foreign DNA contamination, i.e., if only one organism was sequenced).

3. The duplication level, and thus the number of unique reads to be expected (high numbers of identical reads hint at PCR artifacts during library preparation and occur often when the library was not complex enough for the sequencing depth, meaning that there were not enough different DNA pieces in the original sample).
4. The over-representation of sequences, which can indicate if adapters were sequenced along with the reads. This can happen when the initial fragments were too short and the sequencing reaction proceeded into the adapters. In this case, the adapter sequences can be removed computationally.

Mapping Once the quality of the data has been checked and potential problems have been solved (e.g., for adapter contaminations, some mapping tools allow simultaneous trimming of the reads and adapter removal), the sequence reads can be mapped to the reference genome. Examples of tools for this process are BWA (Li and Durbin, 2009, 2010) and Bowtie (Langmead et al., 2009). Several parameters can be set in those tools, such as the number of mismatches allowed. The best parameters depend on the application and on the organism sequenced; thus these parameters have to be found individually. The output of the mapping program is a sequence alignment map (.sam) file, which contains all the mapped reads and their position in the genome.

Duplicate removal, shifting of reads and changing data formats Usually, a .sam file is converted in a compressed, binary .bam file for further processing. Useful tools for conversion of .sam files into .bam files and statistical output generation (e.g., counting the number of reads per chromosome and in total) are SAMtools (Li et al., 2009). As already mentioned, duplicated reads (reads with exactly the same start and end position) are often considered as PCR artifacts during library preparation. There are tools and algorithms which distinguish between duplicates with a high probability of being artifacts and those which might be signal (Moyroud et al., 2011; Muiño et al., 2011), but many work-flows require only one copy or a specified number of copies to be kept. Widely employed tools for this purpose are the Picard tools (http://broadinstitute.github.io/picard/).

Other tools that can be used for file conversions are the BEDTools (Quinlan and Hall, 2010). They can be, for example, employed to convert .bam data into .bed files, which are human-readable files containing the start and stop position of each read.

An operation that is often employed for data resulting in sharp peaks (e.g., TF-binding sites) is the elongation or shifting of reads. As the sequencing machine only sequences the sides of the fragments, only the ends of the submitted DNA pieces are represented by the reads. This can result in one peak on the forward strand representing one end of the fragment and one peak on the reverse strand representing the other end. The results would be double peaks instead of one peak. To overcome this problem, the reads can be elongated in the direction in which their sequencing was performed, while the total length of the reads after elongation will be set to the average fragment length. However, this length is hard to determine and may vary between fragments. The average fragment length can be estimated either from the size selection during library preparation (if performed) or computationally by finding the best correlation between the plus and the minus strand. An alternative to elongating the reads is to shift both plus and minus reads by half of the fragment length towards the middle of the fragment. Routines for this are often included in analysis software for NGS data, for example in ChIP-seq analysis software. A way to reduce the number of duplicated reads that have to be removed from the analysis and to overcome the question of the fragment length is the paired-end sequencing technique. In this technique, both ends of each fragment are sequenced. The two sequences are then saved as a pair and this way the start and stop of each sequenced fragment are known. This technique is especially useful for enhancing the number of sequences to be used and for discovering new sequences as it makes detection of rearrangements like insertion or splice variants easier (Au et al., 2010; Hajirasouliha et al., 2010). However, it is more expensive than single-read sequencing and not necessary for most applications in Arabidopsis because the genome is relatively small.

Visualization   An intuitive way to get a first impression of the results of a NGS experiment is to visualize the data in a genome browser (Stein, 2013; Thorvaldsdóttir et al., 2013). A genome browser displays the chromosomal position in the genome on the x-axis and any given feature (binding sites, reads, genes, etc.) in so-called tracks (Fig. 3.12). To save memory space, there are several compacted formats for the data, for example a .bedgraph in which the number of reads per position is stored but not the single reads themselves. Furthermore, consecutive positions with the same value are summarized. An even smaller file format is provided by bigwig (.bw) files that contain the information from a .bedgraph file in binary format. Such file conversions can be performed with BEDTools.

Normalization   When two ChIP-seq runs are to be compared and quantitative analysis needs to be performed, the data have to be normalized. The normalization method must be chosen depending on the application. One simple way of normalizing NGS data is to specify the coverage at a given position in reads per million, thus normalizing by the number of reads. This linear normalization method yields good results when read numbers in the compared libraries are similar. Other normalization methods are similar to the normalization performed for MAs, like quantile normalization or LOWESS (locally weighted scatterplot smoothing; Gao et al., 2010; Bailey et al., 2013). For RNA-seq, expressing the values as reads per kilobase per million mapped reads (RPKM) and the use of housekeeping genes as a normalization standard are widely used (Chen et al., 2013). A recently proposed method recommends the addition of a defined amount of foreign DNA to all samples prior to sequencing, which in the end serves as an internal control (Bonhoure et al., 2014).

3.3.3 RNA-seq

Time- and position-dependent gene expression play a key role in plant development, and in particular in flower morphogenesis. In this section we describe the different NGS methods that can be employed to characterize the dynamics of transcriptomes, ranging from mRNA, small or long non-coding RNA (ncRNA) patterns to RNA polymerase (Pol) II activity (Fig. 3.13). Description of up-to-date methods for analyzing the diverse non-coding RNA transcriptomes can be found in a recent review (Spicuglia et al., 2013).

Transcription and chromatin dynamics 67

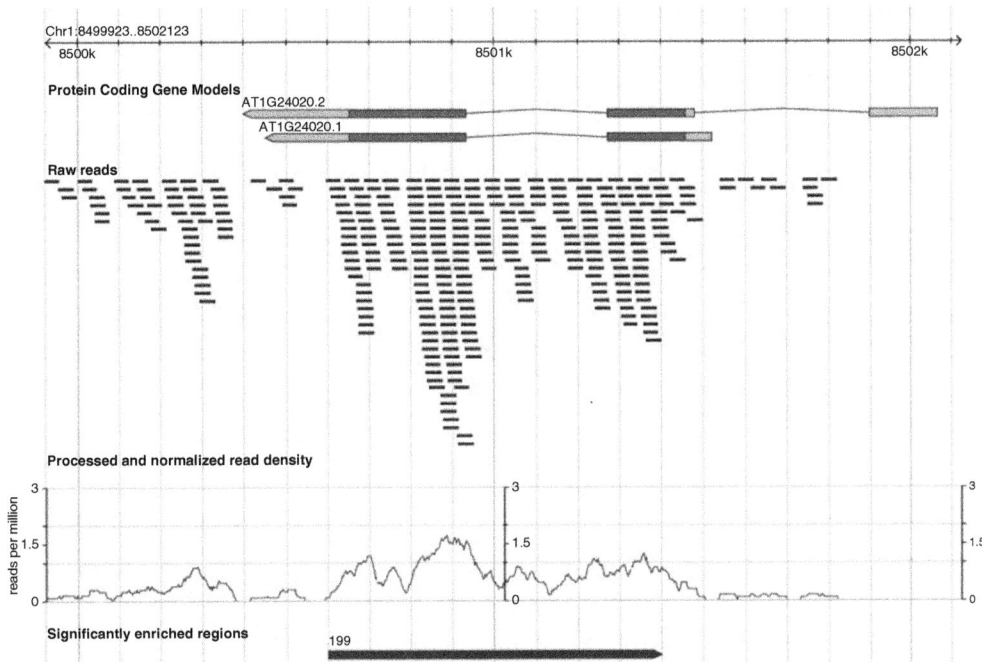

Fig. 3.12 An example of NGS data visualization in a genome browser. The figure shows an example of a view in the genome browser gbrowse (Stein, 2013). The ruler at the top represents the position in the genome, which is visualized (position 8499923 bp to 8502123 bp on chromosome 1 in this example). The next track represents the protein coding gene model, a visualization of the transcripts generated from a locus as an arrow in the direction of transcription. Light gray parts of the arrow represent untranslated regions, dark gray parts represent protein coding parts and thin lines represent introns. Two splice variants for the gene *At1G24020* are represented in this example. The raw reads mapping to the displayed part of the genome are shown beneath. Adding up the number of reads at each position and normalizing by the total number of reads results in the read density shown in the graph. Finally, analysis with a peak-calling program can determine significantly enriched regions. These regions can be for example represented as bars (the lowest dark gray track) and can contain additional information written on the bar (in this case the length of the region in bp).

3.3.3.1 *mRNA*

The most common applications for RNA-seq are the identification of differentially expressed genes (DEGs) between samples or the more general analysis of gene expression levels in various tissues or under different growth conditions. However, mRNA-seq can also be employed to determine the presence of splice variants and to discover novel splice variants. A large fraction of cellular RNA is composed of rRNAs that may impede the detection of less abundant RNAs. Therefore, a first step in RNA-seq typically involves positive selection of polyadenylated RNAs or selective depletion of rRNA. In the case of the standard mRNA-seq procedure, kits are offered for library

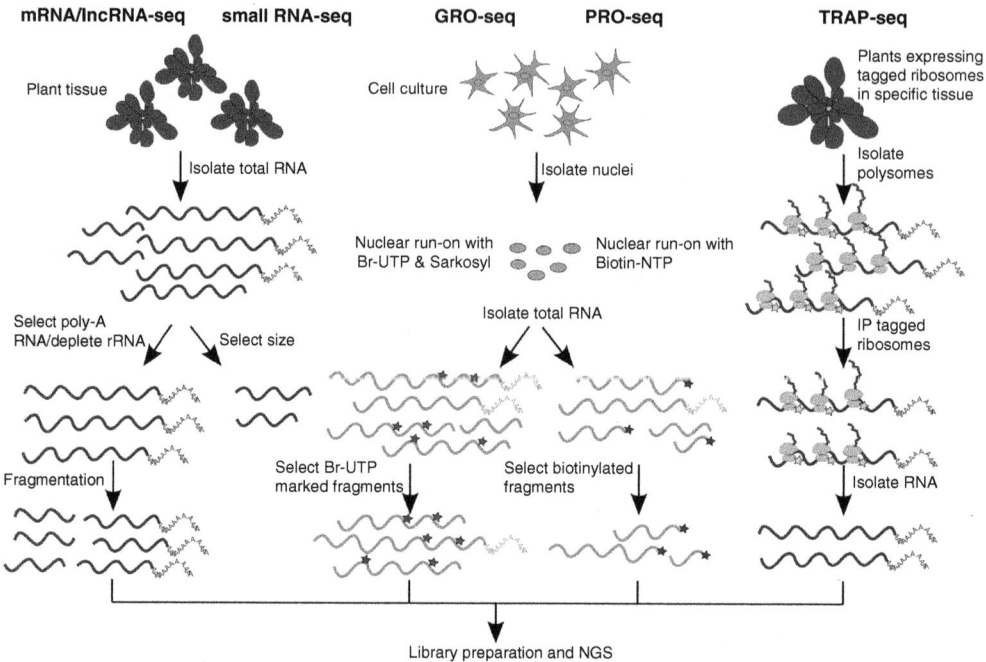

Fig. 3.13 Principles of different RNA-seq techniques. For mRNAs, long non-coding RNAs (lncRNAs) and small RNAs, total RNA can be isolated from the plant tissue of interest and the RNA species of interest is selected in the following steps (e.g., size selection, poly-A selection). Larger RNA molecules have to be fragmented prior to sequencing. To obtain a snap-shot of the RNAs which are synthesized by RNA polymerase at a given moment, nuclear run-on assays can be performed on isolated nuclei (so far only performed in animal cell cultures). In these experiments, a chemical is used to either mark all newly synthesized RNAs starting from the moment the chemical was added (e.g., Br-UTP in the GRO-seq method) or to terminate synthesis of RNAs upon incorporation of the chemical (e.g., biotin-NTP in the PRO-seq method). Afterward, the marked RNAs are isolated for further analysis. To isolate all currently translated RNAs of a certain tissue, plants expressing tagged ribosomes under a tissue-specific promoter can be employed. In this method (TRAP-seq), polysomes are first isolated and then the tagged ribosomes are immunoprecipitated and the RNA contained in these nucleosomes is submitted to NGS. (Figure inspired by Spicuglia et al. (2013).)

preparation, which includes the selection of polyadenylated transcripts or rRNA depletion. RNA adaptors are ligated to the captured transcripts and cDNA synthesized by reverse transcription. The obtained libraries are PCR amplified for high-throughput sequencing (Fig. 3.13).

3.3.3.2 Long non-coding RNA (lncRNA)

With the exception of the Pacific Biosciences RS (PacBio) long read sequencing method (2–3 kb and above 7 kb) (English et al., 2012), long fragments cannot be directly sequenced using current technologies. Thus, a derived library of shorter cDNA (typically 150–200 nucleotides, nt) is usually constructed from chemical or enzymatic fragmentation of lncRNA or of derived cDNA (Fig. 3.13). RNA fragmentation should be preferred for most applications because fragmentation of cDNA favors 3' ends (Spicuglia et al., 2013). An important issue for lncRNA, as for any non-coding RNA, is to be able identify the directionality of transcription. Two main methods can be used for this: the so-called "differential adaptor method", which ligates adaptors in known orientation to 3' and 5' ends of RNA fragments during library construction; or the "differential marking method" that labels the RNA with bisulfite, for conversion of cytidine into uridine, or that incorporates dUTP during second-strand cDNA synthesis (subsequent destruction of the uridine-containing strand allows the identification of transcript orientations during library sequencing) (Levin et al., 2010). In addition, paired-end sequencing provides advantages for unmistakable annotation of lncRNAs.

3.3.3.3 Small RNA

RNA-seq technology can also be used to identify small RNA species (Fig. 3.13). Short non-coding RNA (ncRNA) populations are defined as below 200 nt in size, and represent a significant fraction of the transcriptome. They include mature rRNAs (transcribed by RNA Pol I), tRNAs, 5S, U6 (and many other RNA Pol III-produced transcripts), as well as many RNA Pol II transcripts including, *trans*-acting small interfering RNA (tasiRNA), small nucleolar RNA (snoRNA) and micro RNA (miRNA). These various species are described in numerous reviews. As for lncRNA, it is important to deplete raw RNA extracts from other contaminant species. One approach is excision of the fraction of interest from an acrylamide gel (sizes vary, depending on the species of interest, but are generally shorter than 100 nt), followed by elution and direct cDNA library preparation (without prior fragmentation). An alternative approach for transcripts smaller than 200 nt is enrichment in the short RNA fraction using column-based RNA purification kits. Libraries of cDNA made from this fraction are further size-fractionated on an acrylamide gel to the desired size range. This approach increases the efficiency of adaptor ligation reaction above that of the gel purification method, because it deals with more and larger transcripts (Spicuglia et al., 2013). The obtained cDNA libraries are PCR amplified before sequencing.

3.3.3.4 Nascent RNA for RNA Pol II activity

Another modern development is the discovery of nascent RNA; with this it is hoped to reach time-resolution of the transcriptome during a developmental process and correlate it with the RNA Pol II status for all genes. While the transcriptional status of RNA Pol II can be detected by ChIP-seq (see Section 3.3.4), isolation of nascent RNA is required to verify that the mapped RNA Pol II is engaged in active transcription. This is particularly interesting in connection with the pausing of RNA Pol II at the transcription start site (TSS) (Nechaev et al. 2010), which was recently proposed to facilitate gene expression in animal models, to maintain open chromatin architecture at promoters by preventing nucleosome assembly (Gilchrist and Adelman, 2012). While

70 *Floral development: an integrated view*

not yet reported in Arabidopsis, this function is hypothesized to contribute to fine-tuning of genes involved in development as well as in response to stimuli. In order to identify the transcriptional steps in which RNA Pol II is involved at all genes, ChIP-seq can be performed either (1) on the different isoforms of the C-terminus domain (CTD) of RNA Pol II [commercial antibodies exist that allow the steps pre-initiation complex (PIC) formation–non-phosphorylated CTD, initiation–phosphorylated serine 5 or elongation–phosphorylated serine 2, to be distinguished] or (2) on native or tagged RNA Pol II subunits such as NRPB2, for genome-wide mapping of RNA Pol II.

Genome-wide maps of RNA Pol II can be supplemented by isolation of nascent RNA for which two new RNA-seq procedures have been reported. These are GRO-seq (global run-on library Sequencing) and PRO-seq (precision nuclear run on and sequencing), which allow identification of engaged RNA Pol II (Fig. 3.13). These two methods combine a regular nuclear run-on assay for identification of genes being transcribed at a given time-point, with next generation DNA sequencing as a readout. With GRO-seq, nuclei are incubated with 5-bromouridine 5′-triphosphate (Br-UTP) in the presence of sarkoysl, which prevents the attachment of RNA Pol to the DNA. Thus, only RNA Pol II molecules that were on the DNA before the addition of sarkosyl can produce new Br-U-labeled transcripts. The labeled transcripts are then captured with anti-Br-U antibody-labeled magnetic beads, converted to cDNAs and then submitted to NGS. The sequencing reads are further aligned to the genome, with a resolution of 30–50 bases (Core et al., 2008). GRO-seq thus allows the identification of genes on which RNA Pol II was initially paused.

The more recent PRO-seq approach uses biotin-labeled ribonucleotide triphosphate (biotin-NTP) analogs as a marker in nuclear run-on reactions. On incorporation into the transcript this marker inhibits further elongation, thus ensuring base pair resolution of the position of transcribing RNA Pol at the moment the marker was added (Kwak et al., 2013). The nascent RNAs are then purified on affinity columns and submitted to high-throughput sequencing from their 3′ end. These methods have thus far been validated for *Drosophila*, murine and human embryonic stem cell cultures (Larschan et al., 2011; Min et al., 2011; Kwak et al., 2013). The combination of ChIP-seq for RNA Pol II mapping and PRO-seq for nascent RNA identification should prove very powerful in answering questions about the genome-wide transcriptional status of genes with precise time-resolution, and remains to be tested in plants and in the developing flower.

3.3.3.5 Identification of translating RNA by TRAP-seq in developing flowers

mRNA-seq can also be used to identify the RNA population that is being translated during a developmental process, using the translating ribosome affinity purification (TRAP) methodology (Fig. 3.13). TRAP has been used in Arabidopsis flowers, for which cell-type specific analysis of translating RNAs revealed novel levels of gene regulation (Jiao and Meyerowitz, 2010). Using transgenic lines expressing a tagged ribosomal protein under regulatory sequences of floral master genes (*APETALA1*, *APETALA3* or *AGAMOUS*), the authors immunoprecipitated potential translating mRNAs and submitted them to NGS. Comparison of these translating mRNAs with steady-state RNAs revealed messenger structural features associated with both transcript splicing and translational regulation. Moreover, this experiment allowed the elucidation of novel cell-specific transcripts as well as a new class of ncRNA associated with polysomes, thus bringing new insights in mechanisms regulating the flower translatome.

3.3.3.6 Analysis of RNA-Seq outputs

Computational analysis of RNA-seq outputs as well as available tool-boxes have been extensively described in several reviews and book chapters (Bryant et al., 2012; Garg and Jain, 2013; Pollier et al., 2013; Gulledge et al., 2014; Hehl and Bülow, 2014; Külahoglu and Bräutigam, 2014; Ng et al., 2014).

The determination of DEGs from mRNA-seq data has become a standard procedure in the last few years. Often, such standard analysis is offered as a service along with the sequencing. Several software suites are available for this purpose (e.g., edgeR, DESeq and Cufflinks) (Anders and Huber, 2010; Robinson et al., 2010; Trapnell et al., 2012). Among these, a combination of TopHat and Cufflinks (Trapnell et al., 2012) has become a standard procedure (Külahoglu and Bräutigam, 2014).

For the other methods, analysis procedures are not so standardized, but often the initial descriptions of the methods provide a detailed description of the analysis performed.

3.3.4 ChIP-seq

ChIP is employed to isolate the DNA fragments that are in close contact with the factor of interest in a given sample. The factor of interest can be any protein associated with the chromatin, for example TFs or modified histones (see Section 3.5.3 for a detailed introduction to histone modifications). In combination with NGS, this method provides a tool to demonstrate genome-wide association of a factor and the respective DNA sequences *in vivo*. In this section, we will first describe the general principle of the method with steps common to most protocols, subsequently we will mention some practical considerations which are crucial for a successful experiment and finally we will describe specific steps in the analysis of ChIP-seq data (Fig. 3.14).

Most ChIP experiments include a *cross-linking* step that creates covalent bonds between the DNA, the histone proteins and other factors which are in close contact with the chromatin. This serves to stabilize the interactions to be identified. To isolate the chromatin, first a *nuclear extraction* is performed, involving breaking of the cells by grinding of the material, removal of the cell debris by filtering and subsequent centrifugation in a viscous buffer that allows pelleting of the nuclei while fine components of the cell plasma (like chloroplasts) stay in the solution. Afterward, the purified nuclei are lysed and the chromatin is fragmented into pieces of one to a few nucleosomes. In the subsequent *immunoprecipitation* step, the chromatin is incubated with a protein specifically binding the factor of interest (an antibody in most cases; alternatives to antibodies are ligand–receptor interactions like biotin tagging of the factor of interest and use of streptavidin to precipitate the chromatin of interest; Kolodziej et al.,

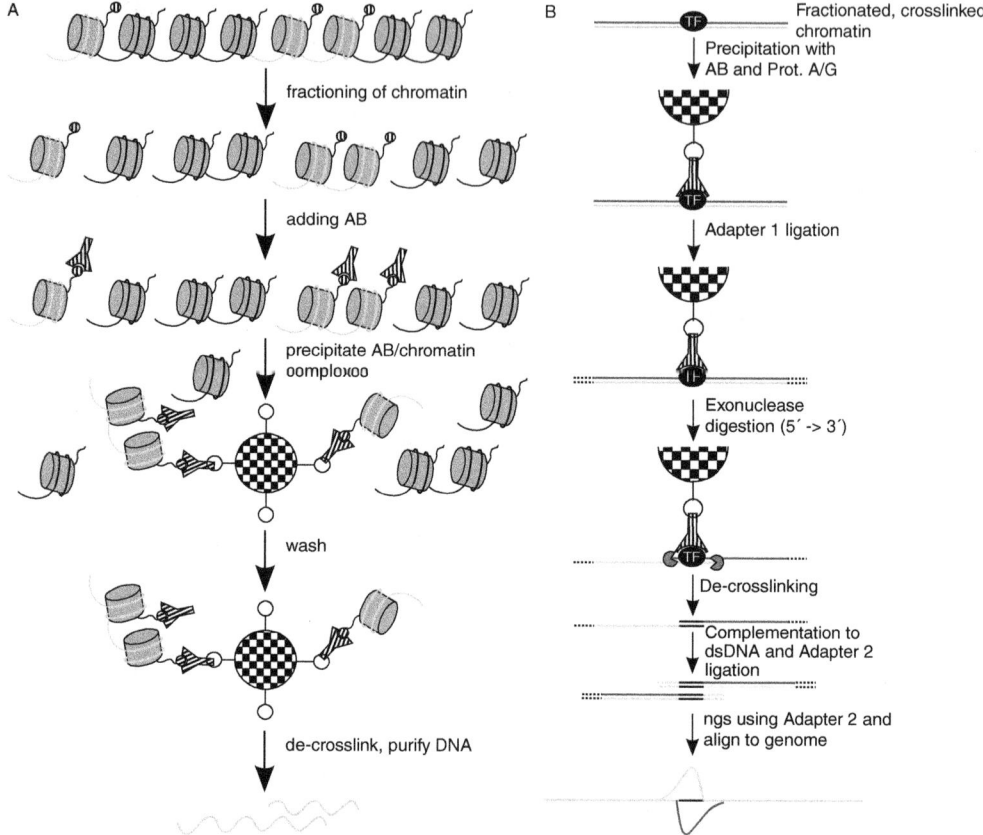

Fig. 3.14 Principle of ChIP (A) and ChIP-exo (B) procedures. (A) The goal of the ChIP procedure is to isolate the DNA in the proximity of a factor of interest, a histone mark (striped) in this case. The DNA fragments of interest are depicted in light gray. After extraction from cross-linked nuclei, the chromatin is fractionated into mono- to tri-nucleosomes and the factor of interest is targeted by a specific antibody (AB, striped). AB/chromatin complexes are precipitated by beads (checkerboard pattern) coated with Protein A/G (white) binding the AB. After repeated washing, the chromatin is eluted, cross-linking is reverted and the DNA of interest is purified for sequencing. (B) The ChIP-exo method is used to enhance the precision of a ChIP experiment, usually for binding of transcription factors (TFs). The first steps are identical to the normal ChIP procedure. In the step when the chromatin is bound to the beads, a first sequencing adapter (dashed black) is ligated to the bound DNA. Then, exonuclease is added to digest both strands in the 5′ to 3′ direction until it reaches the TF which protects the DNA (black). Then, a second adapter (dashed gray) is ligated to the digested ends, NGS is performed using these adapters and the resulting reads are aligned, strand specifically. The results are sharp peaks starting at both end of the protected site. (Figure 3.14B inspired by Mendenhall and Bernstein (2012)). (Figure 3.14B inspired by Mendenhall and Bernstein (2012).)

2009) and the resulting chromatin–antibody complexes are precipitated with beads binding the antibody (mostly beads coated with Protein A or Protein G, which bind immunoglobulins). The beads are washed extensively to remove the chromatin without the factor of interest and then the chromatin associated with the factor of interest is eluted. The obtained sample is enriched for nucleosomes carrying or in association with the factor of interest. To isolate the DNA fragments from the ChIP sample, the *cross-linking is reverted* (chemically, enzymatically and through temperature variation) and the DNA is cleaned up ready for analysis. A detailed protocol for performing ChIP-seq from *Arabidopsis thaliana* tissues can be found in Kaufmann et al. (2010b).

3.3.4.1 Choices to be made

Cross-linking Most ChIP protocols include a cross-linking step, although this step is not absolutely necessary when histone proteins are analyzed, as histone particles are quite stable *per se*. ChIP without cross-linking is termed native ChIP and has already been successfully applied for genome-wide analysis (Bernatavichute et al., 2008).

The most frequently employed reagent for cross-linking of plant tissue is formaldehyde (FA) but other reagents [e.g., di(N-succinimidyl) glutarate (DSG), ethylene glycol bis[succinimidylsuccinate] (EGS), disuccinimidyl suberate (DSS) and dithiobis(succinimidylpropionate) (DSP)] can be employed for factors that are not in direct contact with the DNA (Collas and Dahl, 2008; Kaufmann et al., 2010b). Cross-linking of plant tissue is usually performed directly after harvesting and is enhanced through application of a vacuum; however, cross-linking of already homogenized tissue has also been successfully used for genome-wide studies (e.g., Luo et al., 2012).

Fragmentation of DNA Chromatin fragmentation can be achieved by ultrasound-induced shearing of the DNA or enzymatically using micrococcal nuclease (MNase). Shearing by ultrasound allows one to control the fragment length by the strength of the ultrasound and time for which it is applied. However, the resulting fragments vary in size (e.g., from 200 to 600 bp in our protocol). MNase digestion results in mononucleosomes and thus fragments of around 150 bp. This procedure enhances precision but also renders downstream analysis by PCR difficult due to the short fragment length.

Immunoprecipitation Most ChIP experiments are performed with antibodies that recognizing the factor of interest (Fig. 3.14). In this case, careful analysis of the antibody's specificity is needed to avoid a bias in the results through cross-reactivity. Two systems are frequently employed for the immobilization of Protein A/G for precipitation of the antibody: magnetic beads and agarose beads. Magnetic beads allow a slightly faster processing of the samples but require specific equipment.

Purification of DNA To purify the DNA after de-cross-linking, either classical phenol/chloroform extraction or DNA purification kits can be employed. Phenol/chloroform has the advantage of being independent in the starting and resuspension volume, but residual phenol in the sample might interfere with the library preparation protocol.

Controls One important consideration for the interpretation of ChIP experiments is the choice of the controls. Usually, controls include a so-called input sample containing non-precipitated chromatin and thus all fragments that could be possibly precipitated. A second common control is a negative control. This can be either a no-antibody control that was treated like the ChIP samples but without an antibody, or an IP with an unspecific antibody like anti-rabbit IgG.

Quality control prior to sequencing To avoid wasting resources in the expensive sequencing process, samples have to carefully analyzed prior to submission to NGS. One crucial parameter to be checked is the fragment length. As most library preparation protocols include a size selection step, it has to be ensured that the majority of the submitted fragment is in this size range. This can be assessed by analyzing the ChIP and input samples in a Bioanalyzer machine using a high-sensitivity kit (Agilent) or by de-cross-linking of a fraction of the chromatin and analysis in an agarose gel.

The success of the IP should be analyzed by PCR on a known target *locus* whenever one is known.

Modern preparation kits for bar-coded libraries require high amounts of starting material. The DNA concentration in the samples can be determined using modern quantification tools like the Picogreen technique. To increase the DNA concentration and to minimize experimental variability it widely practiced to pool several IP samples of the same material for sequencing. However, we never reach the required DNA amounts specified by the manufacturers but nevertheless obtain high-quality sequence reads and sufficient numbers of reads when we pool three technical replicates.

3.3.4.2 Analysis of ChIP-seq outputs

The most common task in analyzing a ChIP-seq experiment is peak-calling, meaning the detection of regions with significant enrichment of the ChIPed sample compared with a control sample (or a randomly assumed background). Several peak calling programs exist for this task and they also include functions for pre-processing of the samples, like removal of duplicates and shifting of reads. Depending on the nature of the expected pattern (broad peaks for histone marks, narrow peaks for TFs), different analysis tools can be applied, with MACS (Zhang et al., 2008) being a popular tool for analysis of TF-binding sites and SICER (Zang et al., 2009) being a tool specifically designed for broad histone mark distributions. A review of several peak callers can be found in Kim et al. (2011).

Once peaks have been determined, common tools for NGS analysis like BEDTools can be used to assign genes to the peak regions and downstream analysis tools like motif identifiers (e.g., MEME-ChIP; Machanick and Bailey, 2011) can be employed to find reoccurring features in the targeted DNA sequences. Another task might be the comparison of targeted regions for the same factor in different conditions (e.g., tissues). How this can be achieved is described by Bardet et al. (2012).

A more difficult task is quantitative comparison between ChIP-seq samples. Several tools exist for this task as well, for example SICER, ChIPDiff (Xu and Sung, 2012) and QChIPat (B. Liu et al., 2013). The best-suited tool has to be chosen based on the experimental conditions and the underlying research questions.

One critical parameter for a successful ChIP-seq experiment is the number of unique mapped reads. For human samples, 20 million reads are recommended for TFs and 40 million for histone modifications. For the much smaller Arabidopsis genome, lower numbers are sufficient for peak detection. Several publications have demonstrated successful peak detection with read numbers from 2 to 5 million (Gregis et al., 2013; Xing et al., 2013). Recent considerations indicate that higher read numbers might increase the number of true peaks detected and a reasonable recommendation for fruit flies (with a genome size twice that of Arabidopsis but only half the number of genes) seems to be around 15 million reads (Sims et al., 2014). We found that, especially for quantitative comparisons in Arabidopsis, read numbers of at least 15 million are necessary. The effect of read numbers on peak detection is reviewed in Sims et al. (2014).

3.3.4.3 ChIP-exo

One limitation of ChIP technology is the resolution of the binding sites, which is limited to the fragment size generated during digestion or shearing of the chromatin. This limited resolution can cause difficulties in the exact determination of TF-binding sequences. ChIP-exo technology (Rhee and Pugh, 2011) overcomes this problem, allowing near-single-nucleotide resolution, thanks to the addition of a digestion step with lambda exonuclease during IP (Fig. 3.14). This enzyme digests double-stranded DNA only from 5′ ends. A detailed description of the method, including a protocol, can be found in Rhee and Pugh (2012). In brief, adapters are ligated to the chromatin–antibody complexes on Protein A/G beads during ChIP. Then, exonuclease is added and will digest the DNA fragments starting from both 5′ ends. A protein that is cross-linked to the DNA will protect the DNA and the digestion will stop at this position. Afterward, the chromatin is eluted and processed as a regular ChIP sample for sequencing. A second sequencing adapter is ligated to the digested ends during library preparation, and thus the digested sides (which can be as close to a protein-binding side as 1 bp) will be the sides where sequencing starts. In this way, the start position of the reads mapping to the forward strand will map to the left border (considering the forward strand in the 5′ → 3′ direction) of the binding site and the reads mapping to the reverse strand will mark the right border of the protein-binding site (Rhee and Pugh, 2012).

3.3.5 Chromatin accessibility and nucleosome mapping

If ChIP methods provide important clues as to the sites bound by proteins on chromatin, they give very little information on the actual accessibility of the chromatin. The accessibility of chromatin as the result of its compaction can be probed through several methods using either nuclease sensitivity or mechanical disruption (shearing) of chromatin integrity.

The most widely used methods for measuring chromatin accessibility are based on digestion of DNA by a nuclease followed by deep sequencing. Two nucleases can be used: MNase or DNAseI (Fig. 3.15). The main difference between these nucleases lies in the length of the DNA cleavage products: MNase will ultimately cleave all

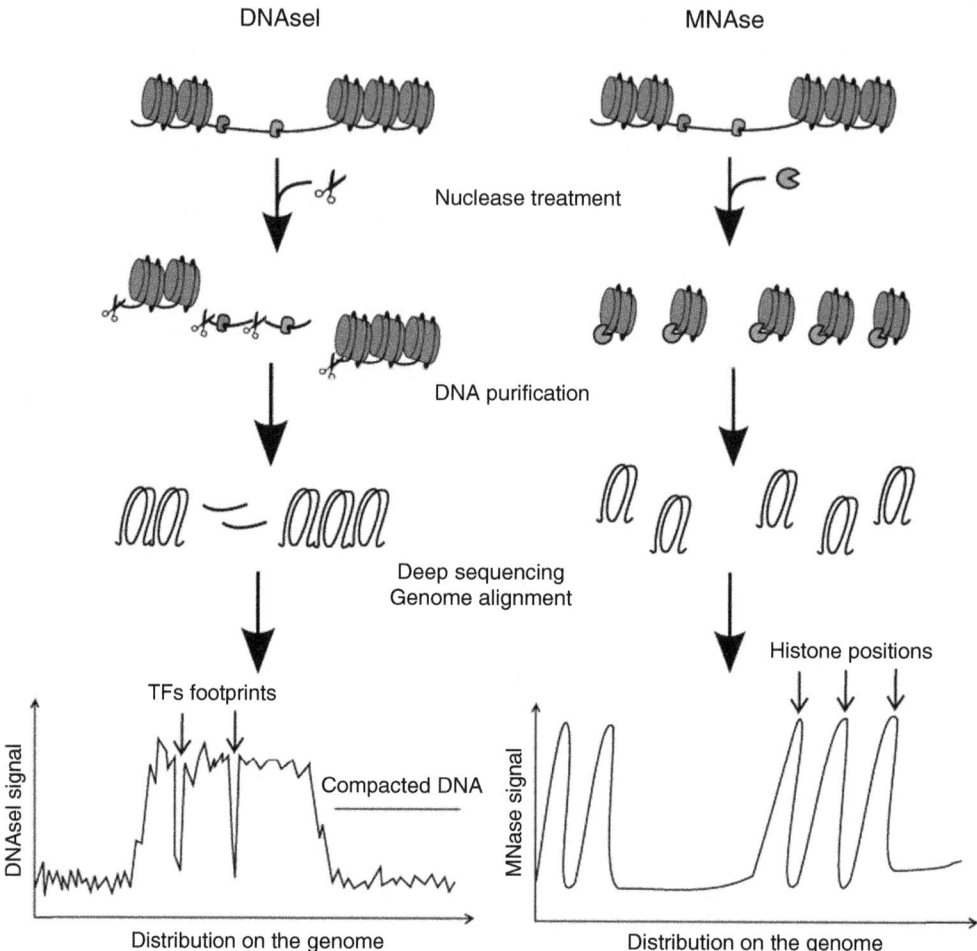

Fig. 3.15 Comparison of the information on chromatin organization obtained by DNAse-seq and MNase-seq. DNAseI treatment (left panel) cleaves accessible DNA into small fragments, DNA is then purified and small sizes (50–100 bp) are selected. From those small fragments, a library is generated and sequenced. Therefore, more accessible chromatin leads to a higher signal while less accessible (e.g., protected by proteins) chromatin leads to lower signal. Local protection of DNA by TFs leaves a footprint in DNAse signal (bottom left). On the other hand, treatment with MNase (right panel) hydrolyzes all accessible DNA (e.g., DNA not protected by histones). After library generation and sequencing, reads are aligned to the genome and mostly map to nucleosome positions (bottom right).

accessible nucleotides while DNAseI will cleave long DNA fragments into shorter ones. Due to this difference in activity, the information obtained from the sequencing of lysis products will be drastically different. Sequencing of MNase products will mostly identify sequences that were protected from the action of the nuclease by nucleosomes, allowing for precise mapping of nucleosomes on the genome (Li et al., 2014). On the other hand, the DNAse workflow involves a size-selection step, selecting fragments of 50–100 bp, significantly enriching the sample in regulatory elements (the size of a nucleosome size is 150 bp) and with a cleaner signal. Therefore, sequencing of the products of DNAseI will identify mainly short DNA fragments, mapping to the most accessible regions of the genome. Also, transcription factors bound on DNA will protect DNA from the action of DNAseI, leaving a "footprint" that can be identified based on the sequencing profile (Zhang et al., 2012). If MNase-seq allows the accurate measure of nucleosome density along the genome, DNAse-seq provides information on occupancy of the genome by proteins—regardless of their function—and can therefore predict occupancy of both nucleosome and TFs on the genome, allowing the identification of putatively transcriptionally active regions (Fig. 3.15). DNAse-seq is therefore very interesting for studying the impact of TFs on chromatin dynamics, as shown in plants by a recent paper focusing on the functions of the MADS domain TFs AP1 and SEP3 during floral organ specification (Pajoro et al., 2014b).

Unlike nuclease-assisted methods, FAIRE (formaldehyde-assisted isolation of regulatory elements) is based on the fact that sonic fractionation of DNA does not occur randomly but has a biological basis (Nagy et al., 2003). This method includes a formaldehyde-assisted cross-linking of the proteins on the sample DNA followed by sonication of the cross-linked DNA (Omidbakhshfard et al., 2014). The main difference between the two methods is the sequence of events following sonication. In FAIRE, DNA preparations of the cross-linked sonicated samples are cleaned with a phenol–chloroform extraction, allowing the purification of non-protein-associated DNA in the aqueous phase, while nucleosome-rich regions are kept in the interphase (see Omidbakhshfard et al., 2014 for a protocol in Arabidopsis). The FAIRE-seq method therefore allows sequencing of nucleosome-depleted regions, overlapping with transcriptionally active regions as shown by the comparison of FAIRE signal and RNA Pol II transcribed sequences (Nagy et al., 2003).

The study of chromatin dynamics is an important field of investigation, and methods are being optimized regularly with the aim of reducing the complexity of the experiment and of the amount of material needed, allowing one to work on smaller tissues. One notable method developed in recent years is the assay for transposase-accessible chromatin using sequencing (ATAC-seq). In this assay the regions of opened chromatin are tagged with sequencing adaptors using a modified version of the Tn5 hyperactive transposon. The purified transposon is loaded *in vitro* with sequencing adapters and integrates itself into active regions of the target genome (Buenrostro et al., 2013). In addition to its simplicity, this method is compatible with small amounts of genetic material. The authors have shown that the method produces data yielding similar information and quality to that of DNAse-seq but with 10^3 to 10^5 times less material.

3.3.6 Physical interaction between distant chromatin regions

If it is possible to predict the sites bound by a single or a set of TFs using immunoprecipitation approaches, it can be difficult to predict the locus potentially regulated by that TF, especially when binding occurs in large intergenic regions. It is possible with the current databases and computational methods to scan the DNA sequence upstream of a gene-coding region to find putative transcription-binding site. This approach, however, is limited to predictions based on TF DNA-binding matrices or consensus on empirically determined regulatory regions and is largely impeded by the great difficulty in predicting promoter regions. One experimental method, so-called chromosome conformation capture (3C and its variations 4C, 5C and Hi-C), allows the identification of physically close chromatin regions independently of their genetic distance, therefore allowing the determination of a putative regulatory sequence for a given locus (Dekker et al., 2013).

In this method, cross-linked DNA is digested by a frequently cutting restriction enzyme (typically *Hind*III or *Nco*I). Chromatin fragments that are physically linked by a chromatin bond are then ligated together based on their physical proximity in a considerably diluted solution (Fig. 3.16). Cross-linking is then reversed and, depending on the technique used, the loci of interest are either PCR amplified (3C or 4C; see Cao et al., 2014 for an example in Arabidopsis or Louwers et al., 2009 in maize) or ligated chromatin fragments are sheared and ligated to adapters for library sequencing (Hi-C; Lieberman-Aiden et al., 2009). While the 3C and 4C methods are focused on one user-specified locus, Hi-C identifies these structures at a chromosome- or genome-wide level. The analysis of the data generated through chromosome conformation capture relies on the detection of co-enriched regions and the generation of a contact matrix, showing physical interactions between genetically distant regions of the genome. A recent study of the *FT* locus in Arabidopsis showed that the merging of regions distant of up to 5 kb from one another in proximity of the transcription start site of FT is necessary for the recruitment of its transcriptional activator CONSTANS (CO) (Cao et al., 2014).

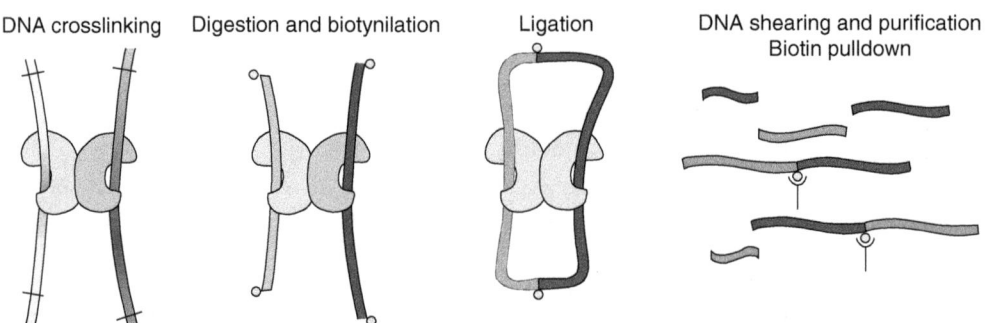

Fig. 3.16 Schematic diagram of a 3C experiment. Native DNA is cross-linked to stabilize protein–protein and protein–DNA interactions. DNA is then digested with a restriction enzyme and sites are filled using biotin-labeled NTPs. DNA fragments are ligated based on their proximity in a highly diluted solution. DNA is then sheared and biotin-labeled fragments are purified. Specific loci can then be tested for co-purification with quantitative reverse transcriptase-PCR (3C) or a library can be generated and sequenced (Hi-C). A genome- (or chromosome-) wide proximity matrix is then generated, allowing the physical interactions between genetically distant elements to be visualized.

3.3.7 Approaches and technologies for space and time resolution

The issue with most developmental genes is that they are tissue- or cell type-specific, and we need to isolate the right cells or nuclei in order to be able to detect, with the best accuracy, any change in their chromatin state, regulator binding or expression dynamics. This is the current challenge in plants, which should allow us (compared with cultures of animal cell lineages) to reach the most accurate and genuine view on the molecular events occurring *in situ*. Several approaches could be used in order to, for instance, separate chromatin profiles for the different developing floral organs (Fig. 3.17).

The first attempts at depicting chromatin marks specific to the SAM versus mature tissue were made using *clavata* mutants—producing a macroscopic SAM—to enrich in nuclei from stem cells versus leaf differentiated cells (Lafos et al., 2011). Other, more technological, methods are laser capture microdissection (LCM), fluorescence activated cell sorting (FACS) and purification of nuclei with the isolation of nuclei tagged in specific cell types (INTACT) system. We will now review these methods.

Technique	Principle	Advantages	Drawbacks
for space-resolution			
LCM 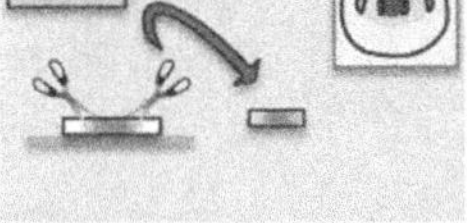	- Section of fixed plant tissue (frozen or paraffin-embedded). - Laser beam dissection of region of interest. - Tissue dissection is based on visual structural features.	- Independence from transgenic markers. - Low alteration in gene expression (because dissection post-fixation). - Use of LCM on small cell populations.	- Low-yield, time-consuming. - Risk of contamination with surrounding cells.
FACS 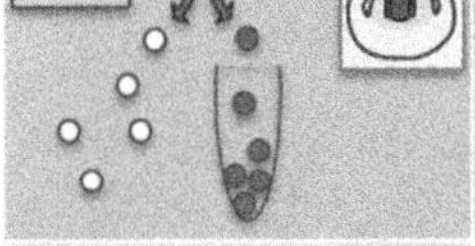	- Flow Cytometry sorting of tissue-specific protoplasts or nuclei. - Based on the expression of a fluorescent reporter type-specific.	- Rapidity of the produre. - Possibility for scaling-up.	- Induction of stress response genes by protoplasting.
INTACT 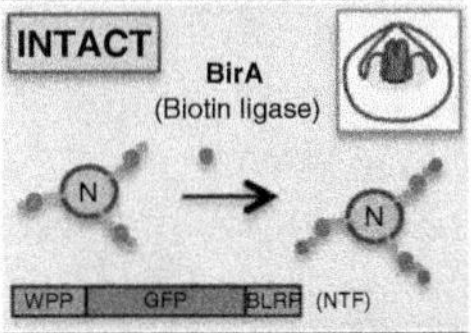	- Affinity-based purification of nuclei (cell or tissue-specific), with streptavidin. - Purification via a biotinylated chimeric GFP Protein, anchored in the nuclear envelope (NTF) and expressed under the control of a cell-or tissue-specific promoter.	- Quick release of nuclei from tissues after homogenization. - Simple, rapid, cost-effective. - Likely high yields and purity.	- Not yet proven to efficiently work on other tissues than root.
for time-resolution			
Inducible ap1cal 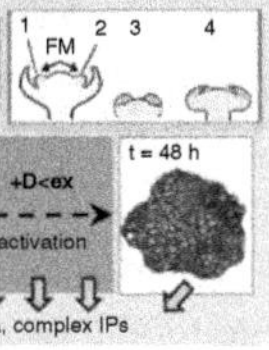	- Growth of arrested ap1cal flower buds is re-induced by Dex treatment - Flowers are harvested at different time points from the apices	- Synchronized development of flowers (for 2 days). - Large amount of tissue. - Time-line to capture dynamic events at the flower.	- Growth condition-dependent (for leaking)

Fig. 3.17 Principle and comparison of space and time resolution technologies for purification of DNA and RNA from Arabidopsis tissues. See main text for more details on the procedures.

3.3.7.1 LCM

With LCM, a region of interest is dissected using a laser beam from fixed and sectioned plant tissue that is frozen or paraffin-embedded (Kerk et al., 2003). The advantages of this method are the independence from transgenic markers and the possibility of collecting samples without much alteration in gene expression because dissection is performed post-fixation. Drawbacks are the low yield and time-consuming procedure and the risk of contamination with surrounding cells because tissue dissection is based on visible structural features. However, recent improvements in NGS technology have allowed the use of LCM on small cell populations. Indeed, RNA-seq on a sample of less than 1 ng of LCM-dissected RNA was successful in resolving the transcriptome of the Arabidopsis female gametophyte central cell (Schmid et al., 2012).

3.3.7.2 FACS

In FACS, protoplasts of a specific tissue are prepared and sorted by flow cytometry and purified based on the expression of a fluorescent reporter driven by a cell type-specific promoter or an enhancer trap in transgenic lines (Birnbaum et al., 2005). Advantages of FACS are the rapidity of the procedure and the possibility for scaling-up. Drawbacks are the induction of stress response genes due to disruption of cell–cell communication by protoplasting, which can be laborious for complex tissues. FACS has been successfully employed to isolate specific cell types from the Arabidopsis root and to profile the transcriptome of cell types in the SAM (Iyer-Pascuzzi and Benfey, 2010). FACS can also be used to sort nuclei, and has been successfully employed to purify vegetative nuclei and perform endosperm-specific chromatin profiling (Borges et al., 2012; Weinhofer and Kôhler, 2014).

3.3.7.3 INTACT

The INTACT system allows the purification of nuclei specific to a cell or tissue type, for both nascent RNA and chromatin landscape analyses. Moreover, because of the significant correlation between nuclear RNA and total cellular RNA, purified nuclei also are a reliable tool for general transcriptome profiling (Deal and Henikoff, 2010). In this system, purification of specific *nuclei* is performed via a biotinylated chimeric green fluorescent protein (GFP), anchored to the outer nuclear envelope (nuclear tagging factor, NTF) and expressed under the control of a cell- or tissue-specific promoter. It uses a two-component transgenic labeling system allowing *in vivo* biotinylation of NTF by the BirA biotin ligase only in cells transgenically expressing both NTF and BirA (Deal and Henikoff, 2011). Biotin-tagged nuclei can thus be efficiently isolated from crude nuclear preparations by direct trapping on streptavidin-coated beads.

The INTACT procedure circumvents many of the limitations associated with LCM and FACS. A major advantage is that nuclei can be quickly released from tissues through homogenization, thus permitting minimal tissue manipulation before fixation and preparation of nuclei. Other advantages are the simplicity, rapidity and cost-effectiveness of the method. Finally, it likely provides higher yield and purity, even though no other INTACT-based analysis has been published since the proof-of-concept study by Deal and Henikoff that reported successful transcriptome and epigenome profiling of Arabidopsis root hair versus non-hair cells (Deal and Henikoff, 2010). Nevertheless, several laboratories are attempting to apply the INTACT method to their favorite tissue, revealing the necessary *de novo* optimization of the purification procedure for each new tissue type (Palovaara et al., 2013).

The TRAP technique described earlier (Section 3.3.3.5) avoids the problem of transcriptome perturbation caused by LCM and FACS, and can probably achieve a higher yield than all the other three techniques (LCM, FACS, INTACT), but it is only suitable for analysis of cell-specific translating mRNA. Comparison between these techniques for transcriptomic analysis in the young developing Arabidopsis embryo is reviewed in Palovaara et al. (2013).

3.3.7.4 Inducible "cauliflower"

The Arabidopsis "cauliflower" inducible system allows circumvention of the spiral phyllotaxy organogenesis that takes place at the Arabidopsis shoot apex (a wild-type Arabidopsis inflorescence is made of flower buds all at different stages of development) and that impedes temporal analysis of molecular events driving differentiation of flower stem cells into organ founder cells. It corresponds to a modified *apetala1 cauliflower* (*ap1cal*) *AP1:GR* inducible genetic background, in which inflorescences over-proliferate a mixture of inflorescence meristems and arrested floral meristems at a very early stage (stage 1), before expression of any of the flower architecture genes has been activated (Wellmer et al., 2006). Development in these inflorescences can be induced by dexamethasone (Dex) treatment, which enables the synchronized development of floral buds (O'Maoiléidigh and Wellmer, 2014). The system has two major advantages. First, it produces a large number of synchronized floral buds that are easy to harvest. Second, the synchronized induction of floral organogenesis allows the molecular dynamics accompanying flower organ development to be monitored. Use of the inducible "cauliflower" system has already proven successful for genome-wide transcriptome profiling and DNA accessibility assessment of TF bound regions, as well as for chromatin marks and chromatin factor-binding analyses at several flower loci (Wellmer et al., 2006; Smaczniak et al., 2012; Pajoro et al., 2014b; Sun et al., 2014).

3.3.8 Summary

In summary, the past decade has seen exceptional technological improvements that have allowed us to gain comprehensive insights into regulatory mechanisms at the genome-wide level. These advances coincided with the full sequencing of the Arabidopsis genome, and subsequently that of other plant species. New approaches permitting purification of DNA and RNA, specific to a developmental time-point, tissue or cell type, are the next challenge. Combined with genome-wide analyses, they will allow us to address the biological questions that arise concerning plant and flower morphogenesis.

82 Floral development: an integrated view

3.4 Modeling TF–DNA binding to predict transcriptional regulation

3.4.1 Introduction

The development of an organism and its response to its environment are determined by gene regulation, which is encoded in the organism's genomic sequence. Gene regulation involves the binding of TFs to their binding sites (TFBS) (provided they are accessible) and the subsequent recruitment of additional factors controlling gene activation (mRNA production) or repression. The state of chromatin (open or compacted; see Section 3.5) has a strong influence on TF accessibility.

One of the major goals of modern biology is to decrypt the information present in genomes in order to predict gene regulation (Fig. 3.18). One of the prerequisites of this complex task is to be able to localize TFBS in genome sequences. This knowledge can then be combined with various types of information such as (1) the expression of TFs in a given tissue, (2) the accessibility of the chromatin and (3) the conservation of binding sites throughout genomes. Understanding how gene expression is controlled is a major challenge: in addition to the multitude of laboratories working on this topic, major international initiatives such as ENCODE (http://www.genome.gov/encode/), modENCODE (http://www.modencode.org/) or FANTOM (http://fantom.gsc.riken.jp/) bring together many groups performing large-scale experiments to crack the code of transcriptional regulation.

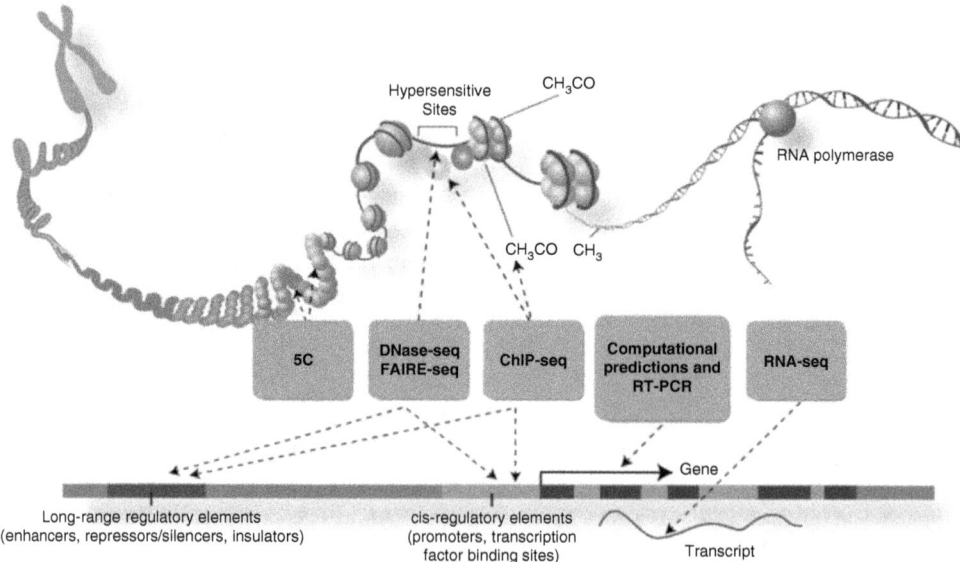

Fig. 3.18 Schematic regulation of gene expression and examples of tools allowing its characterization. Figure from http://www.epigenomebrowser.org/ENCODE/. Credits: Darryl Leja (NHGRI), Ian Dunham (EBI).

In this section we will explain how TFBS can be modeled and how this knowledge can be combined with other types of information to build models of gene regulation.

3.4.2 Generating TF–DNA binding models

3.4.2.1 Basic principles of TFBS models

The basis for building a TF model is to identify as many TFBS as possible. Historically, the most basic tools for describing TF–DNA binding specificity were consensus sequences. They were generated by aligning a few sequences known to be bound by the TF. Binding was mostly established biochemically using sequences identified from the regulatory regions of regulated genes. The predictive power of these consensus sequences was usually quite low. Since they are easy to use and understand, many are still being used. A better tool is called the position weight matrix (PWM) or position-specific scoring matrix (PSSM) (see Wasserman and Sandelin, 2004, for a review). For this, many TFBS have to be identified and aligned, generating a matrix of count or frequency at each position (Fig. 3.19). Frequencies can then be converted into penalties to generate a PSSM or PWM. The rarer a base is at a given position, the stronger the penalty. The score of any DNA sequence can then be computed by

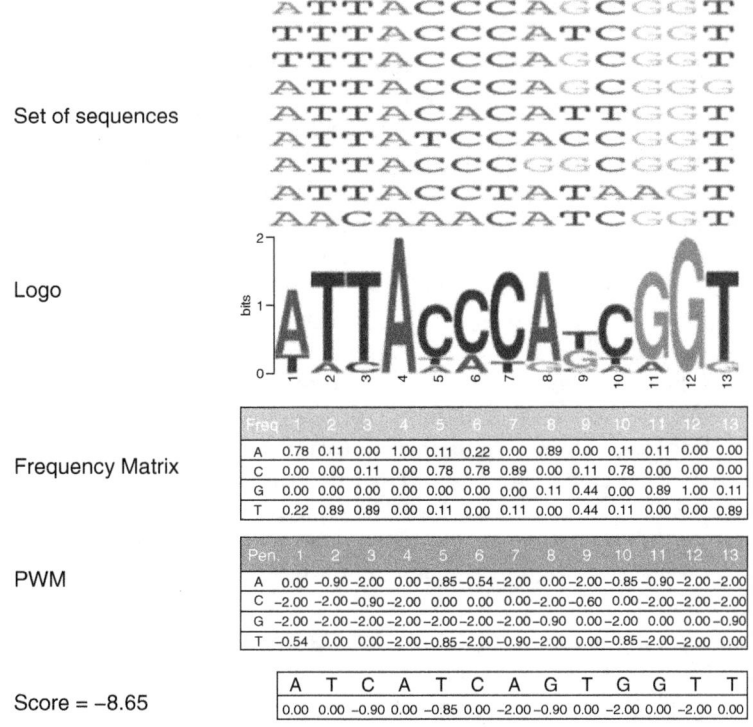

Fig. 3.19 Scheme for building a PWM from a few TFBS.

adding all penalties. The seminal work of Berg and von Hippel Hippel (1987) showed that the resulting score is directly linked to the biochemical affinity of the TF for its TFBS. The PWM can thus be used to scan any genomic sequence and attribute a score at each position. Most scores correspond to background binding sites that have no biological relevance. The highest-scoring positions correspond to high-affinity binding sites, which can be bound by the TF if the region is accessible.

Many improvements to the basic PWM have been proposed to take into account the complexity of protein/DNA binding and a wealth of sophisticated tools are now available (see Mathelier and Wasserman, 2013, and references therein). However, the "basic" PWM is still very commonly used and none of the improved models have emerged as standard.

In addition to identifying individual binding sites, it is often relevant to compute a global occupancy of a large promoter region by a given TF using all the TFBS it contains. To this end, biophysical models have been proposed that integrate multiple TFBS and their affinity and yield a value called predicted occupancy or probability of occupancy (Granek and Clarke, 2005).

3.4.2.2 Identifying a set of TF-bound regions

To build a predictive model, a few tens up to thousands of unique sequences have to be identified that are experimentally bound by the TF of interest. There are several techniques that allow these sequences to be isolated.

1. SELEX (systematic evolution of ligand by exponential enrichment) consists of incubating a recombinant TF with a pool of random oligonucleotides flanked by fixed borders. TF/oligonucleotide complexes are isolated on a gel or by pulling down the protein and isolated DNA is amplified by PCR using the fixed borders. After several rounds of amplification (from three to ten), only high-affinity sequences are present that can be sequenced using NGS (Zhao et al., 2009).
2. ChIP-ChIP or -seq. Sequences isolated using ChIP (see Section 3.3.4) can be processed to identify overrepresented TFBS. Several tools are available, the most popular being MEME (http://meme-suite.org/) and RSAT (http://www.rsat.eu/). The search might be harder than for sequences identified by SELEX because the regions obtained by ChIP are often several hundreds of kilobases in comparison with short SELEX sequences.
3. DIP-ChIP or -seq (DNA immunoprecipitation). Another technique that resembles both SELEX and ChIP is DIP-ChIP (Liu et al., 2005). It uses a naked genome (devoid of histones) as the DNA source and recombinant protein to pull down high-affinity sequences. The advantage of DIP-ChIP is that all potential sites are precisely located on the genome of interest without the further use of a predictive model.
4. Protein binding microarray (PBM). PBMs are efficient and versatile ways to identify TFBS no bigger than 11 bp (Godoy et al., 2011). It consists of a microarray on which tens of thousands of oligonucleotides (50-mers for example) have been spotted that represent several times the diversity of all possible 10-mers or 11-mers. A tagged protein is then hybridized on the chip and detected with a

fluorescent antibody against the tag. Computation then allows the identification of the motifs overrepresented in the most fluorescent spots and the generation of a TF model.

3.4.3 Using TFBS models

3.4.3.1 Databases for TFBS models

Once generated, TFBS are stored in databases that allow one to perform searches with multiple models. The most popular databases are Jaspar (open access) (Mathelier et al., 2014) and Transfac (http://www.gene-regulation.com/pub/databases.html). More specialized databases exist for specific organisms such as AthaMap (http://www.athamap.de/index.php) or DATF (http://datf.cbi.pku.edu.cn/) for Arabidopsis.

3.4.3.2 Combining TFBS model with additional information

The first thing to do with a PWM model is to test its predictive power. This can be done using a receiver operating characteristic area under the curve (ROC-AUC) test that compares the presence of binding sites between the positive set of sequences (from a ChIP experiment) and a negative set (see Moyroud et al., 2011, for an example of use and references on TF ROC-AUC).The main limitation of a TFBS model is that it will identify numerous sites of no biological significance that are present in the DNA just by chance. A major challenge is to distinguish those spurious sites from those involved in regulation. To do this, TFBS models have to be combined with other sources of information.

Combining TFBS models with multiple genome sequences The presence of conserved TFBS is a good indication of their biological relevance for regulation. For this reason, some tools combine the use of TFBS models and sequence conservation to identify meaningful TFBS (e.g., ConSite, http://consite.genereg.net/).

Combining TFBS models with the chromatin landscape The chance of a TFBS being bound is increased if the corresponding genomic region is accessible. Therefore, combining information on TFBS with ChIP with chromatin marks, a DNAseI-seq signal or the presence of nucleosomes can greatly improve the predictive power of the TFBS model (Won et al., 2010).

Identification of regulators of co-expressed genes TFBS models can also be used to identify regulators from a set of co-expressed genes (Villiers et al., 2012). For this, overrepresented *cis*-elements are identified in promoters of co-regulated genes and TFBS databases can be searched for TFBS matching the overrepresented *cis*-elements. However, since many TFs share similar motifs (e.g., bHLH and bZIP share a G-box motif), such analysis often identifies a list of candidate TFs that need to be filtered based on their own expression profile or on genetic evidence.

Using DNAse profiles and TFBS models to guess TF binding As explained in Section 3.3, DNAseI profiling reveals accessible genomic regions. However, within these regions some locations can be partially protected (indicated by the presence of a crease in the DNAseI peak) because they are bound by a TF. A very promising avenue consists of examining the sequence under these creases with TFBS databases to identify the TF present on DNA without the time-consuming generation of one ChIP per TF (Bernstein et al., 2012; Neph et al., 2012; Andersson et al., 2014).

3.4.4 Prospects

Decoding the information present in the genomes of living organisms is a real dream in biology. Whereas this goal appeared unrealistic a few years ago, the advent of massive amounts of genomic data combined with modeling generates hope that it will someday come true. The interplay between experiments and analyses remains key, and the pilot studies on model organisms and model proteins will keep on driving this field forward by generating ideas and hypotheses until the regulatory logic will be as clear for us as it is for the cells of our body.

3.5 Layers of chromatin regulation and their actors

Chromatin is made of basic subunits called nucleosomes. A nucleosome consists of an octamer of histone proteins (H2A, H2B, H3 and H4, each in two copies), around which 147 bp of DNA is wrapped. The level of compaction of the nucleosomes in the chromatin regulates the expression of genes, preventing access to DNA when condensed or permitting access to DNA for the transcription machinery when de-condensed.

Floral development certainly provides a great example of how chromatin dynamics contribute to transcriptional regulation of cell- and stage-specific genes in plants. In this section, we review the mechanisms through which chromatin is brought from a repressed to a permissive state (or vice versa) at specific loci and in a specific cell types, as well as how it is further maintained. We finally report recent results highlighting the interplay that operates between TFs and chromatin regulators for switching the transcriptional status of floral genes.

3.5.1 Nucleosome density and remodeling

Nucleosomes represent direct physical barriers to transcription, namely for the progression of the RNA Pol II complex on target DNA (Bai and Morozov, 2010). Thus, nucleosome density at a given locus influences its transcriptional state and can be employed as a regulatory mechanism for gene expression. As described in Section 3.3, microarray and NGS techniques allow the evaluation of histone density at a genome-wide level, via mapping of histones or more indirectly via an assessment of DNA accessibility. The first plant genome-wide nucleosome map was published by Chodavarapu et al. (2010), for Arabidopsis. It revealed a lower nucleosome density around the TSS of active genes compared with other parts of the genome. The study of Chodavarapu et al. (2010) also showed that nucleosomes are highly enriched at exons, especially in the vicinity of intronic junctions, and that the density of histones was correlated with RNA Pol II positioning. Additionally, heterochromatic regions display a higher nucleosome density than euchromatic regions. These facts strengthened the hypothesis that nucleosome positioning regulates Pol II processivity in plants.

Employing nucleosome density as a means for gene regulation requires the ability to move histones along the DNA, to disrupt DNA–nucleosome contacts or to eject histones from the DNA (Clapier and Cairns, 2009; Hargreaves and Crabtree, 2011). These processes require energy, which is provided by ATP-dependent enzymes of the SWI/SNF-related enzyme family. The exact mechanism of nucleosome repositioning is not completely understood yet, but the process is currently being studied through structural analysis and single-molecule observations. Recent results in this field are reviewed in Narlikar et al. (2013). In Arabidopsis (see Table 3.1), two SWI2/SNF2 ATPases, SPLAYED (SYD) and AtBRAHMA (AtBRM), have been extensively studied. Their function is partially redundant and both are involved in the activation of developmental genes, including meristem and floral organ identity genes like *WUSCHEL (WUS)*, *APETALA2 (AP2)*, *AGAMOUS (AG)*, *APETALA3 (AP3)* and *PISTILLATA (PI)* (Wagner and Meyerowitz, 2002; Kwon et al., 2005; Hurtado et al., 2006). Other ATP-dependent chromatin remodelers involved in floral development in Arabidopsis are CHROMATIN REMODELING12 (CHR12) and CHR23 (both also SWI2/SNF2 ATPases) (Sang et al., 2012), and PICKLE (PKL) and its close homolog PICKLE RELATED2 (PKR2) [chromodomain/helicase/DNA-binding domain (CHD3) ATPases] (Aichinger et al., 2009, 2011). A detailed review of ATP-dependent chromatin remodelers in Arabidopsis can be found in Gentry and Hennig (2014). As will be described in the Section 3.5.4, the ATP-dependent chromatin remodelers are involved in a concerted interplay between other chromatin remodelers and TFs.

3.5.2 Histone variants

Apart from their density, the composition of nucleosomes can also influence the transcriptional status of a gene. As already described, a histone octamer is usually composed of the four canonical histones H2A, H2B, H3 and H4. However, H2A and H3 in particular can be replaced by so-called histone variants, histones that differ from the canonical histone proteins in their amino acid sequence (Kamakaka and Biggins, 2005). The extent of the differences can be very small (e.g., four amino acid residues in H3) and can differ between plants and animals (Ingouff and Berger, 2010). Recent studies suggest that initiation of transcription at the TSS does not require the TSS to be nucleosome free as had been proposed earlier. Rather, it seems that initiation requires the presence of an unstable nucleosome containing the two variants H3.3 and H2A.Z. The detailed findings leading to this conclusion are reviewed in Soboleva et al. (2014).

3.5.2.1 *H3 variants*

In animal cells, the canonical H3.1 is incorporated into chromatin by the chromatin assembly factor 1 (CAF1) chaperone complex after DNA replication. Replacement of H3.1 with its variant H3.3 occurs in association with transcription after the passage of the RNA Pol II complex and is performed by the HIRA chaperone (Mito et al., 2005; Henikoff, 2008).

The association of H3.3 with actively transcribed genes, and thus RNA Pol II (Ahmad and Henikoff, 2002), is conserved in plants. H3.3 often carries active histone modifications (see section 3.5.3), whereas H3.1 is associated with repressive marks (Johnson et al., 2004; Stroud et al., 2012). An example underpinning the role of histone variants in plant development is the changes in H3.3 distribution during leaf differentiation analyzed by Wollmann et al. (2012). Genes that were downregulated in their expression during differentiation displayed a decrease in H3.3, while activated genes displayed an increase in H3.3 abundance. There are thus far only a few illustrations of regulation of floral development by histone H3 variants; for example it was shown that histone H3 variants play a role in male gametophyte development (Okada et al., 2005, 2006).

3.5.2.2 H2A.Z

Changes in the amino acid sequence of H2A.Z compared with H2A alter its interaction interface with H3/H4 and might thus contribute to a less stable nucleosome when H2A.Z is incorporated. Structural differences in the H2A variants and their role in different organisms are reviewed in Weber and Henikoff (2014).

Consistent with the latest hypothesis of H2A.Z being necessary for initiation at the TSS, this variant is found close to the TSS and is associated with activation in Arabidopsis (Zilberman et al., 2008). Furthermore, H2A.Z was recently suggested to aid RNA Pol II transcription in animals (Weber and Henikoff, 2014; Weber et al., 2014). A role for H2A.Z in Arabidopsis development was discovered indirectly by analysis of mutants in the two H2A.Z incorporators ACTIN-RELATED PROTEIN6 (ARP6) and PHOTOPERIOD-INDEPENDENT EARLY FLOWERING1 (PIE1), both SNF2-related proteins. Both are involved in flowering time control and floral development (Noh and Amasino, 2003; Martin-Trillo et al., 2006).

3.5.3 Post-translational modifications (PTMs) of histone tails

Another layer of chromatin-mediated gene regulation is described by the so-called "histone code" hypothesis (Strahl and Allis, 2000). Central to this model are the N-terminal tails of the histone proteins that are not condensed into the core nucleosome particle and therefore accessible for modifying and regulating proteins (Luger et al., 1997). The amino acid residues of the histone tails are subject to various covalent PTMs, including methylation, acetylation, phosphorylation, SUMOylation and ubiquitination. These modifications can be associated with the transcriptional status of the marked locus depending on their nature and the position of the modified amino acid in the tail (Berger, 2007). Histone-modifying enzymes, the so-called "writers" of the histone code, establish cell- and stage-specific histone mark patterns in all higher organisms. The effects of these modifications on the transcriptional status of a gene are currently explained by two mechanisms: either (1) through changes in the net charge of the histones resulting in loosening of DNA–histone interactions and thus enabling removal or replacement of histones or (2) as a signal for so-called "readers" of the histone code. Readers here refer to factors that recognize the histone modifications and alter the chromatin structure as a result of mark recognition (Li et al., 2007; Lauria and Rossi, 2011). The review by Lauria and Rossi (2011) provides a detailed overview of histone mark distributions and functions in plants.

Here we will illustrate the role of histone modifications in plant (flower) development by focusing on two antagonistic groups of chromatin modifiers that play a key role in this process, the repressive Polycomb group (PcG) of proteins and the activating trithorax group (trxG) of proteins. The antagonistic function between PcG and trxG proteins is conserved in plants and animals; however, plants display certain peculiarities differing from the animal model, probably due to their life-long development.

In animals, PcG proteins perform gene repression in two steps (Morey and Helin, 2010): by methylation of lysine 27 of histone 3 (H3K27me3) and subsequent recognition of this mark by readers which remain associated with the repressed locus and catalyze mono-ubiquitination of lysine 119 in H2A (H2Aub). This facilitates compaction of chromatin and prevents transcriptional initiation. Both writing and reading of the marks is performed by the Polycomb repressive complexes (PRCs). Most plant PRC components are encoded by small gene families, allowing the formation of stage-specific writer complexes (PRC2). In most plant tissues (except for the female gametophyte and in the endosperm of the developing seed), the histone methyl transferase (HMT) activity (the actual writing process) is carried out by the partially redundant SET domain proteins CURLY LEAF (CLF) and SWINGER (SWN) (Bemer and Grossniklaus, 2012). The reader complex (PRC1) and its function are not as well conserved in plants as PRC2. The first Arabidopsis PRC1 component to be discovered was LIKE-HETEROCHROMATIN PROTEIN1 (LHP1), which was shown to co-localize with H3K27me3 (Turck et al., 2007). Other putative PRC1 components in Arabidopsis are EMBRYONIC FLOWER1 (EMF1), which also co-localizes with H3K27me3 marks, and homologs of the catalytic subunits for H2Aub deposition, the RING-domain proteins AtRING1a/b and AtBMI1a-c (Calonje et al., 2008). However, the concerted action of PRC2 and PRC1 as reported in animals has not been demonstrated in plants so far. Additionally, PRC1-mediated H2Aub targeting was found to be independent of H3K27me3 at certain loci and H2Aub-independent repression has been reported in plants and in animals. These facts and other recent advances in understanding plant PRC1 function are reviewed in Calonje (2014).

The importance of PcG-mediated gene repression in plant development is impressively demonstrated by loss-of-function mutants in the writer components *CLF* and *SWN*: double *clf/swn* mutants develop into a callus-like structure with somatic embryo formation shortly after germination (Schubert et al., 2005). Furthermore, almost all genes involved in floral development mentioned in this section so far carry the PcG-associated H3K27me3 mark in young seedlings (Farrona et al., 2011) and tissue-specific subsets of H3K27me3 target genes display a high over-representation of developmental functions specific for the respective tissue (Zhang et al., 2007; Engelhorn et al., 2012). This, and the fact that H3K27me3 genome-wide patterns are tissue specific as shown by different meristematic and leaf-specific methylation patterns (Lafos et al., 2011), suggest that PcG repression regulates tissue-specific gene expression during cell differentiation. However, it is not yet clear whether tissue-specific expression is a cause or a consequence of the abundance of H3K27me3 at a locus in a given tissue (Farrona et al., 2011; Engelhorn et al., 2012).

As described in Section 3.2, initiation of floral development requires activation of floral organ and floral meristem identity genes. Thus, repression of these genes by PcG has to be relieved, which is performed by the action of trxG proteins. Most trxG components in Arabidopsis were first discovered by genetic interaction with PcG components, mainly suppression of the *clf* phenotype (curled leaves caused by *AG* ectopic expression). The mechanism of action is still unclear for some trxG components and remains to be clarified. Concerning histone modifications, trxG proteins discovered in Arabidopsis so far include erasers of the H3K27me3 mark, writers of active histone marks (H3K4me3 marks) and factors influencing the abundance of H3K27me3 and H3K4me3 in an as yet unknown manner, maybe as readers and/or recruiters of writers and erasers.

One H3K27me3 eraser described at genome-wide scale in Arabidopsis is RELATIVE OF EARLY FLOWERING 6 (REF6), which catalyzes demethylation of H3K27me3 to H3K27me1 and is involved in the activation of MADS-box floral organ identity genes like *AP1, AP3, PI, AG* and *SEPALLATA3 (SEP3)* (Lu et al., 2011; Figure 3.20, Table 3.1).

A writer of the active H3K4me3 mark that is responsible for activation of homeotic genes during floral development in Arabidopsis is ARABIDOPSIS HOMOLOG OF TRITHORAX 1 (ATX1) (Alvarez-Venegas et al., 2003; Ding et al., 2012; Fig. 3.20). Another H3K4 trimethyl transferase is SET DOMAIN GROUP 2 (SDG2), which is the major H3K4 trimethyl transferase in Arabidopsis involved in various developmental processes including regulation of flowering time and gametophyte development (Guo et al., 2010).

One factor displaying genetic trxG function without being a direct writer or eraser of histone marks is the SAND domain-containing ULTRAPETALA 1 (ULT1) protein (Fig. 3.20, Table 3.1), which induces removal of H3K27me3 at *AG* and *AP3* and displays various defects in floral development when mis-expressed (Carles et al., 2005; Carles and Fletcher, 2009). A possible mechanistic explanation for the action of ULT1 could be through the recruitment of ATX1, with which it interacts (Carles and Fletcher, 2010).

Additionally, trxG components are involved in the recruitment of RNA Pol II and assembly of the PIC, as reported for ATX1 (Ding et al., 2011), and in ATP-dependent chromatin remodeling, as reported for SYD and AtBRM (Wu et al., 2012) (Fig. 3.20).

Although several players in the PcG and trxG regulatory mechanism in Arabidopsis have been identified and characterized in detail, many questions concerning the exact mechanism of gene activation and repression remain open, especially the question of whether deposition and removal of histone marks are the cause or consequence of changes in the transcriptional status of a gene and whether active and repressive histone marks can coexist on a locus (Engelhorn et al., 2014).

A further subject of current research is the interplay between nucleosome density, histone variant deposition and histone marks. As already described, chromatin remodelers that change nucleosome density can be involved in the activation of PcG-repressed genes and certain histone variants can favor active or repressive histone

Layers of chromatin regulation 91

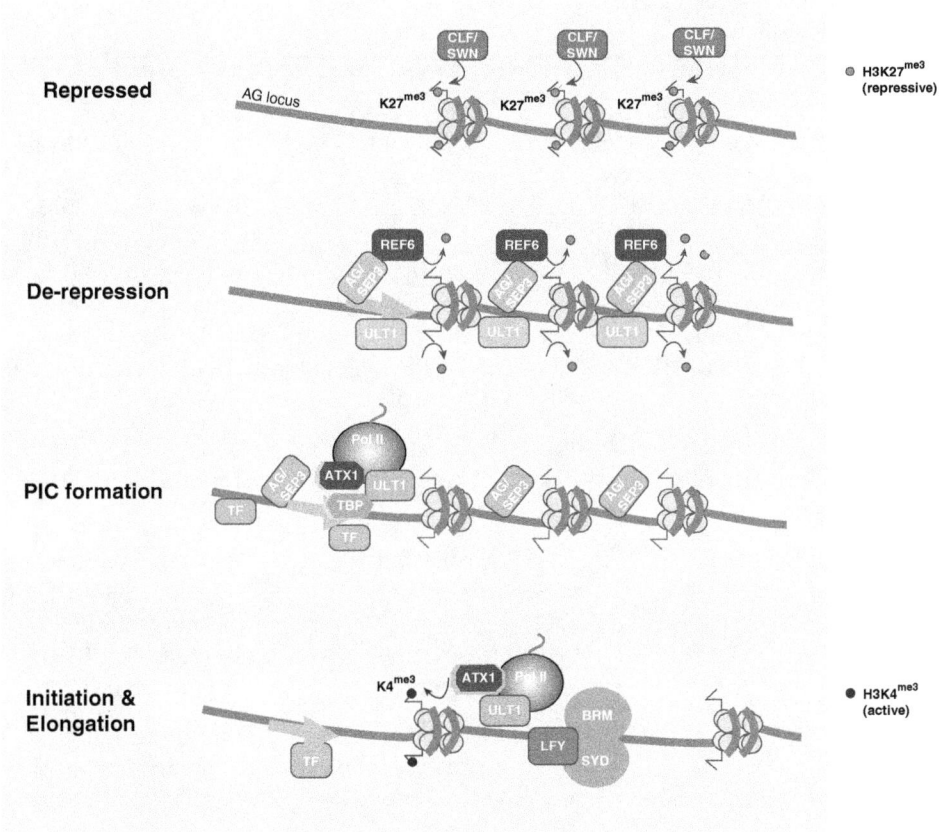

Fig. 3.20 Interactions at the chromatin for floral gene activation in Arabidopsis. The scheme depicts the known molecular events bringing a locus, such as *AG* or *AP3*, from a H3K27me3-mediated repressed state to a H3K4me3-mediated active state. Transcriptional status is indicated on the left. Repressed: the CLF and SWN HMTs function redundantly as components of the PcG complex to maintain the target locus repressed via deposition of repressive H3K27me3 marks. De-repression: AG and AP3 TFs would interact with the REF6 histone demethylase that removes the repressive H3K27me3 marks. PIC formation: the ATX1, ULT1 and possibly AG/AP3 proteins would function in the formation of the transcriptional pre-initiation complex (Pol II, RNA Pol II; TBP, TATA-box-binding protein). Initiation & Elongation: the ATX1 HMT further trimethylates H3K4 within the first 300-bp region of the transcribed gene. The BRM and SYD SWI/SNF2 ATPases may be recruited by the LFY TF for nucleosome disassembly, leading to accessibility of the target region to other TFs and to the general transcriptional machinery. The scheme is not indicative of any relative chronology between the above-described events, which remains to be resolved.

Table 3.1 Actors functioning in chromatin regulation during activation of floral organ identity genes in Arabidopsis

	NAME	FUNCTION	REFERENCE
Snf2-related chromatin remodelers	ATBRM	SWI1/SNF2 ATPase	Hurtado et al. (2006), Wu et al. (2012)
	SYD	SWI1/SNF2 ATPase	Kwon et al. (2005), Wu et al. (2012)
	CHR12, −23	SWI1/SNF2 ATPases	Sang et al. (2012)
	CHR4, −11, −17	SWI1/SNF2 ATPases	Smaczniak et al. (2012)
	PKL, PKR2	CHD3 ATP-dependent chromatin remodelers	Aichinger et al. (2009, 2011)
Histone demethylase	REF6	H3K27me3 demethylase	Lu et al. (2011)
Histone methyl transferases	ATX1	H3K4 methyl transferase	Alvarex-Venegas et al. (2003), Ding et al. (2011)
	SDG2	H3K4 methyl transferase	Berr et al. (2010), Guo et al. (2010)
TFs	AG	MADS domain TF, makes complexes with chromatin remodelers	Sun et al. (2009), Smaczniak et al. (2012)
	AP3, PI	MADS domain TFs, make complexes with chromatin remodelers	Smaczniak et al. (2012)
	LFY	Recruiter of SYD and BRM	Wu et al. (2012),
	SEP3	MADS domain TF, recruiter of SYD and BRM, putative modulator of chromatin accessibility	Wu et al. (2012), Pajoro et al. (2014b)
	AP1	MADS domain TF, putative modulator of chromatin accessibility	Pajoro et al. (2014b)
SAND domain-containing factor	ULT1	Contributing to H3K27me3 changes, interacts with ATX1	Carles and Fletcher (2009)

marks (H3.3 in plants is associated with H3K4me3, H3.1 is associated with repressive marks) (Johnson et al., 2004; Stroud et al., 2012). Several recent results indicate that TFs are involved in this interplay and play a critical role in the recruitment of the chromatin-modifying machinery. The known examples of this interplay are described in Section 3.5.4.

3.5.4 Evidence for the interplay between TFs and chromatin features in the floral development framework

Several recent reports have shown interactions between chromatin and TF dynamics in the context of floral development.

The SYD SWI/SNF chromatin remodeling ATPase factor, initially isolated in a genetic modifier screen in the *leafy* (*lfy*) mutant background (Wagner and Meyerowitz, 2002), and its functionally redundant AtBRM homolog, were shown to physically interact with the LFY TF (Wu et al., 2012). Furthermore, the work of Wu et al. (2012) highlights the role of LFY in the recruitment of SYD to the *AP3* and *AG* regulatory regions and establishes the first link between TF and chromatin remodeling in transcriptional activation (Fig. 3.20). This function might be subsequent to the involvement of LFY in the direct repression and/or eviction of the PRC1 PcG complex that keeps the *AG* and SEP3 *loci* in an inactive state (Wu et al., 2012).

Other studies highlight the interaction of TFs with chromatin remodelers and modifiers or indicate a role in regulation of chromatin accessibility during floral development. Isolation of complexes containing MADS-domain proteins from Arabidopsis inflorescences, followed by systematic identification of the components of these complexes by mass spectrometry, revealed the presence of several chromatin modifiers and remodelers (Smaczniak et al., 2012). Indeed, the MADS-domain-containing proteins AP1, AG and SEP3 interact with the REF6 H3K27me3 demethylase (Smaczniak et al., 2012; Fig. 3.20). Moreover, the ATP-dependent chromatin remodelers CHR4, CHR11 and CHR17 were in found in complexes with AP1, PI, AG, AP3 or SEP3 (Smaczniak et al., 2012; Table 3.1).

Such direct interactions between TFs and chromatin remodelers is interesting in light of the fact that TFs can already sit at loci carrying repressive H3K27me3 marks and lead to progressive removal of repressive marks to further induce gene expression. A good illustration of this is the AG-dependent reduction in H3K27me3 for activation of the *KNUCKLES* (*KNU*) target gene. AG indirectly represses *WUS* via activation of the *KNU* gene (Sun et al., 2009), which encodes a protein composed of a C2H2-type zinc finger and an EAR-like repressor domain (Sun et al., 2009). AG binds directly to the H3K27me3-marked *KNU* promoter, probably contributing to the eviction of the PcG complex (Sun et al., 2014) and leading to a decrease in repressive marks at the *KNU* locus (Sun et al., 2009).

It has been reported in animals that, in the context of compact chromatin, prior binding of a specialized subset of TFs at DNA target sites is required to facilitate accessibility for additional TFs and other regulatory proteins (Magnani et al., 2011; Zaret and Carroll, 2011). These factors are known as pioneering factors. It was recently proposed in plants that MADS-domain-containing TFs could also act as pioneering factors. During floral development in Arabidopsis, binding of AP1 and SEP3 to promoters of flowering time and flower morphogenesis genes precedes increases in chromatin accessibility (Pajoro et al., 2014b).

In summary, gene expression for floral development is tightly regulated by chromatin structure. TFs such as MADS-domain-containing proteins and LFY are probably involved in chromatin de-repression and opening, via interactions with chromatin remodelers and modifiers. MADS-encoding genes are themselves targets for this activation, indicating a multilayered, intricate network of regulation.

3.6 Floral transcription factors: a structural view

3.6.1 Introduction

Gene regulation is essential for determining the complexity and morphology of organisms. One of the most important mechanisms for controlling gene expression is the interaction of TFs with their cognate DNA. Generally, as the complexity of any organism increases, the number and diversity of TFs present in the genome increases. In higher organisms, TFs comprise 5–10% of the genome. For example, *Arabidopsis thaliana* has a genome of 32 835 genes, 1738 of which encode TFs. The *Homo sapiens* genome comprises 46 591 genes, 2886 of which are predicted to be TFs.

3.6.2 Basis for TF–DNA interaction at the structural level

TFs recognize their cognate DNA using a number of different mechanisms (for a review see Rohs et al., 2010, and references therein). The DNA backbone is able to present different shapes depending on the type of DNA (A-, B-, Z- or TA-DNA) and its curvature. B-DNA is the most common form of DNA, present in solution and in cells, which is characterized by a relatively shallow major groove, a deep and narrow minor groove and the base pairs oriented perpendicular to the helical axis. TFs are able to recognize and preferentially bind different overall shapes of DNA. In the case of B-DNA, TF families often use alpha-helices to bind the major groove. However, these interactions are not sufficient to confer specificity. Specificity over more than 5 bp is generally determined by flexible regions of the protein, which are able to interact preferentially with specific bases, present in the minor and the major groves. TFs also recognize the overall shape of the DNA through interaction with the phosphate backbone of the DNA. Such interactions contribute to DNA-binding specificity but to a lesser extent than direct base contacts.

TFs group into over 70 SCOP (Structural Classification of Proteins Database) superfamilies. Common DNA-binding motifs found in multiple TF superfamilies include the HTH (helix-turn-helix) motif (Brennan and Matthews, 1989), the HLH (helix-loop-helix) motif (Massari and Murre, 2000), the leucine zipper motif (Landschulz et al., 1988), the immunoglobulin-like domain (Halaby et al., 1999) and the Zn-finger domain (Laity et al., 2001). While the secondary structural elements within the superfamily are the same, the identity of the amino acid residues (i.e., H-bond acceptors and donors), the presence of flexible regions of the protein and the ability of the protein to oligomerize all play roles in imparting specificity to the interactions between the TF and DNA. Thus, the globular fold of the TF and the presence of structural motifs provide the first level of DNA recognition through shape complementarity. The identity of the amino acids in the TF and the bases in the cognate DNA are responsible for providing a second layer of specificity. The addition of flexible interacting regions of the protein and higher-order protein complex formation imparts additional specificity and the ability of the TF to recognize longer DNA sequences.

Whole-genome sequencing has provided an important overview of the different TFs and TF families present across the kingdoms of life. As TFs often fulfill very specific roles in different organisms, different TF families have undergone lineage or kingdom-specific expansion. In addition, not all TF families are present in all kingdoms. Based on the available genomic data and predicted coding regions, plants have about 40 plant-specific TF families (Richardt et al., 2007). Of these plant-specific families, structural data are currently available for AP2/ERF, B3-ARF-Aux/IAA, NAC, SBP, WRKY and LFY. In this section we will focus on the structure and function of the AP2/ERF, B3-ARF-Aux/IAA, SBP and LFY families and their role in flowering.

3.6.3 Examples of plant-specific floral TFs

3.6.3.1 AP2/ERF

APETALA 2 (AP2) is a floral homeotic class A gene involved in sepal and petal development in whorls 1 and 2 and restricts the expression of the class C gene *AGAMOUS (AG)* in these whorls (Bomblies et al., 1999; Dinh et al., 2012). *AP2* also plays a role in the establishment of floral meristem identity, suppression of floral meristem indeterminacy, the development of the ovule and seed coat and the control of seed mass (Jofuku et al., 1994, 2005; Ohto et al., 2005; Wurschum et al., 2006; Yant et al., 2010). In floral organ development, AP2 functions as a transcriptional repressor by recruiting TOPLESS (TPL) and a histone deacetylase, HDA19. This complex (AP2–TPL–HDA19) associates with the second intron of *AG* during floral development and represses transcription of *AG* in the outer whorls, 1 and 2 (Krogan et al., 2012).

AP2 encodes a transcription factor in the APETALA 2/ethylene response factor (ERF) family. The DBDs of the AP2/ERF family of transcription factors (PDB 3GCC; Fig. 3.21) recognize a 16-bp GCC motif, called a GCC-box. The GCC-box binding domain (GBD) is approximately 60 amino acids long and folds into a three-stranded antiparallel beta-sheet which packs against an alpha-helix. Unlike many DBDs, which are dimeric, the GBD domain binds DNA as a monomer and binds a non-palindromic sequence. The beta-sheet of the GBD interacts with the major groove of DNA via an extensive positively charged patch comprising four highly conserved arginine residues. Arginine and tryptophan residues in the beta-sheet are identified to contact eight of the nine consecutive base pairs in the major groove, and at the same time bind to the sugar phosphate backbones. The cognate DNA is bent to accommodate the beta sheet of the GBD. AP2, unlike the ERF proteins, contains two GBD domains, which may play a role in its regulatory function. In addition, AP2 has an ERF-associated amphiphilic repression (EAR) motif (LxLxL) (Ohta et al., 2001) which is important

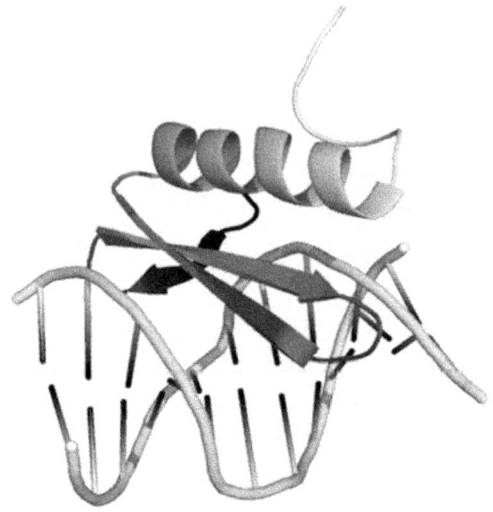

Fig. 3.21 NMR structure of the GCC-box DBD from *Arabidopsis thaliana* ERF1 (PDB 1GCC). The protein and DNA are depicted as cartoons. A three-stranded anti-parallel beta-sheet lies in the major groove and is stabilized by an alpha-helix.

for the direct recruitment of TPL and its role as a repressor of the class C genes in floral organ development (Krogan et al., 2012).

3.6.3.2 B3-ARF-Aux/IAAM

Auxins are a family of plant hormones derived from indole acetic acid (IAA) that are important mediators of different developmental processes including development of roots, shoots and flowers. Auxin indirectly regulates auxin response factor (ARF) family proteins by promoting the degradation of Aux/IAA proteins (Fig. 3.22). Aux/IAA proteins bind to ARFs, forming an inactive complex. When the Aux/IAA proteins are degraded, the ARF family of TFs is able to bind auxin response elements (AuxREs) on promoters of auxin response genes, allowing for their regulation (Fig. 3.23). Degradation of Aux/IAAs occurs via the auxin-binding protein TIR1. TIR1, when bound to auxin, has a high affinity for Aux/IAAs and recruits a ubiquitin ligase to the complex, targeting the Aux/IAAs for degradation and releasing the ARFs (Calderon Villalobos et al., 2012). Structural data for both the DNA-binding domain of ARF1 and ARF5 (Boer et al., 2014) and the Aux/IAA-interacting domain of ARF5 have recently been described (Nanao et al., 2014).

The DBD, called the B3 domain, is present in different plant-specific transcription factors such as RAV1 and the ARF family. The B3 domain is approximately 100 residues long and comprises a seven-stranded open beta-barrel (Fig. 3.22, right). The B3 domain binds the consensus sequence TGTCTC via contacts in adjacent beta-strands. Most contacts between the protein and DNA occur via conserved arginine residues and the major groove of DNA, resulting in a bending of DNA by about 40°.

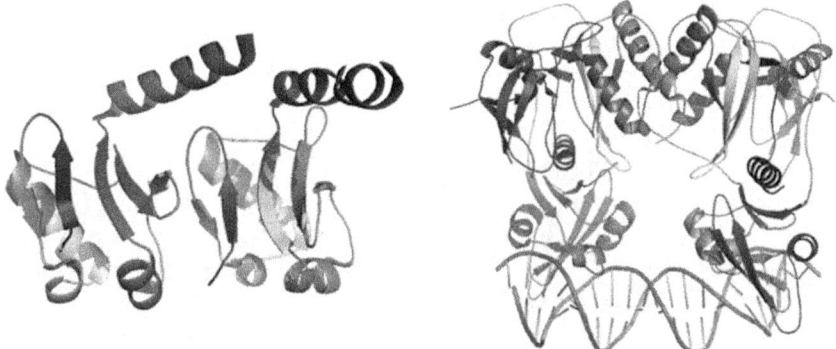

Fig. 3.22 Structures of the plant-specific transcription factor ARF family. Left, oligomerization domain of ARF5 (PDB 4CHK) Nanao et al., 2014. Right, dimeric DBD from ARF5 solved by x-ray crystallography (PDB 4LDX) Boer et al., 2014. Most contacts with the DNA occur via interactions with the beta-strands of the protein in the major groove of DNA. All structures are depicted as cartoons and DNA is shown schematically.

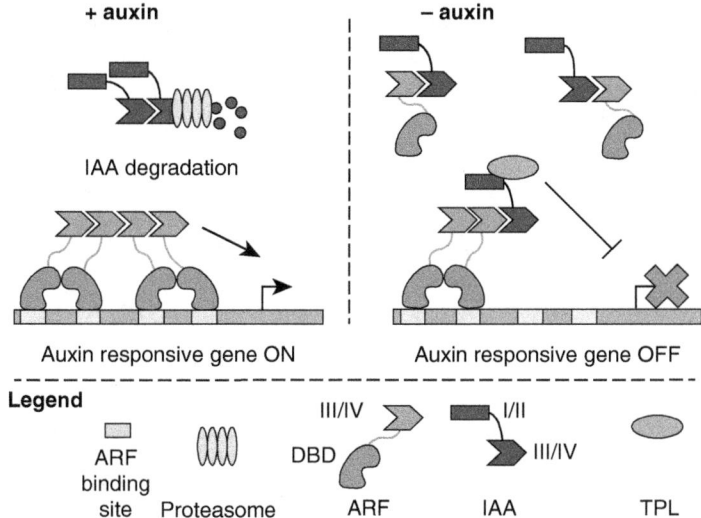

Fig. 3.23 Schematic diagram of auxin response. In the presence of auxin, Aux/IAA proteins are targeted for 26S proteosomal degradation via the formation of a co-receptor complex of auxin-bound Aux/IAAs and TIR1. This frees the ARF transcription factors from their repressive complex and allows them to bind DNA and regulate target genes. In the absence of auxin, the ARFs are part of a repressive complex comprising the Aux/IAAs and co-repressors including TOPLESS (TPL). Figure adapted from Nanao et al. (2014)

Inter-strand loops provide additional interacting residues likely accounting for DNA base recognition (Boer et al., 2014).

3.6.3.3 SBP family

The SQUAMOSA PROMOTER BINDING-LIKE (SBP or SPL) family of TFs promote the transition from vegetative growth to flowering by activating *LFY* as well as a number of MADS-box genes important for flowering time, including *SOC1*, *AP1*, *FUL* and *FT* (Preston and Hileman, 2013). The SBP TFs are often bound in a repressive complex with DELLAs. In the presence of the hormone gibberellin, the soluble gibberellin receptor GID1 binds gibberellin and DELLAs. This complex targets the DELLAs for degradation via the ubiquitination/proteasome pathway, allowing the SBPs to activate the downstream target genes necessary for flowering (Eckardt, 2007).

The SBP family is defined by a highly conserved region of 76–80 amino acids called the SBP domain (Fig. 3.24) (Klein et al., 1996; Yang et al., 2008). The SBP domain is involved in both nuclear import and DNA binding to a GTAC core motif and gene-specific flanking regions (Birkenbihl et al., 2005),(Liang et al., 2008). The SBP domain binds divalent zinc cations via a zinc-finger motif. The zinc-finger motif has the consensus sequences Cys-Cys-His-Cis and Cys-Cys-Cys-His. The SBP domain consists of a three-stranded anti-parallel beta-sheet and two alpha-helices, one N-terminal and the other C-terminal (PBD codes 1UL4, 1UL5 and 1WJ0). The alpha-helices have highly conserved arginine residues which interact with the major groove of DNA in a sequence specific manner and are termed recognition helices (Yamasaki et al., 2004).

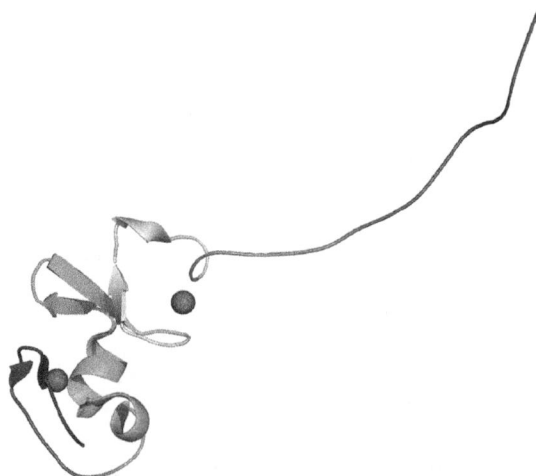

Fig. 3.24 NMR structure of the DBD from *Arabidopsis thaliana* squamosa binding protein-like 7 (PDB 1UL5) Yamasaki et al., 2004, zinc ions are shown as balls. The protein has an unstructured extended C-terminal region. The structure was determined without DNA present.

In any given plant genome, TFs account for between 5 and 10% of all proteins and can span hundreds of structural families. While the majority of folds used to bind DNA are highly conserved across kingdoms, plants have engineered a number of kingdom-specific tertiary and quaternary structures to bind DNA. Helices, beta-strands and random coils all play a role in recognizing specific DNA sequences, as evidenced by the diversity of structures and interaction motifs of the DBDs. This diversity may be due to the complex processes, such as flowering, and the myriad target genes regulated by these TF families. In addition, these DBDs are often part of larger multidomain proteins and/or protein complexes. In many cases, the DBD alone is not sufficient for complete function but requires these additional protein–protein binding domains.

3.6.3.4 LEAFY (LFY)

Origin of LFY As explained in Section 3.2, LFY is a key floral regulator able to convert SAMs or axillary meristems into flowers (Moyroud et al., 2009). Following its identification as a master regulator of floral development, *LFY*-like genes (homologs) were actively sought in various angiosperms. *LFY* homologs have now been isolated in hundreds of flowering plants species (for review see Moyroud et al., 2009, 2010). Interestingly, *LFY*-like genes were also identified in the other divisions of land plants which do not produce flowers: mosses, ferns and gymnosperms (Frohlich and Parker, 2009). Recently, *LFY* homologs were isolated for the first time from charophyte algae, demonstrating that LFY is not a TF specific to terrestrial plants (as previously thought) but that its existence precedes the transition from water to land (Sayou et al., 2014).

LFY is an extremely conserved TF (in particular the two characteristic N- and C-terminal domains) but it does not resemble any other known protein. For this reason, *LFY* seems to have "appeared" suddenly in the charophyte genomes and, to this day, its origins remain obscure. Many TFs were derived from transposons (Breitling and Gerber, 2000; Feschotte, 2008) and this could also be the case for LFY. However, sequence or 3D structural similarities with known transposon elements are too weak to establish LFY's ancestry with confidence.

The LFY DBD The LFY DBD does not resemble any other known DBD. The resolution of its crystallographic structure (Hamès et al., 2008) showed that it is a new fold made of seven alpha-helices including a HTH motif that binds the major groove of DNA and that is commonly found in RNA- or DNA-binding molecules. Direct contact with bases occurs both in the major groove (through the HTH) and in the minor groove through a wing-like extension. The LFY DBD dimerizes upon DNA binding.

Evolution of LFY Unlike most developmental regulators, *LFY* is not part of an extended gene family (Martinez-Castilla and Alvarez-Buylla, 2003; Rokas, 2008; Moyroud et al., 2010) and many species only possess one *LFY* homolog. *LFY* experienced duplication as any other gene but the extra copies seem to have been rapidly eliminated (for review see Moyroud et al., 2009). The reason why copies of *LFY* are not maintained more frequently is not understood: extra copies could be detrimental by affecting plant architecture or by reducing the number of progeny (Weigel and Nilsson,

1995). Alternatively, LFY, as an essential protein occupying a key position in regulatory networks, could be less prone to form an extended family. This phenomenon has been reported for "hub" proteins that contain several distinct interaction surfaces with co-regulators (Kim et al., 2006). Large-scale sequencing projects combined with the ever increasing number of sequenced genomes should provide us with a clearer picture of gene duplication events experienced by *LFY* throughout its evolutionary history and might help us understand why *LFY* did not generate a multicopy family.

Despite the strong conservation of its DBD (all 13 amino acids involved in protein–DNA contact are conserved from mosses to angiosperms), the types of DNA motifs recognized by LFY changed at least twice during evolution: PpLFY, the homolog of LFY in the moss *Physcomitrella patens* is unable to bind to the same DNA motif as LFY and the protein cannot replace Arabidopsis LFY function in planta (Maizel et al., 2005). Similarly, the homolog of LFY in the alga *Klebsormidium subtile* failed to recognize the typical 19-bp DNA motif bound by all angiosperm LFYs and instead interacted with a distinct 12-bp motif (Sayou et al., 2014). Remarkably, in the later case, this change in DNA-binding specificity relies on a couple of amino acids that do not interact directly with the DNA molecule but instead modify the conformation of the DBD and its ability to bind DNA as a dimer (Sayou et al., 2014).

Gene duplication is a major force in the evolution of TFs as it generates "spare copies": following duplication, one copy of the gene is relieved from selective pressure and free to accumulate mutations that can lead to new properties (such as a different DNA-binding specificity) while the other copy still fulfills the original function. Thus, how can an essential but single-copy TF have changed its DNA-binding specificity without any detrimental effect on the organism? A recently identified promiscuous form of LFY, able to bind to all of the three DNA motifs recognized by LFY from angiosperms, mosses or algae could be the key to this mystery: LFY could change its DNA-binding specificity progressively by retaining mutations that make the protein more "flexible" and allow it to interact with new DNA sequences while keeping its ability to bind to its original binding site. This mechanism would enable smooth transitions from one DNA-binding specificity to another in the absence of spare copies of LFY (Sayou et al., 2014).

References

Ahmad, K. and Henikoff, S. (2002). Histone H3 variants specify modes of chromatin assembly. *Proc Natl Acad Sci USA*, 99 (Suppl. 4): 16477–16484.

Aichinger, E., Villar, C.B.R., Farrona, S., Reyes, J.C., Hennig, L. and Köhler, C. (2009). CHD3 proteins and polycomb group proteins antagonistically determine cell identity in Arabidopsis. *PLoS Genet*, 5: e1000605.

Aichinger, E., Villar, C.B.R., Mambro, R.D., Sabatini, S. and Köhler, C. (2011). The CHD3 chromatin remodeler PICKLE and polycomb group proteins antagonistically regulate meristem activity in the Arabidopsis root. *Plant Cell*, 23: 1047–1060.

Alvarez-Venegas, R., Pien, S., Sadder, M., Witmer, X., Grossniklaus, U. and Avramova, Z. (2003). ATX-1, an Arabidopsis homolog of trithorax, activates flower homeotic genes. *Curr Biol*, 13: 627–637.

Anders, S. and Huber, W. (2010). Differential expression analysis for sequence count data. *Genome Biol*, 11, R106.

Andersson, R. et al. (2014). An atlas of active enhancers across human cell types and tissues. *Nature*, 507: 455–461.

Au, K.F., Jiang, H., Lin, L., Xing, Y. and Wong, W.H. (2010). Detection of splice junctions from paired-end RNA-seq data by SpliceMap. *Nucleic Acids Res*, 38: 4570–4578.

Bai, L. and Morozov, A.V. (2010). Gene regulation by nucleosome positioning. *Trends Genet*, 26: 476–483.

Bailey, T., Krajewski, P., Ladunga, I., Lefebvre, C., Li, Q., Liu, T., Madrigal, P., Taslim, C. and Zhang, J. (2013). Practical guidelines for the comprehensive analysis of ChIP-seq data. *PLoS Comput Biol*, 9: e1003326.

Bardet, A. F., He, Q., Zeitlinger, J. and Stark, A. (2012). A computational pipeline for comparative ChIP-seq analyses. *Nature Protoc*, 7: 45–61.

Benková, E., Michniewicz, M., Sauer, M., Teichmann, T., Seifertová, D., Jürgens, G. and Friml, J. (2003). Local, efflux-dependent auxin gradients as a common module for plant organ formation. *Cell*, 115: 591–602.

Berg, O.G. and von Hippel, P.H. (1987). Selection of DNA binding sites by regulatory proteins. Statistical-mechanical theory and application to operators and promoters. *J Mol Biol*, 193: 723–750.

Berger, S.L. (2007). The complex language of chromatin regulation during transcription. *Nature*, 447: 407–412.

Bernatavichute, Y.V, Zhang, X., Cokus, S., Pellegrini, M. and Jacobsen, S.E. (2008). Genome-wide association of histone H3 lysine nine methylation with CHG DNA methylation in *Arabidopsis thaliana*. *PLoS One*, 3, e3156.

Bemer, M. and Grossniklaus, U. (2012). Dynamic regulation of Polycomb group activity during plant development. *Curr Opin Plant Biol*, 15: 523–529.

Bernstein, B.E., Birney, E., Dunham, I., Green, E.D., Gunter, C. and Snyder, M. (2012). An integrated encyclopedia of DNA elements in the human genome. *Nature*, 489: 57–74.

Berr, A., McCallum, E.J., Ménard, R., Meyer, D., Fuchs, J., Dong, A. and Shen, W.-H. (2010). Arabidopsis SET DOMAIN GROUP2 is required for H3K4 trimethylation and is crucial for both sporophyte and gametophyte development. *Plant Cell*, 22: 3232–3248.

Birkenbihl, R.P., Jach, G., Saedler, H. and Huijser, P. (2005). Functional dissection of the plant-specific SBP-domain: overlap of the DNA-binding and nuclear localization domains. *J Mol Biol*, 352: 585–596.

Birnbaum, K., Jung, J.W., Wang, J.Y., Lambert, G.M., Hirst, J.A., Galbraith, D.W. and Benfey, P.N. (2005). Cell type-specific expression profiling in plants via cell sorting of protoplasts from fluorescent reporter lines. *Nat Methods*, 2: 615–619.

Blázquez, M.A., Soowal, L.N., Lee, I. and Weigel, D. (1997). LEAFY expression and flower initiation in Arabidopsis. *Development*, 124: 3835–3844.

Boer, D.R., Freire-Rios, A., van den Berg, W.A., Saaki, T., Manfield, I.W., Kepinski, S., Lopez-Vidrieo, I., Franco-Zorrilla, J.M., de Vries, S.C., Solano, R., Weijers, D. and Coll, M. (2014). Structural basis for DNA binding specificity by the auxin-dependent ARF transcription factors. *Cell*, 156: 577–589.

Bomblies, K., Dagenais, N. and Weigel, D. (1999). Redundant enhancers mediate transcriptional repression of AGAMOUS by APETALA2. *Dev Biol*, 216: 260–264.

Bonhoure, N. et al. (2014). Quantifying ChIP-seq data: a spiking method providing an internal reference for sample-to-sample normalization. *Genome Res*, 24: 1157–1168.

Borges, F., Gardner, R., Lopes, T., Calarco, J.P., Boavida, L.C., Slotkin, R.K., Martienssen, R.A. and Becker, J.D. (2012). FACS-based purification of Arabidopsis microspores, sperm cells and vegetative nuclei. *Plant Methods*, 8: 44.

Bowman, J.L., Smyth, D.R. and Meyerowitz, E.M. (1989). Genes directing flower development in Arabidopsis. *Plant Cell*, 1: 37–52.

Bowman, J.L., Smyth, D.R. and Meyerowitz, E.M. (1991). Genetic interactions among floral homeotic genes of Arabidopsis. *Development*, 112: 1–20.

Breitling, R. and Gerber, J.K. (2000). Origin of the paired domain. *Dev Genes Evol*, 210: 644–650.

Brennan, R.G. and Matthews, B.W. (1989). The helix-turn-helix DNA binding motif. *J Biol Chem*, 264: 1903–1906.

Bryant, D.W., Priest, H.D. and Mockler, T.C. (2012). Detection and quantification of alternative splicing variants using RNA-seq. *Methods Mol Biol*, 883: 97–110.

Buenrostro, J. D., Giresi, P.G., Zaba, L.C., Chang, H.Y. and Greenleaf, W.J. (2013). Transposition of native chromatin for fast and sensitive epigenomic profiling of open chromatin, DNA-binding proteins and nucleosome position. *Nat Methods*, 10: 1213–1220.

Calderon Villalobos, L.I., Lee, S., De Oliveira, C., Ivetac, A., Brandt, W., Armitage, L., Sheard, L.B., Tan, X., Parry, G., Mao, H., Zheng, N., Napier, R., Kepinski, S. and Estelle, M. (2012). A combinatorial TIR1/AFB-Aux/IAA co-receptor system for differential sensing of auxin. *Nat Chem Biol*, 8: 477–485.

Calonje, M. (2014). PRC1 marks the difference in plant PcG repression. *Mol Plant*, 7: 459–471.

Calonje, M., Sanchez, R., Chen, L. and Sung, Z.R. (2008). EMBRYONIC FLOWER1 participates in polycomb group-mediated AG gene silencing in Arabidopsis. *Plant Cell*, 20: 277–291.

Cao, S., Kumimoto, R., Gnesutta, N., Calogero, A., MantovaniHolt, R. and Holt, B.F. (2014). A distal CCAAT / NUCLEAR FACTOR Y complex promotes chromatin looping at the FLOWERING LOCUS T promoter and regulates the timing of flowering in Arabidopsis. *Plant Cell*, 26: 1009–1017.

Carles, C.C. and Fletcher, J.C. (2009). The SAND domain protein ULTRAPETALA1 acts as a trithorax group factor to regulate cell fate in plants. *Genes Dev*, 23: 2723–2728.

Carles, C.C. and Fletcher, J.C. (2010). Missing links between histones and RNA Pol II arising from SAND? *Epigenetics*, 5: 381–385.

Carles, C.C., Choffnes-Inada, D., Reville, K., Lertpiriyapong, K. and Fletcher, J.C. (2005). ULTRAPETALA1 encodes a SAND domain putative transcriptional regulator that controls shoot and floral meristem activity in Arabidopsis. *Development*, 132: 897–911.

Chahtane, H. et al. (2013). A variant of LEAFY reveals its capacity to stimulate meristem development by inducing RAX1. *Plant J*, 74: 678–689.

Chen, Q., Atkinson, A., Otsuga, D., Christensen, T., Reynolds, L. and Drews, G.N. (1999). The Arabidopsis FILAMENTOUS FLOWER gene is required for flower formation. *Development*, 126: 2715–2726.

Chen, C.-M., Shih, T.-H., Pai, T.-W., Liu, Z.-L., Chang, M.D.-T. and Hu, C.-H. (2013). Gene expression rate comparison for multiple high-throughput datasets. *IET Syst Biol*, 7: 135–142.

Chodavarapu, R.K., Feng, S., Bernatavichute, Y.V., Chen, P.-Y., Stroud, H., Yu, Y., Hetzel, J.A., Kuo, F., Kim, J., Cokus, S.J., Casero, D., Bernal, M., Huijser, P., Clark, A.T., Krämer, U., Merchant, S.S., Zhang, X., Jacobsen, S.E. and Pellegrini, M. (2010). Relationship between nucleosome positioning and DNA methylation. *Nature*, 466: 388–392.

Clapier, C.R. and Cairns, B.R. (2009). The biology of chromatin remodeling complexes. *Annu Rev Biochem*, 78: 273–304.

Coen, E.S. and Meyerowitz, E.M. (1991). The war of the whorls: genetic interactions controlling flower development. *Nature*, 353: 31–37.

Cole, M., Chandler, J., Weijers, D., Jacobs, B., Comelli, P. and Werr, W. (2009). DORNROSCHEN is a direct target of the auxin response factor MONOPTEROS in the Arabidopsis embryo. *Development*, 136: 1643–1651.

Collas, P. and Dahl, J.A. (2008). Chop it, ChIP it, check it: the current status of chromatin immunoprecipitation. *Front Biosci*, 13: 929–943.

Core, L.J., Waterfall, J.J. and Lis, J.T. (2008). Nascent RNA sequencing reveals widespread pausing and divergent initiation at human promoters. *Science*, 322: 1845–1848.

Dale, R.K., Pedersen, B.S. and Quinlan, A.R. (2011). Pybedtools: a flexible Python library for manipulating genomic datasets and annotations. *Bioinformatics*, 27: 3423–3424.

Deal, R.B. and Henikoff, S. (2010). Catching a glimpse of nucleosome dynamics. *Cell Cycle*, 9: 3389–3390.

Deal, R.B. and Henikoff, S. (2011). The INTACT method for cell type-specific gene expression and chromatin profiling in *Arabidopsis thaliana*. *Nat Protoc*, 6: 56–68.

Dekker, J., Marti-Renom, M.A. and Mirny, L.A. (2013). interpreting chromatin interaction data. *Nat Rev Genet*, 14: 1–23.

Ding, Y., Avramova, Z. and Fromm, M. (2011). Two distinct roles of ARABIDOPSIS HOMOLOG OF TRITHORAX1 (ATX1) at promoters and within transcribed regions of ATX1-regulated genes. *Plant Cell*, 23: 350–363.

Ding, Y., Ndamukong, I., Xu, Z., Lapko, H., Fromm, M. and Avramova, Z. (2012). ATX1-generated H3K4me3 is required for efficient elongation of transcription, not initiation, at ATX1-regulated genes. *PLoS Genet*, 8: e1003111.

Dinh, T.T., Girke, T., Liu, X., Yant, L., Schmid, M. and Chen, X. (2012). The floral homeotic protein APETALA2 recognizes and acts through an AT-rich sequence element. *Development*, 139: 1978–1986.

Eckardt, N.A. (2007). GA perception and signal transduction: molecular interactions of the GA receptor GID1 with GA and the DELLA protein SLR1 in rice. *Plant Cell*, 19: 2095–2097.

Endress, P. and Doyle, J. (2007). Floral phyllotaxis in basal angiosperms: development and evolution. *Curr Opin Plant Biol*, 10: 52–57.

Engelhorn, J., Reimer, J.J., Leuz, I., Göbel, U., Huettel, B., Farrona, S. and Turck, F. (2012). Development-related PcG target in the apex 4 controls leaf margin architecture in *Arabidopsis thaliana*. *Development*, 139: 2566–2575.

Engelhorn, J., Blanvillain, R. and Carles, C.C. (2014). Gene activation and cell fate control in plants: a chromatin perspective. *Cell Mol Life Sci*, 71: 3119–3137.

English, A.C., Richards, S., Han, Y., Wang, M., Vee, V., Qu, J., Qin, X., Muzny, D.M., Reid, J.G., Worley, K.C., et al. (2012). Mind the gap: upgrading genomes with Pacific Biosciences RS long-read sequencing technology. *PLoS One*, 7: e47768.

Farrona, S., Thorpe, F.L., Engelhorn, J., Adrian, J., Dong, X., Sarid-Krebs, L., Goodrich, J. and Turck, F. (2011). Tissue-specific expression of FLOWERING LOCUS T in Arabidopsis is maintained independently of polycomb group protein repression. *Plant Cell*, 23: 3204–3214.

Ferrándiz, C., Gu, Q., Martienssen, R. and Yanofsky, M.F. (2000). Redundant regulation of meristem identity and plant architecture by FRUITFULL, APETALA1 and CAULIFLOWER. *Development*, 127: 725–734.

Feschotte, C. (2008). Transposable elements and the evolution of regulatory networks. *Nat Rev Genet*, 9: 397–405.

Frohlich, M.W. and Chase, M.W. (2007). After a dozen years of progress the origin of angiosperms is still a great mystery. *Nature*, 450: 1184–1189.

Frohlich, M.W. and Chase, M.W. (2009). The Mostly Male Theory of Flower Evolutionary Origins: From Genes to Systematic Botany (Impact Factor: 1.23). 01/2009; 25(Apr 2000):155–170. DOI:10.2307/2666635

Furey, T.S. (2012). ChIP-seq and beyond: new and improved methodologies to detect and characterize protein-DNA interactions. *Nat Rev Genet*, 13: 840–852.

Gao, D., Kim, J., Kim, H., Phang, T.L., Selby, H., Tan, A.C. and Tong, T. (2010). A survey of statistical software for analysing RNA-seq data. *Hum Genomics*, 5: 56–60.

Garg, R. and Jain, M. (2013). RNA-seq for transcriptome analysis in non-model plants. *Methods Mol Biol*, 1069: 43–58.

Gentry, M. and Hennig, L. (2014). Remodelling chromatin to shape development of plants. *Exp Cell Res*, 321: 40–46.

Gilchrist, D.A. and Adelman, K. (2012). Coupling polymerase pausing and chromatin landscapes for precise regulation of transcription. *Biochim Biophys Acta*, 1819: 700–706.

Godoy, M., Franco-Zorrilla, J.M., Pérez-Pérez, J., Oliveros, J.C., Lorenzo, O. and Solano, R. (2011). Improved protein-binding microarrays for the identification of DNA-binding specificities of transcription factors. *Plant J*, 66: 700–711.

Granek, J. A and Clarke, N.D. (2005). Explicit equilibrium modeling of transcription-factor binding and gene regulation. *Genome Biol*, 6: R87.

Grandi, V., Gregis, V. and Kater, M.M. (2012). Uncovering genetic and molecular interactions among floral meristem identity genes in *Arabidopsis thaliana*. *Plant J*, 69: 881–893.

Grbić, V. and Bleecker, A.B. (2000). Axillary meristem development in *Arabidopsis thaliana*. *Plant J*, 21: 215–223.

Grbic, V. (2005). Comparative analysis of axillary and floral meristem development. *Can J Bot*, 83: 343–349.

Gregis, V. et al. (2013). Identification of pathways directly regulated by SHORT VEGETATIVE PHASE during vegetative and reproductive development in Arabidopsis. *Genome Biol*, 14: R56.

Gulledge, A.A., Vora, H., Patel, K. and Loraine, A.E. (2014). A protocol for visual analysis of alternative splicing in RNA-Seq data using integrated genome browser. *Methods Mol Biol*, 1158: 123–137.

Guo, L., Yu, Y., Law, J.A. and Zhang, X. (2010). SET DOMAIN GROUP2 is the major histone H3 lysine [corrected] 4 trimethyltransferase in Arabidopsis. *Proc Natl Acad Sci USA*, 107: 18557–18562.

Gustafson-Brown, C., Savidge, B. and Yanofsky, M.F. (1994). Regulation of the Arabidopsis floral homeotic gene APETALA1. *Cell*, 76: 131–143.

Hajirasouliha, I., Hormozdiari, F., Alkan, C., Kidd, J.M., Birol, I., Eichler, E.E. and Sahinalp, S.C. (2010). Detection and characterization of novel sequence insertions using paired-end next-generation sequencing. *Bioinformatics*, 26: 1277–1283.

Halaby, D.M., Poupon, A. and Mornon, J. (1999). The immunoglobulin fold family: sequence analysis and 3D structure comparisons. *Protein Eng*, 12: 563–571.

Hamès, C., Ptchelkine, D., Grimm, C., Thevenon, E., Moyroud, E., Gérard, F., Martiel, J.-L., Benlloch, R., Parcy, F. and Müller, C.W. (2008). Structural basis for LEAFY floral switch function and similarity with helix-turn-helix proteins. *EMBO J*, 27: 2628–2637.

Hanano, S. and Goto, K. (2011). Arabidopsis TERMINAL FLOWER1 is involved in the regulation of flowering time and inflorescence development through transcriptional repression. *Plant Cell*, 23: 3172–3184.

Hargreaves, D.C. and Crabtree, G.R. (2011). ATP-dependent chromatin remodeling: genetics, genomics and mechanisms. *Cell Res*, 21: 396–420.

He, J. and Jiao, Y. (2014). Next-generation sequencing applied to flower development: RNA-seq. *Methods Mol Biol*, 1110: 401–411.

Hehl, R. and Bülow, L. (2014). AthaMap web tools for the analysis of transcriptional and posttranscriptional regulation of gene expression in Arabidopsis thaliana. *Methods Mol Biol*, 1158: 139–156.

Heisler, M.G., Ohno, C., Das, P., Sieber, P., Reddy, G.V., Long, J.A. and Meyerowitz, E.M. (2005). Patterns of auxin transport and gene expression during primordium development revealed by live imaging of the Arabidopsis inflorescence meristem. *Curr Biol*, 15: 1899–1911.

Henikoff, S. (2008). Nucleosome destabilization in the epigenetic regulation of gene expression. *Nat Rev Genet*, 9: 15–26.

Holt, A.L., van Haperen, J.M., Groot, E.P. and Laux, T. (2014). Signaling in shoot and flower meristems of *Arabidopsis thaliana*. *Curr Opin Plant Biol*, 17C: 96–102.

Hurtado, L., Farrona, S. and Reyes, J.C. (2006). The putative SWI/SNF complex subunit BRAHMA activates flower homeotic genes in *Arabidopsis thaliana*. *Plant Mol Biol*, 62: 291–304.

Ingouff, M. and Berger, F. (2010). Histone3 variants in plants. *Chromosoma*, 119: 27–33.

Irish, V.F. and Sussex, I.M. (1990). Function of the apetala-1 gene during Arabidopsis floral development. *Plant Cell*, 2: 741–753.

Iyer-Pascuzzi, A.S. and Benfey, P. N. (2010). Fluorescence-activated cell sorting in plant developmental biology. *Methods Mol Biol*, 655: 313–319.

Jiao, Y. and Meyerowitz, E.M. (2010). Cell-type specific analysis of translating RNAs in developing flowers reveals new levels of control. *Mol Syst Biol*, 6: 419.

Jofuku, K.D., den Boer, B.G., Van Montagu, M. and Okamuro, J.K. (1994). Control of Arabidopsis flower and seed development by the homeotic gene APETALA2. *Plant Cell*, 6: 1211–1225.

Jofuku, K.D., Omidyar, P.K., Gee, Z. and Okamuro, J.K. (2005). Control of seed mass and seed yield by the floral homeotic gene APETALA2. *Proc Natl Acad Sci USA*, 102: 3117–3122.

Johnson, L., Mollah, S., Garcia, B.A., Muratore, T.L., Shabanowitz, J., Hunt, D.F. and Jacobsen, S.E. (2004). Mass spectrometry analysis of Arabidopsis histone H3 reveals distinct combinations of post-translational modifications. *Nucleic Acids Res*, 32: 6511–6518.

Kamakaka, R.T. and Biggins, S. (2005). Histone variants: deviants? *Genes Dev*, 19: 295–310.

Kaufmann, K., Muiño, J.M., Jauregui, R., Airoldi, C.A., Smaczniak, C., Krajewski, P. and Angenent, G.C. (2009). Target genes of the MADS transcription factor SEPALLATA3: integration of developmental and hormonal pathways in the Arabidopsis flower. *PLoS Biol*, 7: e1000090.

Kaufmann, K., Wellmer, F., Muiño, J.M., Ferrier, T., Wuest, S.E., Kumar, V., Serrano-Mislata, A., Madueño, F., Krajewski, P., Meyerowitz, E.M., Angenent, G.C. and Riechmann, J.L. (2010a). Orchestration of floral initiation by APETALA1. *Science*, 328: 85–89.

Kaufmann, K., Muiño, J.M., Østerås, M., Farinelli, L., Krajewski, P. and Angenent, G.C. (2010b). Chromatin immunoprecipitation (ChIP) of plant transcription factors followed by sequencing (ChIP-SEQ) or hybridization to whole genome arrays (ChIP-CHIP). *Nat Protoc*, 5: 457–472.

Kempin, S.A., Savidge, B. and Yanofsky, M.F. (1995). Molecular basis of the cauliflower phenotype in Arabidopsis. *Science*, 267: 522–525.

Kerk, N.M., Ceserani, T., Tausta, S.L., Sussex, I.M. and Nelson, T.M. (2003). Laser capture microdissection of cells from plant tissues. *Plant Physiol*, 132: 27–35.

Kim, H., Kim, J., Selby, H., Gao, D., Tong, T., Phang, T.L. and Tan, A.C. (2011). A short survey of computational analysis methods in analysing ChIP-seq data. *Hum Genomics*, 5: 117–123.

Kim, P.M., Lu, L.J., Xia, Y. and Gerstein, M.B. (2006). Relating three-dimensional structures to protein networks provides evolutionary insights. *Science*, 314: 1938–1941.

Klein, J., Saedler, H. and Huijser, P. (1996). A new family of DNA binding proteins includes putative transcriptional regulators of the *Antirrhinum majus* floral meristem identity gene SQUAMOSA. *Mol Gen Genet*, 250: 7–16.

Kolodziej, K.E., Pourfarzad, F., de Boer, E., Krpic, S., Grosveld, F. and Strouboulis, J. (2009). Optimal use of tandem biotin and V5 tags in ChIP assays. *BMC Mol Biol*, 10: 6.

Koornneef, M. and Meinke, D. (2010). The development of Arabidopsis as a model plant. *Plant J*, 61: 909–921.

Krogan, N.T., Hogan, K. and Long, J.A. (2012). APETALA2 negatively regulates multiple floral organ identity genes in Arabidopsis by recruiting the co-repressor TOPLESS and the histone deacetylase HDA19. *Development*, 139: 4180–4190.

Külahoglu, C. and Bräutigam, A. (2014). Quantitative transcriptome analysis using RNA-seq. *Methods Mol Biol*, 1158: 71–91.

Kwak, H., Fuda, N.J., Core, L.J. and Lis, J.T. (2013). Precise maps of RNA polymerase reveal how promoters direct initiation and pausing. *Science*, 339: 950–953.

Kwon, C.S., Chen, C. and Wagner, D. (2005). WUSCHEL is a primary target for transcriptional regulation by SPLAYED in dynamic control of stem cell fate in Arabidopsis. *Genes Dev*, 19: 992–1003.

Lafos, M., Kroll, P., Hohenstatt, M.L., Thorpe, F.L., Clarenz, O. and Schubert, D. (2011). Dynamic regulation of H3K27 trimethylation during Arabidopsis differentiation. *PLoS Genet*, 7: e1002040.

Laity, J.H., Lee, B.M. and Wright, P.E. (2001). Zinc finger proteins: new insights into structural and functional diversity. *Curr Opin Struct Biol*, 11: 39–46.

Landschulz, W.H., Johnson, P.F. and McKnight, S.L. (1988). The leucine zipper: a hypothetical structure common to a new class of DNA binding proteins. *Science*, 240: 1759–1764.

Langmead, B., Trapnell, C., Pop, M. and Salzberg, S.L. (2009). Ultrafast and memory-efficient alignment of short DNA sequences to the human genome. *Genome Biol*, 10: R25.

Larschan, E., Bishop, E.P., Kharchenko, P.V, Core, L.J., Lis, J.T., Park, P.J. and Kuroda, M.I. (2011). X chromosome dosage compensation via enhanced transcriptional elongation in Drosophila. *Nature*, 471: 115–118.

Lauria, M. and Rossi, V. (2011). Epigenetic control of gene regulation in plants. *Biochim Biophys Acta*, 1809: 369–378.

Lee, I., Wolfe, D.S., Nilsson, O. and Weigel, D. (1997). A LEAFY co-regulator encoded by UNUSUAL FLORAL ORGANS. *Curr Biol*, 7: 95–104.

Levin, J.Z., Yassour, M., Adiconis, X., Nusbaum, C., Thompson, D.A., Friedman, N., Gnirke, A. and Regev, A. (2010). Comprehensive comparative analysis of strand-specific RNA sequencing methods. *Nat Methods*, 7: 709–715.

Li, B., Carey, M. and Workman, J.L. (2007). The role of chromatin during transcription. *Cell*, 128: 707–719.

Li, H. and Durbin, R. (2009). Fast and accurate short read alignment with Burrows–Wheeler transform. *Bioinformatics*, 25: 1754–1760.

Li, H. and Durbin, R. (2010). Fast and accurate long-read alignment with Burrows–Wheeler transform. *Bioinformatics*, 26: 589–595.

Li, H., Handsaker, B., Wysoker, A., Fennell, T., Ruan, J., Homer, N., Marth, G., Abecasis, G., Durbin, R. and Subgroup 1000 Genome Project Data Processing (2009). The Sequence Alignment/Map format and SAMtools. *Bioinformatics*, 25: 2078–2079.

Li, G., Liu, S., Wang, J., He, J., Huang, H., Zhang, Y. and Xu, L. (2014). ISWI proteins participate in the genome-wide nucleosome distribution in Arabidopsis. *Plant J.* 78: 706–714.

Liang, X., Nazarenus, T.J. and Stone, J.M. (2008). Identification of a consensus DNA-binding site for the *Arabidopsis thaliana* SBP domain transcription factor, AtSPL14, and binding kinetics by surface plasmon resonance. *Biochemistry*, 47: 3645–3653.

Lieberman-Aiden, E. et al. (2009). Comprehensive mapping of long-range interactions reveals folding principles of the human genome. *Science*, 326: 289–293.

Liljegren, S.J., Gustafson-Brown, C., Pinyopich, A., Ditta, G.S. and Yanofsky, M.F. (1999). Interactions among APETALA1, LEAFY, and TERMINAL FLOWER1 specify meristem fate. *Plant Cell*, 11: 1007–1018.

Liu, B., Yi, J., Sv, A., Lan, X., Ma, Y., Huang, T.H.M., Leone, G. and Jin, V.X. (2013). QChIPat: a quantitative method to identify distinct binding patterns for two biological ChIP-seq samples in different experimental conditions. *BMC Genomics*, 14(Suppl. 8): S3.

Liu, C., Zhou, J., Bracha-Drori, K., Yalovsky, S., Ito, T. and Yu, H. (2007). Specification of Arabidopsis floral meristem identity by repression of flowering time genes. *Development*, 134: 1901–1910.

Liu, C., Xi, W., Shen, L., Tan, C. and Yu, H. (2009). Regulation of floral patterning by flowering time genes. *Dev Cell*, 16: 711–722.

Liu, C., Teo, Z.W.N., Bi, Y., Song, S., Xi, W., Yang, X., Yin, Z. and Yu, H. (2013). A conserved genetic pathway determines inflorescence architecture in Arabidopsis and rice. *Dev Cell*, 24: 612–622.

Liu, X., Noll, D.M., Lieb, J.D. and Clarke, N.D. (2005). DIP-chip: rapid and accurate determination of DNA-binding specificity. *Genome Res*, 15: 421–427.

Lohmann, J.U., Hong, R.L., Hobe, M., Busch, M.A., Parcy, F., Simon, R. and Weigel, D. (2001). A molecular link between stem cell regulation and floral patterning in Arabidopsis. *Cell*, 105: 793–803.

Loman, N.J., Misra, R.V, Dallman, T.J., Constantinidou, C., Gharbia, S.E., Wain, J. and Pallen, M.J. (2012). Performance comparison of benchtop high-throughput sequencing platforms. *Nat Biotechnol*, 30: 434–439.

Long, J. and Barton, M.K. (2000). Initiation of axillary and floral meristems in Arabidopsis. *Dev Biol*, 218: 341–353.

Lord, E.M. and Russell, S.D. (2003). The mechanisms of pollination and fertilization in plants. *Annu Rev Cell Dev Biol*, 18: 81–105.

Louwers, M., Bader, R., Haring, M., Van Driel, R., De Laat, W. and Stam, M. (2009). Tissue- and expression level – specific chromatin looping at maize b1 epialleles. *Plant Cell*, 21: 832–842.

Lu, F., Cui, X., Zhang, S., Jenuwein, T. and Cao, X. (2011). Arabidopsis REF6 is a histone H3 lysine 27 demethylase. *Nat Genet*, 43: 715–719.

Luger, K., Mäder, A.W., Richmond, R.K., Sargent, D.F. and Richmond, T.J. (1997). Crystal structure of the nucleosome core particle at 2.8 Å resolution. *Nature*, 389: 251–260.

Luo, C., Sidote, D.J., Zhang, Y., Kerstetter, R.A., Michael, T.P. and Lam, E. (2012). Integrative analysis of chromatin states in Arabidopsis identified potential regulatory mechanisms for natural antisense transcript production. *Plant J*, 73: 77–90.

Machanick, P. and Bailey, T.L. (2011). MEME-ChIP: motif analysis of large DNA datasets. *Bioinformatics*, 27: 1696–1697.

Magnani, L., Eeckhoute, J. and Lupien, M. (2011). Pioneer factors: directing transcriptional regulators within the chromatin environment. *Trends Genet*, 27: 465–474.

Maizel, A., Busch, M.A., Tanahashi, T., Perkovic, J., Kato, M., Hasebe, M. and Weigel, D. (2005). The floral regulator LEAFY evolves by substitutions in the DNA binding domain. *Science*, 308: 260–263.

Martin-Trillo, M., Lázaro, A., Poethig, R.S., Gómez-Mena, C., Piñeiro, M.A., Martinez-Zapater, J.M. and Jarillo, J.A. (2006). EARLY IN SHORT DAYS 1 (ESD1) encodes ACTIN-RELATED PROTEIN 6 (AtARP6), a putative component of chromatin remodelling complexes that positively regulates FLC accumulation in Arabidopsis. *Development*, 133: 1241–1252.

Mathelier, A. et al. (2014). JASPAR 2014: an extensively expanded and updated open-access database of transcription factor binding profiles. *Nucleic Acids Res*, 42: D142–D147.

Mathelier, A. and Wasserman, W.W. (2013). The next generation of transcription factor binding site prediction. *PLoS Comput Biol*, 9: e1003214.

Martinez-Castilla, L.P. and Alvarez-Buylla, E.R. (2003). Adaptive evolution in the Arabidopsis MADS-box gene family inferred from its complete resolved phylogeny. *Proc Natl Acad Sci USA*, 100: 13407–13412.

Massari, M.E. and Murre, C. (2000). Helix-loop-helix proteins: regulators of transcription in eucaryotic organisms. *Mol Cell Biol*, 20: 429–440.

Melzer, R. and Theissen, G. (2009). Reconstitution of "floral quartets" in vitro involving class B and class E floral homeotic proteins. *Nucl Acids Res*, 37: 2723–2736.

Melzer, R., Verelst, W. and Theissen, G. (2009). The class E floral homeotic protein SEPALLATA3 is sufficient to loop DNA in "floral quartet"-like complexes in vitro. *Nucl Acids Res*, 37: 144–157.

Mendenhall, E.M. and Bernstein, B.E. (2012). DNA–protein interactions in high definition. *Genome Biol*, 13: 139.

Mendes, M.A., Guerra, R.F., Berns, M.C., Manzo, C., Masiero, S., Finzi, L., Kater, M.M. and Colombo, L. (2013). MADS domain transcription factors mediate short-range DNA looping that is essential for target gene expression in Arabidopsis. *Plant Cell*, 25: 2560–2572.

Metzker, M.L. (2010). Sequencing technologies—the next generation. *Nat Rev Genet*, 11: 31–46.

Meyerowitz, E.M., Smyth, D.R. and Bowman, J.L. (1989). Abnormal flowers and pattern formation in floral development. *Development*, 106: 209–217.

Min, I.M., Waterfall, J.J., Core, L.J., Munroe, R.J., Schimenti, J. and Lis, J.T. (2011). Regulating RNA polymerase pausing and transcription elongation in embryonic stem cells. *Genes Dev*, 25: 742–754.

Mito, Y., Henikoff, J.G. and Henikoff, S. (2005). Genome-scale profiling of histone H3.3 replacement patterns. *Nat Genet*, 37: 1090–1097.

Morey, L. and Helin, K. (2010). Polycomb group protein-mediated repression of transcription. *Trends Biochem Sci*, 35: 323–332.

Moyroud, E., Tichtinsky, G. and Parcy, F. (2009). The LEAFY floral regulators in angiosperms: conserved proteins with diverse roles. *J. Plant Biol*, 52: 177–185.

Moyroud, E., Kusters, E., Monniaux, M., Koes, R. and Parcy, F. (2010). LEAFY blossoms. *Trends Plant Sci*, 15: 346–352.

Moyroud, E., Minguet, E.G., Ott, F., Yant, L., Posé, D., Monniaux, M., Blanchet, S., Bastien, O., Thévenon, E., Weigel, D., Schmid, M. and Parcy, F. (2011). Prediction of regulatory interactions from genome sequences using a biophysical model for the Arabidopsis LEAFY transcription factor. *Plant Cell*, 23: 1293–1306.

Muiño, J.M., Kaufmann, K., van Ham, R.C., Angenent, G.C. and Krajewski, P. (2011). ChIP-seq Analysis in R (CSAR): An R package for the statistical detection of protein-bound genomic regions. *Plant Methods*, 7: 11.

Nagy, P.L., Cleary, M.L., Brown, P.O. and Lieb, J.D. (2003). Genomewide demarcation of RNA polymerase II transcription units revealed by physical fractionation of chromatin. *Proc Natl Acad Sci USA*, 100: 6364–6369.

Nanao, M.H., Vinos-Poyo, T., Brunoud, G., Thevenon, E., Mazzoleni, M., Mast, D., Laine, S., Wang, S., Hagen, G., Li, H., Guilfoyle, T.J., Parcy, F., Vernoux, T. and Dumas, R. (2014). Structural basis for oligomerization of auxin transcriptional regulators. *Nat Commun*, 5: 3617.

Narlikar, G.J., Sundaramoorthy, R. and Owen-Hughes, T. (2013). Mechanisms and functions of ATP-dependent chromatin-remodeling enzymes. *Cell*, 154: 490–503.

Nechaev, S., Fargo, D.C., dos Santos, G., Liu, L., Gao, Y. and Adelman, K. (2010). Global analysis of short RNAs reveals widespread promoter-proximal stalling and arrest of Pol II in Drosophila. *Science*, 327: 335–338.

Neph, S. et al. (2012). An expansive human regulatory lexicon encoded in transcription factor footprints. *Nature*, 489: 83–90.

Ng, D.W.-K., Shi, X., Nah, G. and Chen, Z.J. (2014). High-throughput RNA-seq for allelic or locus-specific expression analysis in Arabidopsis-related species, hybrids, and allotetraploids. *Methods Mol Biol*, 1112: 33–48.

Noh, Y.-S. and Amasino, R.M. (2003). PIE1, an ISWI family gene, is required for FLC activation and floral repression in Arabidopsis. *Plant Cell*, 15: 1671–1682.

Ohta, M., Matsui, K., Hiratsu, K., Shinshi, H. and Ohme-Takagi, M. (2001). Repression domains of class II ERF transcriptional repressors share an essential motif for active repression. *Plant Cell*, 13: 1959–1968.

Ohto, M.A., Fischer, R.L., Goldberg, R.B., Nakamura, K. and Harada, J.J. (2005). Control of seed mass by APETALA2. *Proc Natl Acad Sci USA*, 102: 3123–3128.

Okada, K., Ueda, J., Komaki, M.K., Bell, C.J. and Shimura, Y. (1991). Requirement of the auxin polar transport system in early stages of Arabidopsis floral bud formation. *Plant Cell*, 3: 677–684.

Okada, T., Bhalla, P.L. and Singh, M.B. (2005). Transcriptional activity of male gamete-specific histone gcH3 promoter in sperm cells of Lilium longiflorum. *Plant Cell Physiol*, 46: 797–802.

Okada, T., Singh, M.B. and Bhalla, P.L. (2006). Histone H3 variants in male gametic cells of lily and H3 methylation in mature pollen. *Plant Mol Biol*, 62: 503–512.

O'Maoiléidigh, D.S. and Wellmer, F. (2014). A floral induction system for the study of early Arabidopsis flower development. *Methods Mol Biol*, 1110: 307–314.

Omidbakhshfard, M.A., Winck, F.V., Arvidsson, S., Riaño-Pachón, D.M. and Mueller-Roeber, B. (2014). A step-by-step protocol for formaldehyde-assisted isolation of regulatory elements from *Arabidopsis thaliana*. *J Integr Plant Biol*, 56: 527–538.

Otsuga, D., DeGuzman, B., Prigge, M.J., Drews, G.N. and Clark, S.E. (2001). REVOLUTA regulates meristem initiation at lateral positions. *Plant J*, 25: 223–236.

Pajoro, A., Biewers, S., Dougali, E., Valentim, F.L., Mendes, M.A., Porri, A., Coupland, G., Van de Peer, Y., van Dijk, A.D.J., Colombo, L., Davies, B. and Angenent, G.C. (2014a). The (r)evolution of gene regulatory networks controlling Arabidopsis plant reproduction; a two decades history. *J Exp Bot*, 65: 4731–4745.

Pajoro, A. et al. (2014b). Dynamics of chromatin accessibility and gene regulation by MADS-domain transcription factors in flower development. *Genome Biol*, 15: R41.

Palovaara, J., Saiga, S. and Weijers, D. (2013). Transcriptomics approaches in the early Arabidopsis embryo. *Trends Plant Sci*, 18: 514–521.

Parcy, F., Nilsson, O., Busch, M.A., Lee, I. and Weigel, D. (1998). A genetic framework for floral patterning. *Nature*, 395: 561–566.

Pareek, C.S., Smoczynski, R. and Tretyn, A. (2011). Sequencing technologies and genome sequencing. *J Appl Genet*, 52: 413–435.

Pastore, J.J., Limpuangthip, A., Yamaguchi, N., Wu, M.-F., Sang, Y., Han, S.-K., Malaspina, L., Chavdaroff, N., Yamaguchi, A. and Wagner, D. (2011). LATE MERISTEM IDENTITY2 acts together with LEAFY to activate APETALA1. *Development*, 138: 3189–3198.

Pelaz, S., Ditta, G.S., Baumann, E., Wisman, E. and Yanofsky, M.F. (2000). B and C floral organ identity functions require SEPALLATA MADS-box genes. *Nature*, 405: 200–203.

Pidkowich, M., Klenz, J. and Haughn, G. (1999). The making of a flower: control of floral meristem identity in Arabidopsis. *Trends Plant Sci*, 4: 64–70.

Pinyopich, A., Ditta, G.S., Savidge, B., Liljegren, S.J., Baumann, E., Wisman, E. and Yanofsky, M.F. (2003). Assessing the redundancy of MADS-box genes during carpel and ovule development. *Nature*, 424: 85–88.

Pollier, J., Rombauts, S. and Goossens, A. (2013). Analysis of RNA-Seq data with TopHat and Cufflinks for genome-wide expression analysis of jasmonate-treated plants and plant cultures. *Methods Mol Biol*, 1011: 305–315.

Preston, J.C. and Hileman, L.C. (2013). Functional evolution in the plant SQUAMOSA-PROMOTER BINDING PROTEIN-LIKE (SPL) gene family. *Front Plant Sci*, 4: 80.

Quinlan, A.R. and Hall, I.M. (2010). BEDTools: a flexible suite of utilities for comparing genomic features. *Bioinformatics*, 26: 841–842.

Ratcliffe, O.J., Bradley, D.J. and Coen, E.S. (1999). Separation of shoot and floral identity in Arabidopsis. *Development*, 126: 1109–1120.

Reinhardt, D., Mandel, T. and Kuhlemeier, C. (2000). Auxin regulates the initiation and radial position of plant lateral organs. *Plant Cell*, 12: 507–518.

Rhee, H.S. and Pugh, B.F. (2011). Comprehensive genome-wide protein-DNA interactions detected at single-nucleotide resolution. *Cell*, 147: 1408–1419.

Rhee, H.S. and Pugh, B.F. (2012). ChIP-exo method for identifying genomic location of DNA-binding proteins with near-single-nucleotide accuracy. *Curr Protoc Mol Biol*, Chapter 21, Unit 21.24.

Richardt, S., Lang, D., Reski, R., Frank, W, and Rensing, S.A. (2007). PlanTAPDB, a phylogeny-based resource of plant transcription-associated proteins. *Plant Physiol*, 143: 1452–1466.

Robinson, M.D., McCarthy, D.J. and Smyth, G.K. (2010). edgeR: a Bioconductor package for differential expression analysis of digital gene expression data. *Bioinformatics*, 26: 139–140.

Rohs, R., Jin, X., West, S.M., Joshi, R., Honig, B. and Mann, R.S. (2010). Origins of specificity in protein-DNA recognition. *Ann Rev Biochem*, 79: 233–269.

Rokas, A. (2008). The origins of multicellularity and the early history of the genetic toolkit for animal development. *Ann Rev Genet*, 42: 235–251.

Saddic, L.A., Huvermann, B., Bezhani, S., Su, Y., Winter, C.M., Kwon, C.S., Collum, R.P. and Wagner, D. (2006). The LEAFY target LMI1 is a meristem identity regulator and acts together with LEAFY to regulate expression of CAULIFLOWER. *Development*, 133: 1673–1682.

Sang, Y., Silva-Ortega, C.O., Wu, S., Yamaguchi, N., Wu, M.-F., Pfluger, J., Gillmor, C.S., Gallagher, K.L. and Wagner, D. (2012). Mutations in two non-canonical Arabidopsis SWI2/SNF2 chromatin remodeling ATPases cause embryogenesis and stem cell maintenance defects. *Plant J*, 72: 1000–1014.

Sargent, R.D. (2004). Floral symmetry affects speciation rates in angiosperms. *Proc R Soc B: Biol Sci*, 271: 603–608.

Sayou, C., Monniaux, M., Nanao, M.H., Moyroud, E., Brockington, S.F., Thevenon, E., Chahtane, H., Warthmann, N., Melkonian, M., Zhang, Y., Wong, G.K., Weigel, D., Parcy, F. and Dumas, R. (2014). A promiscuous intermediate underlies the evolution of LEAFY DNA binding specificity. *Science*, 343: 645–648.

Schmid, M.W., Schmidt, A., Klostermeier, U.C., Barann, M., Rosenstiel, P. and Grossniklaus, U. (2012). A powerful method for transcriptional profiling of specific cell types in eukaryotes: laser-assisted microdissection and RNA sequencing. *PLoS One*, 7: e29685.

Schubert, D., Clarenz, O. and Goodrich, J. (2005). Epigenetic control of plant development by Polycomb-group proteins. *Curr Opin Plant Biol*, 8: 553–561.

Schultz, E.A. and Haughn, G.W. (1991). LEAFY, a homeotic gene that regulates inflorescence development in Arabidopsis. *Plant Cell*, 3: 771–781.

Shannon, S. and Meeks-Wagner, D.R. (1991). A mutation in the Arabidopsis TFL1 gene affects inflorescence meristem development. *Plant Cell*, 3: 877–892.

Shannon, S. and Meeks-Wagner, D.R. (1993). Genetic interactions that regulate inflorescence development in Arabidopsis. *Plant Cell*, 5: 639–655.

Sims, D., Sudbery, I., Ilott, N.E., Heger, A. and Ponting, C.P. (2014). Sequencing depth and coverage: key considerations in genomic analyses. *Nat Rev Genet*, 15: 121–132.

Smaczniak, C. et al. (2012). Characterization of MADS-domain transcription factor complexes in Arabidopsis flower development. *Proc Natl Acad Sci USA*, 109: 1560–1565.

Smyth, D.R., Bowman, J.L. and Meyerowitz, E.M. (1990). Early flower development in Arabidopsis. *Plant Cell*, 2: 755–767.

Soboleva, T.A., Nekrasov, M., Ryan, D.P. and Tremethick, D.J. (2014). Histone variants at the transcription start-site. *Trends Genet*, 30: 199–209.

Soltis, D.E., Bell, C.D., Kim, S. and Soltis, P.S. (2008). Origin and early evolution of angiosperms. *Ann NY Acad Sci*, 1133: 3–25.

Sparks, E., Wachsman, G. and Benfey, P.N. (2013). Spatiotemporal signalling in plant development. *Nat Rev Genet*, 14: 631–44.

Spicuglia, S., Maqbool, M.A., Puthier, D. and Andrau, J.-C. (2013). An update on recent methods applied for deciphering the diversity of the noncoding RNA genome structure and function. *Methods* 63: 3–17.

Sridhar, V.V., Surendrarao, A. and Liu, Z. (2006). APETALA1 and SEPALLATA3 interact with SEUSS to mediate transcription repression during flower development. *Development*, 133: 3159–3166.

Stein, L.D. (2013). Using GBrowse 2.0 to visualize and share next-generation sequence data. *Brief Bioinform*, 14: 162–171.

Strahl, B.D. and Allis, C.D. (2000). The language of covalent histone modifications. *Nature*, 403: 41–45.

Stroud, H., Otero, S., Desvoyes, B., Ramírez-Parra, E., Jacobsen, S.E. and Gutierrez, C. (2012). Genome-wide analysis of histone H3.1 and H3.3 variants in *Arabidopsis thaliana*. *Proc Natl Acad Sci USA*, 109: 5370–5375.

Sun, B., Looi, L.-S., Guo, S., He, Z., Gan, E.-S., Huang, J., Xu, Y., Wee, W.-Y. and Ito, T. (2014). Timing mechanism dependent on cell division is invoked by Polycomb eviction in plant stem cells. *Science*, 343: 1248559.

Sun, B., Xu, Y., Ng, K.-H. and Ito, T. (2009). A timing mechanism for stem cell maintenance and differentiation in the Arabidopsis floral meristem. *Genes Dev*, 23: 1791–1804.

Sundström, J.F., Nakayama, N., Glimelius, K. and Irish, V.F. (2006). Direct regulation of the floral homeotic APETALA1 gene by APETALA3 and PISTILLATA in Arabidopsis. *Plant J*, 46: 593–600.

Tao, Z., Shen, L., Liu, C., Liu, L., Yan, Y. and Yu, H. (2012). Genome-wide identification of SOC1 and SVP targets during the floral transition in Arabidopsis. *Plant J*, 70: 549–561.

Theissen, G. and Saedler, H. (2001). Plant biology: floral quartets. *Nature*, 409: 469–471.

Thorvaldsdóttir, H., Robinson, J.T. and Mesirov, J.P. (2013). Integrative Genomics Viewer (IGV): high-performance genomics data visualization and exploration. *Brief Bioinform* 14: 178–192.

Trapnell, C., Roberts, A., Goff, L., Pertea, G., Kim, D., Kelley, D.R., Pimentel, H., Salzberg, S.L., Rinn, J.L. and Pachter, L. (2012). Differential gene and transcript expression analysis of RNA-seq experiments with TopHat and Cufflinks. *Nat Protoc*, 7: 562–578.

Turck, F., Roudier, F., Farrona, S., Martin-Magniette, M.-L., Guillaume, E., Buisine, N., Gagnot, S., Martienssen, R.A., Coupland, G. and Colot, V. (2007). Arabidopsis TFL2/LHP1 specifically associates with genes marked by trimethylation of histone H3 lysine 27. *PLoS Genet*, 3: e86.

Vergara Silva, F. et al. (2003). Inside-out flowers characteristic of evolved at least before its divergence from a closely related taxon, *Triuris brevistylis*. *Int J Plant Sci*, 164: 345–357.

Villiers, F., Jourdain, A., Bastien, O., Leonhardt, N., Fujioka, S., Tichtinčky, G., Parcy, F., Bourguignon, J. and Hugouvieux, V. (2012). Evidence for functional interaction between brassinosteroids and cadmium response in *Arabidopsis thaliana*. *J Exp Bot*, 63: 1185–1200.

Wagner, D. and Meyerowitz, E.M. (2002). SPLAYED, a novel SWI/SNF ATPase homolog, controls reproductive development in Arabidopsis. *Curr Biol*, 12: 85–94.

Wang, J.-W., Czech, B. and Weigel, D. (2009). miR156-regulated SPL transcription factors define an endogenous flowering pathway in Arabidopsis thaliana. *Cell*, 138: 738–749.

Wasserman, W.W. and Sandelin, A. (2004). Applied bioinformatics for the identification of regulatory elements. *Nat Rev Genet*, 5: 276–287.

Weber, C.M. and Henikoff, S. (2014). Histone variants: dynamic punctuation in transcription. *Genes Dev*, 28: 672–682.

Weber, C.M., Ramachandran, S. and Henikoff, S. (2014). Nucleosomes are context-specific, H2A.Z-modulated barriers to RNA polymerase. *Mol Cell*, 53: 819–830.

Weigel, D. and Nilsson, O. (1995). A developmental switch sufficient for flower initiation in diverse plants. *Nature*, 377: 495–500.

Weigel, D., Alvarez, J., Smyth, D.R., Yanofsky, M.F. and Meyerowitz, E.M. (1992). LEAFY controls floral meristem identity in Arabidopsis. *Cell*, 69: 843–859.

Weinhofer, I.1, Köhler, C. (2014). Endosperm-specific chromatin profiling by fluorescence-activated nuclei sorting and ChIP-on-chip. *Methods Mol Biol*, 1112: 105–15.

Wellmer, F., Alves-Ferreira, M., Dubois, A., Riechmann, J.L. and Meyerowitz, E.M. (2006). Genome-wide analysis of gene expression during early Arabidopsis flower development. *PLoS Genet*, 2: e117.

Wigge, P.A., Kim, M.C., Jaeger, K.E., Busch, W., Schmid, M., Lohmann, J.U. and Weigel, D. (2005). Integration of spatial and temporal information during floral induction in Arabidopsis. *Science*, 309: 1056–1059.

Winn, A.A. and Moriuchi, K.S. (2009). The maintenance of mixed mating by cleistogamy in the perennial violet *Viola septemloba* (Violaceae). *Am J Bot*, 96: 2074–2079.

Winter, C.M., Austin, R.S., Blanvillain-Baufumé, S., Reback, M.A., Monniaux, M., Wu, M.-F., Sang, Y., Yamaguchi, A., Yamaguchi, N., Parker, J.E., Parcy, F., Jensen, S.T., Li, H. and Wagner, D. (2011). LEAFY Target Genes Reveal Floral Regulatory Logic, cis Motifs, and a Link to Biotic Stimulus Response. *Dev Cell*, 20: 430–443.

Wollmann, H., Holec, S., Alden, K., Clarke, N.D., Jacques, P.-É. and Berger, F. (2012). Dynamic deposition of histone variant H3.3 accompanies developmental remodeling of the Arabidopsis transcriptome. *PLoS Genet*, 8: e1002658.

Won, K.-J., Ren, B. and Wang, W. (2010). Genome-wide prediction of transcription factor binding sites using an integrated model. *Genome Biol*, 11: R7.

Wu, M.-F., Sang, Y., Bezhani, S., Yamaguchi, N., Han, S.-K., Li, Z., Su, Y., Slewinski, T.L. and Wagner, D. (2012). SWI2/SNF2 chromatin remodeling ATPases overcome polycomb repression and control floral organ identity with the LEAFY and SEPALLATA3 transcription factors. *Proc Natl Acad Sci USA*, 109: 3576–3581.

Wuest, S.E., O'Maoileidigh, D.S., Rae, L., Kwasniewska, K., Raganelli, A., Hanczaryk, K., Lohan, A.J., Loftus, B., Graciet, E. and Wellmer, F. (2012). Molecular basis for the specification of floral organs by APETALA3 and PISTILLATA. *Proc Nat Acad Sci USA*, 109: 13452–13457.

Wurschum, T., Gross-Hardt, R. and Laux, T. (2006). APETALA2 regulates the stem cell niche in the Arabidopsis shoot meristem. *Plant Cell*, 18: 295–307.

Xing, D., Wang, Y., Xu, R., Ye, X., Yang, D. and Li, Q.Q. (2013). The regulatory role of Pcf11-similar-4 (PCFS4) in Arabidopsis development by genome-wide physical interactions with target loci. *BMC Genomics*, 14: 598.

Xu, M., Hu, T., McKim, S.M., Murmu, J., Haughn, G.W. and Hepworth, S.R. (2010). Arabidopsis BLADE-ON-PETIOLE1 and 2 promote floral meristem fate and determinacy in a previously undefined pathway targeting APETALA1 and AGAMOUS-LIKE24. *Plant J*, 63: 974–989.

Xu, H. and Sung, W.-K. (2012). Identifying differential histone modification sites from ChIP-seq data. *Methods Mol Biol*, 802: 293–303.

Yamaguchi, A., Wu, M.-F., Yang, L., Wu, G., Poethig, R.S. and Wagner, D. (2009). The MicroRNA-Regulated SBP-Box Transcription Factor SPL3 Is a Direct Upstream Activator of LEAFY, FRUITFULL, and APETALA1. *Dev Cell*, 17: 268–278.

Yamaguchi, N., Wu, M.-F., Winter, C.M., Berns, M.C., Nole-Wilson, S., Yamaguchi, A., Coupland, G., Krizek, B.A. and Wagner, D. (2013). A molecular frame.

Yamaguchi, N., Winter, C.M., Wu, M.F., Kanno, Y., Yamaguchi, A., Seo, M. and Wagner, D. (2014). Gibberellin Acts Positively Then Negatively to Control Onset of Flower Formation in Arabidopsis. *Science*, 344: 638–641.

Yamasaki, K., Kigawa, T., Inoue, M., Tateno, M., Yamasaki, T., Yabuki, T., Aoki, M., Seki, E., Matsuda, T., Nunokawa, E., Ishizuka, Y., Terada, T., Shirouzu, M., Osanai, T., Tanaka, A., Seki, M., Shinozaki, K. and Yokoyama, S. (2004). A novel zinc-binding motif revealed by solution structures of DNA-binding domains of Arabidopsis SBP-family transcription factors. *J Mol Biol*, 337: 49–63.

Yang, Z., Wang, X., Gu, S., Hu, Z., Xu, H. and Xu, C. (2008). Comparative study of SBP-box gene family in Arabidopsis and rice. *Gene*, 407: 1–11.

Yang, F., Wang, Q., Schmitz, G., Müller, D. and Theres, K. (2012). The bHLH protein ROX acts in concert with RAX1 and LAS to modulate axillary meristem formation in Arabidopsis. *Plant J*, 71: 61–70.

Yamaguchi, N., Winter, C.M., Wu, M.F., Kanno, Y., Yamaguchi, A., Seo, M. and Wagner, D. (2014). Gibberellin Acts Positively Then Negatively to Control Onset of Flower Formation in Arabidopsis. *Science*, 344: 638–641.

Yant, L., Mathieu, J., Dinh, T.T., Ott, F., Lanz, C., Wollmann, H., Chen, X. and Schmid, M. (2010). Orchestration of the floral transition and floral development in Arabidopsis by the bifunctional transcription factor APETALA2. *Plant Cell*, 22: 2156–2170.

Zang, C., Schones, D.E., Zeng, C., Cui, K., Zhao, K. and Peng, W. (2009). A clustering approach for identification of enriched domains from histone modification ChIP-Seq data. *Bioinformatics*, 25: 1952–1958.

Zaret, K.S. and Carroll, J.S. (2011). Pioneer transcription factors: establishing competence for gene expression. *Genes Dev*, 25: 2227–2241.

Zhang, X., Clarenz, O., Cokus, S., Bernatavichute, Y.V., Pellegrini, M., Goodrich, J. and Jacobsen, S.E. (2007). Whole-genome analysis of histone H3 lysine 27 trimethylation in Arabidopsis. *PLoS Biol*, 5: e129.

Zhang, Y. et al. (2008). Model-based analysis of ChIP-Seq (MACS). *Genome Biol*, 9: R137.

Zhang, J., Chiodini, R., Badr, A. and Zhang, G. (2011). The impact of next-generation sequencing on genomics. *J Genet Genomics*, 38: 95–109.

Zhang, W., Zhang, T., Wu, Y. and Jiang, J. (2012). Genome-wide identification of regulatory DNA elements and protein-binding footprints using signatures of open chromatin in Arabidopsis. *Plant Cell*, 24: 2719–2731.

Zhang, S.-J. et al. (2014). Evolutionary interrogation of human biology in well-annotated genomic framework of rhesus macaque. *Mol Biol Evol*, 31: 1309–1324.

Zhao, Y., Granas, D. and Stormo, G.D. (2009). Inferring binding energies from selected binding sites. *PLoS Comput. Biol*, 5: e1000590.

Zilberman, D., Coleman-Derr, D., Ballinger, T. and Henikoff, S. (2008). Histone H2A.Z and DNA methylation are mutually antagonistic chromatin marks. *Nature*, 456: 125–129.

Part 2

Concepts in physics and emerging methods

4
Thermodynamics (a reminder)

Giuseppe (Joe) ZACCAI

CNRS, Institut de Biologie Structurale, and Institut Laue Langevin, Grenoble, France

Abstract

Biological processes take place in an aqueous environment, and the discovery and characterization of biological macromolecules is tightly interwoven with the development of physical chemistry and solution thermodynamics. Following a recap of relations in calorimetry and fundamental thermodynamics, this chapter examines their extension to aqueous solutions and macromolecular hydration and solvation. The thermodynamics approach to analytical ultracentrifugation and small-angle X-ray and neutron scattering is presented. Finally, work is discussed in which NMR and inelastic neutron scattering experiments are used to explore a molecular dynamics interpretation of thermodynamics functions.

Keywords

Aqueous solution, macromolecule, specific heat, pressure, enthalpy, entropy, X-ray small-angle scattering, neutron small-angle scattering, SAXS, SANS, solvation, hydration, inelastic neutron scattering, NMR, intrinsically disordered protein, IDP, density of states, molecular dynamics

From Molecules to Living Organisms: An Interplay Between Biology and Physics. First Edition. Eva Pebay-Peyroula et al. © Oxford University Press 2016.
Published in 2016 by Oxford University Press.

Chapter Contents

4 Thermodynamics (a reminder) 119
Giuseppe (Joe) ZACCAI

4.1 Brief historical review	121
4.2 Why structural biologists need to be reminded about thermodynamics	121
4.3 Calorimetry	122
4.3.1 Differential scanning calorimetry	122
4.3.2 Isothermal titration calorimetry	122
4.3.3 Definitions	123
4.4 Current methods for the study of biological macromolecules in solution	123
4.5 Thermodynamics of aqueous solutions	124
4.5.1 Solute concentration	124
4.5.2 Partial volume	124
4.5.3 Colligative properties	124
4.5.4 Chemical potential and activity	125
4.5.5 Osmotic pressure	125
4.5.6 Virial coefficients	126
4.6 Macromolecular solutions	127
4.6.1 Ionic strength	127
4.6.2 Polyelectrolytes	127
4.6.3 Water, salt and the hydrophobic effect	128
4.7 A thermodynamics approach to scattering methods: macromolecule–solvent interactions (hydration and solvation)	129
4.7.1 Sedimentation equilibrium	130
4.7.2 Light scattering	130
4.7.3 X-ray scattering	130
4.7.4 Neutron scattering	130
4.8 Relating the thermodynamics and particle approaches	130
4.9 NMR and neutrons give molecular meaning to thermodynamics parameters	131
References	132

4.1 Brief historical review

The discovery of biological macromolecules is tightly interwoven with the history of physical chemistry, which formally emerged as a discipline in the nineteenth century [1]. It was evident that biological processes take place in an aqueous environment and the first papers published in this field were concerned with reactions in solution. Osmotic pressure was described from experiments with animal bladder membranes separating alcohol and water, and its importance was recognized for living systems. The freezing-point depression law made it possible to determine the molecular weight of dissolved substances. Brownian motion was discovered. Fick described diffusion in concentration gradients from a law based on heat flow. A study of the diuretic and laxative effects of salts led to their classification in the Hofmeister series according to how they modified the solubility of protein. The concepts of activity, ionic strength and Donnan membrane potential were introduced and the Debye–Hückel theory for electrolyte solutions was proposed at the beginning of the twentieth century. Emil Fischer had discovered proteins as agents of biochemical activity, and solution experiments strongly suggested that they were constituted of discrete macromolecules. Theodor Svedberg proved the molecular nature of proteins in 1925 when he made the first "direct observation" of hemoglobin molecules with an analytical centrifuge.

In the second half of the twentieth century, modern biophysical methods developed for the characterization of polymers, and especially polyelectrolytes, contributed significantly to our current understanding of biological macromolecules in solution. Proteins have evolved in the presence of water and cannot be considered separately from their hydration and solvation shells [2]. We should also recall that crystals of biological macromolecules contain an appreciable amount of solvent and are really organized macromolecular solutions.

4.2 Why structural biologists need to be reminded about thermodynamics

Someone once wrote that the three secrets behind a good structural biology experiment were (1) biochemistry, (2) biochemistry and (3) biochemistry. In other words, if the sample is not what one assumes it to be the most beautiful biophysics experiments will be meaningless. Electrons, X-rays, neutrons, light or gravitational fields in ultracentrifugation are incapable of judging the biological quality of a sample. They just "look" at what happens to be there. The biochemical approach, however, is mainly qualitative, while biophysics is quantitative and based on the physical chemical principles of thermodynamics.

Consider the example of a small-angle scattering (SAS) experiment on a protein solution. Biochemical characterization is essential for ensuring sample purity and appropriate conditions for the buffer solvent in which the protein is likely to be in an active state. A normal biochemical determination will also yield a relative protein concentration in the solution that may differ from the absolute value by as much as 30%. A SAS experiment will be planned with a simple picture in mind of identical particles in an otherwise empty volume, not interacting with each other, and occupying all orientations. It is a nice mental picture, like Enrico Fermi's observation that he used to think of electrons as little red balls, but it is severely limited with respect to scientific reality and contains no information about interactions that could lead to aggregation or repulsion, solvation interactions with small solvent solutes or hydration interactions with solvent water. The scientific reality of solutions is what we have discovered from more than 150 years of physical chemistry and thermodynamics.

Thermodynamics is also essential for structural biology in order to characterize macromolecular sample states, such as, for example, transitions in lipid membranes, protein unfolding, protein–nucleic acid interactions, binding states, etc.

4.3 Calorimetry

It is not possible to write a chapter on thermodynamics without mentioning calorimetry, which was developed in the twentieth century for the careful study of the physical chemistry of solutions. Classical thermodynamics is a phenomenological science, concerned with precise observations on a system such as a macromolecular solution.

4.3.1 Differential scanning calorimetry

Differential scanning calorimetry (DSC) measures heat exchange in a system during temperature-induced changes. Temperature (T) and enthalpy (H) are conjugate extensive and intensive variables, respectively. When a change in the state of a system (e.g., protein unfolding, ligand binding, transitions in lipid bilayers) results from a change in temperature, there is a corresponding change in enthalpy. Heat is absorbed ($\Delta H > 0$) if the change results from an increase in temperature or heat is released ($\Delta H < 0$) if the change results from a decrease in temperature:

$$\Delta H = \int \Delta C_p(T) \mathrm{d}T$$
$$\Delta S = \int \frac{\Delta C_p(T)}{T} \mathrm{d}T \tag{4.1}$$

where ΔS is entropy change (see Section 4.3.3) and ΔC_p is the change in partial specific heat capacity at constant pressure as a function of temperature due to a given macromolecular process (e.g., protein unfolding).

4.3.2 Isothermal titration calorimetry

Isothermal titration calorimetry (ITC) is unique in providing not only the magnitude of the enthalpy change upon binding but also, in favorable experimental conditions, values for the binding affinity and entropy changes. Because these parameters fully define the *energetics* of the binding process, ITC is playing an increasingly important role in the detailed study of protein–ligand interactions and associated molecular design approaches, in particular with respect to drug design.

4.3.3 Definitions

A complete set of thermodynamics equations describing a system can be calculated from the functional dependence of enthalpy on temperature, which can be obtained, in principle, from careful calorimetric measurements. A non-exhaustive list of useful thermodynamics definitions is given in the following.

Enthalpy: the change in enthalpy, ΔH, during a chemical reaction is the heat absorbed or released in the breaking and formation of bonds. ΔH represents the experimentally measured quantity in calorimetry.

Entropy, S, is a function of state which increases with time until it reaches a maximum at equilibrium. The second law of thermodynamics postulates that an isolated system evolves in order to maximize its entropy. The equilibrium state is a disorganized state, in which the system components sample all possible configurations available to them. Boltzmann, in his development of statistical thermodynamics, expressed the concept quantitatively and related entropy to the number of configurations, p, available to the system:

$$S = k_B \ln p. \tag{4.2}$$

Gibbs free energy, G, represents the part of the energy in a system that is not used for the population of the different entropy configurations. Because G is not involved in maintaining thermal agitation, it can be transformed into useful work, hence its name *"free energy"*. At constant temperature, the free energy released by a reaction is given by:

$$\Delta G = \Delta H - T\Delta S. \tag{4.3}$$

A system that exchanges energy with its surroundings evolves in order to maximize its free energy. The spontaneous sense of a reversible reaction is in the direction for which $\Delta G < 0$. When two states of a system are in equilibrium, the difference in their free energy is zero, $\Delta G = 0$.

4.4 Current methods for the study of biological macromolecules in solution

X-ray crystallography and NMR dominate structural biology for the determination of three-dimensional structures to atomic resolution. But crucial questions would still remain even if all the protein and nucleic acid structures in genomes were to be solved. How do macromolecules interact with each other and what are their structures and dynamics, and how do they change during biological activity in the crowded environment of a living cell? These questions are addressed by the complementary application of a variety of methods. In order to have a decent signal to noise ratio, NMR, small-angle X-ray and neutron scattering (SAXS, SANS) and analytical ultracentrifugation (AUC) samples comprise solutions containing large molecular ensembles for which the laws of thermodynamics are readily applicable. For example, 100 µl of a 1 mg/ml solution of hemoglobin contains 10^{18} protein molecules.

4.5 Thermodynamics of aqueous solutions

An aqueous solution is a homogenous mixture at the molecular level of a majority component, the solvent (water), and solutes (small molecules, salt ions, etc.).

4.5.1 Solute concentration

Solute concentration can be defined as the mass of solute per unit mass of solvent or unit volume of solution, or as moles of solute per liter of solution (*molarity*) or *molality* (moles of solute per kilogram of solvent). Mass per mass and molality units have the advantage that they are based on measuring invariant masses, whereas the volume of a solution depends on temperature and pressure.

Measuring the concentration of a protein or nucleic acid solution is not trivial. "Dry" material obtained from lyophilization or precipitation always contains an unknown quantity of hydration water and salt ions necessary for the maintenance of an active structure. Very low concentrations may be sufficient for many experiments, and it is not possible to weigh precisely micrograms or less of material. Relative protein concentrations can be measured by colorimetric assays (e.g., the Bradford assay) or by spectrophotometry. Absorbance at 280 nm is sensitive to the presence of tryptophan, tyrosine and cysteine amino acid residues. The exact value of the extinction coefficient, however, varies with the number of sensitive residues in the protein. Nucleic acids show a strong absorbance at 260 nm (one absorbance unit corresponds to about 40 µg/ml). Unless they have been calibrated (e.g., by quantitative amino acid analysis), colorimetric and spectrophotometric measurements yield relative values. However, absolute concentration values are essential, for example, in the absolute scale interpretation of SANS and SAXS.

4.5.2 Partial volume

Partial volume is the change in volume of a solution when a solute is added under given conditions (the *partial specific volume* is the change in volume per added mass). It is not simply the volume occupied by the solute, because its presence may lead to a change in volume of the solvent. The case of charged solutes is an interesting illustration of this. Water molecules behave like electric dipoles. In the presence of ions, water molecules will orient around the charges, effectively occupying a smaller volume than in the bulk liquid; this effect is called *electrostriction*. The partial volume of a salt ion, therefore, can be positive, negative or even close to zero. The decrease in volume due to electrostriction for Na^+ or Mg^{2+}, for example, is greater than the volume they effectively occupy in the solution, and their partial volumes are negative. The partial volume of K^+ is slightly positive and close to zero; the ion fits into the liquid water structure with very little disturbance. In general, values of solute partial volumes depend on solvent composition and concentration.

4.5.3 Colligative properties

Colligative properties depend only on the number of solute molecules per unit volume and not on the mass or nature of those molecules. The colligative properties of solutions provide essential evidence for the very existence of atoms and molecules. Concentration in molar or molal units, therefore, is more relevant in experiments that involve colligative properties.

In an *ideal solution* at constant temperature, the partial pressure of a component is proportional to its mole fraction. This is *Raoult's law*. The rise in boiling point and decrease in freezing point that result when a solute is dissolved in a solvent are colligative properties related to Raoult's law and applicable to dilute solutions of non-volatile molecules. The temperature difference is proportional in each case to the number of solute molecules present. The laws can be adapted to non-ideal solutions by applying the concepts of *chemical potential* and *activity*.

4.5.4 Chemical potential and activity

The chemical potential, μ, of a solute is the gain in free energy upon addition of 1 mole of solute to the solution:

$$\mu = \Delta G / \Delta C. \tag{4.4}$$

The difference in free energy between two solutions with concentrations C_A and C_B, respectively (moles per volume of solution), is given by:

$$G_A - G_B = \Delta G = -RT \ln \frac{C_A}{C_B}. \tag{4.5}$$

Combining equations (4.4) and (4.5) we get:

$$\Delta G = \mu(C_A - C_B) = -RT \ln \frac{C_A}{C_B}. \tag{4.6}$$

(Equation 4.6) is valid for an *ideal solution*. The physicochemical concept of *activity* was introduced by Gilbert Lewis in 1908 to account for deviations from ideal behavior (equation 4.5):

$$\Delta G = -RT \ln \frac{a_A}{a_B} \tag{4.7}$$

where $a_J = \gamma_J C_J$ is the activity of solute J and γ_J is the activity coefficient, equal to 1 for an ideal solution. Activity coefficients are obtained experimentally from deviations from ideal solution laws: in order to apply equations derived in the ideal case to other solutions concentration C is replaced by activity a.

4.5.5 Osmotic pressure

Osmotic pressure is a colligative property that involves small-molecule solutes as well as macromolecular components in a solution. When two aqueous solutions with different concentrations are separated by a semi-permeable membrane (a membrane that is permeable to water but not to solutes), water will flow across the membrane from the more dilute solution increasing the pressure on the side with the more concentrated solution according to equation (4.8).

The osmotic pressure associated with a volume V of solution containing N moles of solute is given by:

$$\Pi V = NRT \tag{4.8}$$

where Π is osmotic pressure, R and T are the gas constant and absolute temperature, respectively, and N is the number of moles of solute in the volume V. The molar concentration $C(= N/V)$ of solute is simply related to the mass concentration C_M and molar mass M by:

$$\frac{C_M}{M} = C. \tag{4.9}$$

The difference between the osmotic pressure in the two compartments is given by:

$$(\Pi_A - \Pi_B) = (C_A - C_B)RT. \tag{4.10}$$

Note that equation (4.8) appears to be identical to the ideal gas equation. The mechanism, however, is fundamentally different. In an ideal gas, particles are moving around and the pressure is due to their collisions with the walls restricting the volume. The difference in osmotic pressure between the two sides does *not* arise from solute molecules "pressing" against the membrane, like gas molecules on the high-concentration side, it is the *hydrostatic pressure* due to the water that develops on the more concentrated side of the membrane—it is due to water trying to move from the dilute to the concentrated side of the membrane in an attempt to equalize its chemical potential.

For more than one type of solute, equation (4.8) becomes:

$$\Pi V = \sum_j N_j RT \tag{4.11}$$

where the sum is over all solutes for which the membrane is impermeable.

If only one of the compartments contains a solute (e.g., C_B is zero, the compartment only contains water), equation (4.8) provides an especially sensitive measure of the molar mass of solute A: this has been used extensively in the field of polymers. Note that the method is *not* useful for the determination of molecular weights of biological macromolecules such as proteins or nucleic acids because the molar concentration is usually very low and biological macromolecules are not stable in a pure water solvent; the much higher molar concentrations of buffer solutes and ions strongly dominate the osmotic pressure.

4.5.6 Virial coefficients

Equations derived for the ideal case could be considered as good approximations in the case of dilute solutions, provided interparticle interactions are negligible. Although such interactions can be accounted for by an analysis in terms of activity coefficients and chemical potentials the *virial coefficient* approach is more widely used for solutions of biological macromolecule. For a non-ideal solution equation (4.7) (using also 4.8) can be expanded in powers of mass concentration:

$$\frac{\Pi}{CRT} = \frac{1}{M} + A_2 C_M + A_3 C_M^2 + \cdots \qquad (4.12)$$

where $A_2, A_3 \ldots$ are the second, third $\ldots$ virial coefficients, respectively. They provide quantitative information on particle–particle interactions. Positive coefficients correspond to repulsion and negative coefficients to attraction between the particles.

4.6 Macromolecular solutions

Biological macromolecules require special solvent conditions in order to be stable and active, and salts, in particular, have important specific and non-specific effects due to ion association or binding. These effects depend on salt type and concentration and on the macromolecule. Specific salt effects are used in biochemistry and biophysics for fractionation, solubilization, crystallization, etc. of macromolecules. The thermodynamic concept of *ionic strength* is valid at very low ionic concentrations (in the mM range) and accounts for non-specific effects due to electric charge.

4.6.1 Ionic strength

Ionic solutions are not *ideal* because of the electrostatic interaction between charges. Lewis and Randall first introduced the concept of activity for strong electrolytes to account for the effects of ions of different valency. They postulated that the mean activity of a completely dissociated electrolyte in *dilute* solution depends only on what they called the ionic strength, i, of the solution:

$$i = \frac{1}{2} \sum_j C_j z_j^2 \qquad (4.13)$$

where C_j and z_j are, respectively, the molar concentration and charge of ion j.

Note that the concept of ionic strength is applicable only to (very) dilute solutions, in which the "nature" of the ions and their interactions with water are neglected so that they can be considered to act as point charges.

4.6.2 Polyelectrolytes

Solutions are always electrically neutral (otherwise they would explode!) with equal numbers of positive and negative ions. Negatively charged polyelectrolytes, like DNA, will have associated positive counterions which will dissociate from the macromolecule in solution.

Consider, for example, a dilute solution of NaCl in a vessel separated into two compartments by a membrane that is permeable to water and to the Na^+ and Cl^- ions.

A 70-nucleotide Na·tRNA salt, which dissociates into a tRNA polyion with 70 negative charges and 70 Na$^+$ ions, is dissolved in the left-hand compartment. The polyion cannot cross the membrane. Its negative charges are restricted to the left-hand compartment so that in order to maintain electroneutrality on either side of the membrane, there will be more small positive ions (Na$^+$) than unbound negative ions (Cl$^-$) in that compartment. Since there are also 70 positive counter-ions added per macromolecule, the result is an outflow of salt from the left-hand compartment to the right-hand compartment.

The phenomenon is called the *Donnan effect* and the distribution of ions across the membrane to establish chemical potential equilibrium between the left- and right-hand compartments is termed the Donnan distribution.

4.6.3 Water, salt and the hydrophobic effect

A water molecule can participate in four hydrogen bonds (two H donors and the O atom accepting two). In the current model, liquid water is regarded as a highly dynamic system of fluctuating directional hydrogen bonds, constantly rearranging between different partners. The formation of a hydrogen bond leads to a decrease in enthalpy (negative ΔH), while the configurational freedom of having different partners with which to form bonds leads to an increase in entropy (positive ΔS). Both features are thermodynamically favored because they contribute to a decrease in free energy ($\Delta G = \Delta H - T\Delta S$). At high salt concentrations, the structure of liquid water and its specific interactions with solutes have a significant effect on the properties of macromolecules in solution. In general, a solute molecule will perturb the dynamics of the water molecules with which it is in contact. Polar solutes will orient the surrounding water molecules in a preferential way compared to when they are in the bulk liquid. Recall how an ion with negative partial volume exerts electrostriction by "pulling" water molecules towards itself. Apolar solutes perturb the structure and dynamics of liquid water because solvent molecules in their vicinity have fewer hydrogen bonding possibilities, leading to a decrease in entropy. Exposure of apolar surfaces to water is therefore not favored thermodynamically. Apolar solutes are poorly soluble in water, displaying an anomalous solubility curve—their solubility decreases with increasing temperature.

The *hydrophobic effect* results from the low solubility of apolar solutes in water. Because contact between the solute and water is entropically unfavorable, apolar solutes will tend to aggregate to minimize their surface of interaction with the surrounding water molecules.

Salt ions perturb the structure of water according to specific properties, especially charge and volume. *Salting-out* ions reduce the solubility of apolar solutes in water even further and favor their precipitation. Biological macromolecules possess heterogeneous surfaces, with charged as well as apolar patches, leading to complex hydration interactions, folding and solubility behavior. The *Hofmeister series* is a phenomenological classification of different ions according to how they stabilize and reduce the solubility of proteins, in one direction, or on the contrary destabilize structures and increase solubility, in the other (Table 4.1) [1].

Table 4.1 The Hofmeister series. Anions: phosphate is the most salting-out, while chloride is neutral and thiocyanide the most salting-in. Cations: ammonium, potassium and sodium are neutral, while calcium and magnesium are more salting-in than lithium

---
Cations:
Phosphate > sulfate > acetate > chloride < bromide < chlorate < thiocyanide

Anions:
Ammonium, potassium, sodium < lithium < calcium, magnesium

4.7 A thermodynamics approach to scattering methods: macromolecule–solvent interactions (hydration and solvation)

Consider a solution made up of three components: water (component 1), a macromolecule (component 2) and a small solute such as salt (component 3). In order to have well-defined solvent conditions, the solution is dialyzed against a large volume of solvent [3].

Density increments are bulk properties of the solution. The mass density increment accounts for the presence of the macromolecule itself and also for its interactions with diffusible solvent components:

$$(\partial \rho / \partial C_2)_\mu = (1 + \xi_1) - \rho^\circ (\bar{v}_2 + \xi_1 v_1) \tag{4.14}$$

where ξ_1 is an interaction parameter in grams of water per gram of macromolecule and v_x is the partial specific volume (in ml/g) of component x. The parameter ξ_1 represents the water that flows into the dialysis bag to compensate for the change in solvent composition caused by the association of (or repulsion by) water and small solute components with the macromolecule. The mass density increment is essentially equivalent to the buoyancy term arising from Archimedes' principle.

In X-ray and neutron scattering, electron and neutron scattering density increments, respectively, are defined analogously:

$$(\partial \rho_{el} / \partial C_2)_\mu = (l_2 + \xi_1 l_1) - \rho^\circ_{el}(\bar{v}_2 + \xi_1 \bar{v}_1) \tag{4.15}$$

$$(\partial \rho_{n} / \partial C_2)_\mu = (b_2 + \xi_1 b_1) - \rho^\circ_{n}(\bar{v}_2 + \xi_1 \bar{v}_1) \tag{4.16}$$

where l_x and b_x are the electron and neutron scattering amplitudes, respectively, per gram of component x. They can be calculated readily from chemical compositions.

In the same way as for the mass density increment, equations for X-ray and neutron scattering are equivalent to buoyancy terms where mass is replaced by scattering amplitudes. The equations can be used jointly to solve for unknown parameters, for example, ξ_1 and v_2. Combining the mass and X-ray equations is not very useful, because mass and electron density are close to being proportional to each other. On the other hand, atomic neutron scattering amplitudes (which can be positive or negative) are completely independent of mass or electron content, so that combining the neutron equation with mass and/or X-ray results is extremely useful.

Relations between *density increments* and *measurable parameters* are now set out.

4.7.1 Sedimentation equilibrium

The condition for sedimentation equilibrium is given by;

$$\mathrm{d}\ln C_2/\mathrm{d}r^2 = (\omega^2/2RT)\,(\partial\rho/\partial C_2)_\mu M_2 \qquad (4.17)$$

where r is the distance to the center of the rotor, ω is the angular velocity and $(\partial\rho/\partial C_2)_\mu$ is the mass density increment at constant chemical potential of the diffusible solvent components (equation (4.14)).

4.7.2 Light scattering

Light scattering is given by:

$$\Delta R(0) - K\,(\partial n/\partial C_2)_\mu^2\,C_2 M_2 \qquad (4.18)$$

where $\Delta R(0)$ is the light scattering in the forward direction (zero angle) in excess of solvent scattering and K is an optical constant. n is the refractive index.

4.7.3 X-ray scattering

For forward scattering of X-rays:

$$I_{\mathrm{el}}(0) = K_{\mathrm{el}}(\partial\rho_{\mathrm{el}}/\partial C_2)_\mu^2 C_2 M_2/N_A \qquad (4.19)$$

where $I_{\mathrm{el}}(0)$ is forward scattering of X-rays, K_{el} is a calibration constant and N_A is Avogadro's number.

4.7.4 Neutron scattering

Neutron scattering is given by:

$$I_{\mathrm{n}}(0) = K_{\mathrm{n}}(\partial\rho_{\mathrm{n}}/\partial C_2)_\mu^2 C_2 M_2/N_A \qquad (4.20)$$

where $I_{\mathrm{n}}(0)$ is forward scattering of neutrons and K_{n} is a calibration constant.

4.8 Relating the thermodynamics and particle approaches

The interaction parameter can be expressed in terms of B_1 grams of water and B_3 grams of salt bound by the particle [3]:

$$\xi_{11} = B_1 - B_3/w_3. \qquad (4.21)$$

When B_1 and B_3 are constant a plot of ξ_1 against $1/w_3$ gives a straight line. The "bound" water and salt values may be positive or negative (when the component is excluded from the solvation shell). Applying the model to the interaction parameters, we obtain:

$$(\partial\rho/\partial C_2)_\mu = (1 + B_1 + B_3) - \rho^\circ(\bar{v}_2 + B_1\bar{v}_1 + B_3\bar{v}_3) \qquad (4.22)$$

$$(\partial\rho_{\mathrm{el}}/\partial C_2)_\mu = (l_2 + l_1 B_1 + l_3 B_3) - \rho^\circ(\bar{v}_2 + B_1\bar{v}_1 + B_3\bar{v}_3) \qquad (4.23)$$

$$(\partial\rho_{\mathrm{n}}/\partial C_2)_\mu = (b_2 + b_1 B_1 + b_3 B_3) - \rho^\circ(\bar{v}_2 + B_1\bar{v}_1 + B_3\bar{v}_3). \qquad (4.24)$$

Note that whereas a straight-line dependence of ξ_1 versus $1/w_3$ is consistent with a particle that is invariant in composition, a straight-line dependence of density increment versus solvent density is consistent with a particle in solution that is invariant in both composition and volume. Equations (4.22)–(4.24) are the contrast equations in the case of a three-component solution. Note that where the particle composition does not depend on the solvent the slope of the density increment versus solvent density is identical for the three methods and equal to the volume of the solvated particle.

A reduced density increment for an invariant particle was proposed in order to plot mass, X-ray and neutron data together:

$$(\partial \rho / \partial C_2)^*_\mu = (\partial \rho_x / \partial C_2)_\mu / (x_2 + x_1 B_1 + x_3 B_3)$$

$$(\rho^0_x)^* = \rho^0_x / (x_2 + x_1 B_1 + x_3 B_3) \quad (4.25)$$

$$(\partial \rho / \partial C_2)^* = 1 - (\rho^0_\mu)^* V_T$$

where x refers to mass, electron or neutron scattering length, respectively; x_2 is for macromolecules, x_1 is for water and x_3 is for salt. ρ^0_μ is the solvent scattering density and V_T is total volume of macromolecules plus B_1 water and B_3 salt.

Bonneté et al. [4] used equation (4.25) to make a single plot of neutron, X-ray and mass data to reveal and quantify the hydration and solvation of malate dehydrogenase from an extreme halophile in terms of B_1 grams of bound water and B_3 grams of salt per gram of protein.

The thermodynamics density increment analysis of AUC and scattering data is rigorous and provides absolute scale information on hydration and solvation under given solvent conditions. However, a model approach to the scattering curve beyond forward scattering is required in order to place hydration within a structure. This was done in a combined contrast variation approach using SANS in H_2O and D_2O, and with SAXS for four soluble proteins of known structure [5]. In the four cases, the SAXS curve indicated the largest particle in solution (X-rays are sensitive to a hydration shell of higher density than bulk water), the SANS/D_2O curve indicated the smallest particle (in D_2O, the hydration shell has opposite SANS contrast to that of the protein) and the SANS/H_2O curve indicated an intermediate particle size (in H_2O the contrast of the hydration shell is close to zero). The results clearly established the presence of a hydration shell denser than the bulk water surrounding each protein. Its contrast is high and positive for SAXS, close to zero (practically invisible) for SANS/H_2O and of opposite contrast to that of the protein for SANA/D_2O (leading to an apparently smaller particle than the protein alone). The results revealed a first hydration shell with an average density about 10% larger than that of bulk water [5].

4.9 NMR and neutrons give molecular meaning to thermodynamics parameters

Neutron wavelengths and energies are of the same order as the fluctuations and energies, respectively, of thermal motions in macromolecules. Since the 1950s, neutron inelastic scattering has played a fundamental role in condensed matter physics through the experimental interpretation of specific heat in terms of phonon dispersion relations and density of states at the level of atomic motions. These concepts are now being applied to macromolecules and their interactions. In a combined study using neutron scattering and molecular dynamics simulations, the change in the vibrational density of states of a protein on binding a ligand was determined and analyzed. The vibrations of the complex soften significantly relative to those of the unbound protein. The resulting change in free energy was directly determined by the change in density of states and found to contribute significantly to the binding equilibrium [6].

Molecular recognition by proteins is fundamental to almost every biological process, particularly the protein interactions underlying signal transduction. Understanding the basis of protein–protein interactions, however, requires a full characterization of the thermodynamics. NMR has emerged as a powerful tool for characterizing protein dynamics and has been used to estimate changes in conformational entropy from corresponding changes in conformational dynamics in calmodulin binding to different ligands [7]. The squared order parameters for different motion classes were found to be linearly correlated to the contribution of entropy to binding free energy. Large-amplitude side-chain rotamer sampling (jumping between potential wells) displayed an inverse correlation with the entropy contribution (a possible explanation is that rotamer sampling facilitates binding leading to an overall reduction in entropy), whereas smaller-amplitude motions, within a well, were found to be positively correlated with the contribution of entropy to binding free energy (a possible explanation being that the reduction in entropy due to binding is compensated by an increase in local mobility outside the active site).

Intrinsically disordered proteins (IDP) and their order–disorder transitions are likely to play an important role in certain biological systems. Their structural characterization, however, has required the development of specific approaches adapted to the disordered state. Ensemble approaches are particularly well adapted for mapping the conformational energy landscape sampled by a protein at atomic resolution.

Significant advances have also been achieved in the development of calibrated NMR approaches to the statistical representation of the conformational behavior of IDP, which also allow the observation of flexible IDP even within large molecular complexes—as shown by a recent study of the disordered domain of measles virus protein [8].

References

[1] Serdyuk IN, Zaccai N, Zaccai G (2007) *Methods in Molecular Biophysics: Structure, Dynamics, Function*. Cambridge, UK: Cambridge University Press.
[2] Zaccai G (2013) Hydration shells with a pinch of salt. *Biopolymers* **99**: 233–238.
[3] Eisenberg H (1981) Forward scattering of light, X-rays and neutrons. *Q Rev Biophys* **14**: 141–172.
[4] Bonneté F, Ebel C, Zaccai G, Eisenberg H (1993) Biophysical study of halophilic malate dehydrogenase in solution: revised subunit structure and solvent

interactions of native and recombinant enzyme. *J Chem Soc, Faraday Trans* **89**: 2659–2666.

[5] Svergun DI, Richard S, Koch MH, Sayers Z, Kuprin S, Zaccai G (1998) Protein hydration in solution: experimental observation by x-ray and neutron scattering. *Proc Natl Acad Sci USA* **95**: 2267–2272.

[6] Balog E, Becker T, Oettl M, Lechner R, Daniel R, Finney J, and Smith JC (2004) Direct determination of vibrational density of states change on ligand binding to a protein. *Phys Rev Lett* **93**: 028103.

[7] Frederick KK, Marlow MS, Valentine KG, Wand AJ (2007) Conformational entropy in molecular recognition by proteins. *Nature* **448**: 325–329.

[8] Jensen MR, Ruigrok RWH, Blackledge M (2013) Describing intrinsically disordered proteins at atomic resolution by NMR. *Curr Opin Struct Biol* **23**: 426–435.

5
NMR spectroscopy: from basic concepts to advanced methods

Enrico LUCHINAT

Magnetic Resonance Center – CERM and Department of Biomedical, Clinical and Experimental Sciences, University of Florence, Italy

Abstract

This chapter aims to give a general overview of the field of NMR spectroscopy as applied to structural biology. In the first section the basic concepts of NMR theory are outlined, together with some technical aspects, from the nuclear spin to multidimensional experiments. In the second section more advanced applications of protein NMR are presented, including methods for studying large proteins, fast-pulsing techniques, carbon-detected NMR, paramagnetic NMR and solid-state NMR. The final section focuses on a recent advance of NMR applied to the structural biology of proteins: in-cell NMR. With this method, high-resolution NMR is used to investigate proteins in living cells at atomic detail. The main effects of the intracellular environment on protein structure and dynamics are explained. Several in-cell NMR approaches are shown: protein expression in bacterial cells, protein insertion or expression into higher eukaryotes. Some applications of protein expression by transient transfection in human cells are given.

Keywords

Nuclear magnetic resonance, NMR, in-cell NMR, NMR theory, nuclear spin, NMR basics, structural biology, NMR methods, atomic resolution, living cells

From Molecules to Living Organisms: An Interplay Between Biology and Physics. First Edition.
Eva Pebay-Peyroula et al. © Oxford University Press 2016.
Published in 2016 by Oxford University Press.

Chapter Contents

5 NMR spectroscopy: from basic concepts to advanced methods   134
    Enrico LUCHINAT

  5.1 Basic concepts of NMR spectroscopy   136
     5.1.1 The nuclear spin   136
     5.1.2 Spins in a magnetic field   137
     5.1.3 The chemical shift   138
     5.1.4 Radiofrequency pulses and Fourier transform   138
     5.1.5 Vectorial description, the rotating frame   139
     5.1.6 Spin relaxation   141
     5.1.7 J-coupling, dipolar coupling and the nuclear Overhauser effect   142
     5.1.8 Multidimensional NMR   143
     5.1.9 Heteronuclear 2D and 3D experiments   145
     5.1.10 Protein NMR: from backbone assignment to structure calculation   147
  5.2 Advances in protein NMR   148
     5.2.1 Increasing the size limit in solution NMR   148
     5.2.2 Increasing the speed of NMR   149
     5.2.3 Carbon-detected NMR   151
     5.2.4 NMR of paramagnetic molecules   153
     5.2.5 Solid-state NMR: a brief overview   155
  5.3 In-cell NMR: towards integrated structural cell biology   157
     5.3.1 Why in-cell NMR?   157
     5.3.2 In-cell versus *in vivo* NMR   158
     5.3.3 Effects of the intracellular environment on the protein   158
     5.3.4 Measurable effects on the NMR spectra   158
     5.3.5 Isotopic labeling strategies   159
     5.3.6 Overview of the existing techniques   161
     5.3.7 Protein expression in mammalian cells: applications   162
     5.3.8 Pushing the limits: in-cell solid-state NMR   164

  References   165

5.1 Basic concepts of NMR spectroscopy

5.1.1 The nuclear spin

Nuclear magnetic resonance (NMR) spectroscopy exploits the magnetic properties of atomic nuclei (for an overview of the basic concepts of NMR spectroscopy see references by Hore and colleagues[1,2]). The nuclei of certain isotopes possess a magnetic moment which is extremely sensitive to their chemical surroundings, and as such can act as an excellent probe for investigating the molecular structure of matter. The magnetic properties of every nucleus are given by its *spin*, i.e., its intrinsic angular momentum, the magnitude of which is determined by the *spin quantum number* I (Fig. 5.1), following the relationship

$$\bar{I} = \sqrt{I(I+1)}\hbar \quad \text{with} \quad I = 0, 1/2, 1, 3/2, 2, \ldots$$

The spin quantum number is determined by the number of unpaired protons and neutrons. Therefore, different isotopes of the same element will have different spin quantum numbers. A nucleus with spin I will have $2I+1$ possible spin states, and the projections onto an arbitrarily chosen axis (Fig. 5.1) will be

$$I_z = m\hbar \quad \text{with} \quad m = I, I-1, I-2, \ldots, -I+1, -I.$$

The magnetic moment $\bar{\mu}$ of a nucleus is given by

$$\bar{\mu} = \gamma \bar{I}$$

where the constant of proportionality γ is the gyromagnetic ratio, which has a unique value for each nucleus.

For several reasons, it is often preferred in NMR spectroscopy to exploit nuclei with spin quantum number $I = 1/2$. In organic and biological molecules, the most abundant nucleus with such a spin quantum number is ^{1}H, which has very high natural abundance (99.99%) and the highest gyromagnetic ratio. When studying complex systems, it is often necessary to probe different nuclei as well. Unfortunately, the second most abundant nucleus, ^{12}C, has $I = 0$, and is magnetically silent. ^{13}C, on the other hand, has $I = 1/2$, $\gamma_C \cong \frac{1}{4}\gamma_H$, and has a natural abundance of 1.1%. For ^{14}N, $I = 1$, so it is technically non-silent, but difficult to detect for other reasons, while ^{15}N has $I = 1/2$, $\gamma_N \cong \frac{1}{10}\gamma_H$, but has a natural abundance of only 0.37%.

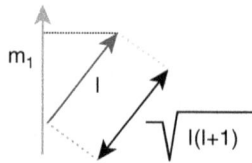

Fig. 5.1 The nuclear spin quantum number I.

5.1.2 Spins in a magnetic field

In the absence of an external magnetic field, all $2I+1$ spin states of a nucleus with spin I have the same energy. When a nucleus with spin I is put in an external magnetic field $\bar{B}$, along the z-axis, the degeneracy is removed, and the energy of each spin state will be:

$$E = -\bar{\mu} \cdot \bar{B} = -\mu_z B = -m\hbar\gamma B.$$

Therefore, the spin states of the nucleus will be separated from each other by an energy of $\Delta E = -\hbar\gamma B$ (Fig. 5.2). If we irradiate the nucleus with an electromagnetic radiation of energy $h\upsilon$ that matches the separation between the spin states, we have the resonance condition:

$$\Delta E = h\upsilon = -\hbar\gamma B, \quad \upsilon = -\frac{\gamma B}{2\pi} \quad \text{or} \quad \omega_0 = -\gamma B$$

where ω_0 is the Larmor frequency of that nucleus in the magnetic field B. Since ^{1}H is the nucleus most often used in NMR, the magnetic field of the NMR spectrometer is indicated by the ^{1}H Larmor frequency (in hertz) instead of in the correct SI units for magnetic field (tesla). For example, a NMR spectrometer with a magnetic field of 9.4 T is usually referred to as a 400 MHz instrument.

When the resonance condition is met, the electromagnetic radiation will induce transitions between the spin states of the irradiated nuclei. A short time after irradiation at the Larmor frequency, the excited nuclei in the sample will transition back to the ground state, and in so doing will emit electromagnetic radiation at the same frequency. It is by detecting this emitted radiation that NMR spectroscopy can obtain information about the structure of matter, from small molecules to huge macromolecular assemblies. The intensity of the emitted radiation is directly proportional to the population difference between two spin energy levels. This difference is extremely small, as the energy gap is small compared with the thermal energy at room temperature, and is given by an (approximated) Boltzmann equation:

$$\frac{\Delta n}{n} \approx \frac{\Delta E}{2kT}$$

which for ^{1}H at 400 MHz and 300 K gives a population difference of roughly 3×10^{-5}. Therefore, only one ^{1}H nucleus in 30 000 will be excited by the electromagnetic radiation. Nuclei with a lower gyromagnetic ratio will have an even smaller population

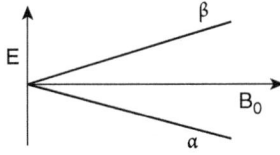

Fig. 5.2 Energy separation between two spin states as a function of the magnetic field B_0.

difference. For this reason NMR is an intrinsically insensitive technique. However, at low energies NMR has a big advantage over all other spectroscopic techniques: indeed, electromagnetic waves in the radiofrequency (RF) spectrum leave the system almost unperturbed. This allows us to investigate the properties of biological molecules in conditions very close to the physiological environment, i.e., at room temperature, in aqueous buffers, at neutral pH, etc. without any risk to the stability of the sample during the NMR experiment. Sensitivity can be significantly improved by increasing the strength of the external magnetic field. This is one of the reasons why higher-field spectrometers being increasingly requested.

5.1.3 The chemical shift

As stated in Section 5.1.2, the resonance frequency of each nucleus is exactly determined by the gyromagnetic ratio and the magnetic field strength. How would we be able to obtain any information on the structure and dynamics of matter if all nuclei of a given type (e.g., ^{1}H) had only one Larmor frequency (e.g., 400 MHz) at a given magnetic field (e.g., 9.4 T)? In this case, the devil (and the beauty!) is in the detail. Indeed, the actual field B experienced by any given nucleus is influenced by the local electron distribution, which will contribute to the external field B_0 with a small *shielding* magnetic field B' of opposite sign ($B = B_0 - B'$). Therefore, nuclei of the same type in different positions within a molecule will experience a slightly different B, and will have slightly different Larmor frequencies. B' itself is proportional to B_0, therefore these differences are independent of the strength of B_0. Commonly, these small differences in frequency are defined relative to a reference nucleus ($\nu - \nu_{\text{ref}}$), and expressed by a dimensionless parameter δ called the *chemical shift*:

$$\delta = 10^6 \frac{\nu - \nu_{\text{ref}}}{\nu_{\text{ref}}}$$

which is scaled by 10^6 and is therefore expressed in parts per million (ppm) of the reference Larmor frequency. As they are usually represented, NMR spectra go from less shielded nuclei on the left (high ppm) to more shielded nuclei on the right (low ppm).

These comparatively small differences in frequency are an extremely valuable source of information, as they reflect the nature of the chemical bonds between atoms and are sensitive to other, even minor, effects (Fig. 5.3). With sufficient resolution, thousands of unique atoms can in principle be resolved by virtue of their chemical shift. Higher magnetic fields provide better resolution, together with increased sensitivity, and this is critical for investigating larger and more complex systems.

5.1.4 Radiofrequency pulses and Fourier transform

Given that in any molecule each nucleus of the same kind (e.g., ^{1}H) resonates at a slightly different frequency, how can the NMR spectrometer excite—and detect—all of them at once? The old, first generation of NMR spectrometers used to achieve this by irradiating the sample with a RF at a fixed wavelength and, by varying the strength of the static magnetic field in a linear manner, they could excite all different nuclei one

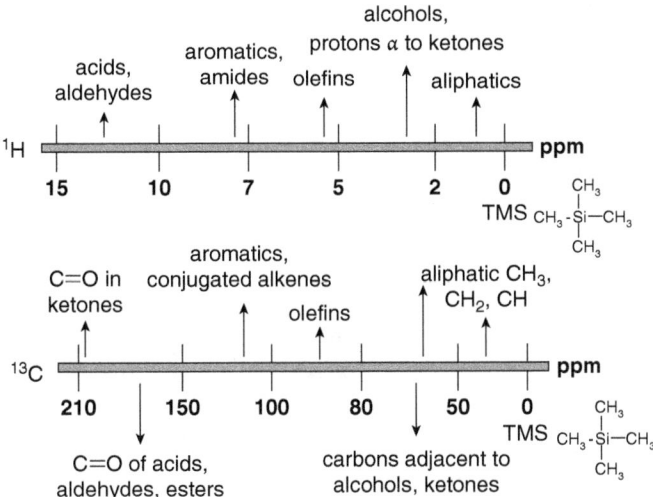

Fig. 5.3 ^{1}H and ^{13}C chemical shift ranges of the most common functional groups in organic molecules.

by one, starting from the less shielded ones on the left (which is why the NMR spectra are represented in that way). Modern NMR spectrometers work in a different way. The static magnetic field is created by a superconducting magnet, which is maintained at a few kelvin by liquid helium, and is kept at constant strength. The spectrometer, instead of doing a frequency sweep to excite all nuclei, irradiates the sample with a single, short RF pulse at a fixed frequency. An important property of pulses is that they uniformly excite a broad region of the frequency spectrum, centered around the frequency of the pulse itself. The shorter the pulse, the broader the excitation window will be. Once all the nuclear spins have been excited, the RF is switched off and the spins are allowed to evolve in the so-called *free precession*. During this period, each spin will emit a RF signal at its Larmor frequency, which is then detected. The period during which this signal is recorded is called *free induction decay* (FID). The recorded signal is a measure of amplitude as a function of time. To produce a frequency spectrum, the FID is processed by means of a discrete Fourier transform, which converts the time-dependent information, $s(t)$, to a frequency-dependent intensity plot, $S(\omega)$, which is the NMR spectrum:

$$S(\omega) = \int_{-\infty}^{\infty} s(t)e^{-i\omega t} dt.$$

5.1.5 Vectorial description, the rotating frame

The *vector model* of NMR spectroscopy is a convenient means of describing in an intuitive way what is happening to the nuclear magnetization in the sample during a NMR experiment.

140 NMR spectroscopy

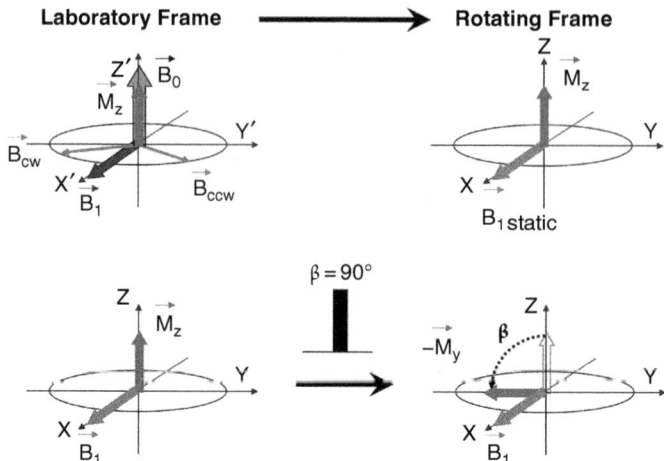

Fig. 5.4 Change of coordinates between the laboratory frame and the rotating frame; the effect of a RF pulse on the net magnetization in the rotating frame (cw, clockwise; ccw, counterclockwise).

When the nuclei are put in a strong magnetic field B_0, a net, or *bulk*, magnetization $\bar{M}$ is created along the magnetic field, which results from the population difference at the thermal equilibrium between the nuclei in the ground spin state and those in the excited spin state. The direction of B_0 defines the z' axis of the x', y', z' coordinate system called the *laboratory frame* (Fig. 5.4). If we irradiate the sample with a RF pulse of frequency ω_{rf}, close to the Larmor frequency of the nuclei ω_0, the bulk magnetization M will be tilted from the z' axis and will immediately start to precess along z' at the Larmor frequency. These motions are difficult to visualize in this frame as they occur at very high frequencies. However, if we view the system in a *rotating frame*, represented by a x, y, z coordinate system, which rotates about the z' axis at the frequency ω_{rf}, the clockwise rotating component of the oscillating magnetic field created by the RF pulse will appear to be static (let's call it B_1), and will be oriented (say) along x, while the static magnetic field B_0 will be replaced by a small effective (residual) field $\Delta B_0 = -(\omega_0 - \omega_{rf})/\gamma$, which is usually negligible with respect to B_1. Therefore, in the rotating frame, the RF pulse will start tilting the vector $\bar{M}$ along the x axis at the angular frequency $\omega_1 = -\gamma B_1$, until it is turned off (Fig. 5.4). Longer RF pulses will tilt $\bar{M}$ by larger angles, and by adjusting the pulse length we can tilt $\bar{M}$ by any *flip angle*. A 90° pulse will put the magnetization along the plane (*transverse* magnetization), while a 180° pulse will invert the equilibrium magnetization (i.e., put it along $-z$). The detector, which is the same coil which generated the RF pulse, can only register RF signals in the x–y plane that are generated by the transverse magnetization during its *free precession*.

5.1.6 Spin relaxation

When it is in-plane, the magnetization $\bar{M}$ will not precess forever in the x–y plane. Given enough time (which can vary between microseconds and seconds, or even more) the spin population will return to thermal equilibrium. This process is known as *spin–lattice* or *longitudinal* relaxation, and is often characterized by (or approximated to) a single exponential function with time constant T_1. In the rotating frame it can visualized as the time during which $\bar{M}$ returns along the z axis. Transverse magnetization will also experience a different relaxation mechanism, called *spin–spin* or *transverse* relaxation, which is a loss of *phase coherence* between the individual spins. It is approximated to a single exponential function with time constant T_2. In the rotating frame, it can be seen as the disappearance of the x–y component of the net magnetization. The minimum linewidth of a signal in the NMR spectrum is given by the transverse relaxation rate: the faster it is, the broader the signal we will obtain.

The mechanisms governing the relaxation behavior of spins in molecules can be quite complex, and are a function of the properties of the nuclei, the chemical bonds, the molecular tumbling rate in solution and the external magnetic field, and their detailed description is outside of the scope of this chapter. However, there are some simple rules which are important when applying NMR to biological macromolecules: T_2 is always shorter than T_1; T_2 always decreases with decreasing molecular tumbling rate (which in turn decreases with increasing molecular weight); for small molecules, T_1 is close to T_2 and decreases with increasing molecular weight, while for large molecules (e.g., proteins) T_1 starts to increase with increasing molecular weight, and becomes larger than T_2 (Fig. 5.5).

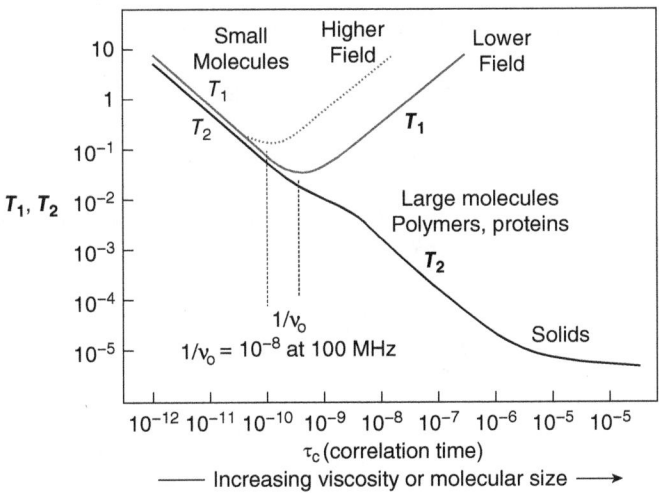

Fig. 5.5 Variation of the relaxation time constants T_1 and T_2 as a function of the rotational correlation time. Copyright 1948 by the American Physical Society. http://www.chem.wisc.edu/areas/reich/nmr/

At increasing magnetic fields, the tipping point where T_1 diverges from T_2 occurs at smaller molecular weights, so when we are working with proteins at high magnetic fields we are in the $T_1 \gg T_2$ regime. In practice, this means that after the detected signal has decayed to zero in the FID (say in a few milliseconds) we still have to wait for a longer times (usually 1 s or more) for the system to reach thermal equilibrium so that we can repeat the experiment and record another FID.

5.1.7 J-coupling, dipolar coupling and the nuclear Overhauser effect

The nuclei in a molecule experience coupling effects, which arise from magnetic interactions between them and between nuclei and electrons. There are mainly two kinds of coupling: *J-coupling*, also called *scalar* or *spin–spin* coupling, and dipolar coupling. J-coupling between two nuclei arises from the presence of one (or more) chemical bond(s) connecting them. J-coupling is the consequence of the contact (or Fermi) interaction between electron and nuclear spins, and does not have a classical description. The presence of a chemical bond causes a splitting in the energy levels of each nucleus, as a function of the spin state of the other nucleus. This in turn causes line splitting in the NMR spectrum, which is independent of the strength of the external magnetic field (Fig. 5.6).

J-coupling is useful not only because it can deliver information on the network of chemical bonds in a molecule but also because it can be exploited to transfer magnetization from one nucleus to a J-coupled neighbor, in a coherent manner. Therefore, J-coupling is the basic principle upon which many NMR experiments for protein resonance assignment and structure determination are designed.

Dipolar coupling arises from the direct magnetic interaction between nuclei. A magnetic dipole is associated with each non-zero nuclear magnetic moment μ, which in turn generates a small magnetic field B_μ, which has the geometry shown in Fig. 5.7 and falls off from the position of μ with $1/r^3$. This small magnetic field will affect the magnetic moments of nearby nuclei in a way that is both dependent on the distance and on the angle θ between the vector connecting the two nuclei and the axis (z) of the external magnetic field. The component along z, in particular, is positive for certain angles and negative for others (and null at $\theta = \theta_m \approx 54.74°$ for which $3\cos^2 - 1 = 0$; θ_m is called the *magic angle*, and will be important when we introduce solid-state NMR). Depending on the spin state of the first nucleus, its magnetic moment will be

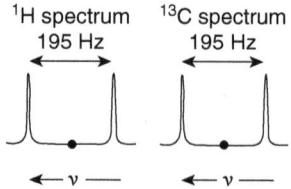

Fig. 5.6 J-coupling between chemically bonded ^{1}H and ^{13}C. Reproduced from Hore,[1] by permission of Oxford University Press.

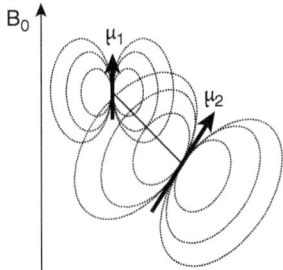

Fig. 5.7 Two nuclear magnetic moments in a magnetic field undergoing dipolar interaction.

oriented along either $+z$ or $-z$, and will give rise to a splitting of the energy levels of the second nucleus. This splitting has a large effect on solids, but in solution, where molecules are freely tumbling at a rate much faster than the frequency relative to the dipolar interaction energy, it averages to zero. Therefore in solution, dipolar couplings do not produce splitting in the NMR spectra. However, they are one of the main causes of spin–lattice relaxation.

An important consequence of dipolar coupling is the so-called *nuclear Overhauser effect* (NOE). The homonuclear NOE can be described by considering two non-equivalent protons (i.e., with two different Larmor frequencies), close in space, with no scalar coupling but only dipolar coupling between them. The NOE shares the nature of the dipolar effect: it is through-space, not through-bond. If the populations of the two spin states of the first nucleus are equalized (by applying a continuous RF centered at its Larmor frequency), the NMR signal of that nucleus will disappear. At the same time, however, the intensity of the second NMR signal will be affected (i.e., it could get stronger, weaker or even change sign). The NOE can be quantified by defining the parameter η so that:

$$\eta = \frac{i - i_0}{i_0}$$

where i is the perturbed signal intensity and i_0 is the unperturbed intensity. The NOE arises from relaxation pathways occurring in the two-spin system in which the population differences of both nuclei change simultaneously, i.e., from *cross-relaxation* processes. The probability of these relaxation pathways depends on the difference in Larmor frequency, the tumbling rate of the protein and, importantly, on the internuclear distance. Specifically, NOEs decrease proportional to $1/r^6$, so for example ^{1}H–^{1}H distances lower than about 5 Å usually give rise to measurable NOEs. This dependence renders the NOE an extremely useful tool for obtain through-space distance restraints in a molecule, and in fact NOEs are the basis for the determination of protein structure.

5.1.8 Multidimensional NMR

NMR spectra with more than one dimension are increasingly necessary when dealing with large molecules such as proteins or nucleic acids. By using *pulse sequences*, in which the magnetization is manipulated by a series of pulses and time delays, it is possible to encode one or more additional frequency dimensions into a series of FIDs (which are intrinsically one-dimensional, 1D), which are then processed to obtain the final n-dimensional spectrum. Traditionally, a generic 2D pulse sequence can be divided into these sequential steps:

$$\text{preparation} \rightarrow \text{evolution}\,(t_1) \rightarrow \text{mixing} \rightarrow \text{detection}\,(t_2).$$

To illustrate these steps, let us take the 2D ^{1}H–^{1}H NOE spectroscopy (NOESY) experiment. The NOESY experiment correlates spins which are close together in space, by means of the NOE between them. The NOESY pulse sequence starts with a 90° pulse (preparation), which brings the magnetization vector of all spins in the x–y plane. Next, free precession is allowed to take place during time t_1 (evolution). During t_1, all spins will precede in the x–y plane according to their Larmor frequency. Then, a second 90° pulse is applied, which will again create some net magnetization oriented along the z axis. The extent of this magnetization will depend on the position on the x–y plane that each spin had reached during t_1 (i.e., the evolution of each spin is now *encoded* in the z-axis component). The system is now allowed to evolve for a fixed delay, τ_m (mixing), during which part of the z-oriented magnetization of one spin will transfer to other spins via the NOE. Finally, a third 90° pulse is applied which puts these components of the magnetization back on the x–y plane, and the FID is recorded. The experiment is repeated several times, and the evolution time t_1 is gradually incremented. In the end, a series of FIDs is recorded. If we transform each FID, we will obtain a series of 1D spectra. Each spectrum contains all the expected ^{1}H signals, but their intensity will vary with increasing t_1. Specifically, each signal intensity will be *modulated* both by its own Larmor frequency and, to a lesser extent, by the Larmor frequency of a second nucleus which had transferred part of its magnetization via the NOE. If we look at each signal separately, we can appreciate how the intensity fluctuations as a function of t_1 resemble a FID, as if it was recorded in the *indirect dimension*. Thus, if we take the array of 1D spectra and transform each point along t_1, we will obtain a 2D spectrum, which has the ^{1}H frequency in both direct (F2) and indirect (F1) dimensions. The spectrum will have a diagonal, which arises from the magnetization which evolved with the frequency of the same nucleus during t_1 and t_2, and some *crosspeaks*, which arise from the magnetization which evolved with the frequency of one nucleus during t_1, was transferred to another nucleus during the mixing time via the NOE and evolved with the frequency of the second nucleus during t_2 (Fig. 5.8). The 2D NOESY spectrum thus contains the pairwise correlations between all nuclei close in space, and each crosspeak intensity will be (in principle) proportional to $1/r^6$.

Using the same principle, other 2D experiments have been designed which correlate nuclei via J-coupling: correlation spectroscopy (COSY) and total correlation spectroscopy (TOCSY) for nuclei of the same type and heteronuclear single quantum correlation (HSQC) and heteronuclear multiple quantum correlation (HMQC) for nuclei of different types (*heteronuclear* experiments). By adding further evolution times, spectra of three or more dimensions are created, and between two evolution times

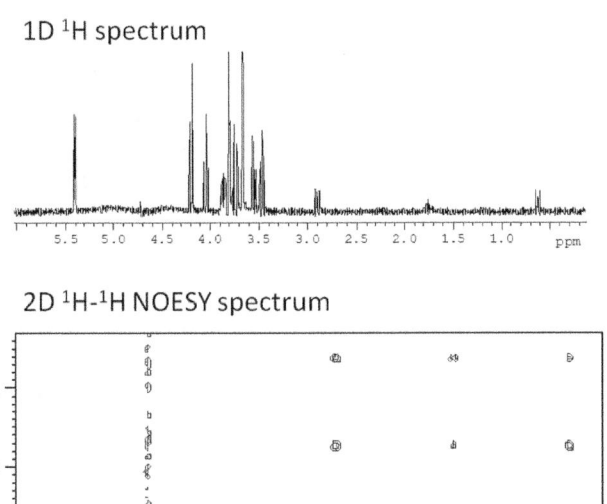

Fig. 5.8 1D ^{1}H spectrum (top) and 2D ^{1}H–^{1}H NOESY spectrum.

the nuclei can be mixed together either via J-coupling or via NOE, encoding different types of information in each dimension.

5.1.9 Heteronuclear 2D and 3D experiments

When studying a system such as a protein made up of thousands of non-equivalent atoms, signal overlap in a 1D spectrum becomes a serious issue and multidimensional experiments are necessary to provide sufficient separation between all the signals. To date, the basic tools of protein NMR spectroscopy are heteronuclear experiments, which have the advantage of providing information on different types of atoms in the same spectrum and establishing sequential correlation between them. Their success is also partly due to the relative ease nowadays of producing by recombinant techniques a sample of protein enriched in (*labeled with*) NMR-active isotopes, ^{13}C and ^{15}N. The most commonly performed protein heteronuclear experiments are constructed in a modular fashion, in which building blocks provide a simple way of controlling the magnetization transfer between nuclei, and its evolution, and can be combined to create increasingly complex combinations.

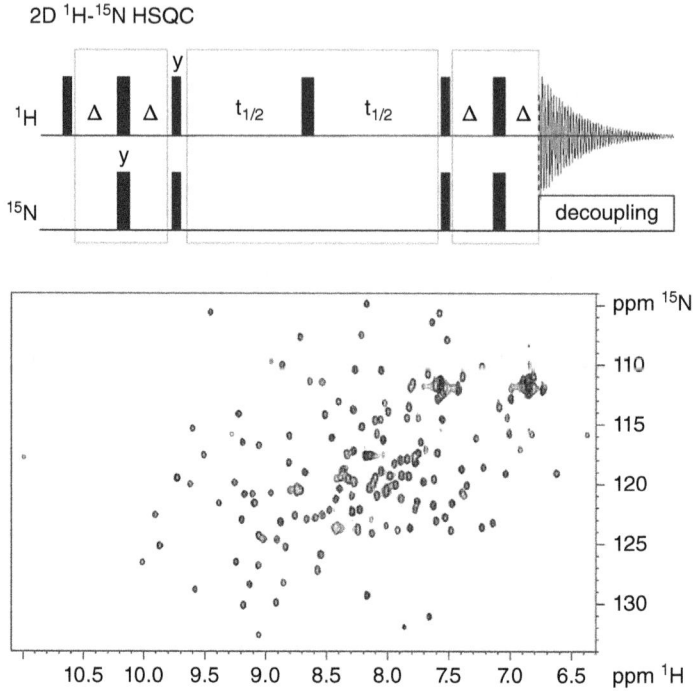

Fig. 5.9 Pulse scheme of the ^{1}H–^{15}N HSQC experiment (top) and the ^{1}H–^{15}N HSQC spectrum of a globular protein (bottom).

The simplest heteronuclear experiment which makes use of these building blocks is the HSQC experiment (Fig. 5.9). It produces a 2D spectrum which correlates ^{1}H resonances with those of the heteronuclei chemically bonded to them (^{1}H–^{15}N HSQC correlates amide protons and nitrogens, ^{1}H–^{13}C correlates all organic protons and carbons). The pulse sequence is made of three building blocks, which can somehow still be thought of as the preparation → evolution → mixing parts of the spectrum. The first block is called INEPT (insensitive nuclei enhancement by polarization transfer); this exploits the J-coupling between ^{1}H and X (X = ^{15}N or ^{13}C) to coherently transfer the net ^{1}H magnetization to X. The second block is the evolution period (t_1), in which the magnetization evolves on the x–y plane with the frequency of X. The third block is an *inverse* INEPT, which, as the name implies, does the opposite work of the first INEPT, and transfers the magnetization back to the ^{1}H, which is then detected. The reason for this double transfer (^{1}H → X → ^{1}H, also called *out-and-back*) is to provide sensitivity enhancement. Indeed, direct excitation or detection of the heteronucleus would result in much lower sensitivity due to the lower γ (this is especially true for ^{15}N, as $\gamma_N \cong \frac{1}{10}\gamma_H$).

Building blocks analogous to those of the HSQC experiment can be combined to generate a series of so-called *triple-resonance* experiments. Provided the protein is labeled with both ^{13}C and ^{15}N, and the NMR probe can work in three-channel mode (i.e., it is able to simultaneously excite ^{1}H, ^{13}C and ^{15}N), triple-resonance experiments, usually in three dimensions, can be acquired, which provide correlation between the atoms of the protein *backbone* (i.e., H^N, C′, Cα, Cβ). These correlations are both *intra-residue* and *inter-residue*, and therefore allow *sequential connectivity* to be established between all the amino acids. This sequential connectivity is the first important step towards the complete *resonance assignment* of a protein, which is necessary for structure calculation. Even if the final goal is not to calculate a protein structure, the backbone assignment itself is extremely useful as it provides a key to understanding protein conformational changes (induced, for example, by protein–protein interaction, drug or metal binding, etc.), which can be mapped at the residue level by means of chemical shift changes.

5.1.10 Protein NMR: from backbone assignment to structure calculation

Solution NMR can provide several types of experimental data which can be used as restraints to calculate the structure of a protein.[3] Classically, the restraints are mostly derived from three distinct phenomena: chemical shifts, J-coupling constants and NOEs. Several other complementary sources of structural information exist, such as residual dipolar couplings (RDCs), and paramagnetic effects such as paramagnetic relaxation enhancement (PRE) and pseudo-contact shifts (PCSs), which will be further discussed in Section 5.2.

Chemical shifts reflect the chemical environment of each nucleus, and therefore contain information on the 3D structure of the protein. Although there are methods for calculating the structure of a protein solely by relying on chemical shift information, this information is often difficult to extract as there are infinite possible chemical environments which would produce the same chemical shift. In some specific cases, however, chemical shifts can provide accessible information. The Hα, Cα and C′ chemical shifts, for example, are influenced by the local secondary structure, while the Cβ chemical shift is mainly indicative of the type of residue. These data alone can be used to determine the secondary structure of the protein without recurring to any structure calculation.

More specific structural data are obtained by measuring variations in the three-bond (^{3}J)-coupling between ^{1}HN and ^{1}Hα. The frequency of this J-coupling is indeed a function of the dihedral angle between these two nuclei. Analogously, other ^{1}H ^{3}J-couplings give information on the side-chain dihedral angles.

Finally, the usually most important source of structural restraints is the ^{1}H–^{1}H NOE, which was described in Section 5.1.7.

The classical workflow for protein structure calculation by NMR can be briefly summarized as:

backbone resonance assignment → side-chain resonance assignment →

measurement of backbone and side-chain J-couplings

(dihedral angle restraints) → ^{1}H–^{1}H NOE measurements

(distance restraints) →(collection of other restraints) →

iterative structure calculation.

Once all the experimental data have been extracted from the spectra, the actual structure calculation is made by a simplified molecular dynamics simulation in which the protein dihedral angles are moved and a target function is minimized to satisfy as many experimental restraints as possible. Usually, the software suites for protein structure calculation operate in a semi-automatic, back-and-forth iterative procedure, so that errors or ambiguities in the resonance assignment can be progressively corrected, decreasing the number of non-satisfied restraints until the simulation converges to a set of best-possible structures.

5.2 Advances in protein NMR

5.2.1 Increasing the size limit in solution NMR

One of the limits of protein NMR is given by the tumbling rate of the molecule in solution. Bigger proteins have slower tumbling rates (i.e., a higher rotational correlation time), which cause an increase in transverse relaxation rates ($R_2 = 1/T_2$) for all nuclei. This causes severe line broadening effects, above 30–40 kDa for ^{1}H–^{15}N HSQC spectra and even worse for triple resonances. There are several ways to mitigate this effect, which together allow spectra to be recorded on systems up to 1 MDa. (For an overview of the topics described in this section see Bertini et al.[3])

Above 30 kDa, a strategy that is generally good is to produce deuterated samples, i.e., in which a sizable fraction of the non-exchangeable ^{1}H have been replaced with ^{2}H. ^{13}C, ^{15}N, ^{2}H labeling usually provides an improved signal to noise (S/N) ratio for systems above 30 kDa.

There are also pulse sequences specifically designed to improve the linewidth at high molecular weights. TROSY (transverse relaxation optimized spectroscopy) has become a very useful building block which has been added to the set of triple-resonance experiments to reduce the extent of transverse relaxation of the ^{1}H–^{15}N crosspeaks.[4] The TROSY effect arises from the cross-correlation of two distinct relaxation mechanisms which are modulated by the same correlation time (e.g., molecular reorientation), and its effect increases with the magnetic field, reaching a maximum around 1 GHz. The TROSY effect causes the two components of a H–N J doublet to relax at different rates: for one component, the two relaxation mechanisms add; for the other, they cancel out. The simplest implementation is ^{1}H–^{15}N TROSY-HSQC. Normally, HSQC transfers the ^{1}H magnetization to ^{15}N via J-coupling. In HSQC, the two components of the J doublet are mixed together (in a process called *decoupling*), during both t_1 and t_2. In TROSY-HSQC the two components are treated separately, and the sequence is built to select only one of the two at the end: the slower-relaxing one. In this way, ^{15}N-labeled proteins up to 100 kDa can be observed.

The CRINEPT (cross-correlated relaxation-enhanced INEPT) building block pushes this limit further.[5] CRINEPT is a modified version of INEPT which exploits the cross-correlated relaxation, instead of J-coupling, to achieve polarization transfer. The CRINEPT delay is much shorter than the INEPT one, and allows enough magnetization to survive be transferred to the ^{15}N and back, despite the short T_2 of the

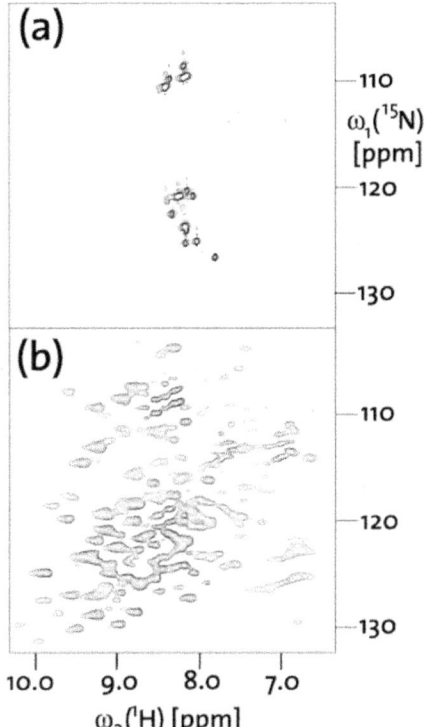

Fig. 5.10 Comparison between a conventional ^{1}H–^{15}N HSQC (a) and a ^{1}H–^{15}N CRINEPT–TROSY (b) acquired on the 800-kDa GroEL protein. Reproduced with permission from Riek et al.[6] Copyright 2002 American Chemical Society.

system. CRINEPT is usually used in combination with the TROSY effect to detect signals of systems up to 1 MDa in size (Fig. 5.10).[6]

Heteronuclear-only experiments can also be applied to overcome the size limit. This is because R_1 and R_2 relaxation rates are a function of γ^2. As an example, a 2D ^{13}C-^{13}C NOESY spectrum of ferritin, a 480 kDa 24-mer, provides good S/N ratio and sharp lines, whereas its ^{1}H counterpart would be totally empty.[7]

5.2.2 Increasing the speed of NMR

A drawback of multidimensional NMR experiments is the long experimental time needed to record them. As NMR is intrinsically not that sensitive, multiple scans of 1D experiments have to be recorded and combined. A delay has to be put between two scans to allow longitudinal relaxation to re-establish the equilibrium population. For larger systems, the T_1 constant which mediates longitudinal relaxation increases and longer *inter-scan delays* (up to several seconds) become necessary. To make things worse, multidimensional experiments are constructed by a series of 1D scans, at incremental indirect evolution delays. Recording a 1D spectrum may take a few seconds, but conventional 2D spectra require hours and 3D spectra days. 4D spectra would require weeks, 5D years and so on...! Several strategies have been developed to circumvent these limitations, and can be combined.

A conventional 3D experiment requires a FID to be recorded (t_3) for each pair of increments in the indirect dimensions (t_1, t_2). These pairs form a grid in the 2D space of indirect dimensions. A 4D experiment requires a lattice in the 3D space of (t_1, t_2, t_3) indirect dimensions to be sampled. This leads to an exponential increase in experimental time. A strategy for reducing this experimental time is to decrease the number of sampled points in the indirect dimensions. There are several methods for achieve this, which are generally called *sparse-sampling* methods. One way is to radially sample the indirect dimension. Doing this for a 3D experiment would produce a series of 2D spectra, which are literally *projections* of the 3D spectrum along different angles that can be used to reconstruct the original spectrum.[8-11] Another method is to do *random sampling* (or *non-uniform sampling*, NUS), in which a series of randomly distributed points is sampled during the experiment. The full spectrum can then be directly reconstructed using appropriate algorithms (Fourier transform is not the most suitable in this case) such as MaxEnt[12,13] or MDD.[14,15]

A conceptually different strategy, which has turned out to be extremely valuable for real-time applications and diluted samples, aims to increase the sensitivity per time unit of a NMR experiment by reducing the inter-scan delays. To achieve high sensitivity, the so-called *fast-pulsing* methods mainly rely on two effects: *longitudinal relaxation enhancement* (LRE) and *Ernst angle excitation*. LRE is achieved by reducing the effective T_1 of the detected spins. A lower T_1 means that the spins will reach equilibrium earlier and the pulse sequence can be repeated at a higher frequency. Dipolar couplings are the main mechanism for longitudinal relaxation of ^{1}H, and as a consequence the relaxation rate of each nucleus will depend on the spin state of the neighboring nuclei. If the neighbors are kept at thermal equilibrium, the nucleus will dissipate its energy more efficiently and will relax faster. Therefore, T_1 can be effectively reduced by selectively exciting only the spins of interest. This is easily done in practice by replacing *hard* excitation pulses (e.g., short square pulses) with longer, *soft* pulses properly *shaped* (i.e., with a given amplitude modulation which translates in a uniform narrow excitation window). Amide-selective excitation schemes have been implemented in the triple resonances: the so-called BEST (band-selective excitation short-transient) sequences can substantially improve the S/N ratio for folded proteins at high magnetic fields compared with standard experiments. When working at high repetition rates, another trick can be added to further increase the S/N ratio: Ernst angle excitation. It can be shown that, for maximum sensitivity, the optimal flip angle of the excitation pulse is a function of the recovery time T_{rec} and of the effective T_1:

$$\cos \beta^{\mathrm{opt}} = e^{-T_{\mathrm{rec}}/T_1}.$$

Therefore, proper adjustment of the excitation pulse (this is especially true for simple pulse sequences) will result in greater sensitivity. The SOFAST–HMQC sequence

(HMQC is a simpler version of HSQC) combines Ernst angle excitation with LRE and it is now broadly applied to highly diluted solutions, in-cell NMR experiments and for real-time monitoring of conformational changes on the seconds time scale.

5.2.3 Carbon-detected NMR

The classical sets of heteronuclear NMR experiments used for protein resonance assignment and/or structure calculation all rely on excitation and detection of ^{1}H resonances. Setting aside historical reasons, this is mainly due to the higher sensitivity of ^{1}H, which is usually looked for when dealing with protein samples. However, in some cases the detrimental effect of γ_H on the relaxation properties of the system override the benefit of higher polarization. Thanks to technological advancements—higher magnetic fields and more sensitive electronics (cryogenically cooled probes)—the sensitivity of ^{13}C NMR has increased by an order of magnitude, allowing the use of carbon-detected experiments on ^{13}C-labeled proteins. In the last decade many sequences have been developed which are based on ^{13}C detection, and some of them do not excite ^{1}H spins at all (so-called *protonless* experiments).

^{13}C-based experiments offer several advantages when studying biological molecules. ^{13}C nuclei suffer less from dipolar relaxation, thanks to the lower γ. This translates into sharp linewidths even for large systems for which ^{1}H resonances start to broaden out; for the same reason, ^{13}C resonances close to paramagnetic centers relax more slowly than ^{1}H and can be used to study paramagnetic centers in metalloproteins.[16,17] Carbon atoms are unaffected by chemical exchange, unlike many hydrogen atoms such as amide ^{1}H. Finally, ^{13}C chemical shifts have a larger dispersion in frequency than ^{1}H (not only in ppm, but also in Hz). Importantly, some types of ^{13}C nuclei in a protein are closely grouped together, and are well separated from other ^{13}C resonances (e.g., C' resonances lie between 170 and 185 ppm, very far from Cα, Cβ and aliphatic carbons). As a consequence of this, NMR experiments have been designed which selectively manipulate the different types of ^{13}C nuclei, by means of shaped pulses and appropriate J-transfer delays, so that these ^{13}C can be *almost* considered as independent channels in the spectrometer.

A field in structural biology where ^{13}C-detected NMR experiments are ideally employed is the study of *intrinsically disordered proteins* (IDPs).[18–21] IDPs are proteins, or parts of longer constructs, which do not have a stable tertiary structure in their native state. IDPs in eukaryotic cells often harbor linear recognition motifs that are recognized by many cellular partners and can have regulatory functions. Due to the dynamic nature of such proteins, solution NMR is probably the most suitable technique for atomic-resolution studies of IDPs. However, the classical ^{1}H-centered approach for protein NMR often fails when working with IDPs. The main reason for this is the severe signal overlap: in unstructured polypeptides, the ^{1}H$_N$ chemical shift dispersion is drastically decreased, and the Cα and Cβ chemical shifts depend almost entirely on the residue type. Another reason is the higher rate of H$_N$ chemical exchange in polypeptides. ^{13}C-based experiments alleviate this problem; so, for example, a ^{13}C-detected CON experiment (which correlates backbone ^{13}C' and ^{15}N; see Fig. 5.11) has much higher resolution and chemical shift dispersion than the corresponding ^{1}H–^{15}N HSQC

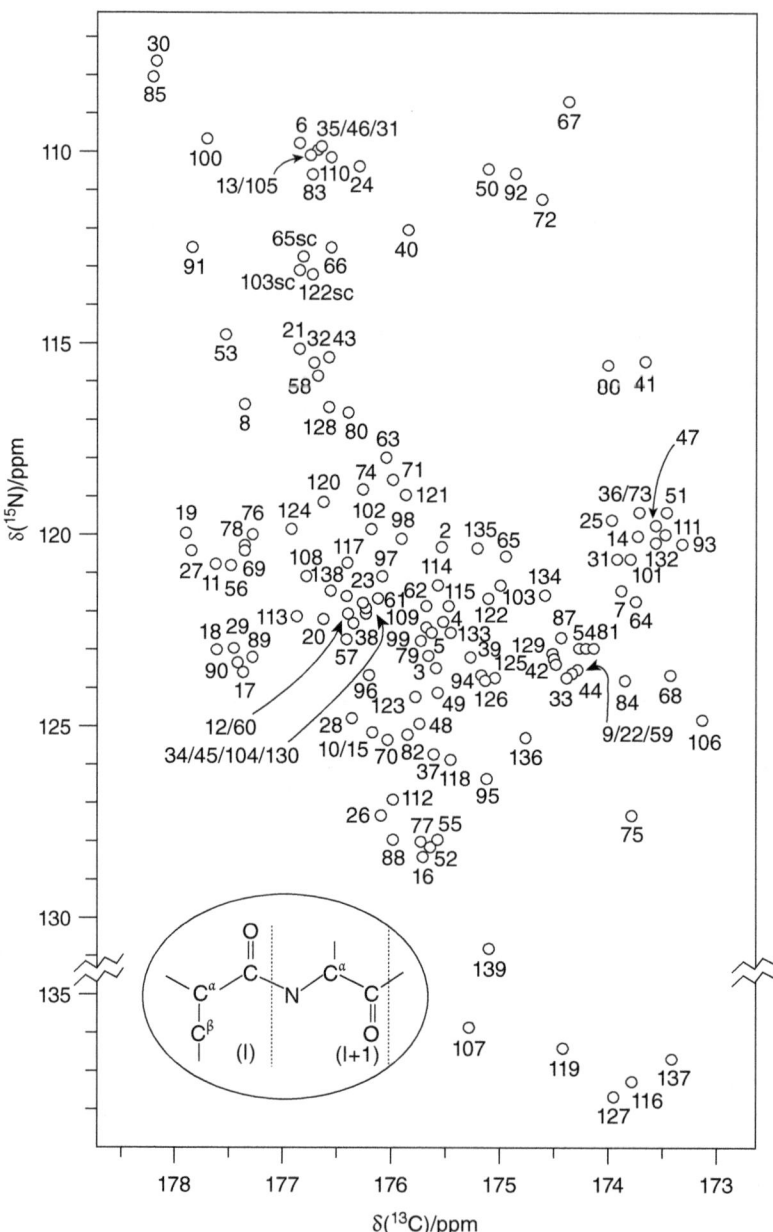

Fig. 5.11 Chemical shift plot reconstructed from a ^{13}C-detected ^{13}C–^{15}N correlation spectrum of an IDP. Adapted with permission from Bermel et al.[18] Copyright 2006 American Chemical Society.

of the same protein. A large set of multidimensional ^{13}C-detected experiments has been developed, allowing protonless assignment strategies for backbone and side-chain ^{13}C and ^{15}N.

The sensitivity lost in protonless experiments can be partially recovered by implementing *H-start* variants of ^{13}C-detected experiments.[22,23] In this case, ^{1}H (which can be either amide or aliphatic) is used as a source of magnetization, and is transferred to ^{13}C at the beginning of the pulse sequence, providing an increase in S/N noise ratio. This strategy is usually best suited for small, non-deuterated proteins.

5.2.4 NMR of paramagnetic molecules

Many proteins coordinate one or more metal ions. Metal ions in living organisms are essential for ensuring the correct folding of proteins, providing the catalytic activity of some classes of enzymes and modulating complex electron transfer reactions. Some metal ions are paramagnetic, i.e., they have one or more unpaired electrons in their outer shell. Paramagnetic metal ions can have profound effects on the NMR spectra of proteins. While these effects may seem disruptive at a first glance, they actually provide several unique restraints for characterizing protein structure and dynamics. To observe paramagnetic effects the native metal cofactor can be used if it is paramagnetic (e.g., Cu^{2+} in Cu,Zn-superoxide dismutase 1), or it can be substituted with a chemically analogous metal ion which is paramagnetic (e.g., diamagnetic Ca^{2+} can be substituted with many paramagnetic lanthanide ions). The latter strategy provides two structurally identical samples that can be compared to accurately measure the paramagnetic effects.[24,25]

The main effects of a paramagnetic metal ion, compared with a diamagnetic control, are: a change in the position of some peaks, which are usually more dispersed in the spectrum; a variable broadening effect on the peaks; and the disappearance of some other peaks. The magnitude of these effects depends on the distance of each atom from the metal ion, and from the latter's paramagnetic properties. Therefore, different paramagnetic ions will have different effects, and may provide non-redundant information, thus complementing each other.[26]

The change in chemical shift due to a paramagnetic center is called the *hyperfine shift*, and is composed of two contributions: the *contact shift* and the *pseudocontact shift* (PCS). The hyperfine shift is caused by interactions between the nuclear magnetic moment and the average magnetic moment of the unpaired electron, induced by the external magnetic field. The contact shift is caused by the presence of non-zero unpaired electron spin density at the nucleus. This is the case if the molecular orbital occupied by the electron has partial s character. Furthermore, molecular orbitals with two paired electrons can also be influenced by the unpaired electron, by the *spin polarization* effect. In either case, the presence of a contact shift implies a through-bond connectivity between the atom and the paramagnetic ion. A PCS is caused when the dipolar coupling between the nuclear magnetic moment and the electron magnetic moment does not average to zero upon rotation of the molecule in solution. The non-zero averaging is due to anisotropy of the electron magnetic moment. The magnetic susceptibility tensor χ describes how the electron magnetic moment changes when the molecule rotates with respect to the magnetic field. The PCS of a given nucleus therefore depends on its distance from the paramagnetic metal and its angle with respect to the axes of the tensor χ:

$$\delta^{pcs} = \frac{1}{12\pi r^3}\left[\Delta\chi_{ax}(3\cos^2\theta - 1) + \frac{3}{2}\Delta\chi_{rh}\sin^2\theta\cos 2\varphi\right]$$

where the elements $\Delta\chi_{ax} = \chi_{zz} - (\chi_{xx} + \chi_{yy})/2$ and $\Delta\chi_{rh} = \chi_{xx} - \chi_{yy}$ reflect the anisotropy of the tensor χ. PCSs do not depend on the magnetic field, nor on the type of nucleus. Plotting the iso-PCS surfaces (Fig. 5.12) for some tensors shows that infinite atomic coordinates will give the same PCS. However, in principle, PCSs alone from two or more different paramagnetic metals ions would provide a unique molecular structure. In reality, PCSs are more usefully employed as additional restraints when solving a structure, or to refine an existing one. They also provide a valuable source of data when studying the relative orientation of two protein domains connected by a flexible linker.

The line broadening effect in the paramagnetic spectrum and the disappearance of crosspeaks arise from increased nuclear relaxation rates induced by the paramagnetic ion. This effect is measured as paramagnetic relaxation enhancement (PRE). The PRE depends on the metal-nucleus distance as $1/r^6$ and is caused by the dipolar interaction of a nucleus with the unpaired electron and with its associated averaged induced electron magnetic moment. The interaction is modulated by molecular tumbling in solution and by the electron relaxation time. Due to the strong dependence on the distance from the paramagnetic metal ion, the PRE in practice causes the disappearance of signals from nuclei in a "blind sphere" around the metal ion, and is large enough to be measurable within a spherical shell centered around the metal ion (Fig. 5.13).

Finally, additional structural restraints come from residual dipolar couplings (RDCs). RDCs are "classical" restraints which become available when the molecule in solution is partially oriented by the use of orienting media (such as bicelles or stretched gels). The partial orientation gives rise to a non-zero averaging of the dipolar coupling between neighboring nuclei, and causes a net change in frequency of the J-splitting (which is normally caused by the scalar coupling only). By recording non-decoupled variants of heteronuclear spectra such as the HSQC, the J-splitting can be measured for each crosspeak and the RDCs can be calculated by difference from a non-oriented control sample. Paramagnetic RDCs (PRDCs) are caused by the

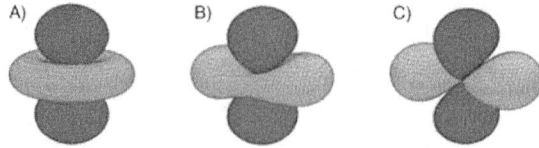

Fig. 5.12 Iso-PCS surfaces calculated from the equation above for tensors with different anisotropy. Reproduced from Bertini et al.,[26] with permission from John Wiley and Sons.

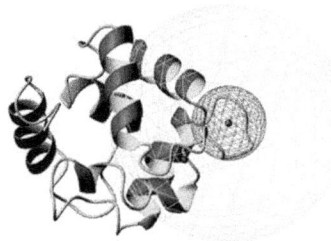

Fig. 5.13 Blind spheres centered around a paramagnetic metal ion for ^{1}H and ^{13}C-detected NMR experiments. Reproduced from Bertini et al.,[26] with permission from John Wiley and Sons.

same magnetic susceptibility anisotropy responsible for PCSs. Indeed, anisotropy in the magnetic susceptibility causes a partial orientation of the molecule in strong magnetic fields, without an external orienting medium. PRDCs for two coupled nuclei (e.g., amide ^{1}H–^{15}N) are given by a formula which has the same functional form as that for PCSs, but PRDCs are independent of the distance from the metal, and the angular dependence is given by the orientation of the H–N vector with respect to the tensor frame:

$$\Delta v^{\mathrm{PRDC}} \propto -B_0^2 \left[\Delta \chi_{ax} \left(3\cos^2 \Theta - 1 \right) + \frac{3}{2} \Delta \chi_{rh} \sin^2 \Theta \cos 2\Phi \right].$$

5.2.5 Solid-state NMR: a brief overview

The principal limit of solution NMR is the molecular weight. That is because the larger a molecule is, the slower the tumbling rate it will have in solution (unfolded proteins such as IDPs are an exception). Therefore, large systems such as multiprotein complexes, membrane-associated proteins, oligomers and fibrils cannot be investigated by solution NMR. In recent years, huge technical advances have made it possible to characterize these systems by solid-state NMR. Here, the effects on the nuclear spins in the solid state are summarized, and the basic steps in magic angle spinning solid-state NMR are summarized.

Following from Section 5.1, nuclear spins ($I = 1/2$) of a molecule in a magnetic field undergo many effects: interaction with the external field, chemical shift (i.e., chemical shielding), scalar coupling and dipole–dipole coupling. The chemical shift and the dipolar coupling interactions are not isotropic and depend on the orientation of the molecule with respect to the magnetic field (i.e., they can be described with *tensors*). If the molecule tumbles rapidly, as in a solution, these quantities are averaged over the whole solid angle: the chemical shift becomes a scalar quantity, and the dipolar interaction averages to zero. Instead, if the molecule does not rotate with respect to the magnetic field, as in a solid, both chemical shift and dipolar coupling will contribute to the Larmor frequency of each nucleus in an orientation-dependent way. Thus, in a powder sample, the lineshape of each spin will be the sum of the resonances at all possible orientations. The magnitude of the *chemical shielding anisotropy* (CSA) and of the dipolar coupling are such that each lineshape will cover most of the NMR spectrum, in fact preventing data interpretation.

Magic angle spinning (MAS) is a method that is now widely used for protein solid-state NMR, and partly recovers the narrow spectral lines typical of solution NMR. In practice, the sample inside the magnetic field is rotated rapidly about an axis tilted by the *magic angle*:

$$\beta = \theta_m = \arccos\left(\frac{1}{\sqrt{3}}\right) \approx 54.74°$$

with respect to the axis of the external magnetic field. If the spinning rate is fast enough, with respect to the magnitude of the anisotropic interactions, the CSA and the dipolar couplings are averaged out, in analogy to what happens in solution. Under MAS, the static lineshapes break up into a series of sharp peaks separated by the spinning frequency. The larger the MAS frequency (v_R), the larger the separation; when v_R is higher than the static linewidth, a single peak is obtained.

MAS solid-state NMR is especially powerful when studying protein samples which cannot be crystallized. Fibrillar structures are one example (Fig. 5.14).[27–30] Additionally, it was recently found that sedimented proteins can be successfully studied; these have the main advantage that the rotor can be easily packed starting from a solution of the protein.[31,32] Packing can be done either directly in the instrument under MAS or by means of a special device which fits the rotor into an ultracentrifuge bucket.[33]

Due to the large dipolar coupling between ^1H nuclei, proton-detected experiments are only useful at ultra-high MAS ($v_R \geq 60\,kHz$), and at lower MAS, NMR studies of proteins usually rely on heteronuclear ^{13}C-detected experiments. Many sequences have

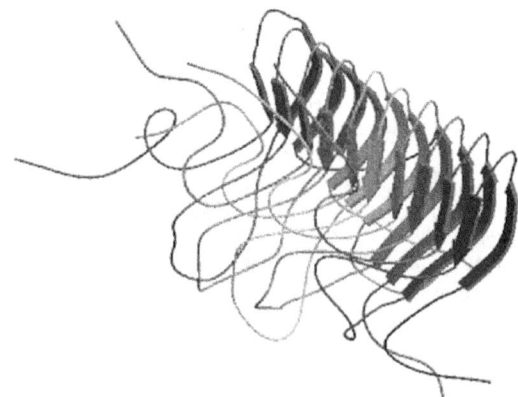

Fig. 5.14 Fibrillar structure of the HET-s prion solved by solid-state NMR (PDB: 2RNM).[28] Reproduced with permission from Wang et al.[42] Copyright 2011 American Physical Society.

been designed for MAS solid-state NMR that achieve polarization transfers between nuclei and allow the recording of multidimensional spectra conceptually similar to those used in solution NMR. However, they are made of building blocks which exploit different phenomena from their solution counterparts. To achieve initial polarization transfer between ^{1}H and ^{13}C, for example (for sensitivity enhancement, equivalent to an INEPT block in solution), the *cross-polarization* (CP) technique is used. CP exploits heteronuclear dipolar coupling and employs simultaneous RF irradiation on both ^{1}H and ^{13}C, so that the nutation frequency along the RF field axis is the same for both nuclei. In this condition, polarization transfer via simultaneous spin transitions of the two nuclei is achieved. The mixing part of these heteronuclear spectra can exploit different effects, and several implementations exist. Among the most commonly used are *proton-driven spin diffusion* (PDSD) and *dipolar-assisted rotational resonance* (DARR), which both allow polarization exchange between ^{13}C nuclei by means of the dipolar couplings with ^{1}H nuclei that are close in space. These mixing sequences allow the identification of the ^{13}C spin system of each residue; the resulting 2D spectrum is conceptually similar to a ^{13}C–^{13}C TOCSY in solution.

With the same principles, CP blocks and mixing sequences can be combined to transfer polarization between ^{13}C and ^{15}N. Thus, pulse sequences have been designed which resemble 2D or 3D carbon-detected solution experiments, such as N–C′, N–Cα, and more complex transfers (e.g., N–Cα–Cβ, N–C′–Cα–Cβ) which can be successfully used to obtain residue-specific resonance assignments that are the basis for protein structure characterization by solid-state NMR.

5.3 In-cell NMR: towards integrated structural cell biology

5.3.1 Why in-cell NMR?

The classical approaches of structural biology, including solution NMR of purified proteins, are highly reductionist. Usually, complex biological systems are decomposed into their minimal components (most often macromolecules), the underlying mechanisms are reconstructed, usually by analyzing pair-wise interactions, and finally the information is added up to try to explain the emergent properties observed in the whole system. Such approaches, albeit extremely powerful and still largely necessary, struggle to understand how macromolecules really behave—in a structural and dynamic sense—and exert their function inside the cellular environment.

In-cell NMR spectroscopy tries to answer some of these open questions. It is a one-of-a-kind approach in that it combines the ability of solution NMR to obtain atomic-resolution data with the possibility of investigating biological macromolecules within the cell, i.e., very close to their true physiological condition. The approach is relatively young, and it has shown a lot of potential for understanding how the cellular environment affects protein folding and function and for investigating critical steps in protein maturation, e.g., protein phosphorylation, metal binding and redox-induced folding events, many of which are dependent on the cellular compartment and may rely on specific partners and molecular chaperones.

5.3.2 In-cell versus *in vivo* NMR

It is worth mentioning that other intracellular NMR studies had been carried out before the in-cell NMR technique was developed. These studies usually relied on isotopic labeling of delivered molecules, and were principally aimed at understanding specific metabolic pathways, without altering the concentration of any particular metabolite. Therefore, these approaches—which can be generally referred to as *in vivo* NMR—are not oriented to structural biology. Instead, in-cell NMR is a structural technique focused on one macromolecule (e.g., a protein), which is increased in concentration and selectively labeled with NMR-active isotopes so that it can be characterized at high resolution within a cell.[34]

Exclusively in-cell NMR-based protein characterization studies can be pursued, and in-cell protein structure determination has been demonstrated from scratch.[35] However, a hybrid approach is sometimes preferred for in-cell NMR, which consists in doing comparative analysis of the data collected in-cell with *in vitro* samples of the same protein(s) in various well-defined conditions. In this way, the structural features can be conveniently extracted by conventional solution NMR techniques and are then used to complement and interpret the intracellular data.

5.3.3 Effects of the intracellular environment on the protein

The intracellular environment can affect the protein in several non-trivial ways. The cytosol, for example, is a complex mixture containing around 300 g/l of proteins, inorganic ions such as K^+ and Mg^{2+}, glutamate, glutathione and a plethora of less abundant molecules. It has pH close to 7.0, and has reducing properties ($E_0 \cong -290\,\mathrm{mV}$). While the cytosolic buffering effects can be simulated *in vitro* to some extent, the high concentration of proteins gives rise to *macromolecular crowding* effects, which are difficult to reproduce *in vitro*.[36] Thermodynamically, the crowding is generally thought to have an excluded-volume effect on proteins, which should stabilize the folded state with an entropic contribution. However, studies have shown that an enthalpic contribution can also occur, which destabilizes the folding state due to favorable interactions between an exposed hydrophobic core and the crowders.[37,38]

On top of these effects, which are generally sequence-independent, there are protein-dependent interactions with cellular components (e.g., other proteins, membranes, DNA, etc.). The affinities for all these interacting molecules are likely to differ by many orders of magnitude: interactions with specific partners may by stronger, while other interactions could be extremely weak. However, the sum of many weak interactions is likely to cause drastic effects when observing a protein by in-cell NMR.[39]

5.3.4 Measurable effects on the NMR spectra

The measurable effects of the intracellular environment on the protein NMR resonances can be changes in chemical shift, changes in signal intensity/linewidth or complete signal loss.[40]

Changes in chemical shift reflect conformational changes of the protein, and are usually the effects most often studied. Changes in chemical shift can be mapped residue-by-residue on the protein structure, and may arise from effects of the environmental properties such as pH and ionic strength, as well as from interactions with soluble (i.e., free-tumbling) partners, small molecules or metal binding, post-translational modifications.

Changes in signal linewidth point to different relaxation properties of the resonances. Commonly, a general line broadening is observed for all the signals of a folded protein, and is a direct consequence of a decreased tumbling rate (Fig. 5.15). The extent of this phenomenon is strongly protein-dependent, and does not clearly correlate with the molecular size.[41,42] Transient interactions with other cellular components are thought to cause the reduced rotational mobility. In the worst cases, the tumbling rate of the protein will decrease to a point that all its signals will be broadened beyond detection. This effect, sometimes referred to as protein *stickiness*, has by now been observed for many globular proteins and is not yet clearly explained. In most cases, cell lysis (and consequent dilution of the intracellular solution) breaks those interactions, and the protein tumbles freely as in pure buffer. Some hypothesize that the protein *quinary structure* is responsible of this effect, and that it has biological significance as a means to regulate protein function.[43]

The line broadening effects for IDPs are less drastic, due to their unstructured nature. In IDPs, weak interactions would cause line broadening or disappearance of the signals directly involved in the interactions without affecting the rest of the polypeptide chain, thereby providing valuable information on the protein-interacting sites.

5.3.5 Isotopic labeling strategies

Isotopic labeling is usually necessary for in-cell NMR as a way of creating *contrast* between the protein of interest and the other cellular components. The feasibility of many labeling schemes for in-cell NMR has been evaluated, and some strategies were found to perform better in the cellular environment.[44,45] Uniform labeling of the protein is usually desirable to easily characterize protein conformation and dynamics from heteronuclear correlation spectra. However, uniform ^{13}C labeling is not amenable to in-cell studies, for two main reasons: the high enough natural abundance of ^{13}C and the high carbon content of all biomolecules. Those two effects cause signals of abundant cellular components (e.g., membranes) to mask out any useful signal from our ^{13}C-enriched protein. Unlike ^{13}C, ^{15}N uniform labeling is usually preferred, as the natural abundance of ^{15}N is much lower, and nitrogen is not so widely distributed in biomolecules other than proteins. Therefore, ^{1}H–^{15}N NMR spectra are quite clean and can be easily analyzed. Amino acid-selective labeling strategies can also be successfully employed. ^{15}N labeling on selected amino acid types is often useful to facilitate the interpretation of in-cell NMR data.[46] Methyl–^{13}C methionine labeling has also been successfully employed, thanks to the spectral distribution and the low isotopic scrambling of the methyl group.[47]

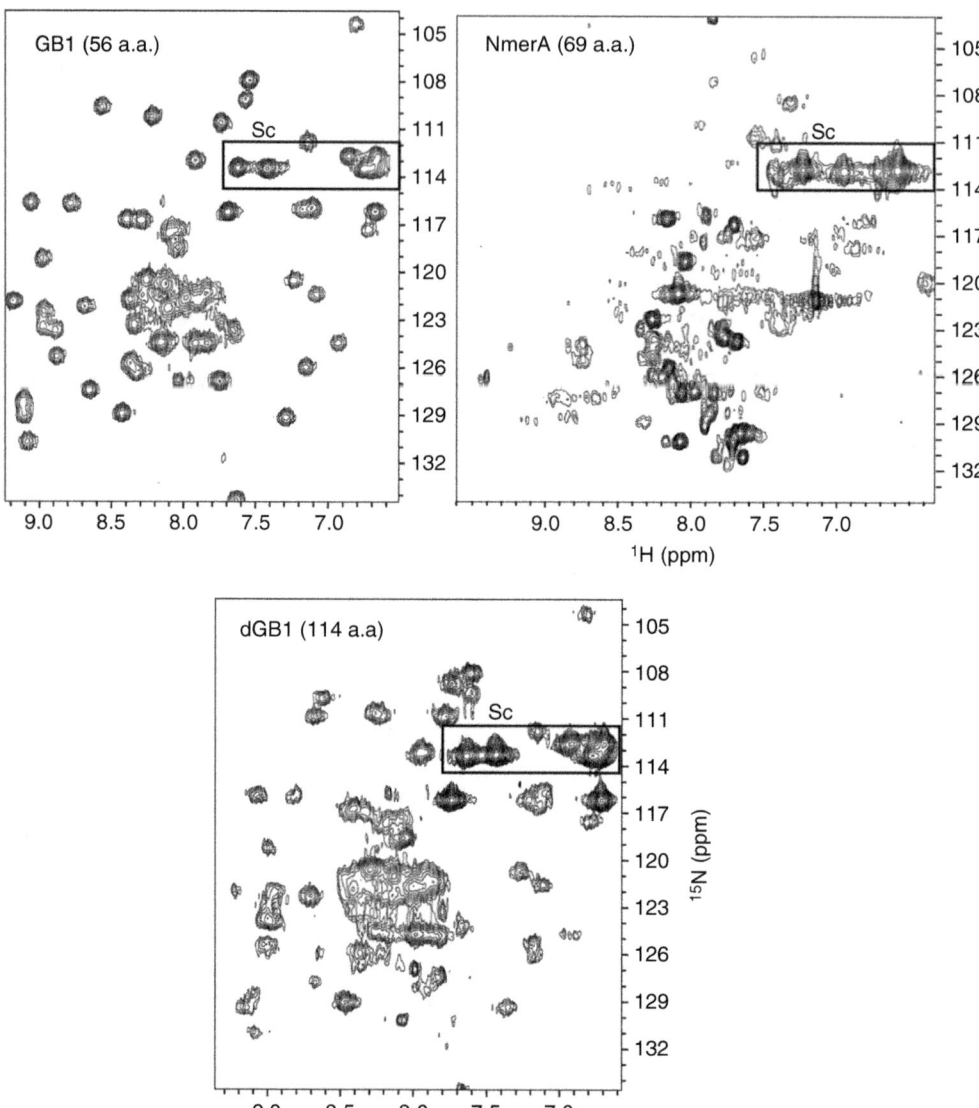

Fig. 5.15 Globular proteins can have different in-cell NMR relaxation properties: the signals of NmerA (center) are broader than those of dGB1 (right). Reproduced with permission from Wang et al.[42] Copyright 2011 American Physical Society.

5.3.6 Overview of the existing techniques

Several techniques have been developed for performing in-cell NMR experiments that allow the study of proteins in different organisms and use different approaches to insert a labeled protein inside the cells.

The first technique to be developed was NMR on bacterial cells.[34] Strains of *Escherichia coli* are commonly used, and the protein is overexpressed within the cells by means of recombinant DNA technology. Selective isotopic labeling of the protein is achieved by replacing the unlabeled medium used to grow the cells with isotopically labeled minimal medium at the time of induction of protein expression. This approach has the advantage that is relatively easy to implement in a laboratory for recombinant protein expression, which is the most common way of producing *in vitro* protein NMR samples. Furthermore, many advanced biotechnological techniques for protein expression can be exploited for complex in-cell NMR studies. As an example, a strategy was developed to sequentially express two (or more) proteins in a controlled manner, using independent induction systems so that only one protein at a time can be selectively labeled and easily observed in the NMR experiments. This strategy, called STINT-NMR, has been successfully employed to investigate protein–protein interactions in *E. coli* cells.[48–50]

Many approaches were developed for use in eukaryotic cells. Yeast cells, for example, can be used for in-cell NMR in a similar way to *E. coli*, employing the existing technology for protein expression in yeast cells.[51] For higher eukaryotes, the approaches are different. A first strategy is to use *Xenopus laevis* (African clawed frog) oocytes.[52] These oocytes are quite large (1 mm in diameter), and can be microinjected with a concentrated solution of labeled protein (Fig. 5.16). The protein inside the oocytes can then be observed by NMR. It has been reported that also protein expression in insect cells can be used for in-cell NMR purposes. DNA is delivered to insect cells by a baculovirus infection system.[53]

Other strategies have been developed to perform in-cell NMR on cultured mammalian cells. A few of them rely on protein insertion, while protein overexpression in cultured human cells was recently successfully employed for in-cell NMR. A first approach for protein insertion relied on the HIV1-TAT *cell-penetrating peptide* (CPP) N-terminally fused to the protein (or chemically linked via a S–S bond to a cysteine residue). The CPP fuses with the plasma membrane and allows the fused protein to

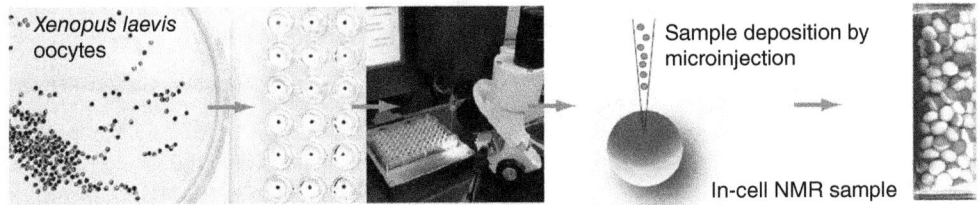

Fig. 5.16 In-cell NMR in *X. laevis* oocytes. Reproduced by permission from Macmillan Publishers Ltd: D.S. Burz, A. Shekhtman, *Nature*, 458, 2009. Copyright 2009.

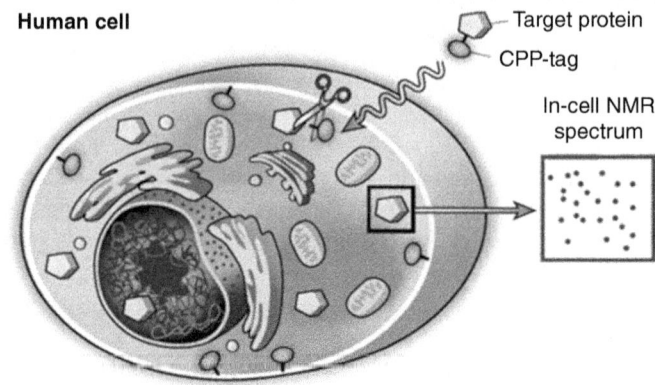

Fig. 5.17 Protein insertion via a CPP-tag. Reproduced by permission from Macmillan Publishers Ltd: D.S. Burz, A. Shekhtman, *Nature*, 458, 2009. Copyright 2009.

translocate inside the cell. A few examples exist where a CPP tag successfully has delivered a protein inside human cells (Fig. 5.17).[54,55] However, the import efficiency is strongly protein-dependent. Labeled proteins can also be delivered to human cells by means of pore-forming toxins. In one example, Streptolysin O was employed to reversibly permeabilize the plasma membrane, so that the protein could translocate inside the cells.[56] Recently, it has been shown that proteins can be efficiently delivered to different cell lines by means of *cell electroporation*. Electroporation is commonly used for *cell transfection*, i.e., to deliver exogenous DNA to the cells. A strong pulsed electric potential is generated in the cell suspension, and holes are formed on the cell surface. By tuning the strength and number of the pulses, proteins can be also efficiently delivered while preserving cell viability.

5.3.7 Protein expression in mammalian cells: applications

Protein expression in cultured human cells can also be applied to in-cell NMR.[57] This approach is conceptually more similar to how in-cell NMR is performed in bacterial and yeast cells. In human cells, protein expression is achieved by transiently transfecting cells with the exogenous DNA encoding the protein of interest. An efficient and cost-effective transfection system is used, which relies on a DNA:polyethylenimine complex, so that a high number of cells ($>10^7$) can be transfected with high DNA copy number per cell (Fig. 5.18). The DNA vector contains a strong constitutive promoter, so that protein synthesis quickly starts once the DNA is internalized. As with *E. coli* and yeast cells, protein labeling is achieved by switching from an unlabeled growth medium to an isotopically enriched one. In this case, however, a complete labeled medium is necessary (i.e., one containing all the essential amino acids). The main strength of protein expression for NMR in human cells is that folding and protein maturation events can be monitored starting from the newly synthesized polypeptide.

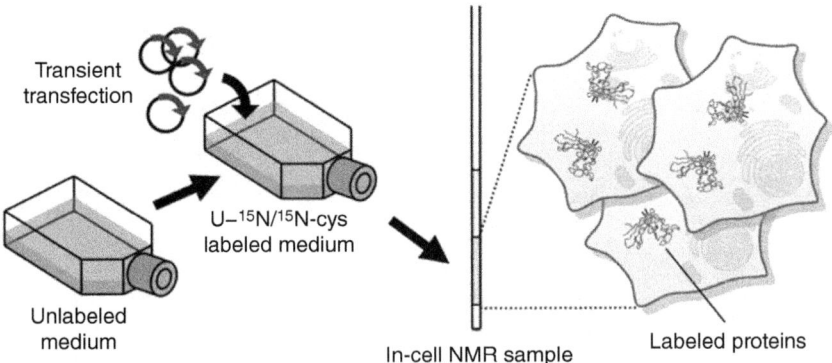

Fig. 5.18 Protein expression and labeling in human cells for in-cell NMR.

With this approach, the complete sequence of events leading to mature human copper, zinc superoxide dismutase 1 (Cu,Zn-SOD1) was characterized in human cells.[57] SOD1 is a 32-kDa homodimeric enzyme which protects cells from oxidative stress. To reach the mature active form, apo-SOD1 has to bind one zinc and one copper ion, dimerize and form an intrasubunit disulfide bond. Mutations of the *SOD1* gene have been linked to the familial form of amyotrophic lateral sclerosis, a degenerative motor neuron disease, and are thought to cause the incomplete maturation of SOD1 by destabilizing the apo state. The wild-type protein was expressed in HEK293T cells and the intermediate steps of maturation could be observed by treating the cells with different quantities of zinc and copper ions. Each step was characterized by comparing the in-cell NMR data, obtained with both uniform ^{15}N and [^{15}N]-cysteine labeling, with *in vitro* samples of SOD1 in various metallation and redox states. Zinc treatment caused apo-SOD1 to bind one zinc ion stoichiometrically (with higher-than-*in vitro* selectivity towards the zinc-binding site), and to dimerize. Copper treatment alone was not sufficient to produce the mature protein, due to the tightly controlled copper homeostasis of the cell. The role of the specific copper chaperone for SOD1 (CCS) was then investigated in-cell, by simultaneous expression of both proteins. Upon CCS co-expression, complete maturation of SOD1 was observed (i.e., Cu-CCS-dependent copper delivery and cysteine oxidation). Additionally, a previously unreported effect of Cu-independent, CCS-dependent cysteine oxidation was observed.

Redox-dependent protein folding can also be investigated by NMR of intracellularly expressed proteins. The folding and redox state of the mitochondrial protein Mia40 was characterized in the cytoplasm prior to mitochondrial import.[58] Mia40 is a small oxidoreductase of the intermembrane space (IMS) of mitochondria. To reach the IMS, Mia40 has to cross the outer mitochondrial membrane in an unfolded, reduced state; in the IMS, it obtains its final structure with the formation of two structural disulfide bonds. When overexpressed in human cells, Mia40 accumulated in the cytoplasm. By in-cell NMR, we found that, remarkably, this cytoplasmic form of Mia40 was fully folded and oxidized. Additionally, Mia40 folding and its redox state were found

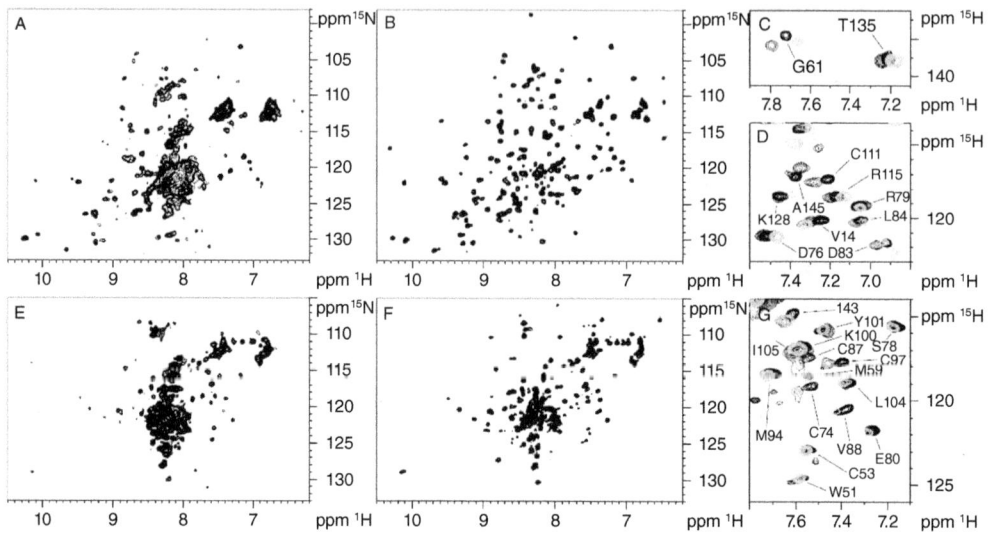

Fig. 5.19 In-mitochondria NMR spectra of proteins localized in the intermembrane space of intact mitochondria. Reproduced from Barbieri et al.,[59] with permission from Elsevier.

to be dependent on the levels of the cytoplasmic thiol-disulfide regulating proteins, glutaredoxin 1 and thioredoxin 1. Indeed, increased levels of either of these proteins allowed Mia40 to remain mostly in the unfolded, reduced state, which is import-competent.

Finally, intracellular expression can be used to target proteins to specific organelles. Intact organelles containing sufficient amounts of labeled protein can be then isolated to perform *in organello* NMR experiments. This approach was applied to characterize SOD1 and Mia40, previously observed in the cytoplasm, in the IMS of intact mitochondria (Fig. 5.19).[59] For this purpose, the desired protein localization was obtained by fusing the proteins to a N-terminal mitochondrial targeting sequence (MTS), specific for IMS localization. MTS-tagged proteins are mainly localized in the mitochondria after protein synthesis, and do not accumulate in the cytosol. In principle, analogous approaches would allow characterizing proteins in the endoplasmic reticulum, or in the Golgi network.

5.3.8 Pushing the limits: in-cell solid-state NMR

One of the intrinsic limits of in-cell NMR is the unexpected behavior of many soluble, globular proteins, which in the cellular environment are "sticky" and tumble so slowly that they cannot be detected by solution NMR. For these proteins, intracellular characterization may still be feasible by means of MAS solid-state NMR. It is a challenge to achieve proper in-cell solid-state NMR as many technical problems arise. Mainly, cell integrity has to be preserved under MAS: the radial acceleration inside the rotor at high MAS can be enough to destroy the cells. There are additional problems of sensitivity due to the maximum effective concentration of the labeled protein inside the rotor being much lower than that of a pure sample. So far there are just two examples of cellular solid-state NMR, both of which make use of *E. coli* cells, which better resist high acceleration, and carried out at very low temperatures thus cryoprotecting the cells.[60,61] For eukaryotic cells, sample preparation techniques similar to those used in electron microscopy need to be developed to preserve sample integrity. For example, chemical or cryo-fixation methods may need to be employed, even though such techniques are not classed as "living cell" techniques.

References

1. Hore, P. J. *Nuclear magnetic resonance*. Oxford University Press (1995).
2. Hore, P. J., Jones, J. J. A. and Wimperis, S. *Nmr: the toolkit*. Oxford University Press (2000).
3. Bertini, I., McGreevy, K. S. and Parigi, G. *NMR of biomolecules: towards mechanistic systems biology*. John Wiley and Sons (2012).
4. Pervushin, K., Riek, R., Wider, G. and Wüthrich, K. Attenuated T2 relaxation by mutual cancellation of dipole-dipole coupling and chemical shift anisotropy indicates an avenue to NMR structures of very large biological macromolecules in solution. *Proc. Natl. Acad. Sci. USA* **94**, 12366–12371 (1997).
5. Riek, R., Wider, G., Pervushin, K. and Wüthrich, K. Polarization transfer by cross-correlated relaxation in solution NMR with very large molecules. *Proc. Natl. Acad. Sci. USA* **96**, 4918–4923 (1999).
6. Riek, R., Fiaux, J., Bertelsen, E. B., Horwich, A. L. and Wuthrich, K. Solution NMR techniques for large molecular and supramolecular structures. *J. Am. Chem. Soc.* **124**, 12144–12153 (2002).
7. Matzapetakis, M., Turano, P., Theil, E. C. and Bertini, I. ^{13}C–^{13}C NOESY spectra of a 480 kDa protein: solution NMR of ferritin. *J. Biomol. NMR* **38**, 237–242 (2007).
8. Szyperski, T., Wider, G., Bushweller, J. H. and Wüthrich, K. 3D ^{13}C–^{15}N-heteronuclear two-spin coherence spectroscopy for polypeptide backbone assignments in ^{13}C–^{15}N-double-labeled proteins. *J. Biomol. NMR* **3**, 127–132 (1993).
9. Simorre, J. P., Brutscher, B., Caffrey, M. S. and Marion, D. Assignment of NMR spectra of proteins using triple-resonance two-dimensional experiments. *J. Biomol. NMR* **4**, 325–333 (1994).
10. Kim, S. and Szyperski, T. GFT NMR, a new approach to rapidly obtain precise high-dimensional NMR spectral information. *J. Am. Chem. Soc.* **125**, 1385–1393 (2003).
11. Kupce, E. and Freeman, R. Projection-reconstruction of three-dimensional NMR spectra. *J. Am. Chem. Soc.* **125**, 13958–13959 (2003).
12. Stern, A. S., Donoho, D. L. and Hoch, J. C. NMR data processing using iterative thresholding and minimum l(1)-norm reconstruction. *J. Magn. Reson.* **188**, 295–300 (2007).

13. Mobli, M., Maciejewski, M. W., Gryk, M. R. and Hoch, J. C. An automated tool for maximum entropy reconstruction of biomolecular NMR spectra. *Nat. Methods* **4**, 467–468 (2007).
14. Orekhov, V. Y., Ibraghimov, I. and Billeter, M. Optimizing resolution in multidimensional NMR by three-way decomposition. *J. Biomol. NMR* **27**, 165–173 (2003).
15. Jaravine, V., Ibraghimov, I. and Orekhov, V. Y. Removal of a time barrier for high-resolution multidimensional NMR spectroscopy. *Nat. Methods* **3**, 605–607 (2006).
16. Bermel, W., Bertini, I., Felli, I. C., Kümmerle, R. and Pierattelli, R. 13C direct detection experiments on the paramagnetic oxidized monomeric copper, zinc superoxide dismutase. *J. Am. Chem. Soc.* **125**, 16423–16429 (2003).
17. Babini, E., Felli, I. C., Lelli, M., Luchinat, C. and Pierattelli, R. Backbone and side-chains ^{1}H, ^{13}C and ^{15}N NMR assignment of human beta-parvalbumin. *J. Biomol. NMR* **33**, 137 (2005).
18. Bermel, W. et al. Protonless NMR experiments for sequence-specific assignment of backbone nuclei in unfolded proteins. *J. Am. Chem. Soc.* **128**, 3918–3919 (2006).
19. Bertini, I., Felli, I. C., Gonnelli, L., Vasantha Kumar, M. V. and Pierattelli, R. High-resolution characterization of intrinsic disorder in proteins: expanding the suite of (13)C-detected NMR spectroscopy experiments to determine key observables. *Chembiochem* **12**, 2347–2352 (2011).
20. Bermel, W. et al. Speeding up sequence specific assignment of IDPs. *J. Biomol. NMR* **53**, 293–301 (2012).
21. Felli, I. C. and Pierattelli, R. Novel methods based on (13)C detection to study intrinsically disordered proteins. *J. Magn. Reson.* **241**, 115–125 (2014).
22. Bermel, W. et al. H-start for exclusively heteronuclear NMR spectroscopy: the case of intrinsically disordered proteins. *J. Magn. Reson.* **198**, 275–281 (2009).
23. Bermel, W., Bertini, I., Felli, I. C. and Pierattelli, R. Speeding up (13)C direct detection biomolecular NMR spectroscopy. *J. Am. Chem. Soc.* **131**, 15339–15345 (2009).
24. Fragai, M., Luchinat, C. and Parigi, G. "Four-dimensional" protein structures: examples from metalloproteins. *Acc. Chem. Res.* **39**, 909–917 (2006).
25. Bertini, I., Luchinat, C., Parigi, G. and Pierattelli, R. Perspectives in paramagnetic NMR of metalloproteins. *Dalton Trans* 3782–3790 (2008). doi:10.1039/b719526e
26. Bertini, I., Luchinat, C., Parigi, G. and Pierattelli, R. NMR spectroscopy of paramagnetic metalloproteins. *Chembiochem* **6**, 1536–1549 (2005).
27. Siemer, A. B., Ritter, C., Ernst, M., Riek, R. and Meier, B. H. High-resolution solid-state NMR spectroscopy of the prion protein HET-s in its amyloid conformation. *Angew. Chem. Int. Ed. Engl.* **44**, 2441–2444 (2005).
28. Wasmer, C. et al. Amyloid fibrils of the HET-s(218–289) prion form a beta solenoid with a triangular hydrophobic core. *Science* **319**, 1523–1526 (2008).
29. Daebel, V. et al. β-sheet core of tau paired helical filaments revealed by solid-state NMR. *J. Am. Chem. Soc.* **134**, 13982–13989 (2012).
30. Banci, L. et al. Solid-state NMR studies of metal-free SOD1 fibrillar structures. *J. Biol. Inorg. Chem.* **19**, 659–666 (2014).

31. Bertini, I. et al. Solid-state NMR of proteins sedimented by ultracentrifugation. *Proc. Natl. Acad. Sci. USA* **108**, 10396–10399 (2011).
32. Bertini, I. et al. NMR properties of sedimented solutes. *Phys Chem Chem Phys* **14**, 439–447 (2012).
33. Bertini, I. et al. On the use of ultracentrifugal devices for sedimented solute NMR. *J. Biomol. NMR* **54**, 123–127 (2012).
34. Serber, Z. et al. High-resolution macromolecular NMR spectroscopy inside living cells. *J. Am. Chem. Soc.* **123**, 2446–2447 (2001).
35. Sakakibara, D. et al. Protein structure determination in living cells by in-cell NMR spectroscopy. *Nature* **458**, 102–105 (2009).
36. Dedmon, M. M., Patel, C. N., Young, G. B. and Pielak, G. J. FlgM gains structure in living cells. *Proc. Natl. Acad. Sci. USA* **99**, 12681–12684 (2002).
37. Schlesinger, A. P., Wang, Y., Tadeo, X., Millet, O. and Pielak, G. J. Macromolecular crowding fails to fold a globular protein in cells. *J. Am. Chem. Soc.* **133**, 8082–8085 (2011).
38. Benton, L. A., Smith, A. E., Young, G. B. and Pielak, G. J. Unexpected effects of macromolecular crowding on protein stability. *Biochemistry* **51**, 9773–9775.
39. Crowley, P. B., Chow, E. and Papkovskaia, T. Protein interactions in the *Escherichia coli* cytosol: an impediment to in-cell NMR spectroscopy. *Chembiochem* **12**, 1043–1048 (2011).
40. Selenko, P. and Wagner, G. Looking into live cells with in-cell NMR spectroscopy. *J. Struct. Biol.* **158**, 244–253 (2007).
41. Barnes, C. O., Monteith, W. B. and Pielak, G. J. Internal and global protein motion assessed with a fusion construct and in-cell NMR spectroscopy. *Chembiochem* **12**, 390–391 (2011).
42. Wang, Q., Zhuravleva, A. and Gierasch, L. M. Exploring weak, transient protein–protein interactions in crowded in vivo environments by in-cell nuclear magnetic resonance spectroscopy. *Biochemistry* **50**, 9225–9236 (2011).
43. Wirth, A. J. and Gruebele, M. Quinary protein structure and the consequences of crowding in living cells: leaving the test-tube behind. *Bioessays* **35**, 984–993 (2013).
44. Reckel, S., Löhr, F. and Dötsch, V. In-cell NMR spectroscopy. *Chembiochem* **6**, 1601–1606 (2005).
45. Xu, G. et al. Strategies for protein NMR in *Escherichia coli*. *Biochemistry* **53**, 1971–1981 (2014).
46. Banci, L., Barbieri, L., Bertini, I., Cantini, F. and Luchinat, E. In-cell NMR in *E. coli* to monitor maturation steps of hSOD1. *PLoS ONE* **6**, e23561 (2011).
47. Serber, Z. et al. Methyl groups as probes for proteins and complexes in in-cell NMR experiments. *J. Am. Chem. Soc.* **126**, 7119–7125 (2004).
48. Burz, D. S., Dutta, K., Cowburn, D. and Shekhtman, A. Mapping structural interactions using in-cell NMR spectroscopy (STINT-NMR). *Nat. Methods* **3**, 91–93 (2006).
49. Burz, D. S. and Shekhtman, A. In-cell biochemistry using NMR spectroscopy. *PLoS ONE* **3**, e2571 (2008).

50. Maldonado, A. Y., Burz, D. S., Reverdatto, S. and Shekhtman, A. Fate of pup inside the *Mycobacterium* proteasome studied by in-cell NMR. *PLoS ONE* **8**, e74576 (2013).
51. Bertrand, K., Reverdatto, S., Burz, D. S., Zitomer, R. and Shekhtman, A. Structure of proteins in eukaryotic compartments. *J. Am. Chem. Soc.* **134**, 12798–12806 (2012).
52. Selenko, P., Serber, Z., Gadea, B., Ruderman, J. and Wagner, G. Quantitative NMR analysis of the protein G B1 domain in *Xenopus laevis* egg extracts and intact oocytes. *Proc. Natl. Acad. Sci. USA* **103**, 11904–11909 (2006).
53. Hamatsu, J. et al. High-resolution heteronuclear multidimensional NMR of proteins in living insect cells using a baculovirus protein expression system. *J. Am. Chem. Soc.* **135**, 1688–1691 (2013).
54. Inomata, K. et al. High-resolution multi-dimensional NMR spectroscopy of proteins in human cells. *Nature* **458**, 106–109 (2009).
55. Danielsson, J. et al. Pruning the ALS-associated protein SOD1 for in-cell NMR. *J. Am. Chem. Soc.* **135**, 10266–10269 (2013).
56. Ogino, S. et al. Observation of NMR signals from proteins introduced into living mammalian cells by reversible membrane permeabilization using a pore-forming toxin, streptolysin O. *J. Am. Chem. Soc.* **131**, 10834–10835 (2009).
57. Banci, L. et al. Atomic-resolution monitoring of protein maturation in live human cells by NMR. *Nat. Chem. Biol.* **9**, 297–299 (2013).
58. Banci, L., Barbieri, L., Luchinat, E. and Secci, E. Visualization of redox-controlled protein fold in living cells. *Chem. Biol.* **20**, 747–752 (2013).
59. Barbieri, L., Luchinat, E. and Banci, L. Structural insights of proteins in sub-cellular compartments: In-mitochondria NMR. *Biochim. Biophys. Acta* **1843**, 2492–2496 (2014).
60. Reckel, S., Lopez, J. J., Löhr, F., Glaubitz, C. and Dötsch, V. In-cell solid-state NMR as a tool to study proteins in large complexes. *Chembiochem* **13**, 534–537 (2012).
61. Renault, M. et al. Cellular solid-state nuclear magnetic resonance spectroscopy. *Proc. Natl. Acad. Sci. USA* **109**, 4863–4868 (2012).

6
Size exclusion chromatography with multi-angle laser light scattering (SEC-MALLS) to determine protein oligomeric states

Albert GUSKOV and Dirk Jan SLOTBOOM

University of Groningen, The Netherlands

Abstract

Proteins are natural polymers that have evolved the ability to oligomerize to enhance their stability and functions. There are different ways to determine the oligomerization state of proteins, and multi-angle laser light scattering (MALLS) is one of the most powerful. It allows for fast and reliable determination of the molar mass, and therefore the oligomerization state, of polymers in solution. It is also widely used to check the homogeneity and monodispersity of purified protein samples. Furthermore, MALLS allows quantification of the number of molecules (detergents, lipids, oligosaccharides) bound to the protein of interest. It is one of the few techniques that can be used to determine the detergent micelle size of purified membrane proteins.

Keywords

Oligomerization, oligomeric proteins, isologous oligomerization, oligomeric state, obligate oligomer, oligomerization interface, multi-angle light scattering, MALLS, molar mass determination, micelle size

Chapter Contents

6 Size exclusion chromatography with multi-angle laser light scattering (SEC-MALLS) to determine protein oligomeric states 169
    Albert GUSKOV and Dirk Jan SLOTBOOM

  6.1 Introduction 171
  6.2 Multi-angle light scattering 174
      6.2.1 Mathematical apparatus 174
  6.3 Applications 178
      6.3.1 Determination of detergent micelle size 178
      6.3.2 Determination of the oligomeric state of membrane proteins 179
  6.4 Outlook 180

    References 180

6.1 Introduction

Proteins are natural heteropolymers of amino acid residues. They can be of very different lengths, but more importantly they can adopt (or fold into) very different spatial structures. The latter allows the general classification of proteins into three main categories (Fig. 6.1): (1) fibrous, (2) membrane and (3) globular (sometimes termed water-soluble) proteins. In addition to these well-structured proteins there is a class of so-called intrinsically disordered (or unstructured) proteins (see Uversky 2013 for a review), which prefer an unfolded state to the stable tertiary structure. However, it must be noted that many of these unstructured proteins are able to organize into more stable forms upon interaction with their targets/cofactors.

Membrane and globular proteins typically fold into a compact shape. Fibrous proteins are usually composed of large repetitive blocks of either secondary structural elements or complete globular subunits held together by inter-polypeptide hydrogen bonds (and other bonds such as disulfide bridges). Such a spatial arrangement allows for the formation of long and durable structural elements, for example hair, silk, muscle fibers, cytoskeleton, etc.

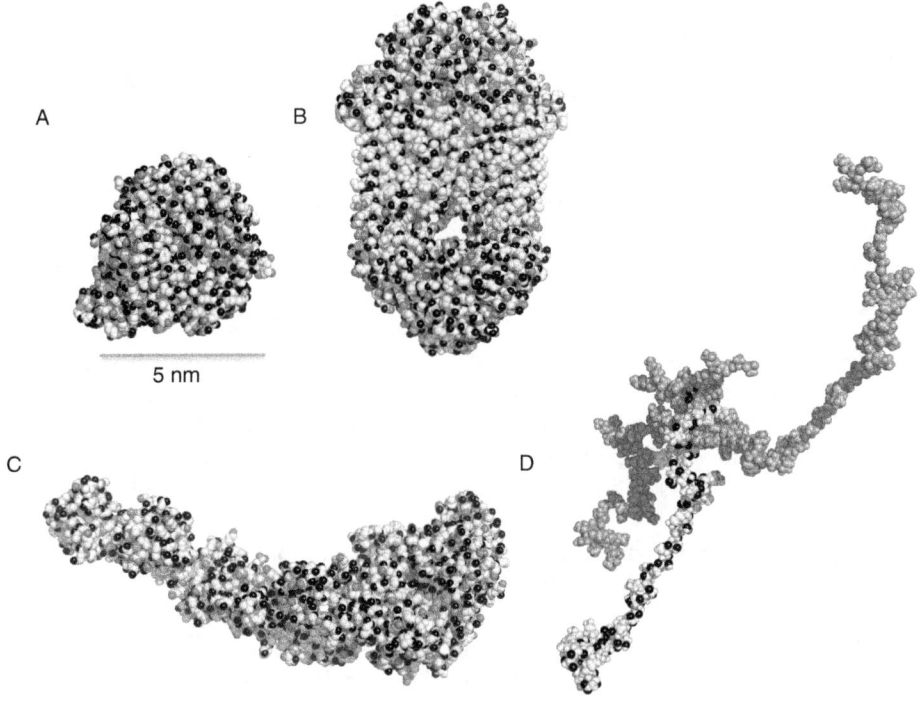

Fig. 6.1 Different classes of proteins: (A) globular, (B) membrane, (C) fibrous, and (D) intrinsically disordered (additional conformations are shown in gray to emphasize the flexibility of this type of protein). All proteins are depicted at the same scale.

Globular proteins—as might be deduced from their name—have a roughly globular or spherical shape. These proteins fold in a way that minimizes exposure of their hydrophobic residues to the solvent, thereby forming a hydrophobic core, which is shielded by a layer containing polar residues. Such an arrangement—apolar residues *in* and polar ones *out*—explains the water-solubility of these proteins. Globular proteins have a large range of different functions—they may serve as enzymes, messengers, regulators, transporters, and as part of the immune system, etc.

Membrane proteins are embedded into or associated with a cellular or organellar membrane, and have surface-exposed hydrophobic regions that interact with the lipid bilayer. Membrane proteins have important roles in signal transduction and transport of a vast variety of molecules through the semi-impermeable lipid bilayer.

Many proteins self-organize into supramolecular assemblies to perform their functions. This process is termed oligomerization, and it has been estimated that about 80% of proteins can form functional oligomeric proteins (Klotz et al. 1970; Traut 1994). Oligomeric proteins can have a compact shape, or can be extended (fibrous-like). The interactions between monomers in an oligomer can be via identical surfaces on each monomer (so-called isologous oligomerization) or via non-identical surfaces (heterologous oligomerization) (Monod et al. 1965; Cornish-Bowden and Koshland 1971; Jones and Thornton 1995; Nooren and Thornton 2003a; Hashimoto et al. 2011) (Fig. 6.2).

The residues at the oligomerization interfaces tend to be more conserved than other surface residues; moreover the conserved residues provide the larger part of binding energy (Clackson and Wells 1995; Ali and Imperiali 2005). Furthermore, in many cases these oligomerization hot spots have evolved in such a way as to enhance the necessary interactions for oligomer formation while simultaneously preventing or diminishing unwanted interactions. As discussed in Nishi et al. (2013), the evolutionary conservation of oligomerization hot spots within subfamilies might be advantageous for the separation of functional pathways of close paralogs. Intriguingly, different oligomeric states of the same protein can play different roles; for example, the inactive dimer of pyruvate kinase M2 triggers the accumulation of phosphometabolites in tumors, whereas the active tetramer interacts with the glycolytic enzyme complex in normal cells (Mazurek et al. 2005). Another more recent example is the dual-affinity nitrate transporter (Sun et al. 2014). This transporter switches between monomeric and dimeric states for high- and low-affinity transport, respectively. See Nishi et al. (2013), for more examples of functional differences between different oligomerization states.

In many cases oligomerization interfaces are more or less planar (Jones and Thornton 1996, 1997; Marianayagam et al. 2004) and proportional to the size of a protein—the larger the interacting subunits, the larger the interface (Brooijmans et al. 2002; Nooren and Thornton 2003a, b). Protomers that are not stable individually *in vivo* form so-called obligate oligomers, in contrast to the stable ones that form non-obligate complexes (Nooren and Thornton 2003a). For proteins that form non-obligate oligomers (with transient or permanent interactions), the oligomerization interfaces are generally smaller and more polar/charged (Jones and Thornton 1996, 1997; Lo Conte et al. 1999; Zhanhua et al. 2005; Dey et al. 2010) in contrast to the stable (obligate) oligomers (usually with permanent interactions) in which interfaces

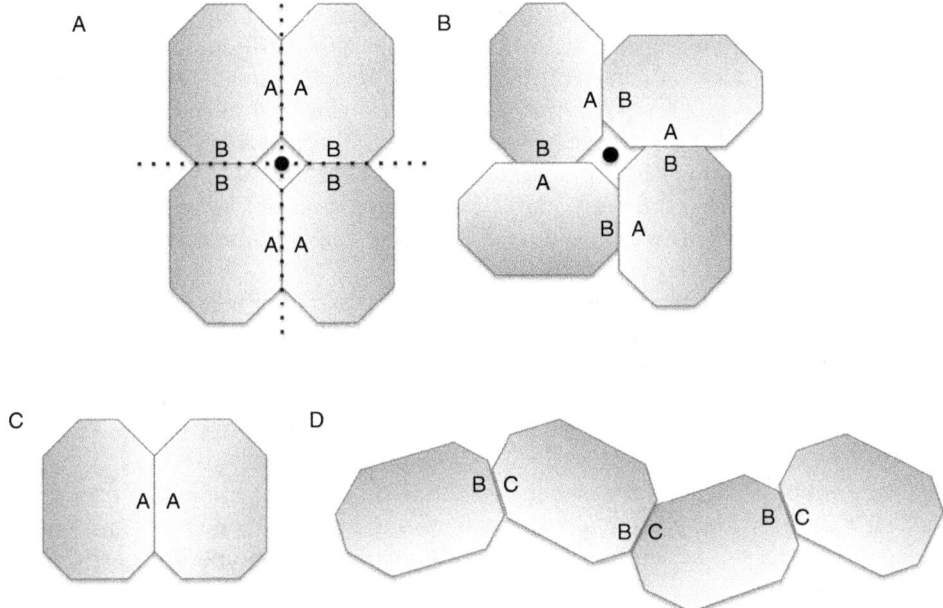

Fig. 6.2 Different type of oligomers (modified after Griffin and Gerrard 2012). (A) Homotetramer (dimer of dimers) with isologous interfaces. Two-fold symmetry axes are shown as dashed lines and a large black dot for the one pointing towards reader. (B) Homotetramer with heterologous interfaces. The four-fold symmetry axis is shown by the large black dot. (C) Isologous dimer. (D) Heterologous interfaces allow unlimited oligomerization.

are large and predominantly hydrophobic (Bahadur et al. 2003; Nooren and Thornton 2003a; Ofran and Rost 2003; Zhanhua et al. 2005).

There is no single answer to why proteins evolved the ability to oligomerize. The basic paradigm is that the structure of a protein is fine-tuned to its function(s). Thus, in principle, protein oligomerization should be beneficial for a protein function. Indeed, the formation of shared active sites and the cooperativity effect are the most obvious benefits that come with the oligomerization (Creighton 1997; Marianayagam et al. 2004; Ali and Imperiali 2005). The other benefit is the formation of a larger, more energetically favorable and more stable oligomeric structure, resistant to denaturation (Miller et al. 1987; Jones and Thornton 1995; Goodsell and Olson 2000). However, for some proteins it has been shown that lone monomers have the same (or an increased) level of stability (Franke et al. 1998; Mossing and Sauer 1990; Borchert et al. 1994). Additionally, oligomerization may be beneficial for osmotic regulation in cells, as the large oligomers have a considerably smaller solvent-exposed area compared with individual monomers, thus decreasing the amount of solvent required for protein hydration, which in turn reduces osmotic pressure (Creighton 1997; Goodsell and Olson 2000). Among other possible benefits are an increase in the thermal stability (Ahern et al. 1987; Gerk et al. 2000); minimization of the amount of genetic material required (at least in the prokaryotes) and improvement in translation error management (discarding one monomer instead of the whole structure); as well as protection against misfolding, which might lead to the formation of perilous aggregates such as amyloid fibrils (Dobson 2002) that are involved in numerous human diseases, such as Alzheimer's, Parkinson's and Huntington's diseases. Amyloid fibrils are long, insoluble, extremely stable and rich in β-sheets and usually accumulate in the extracellular space (Hammer et al. 2008). It must be noted though, that not all amyloid fibrils are detrimental; it has been shown, that human hormones may self-assemble into amyloids for storage purposes (Maji et al. 2009).

Several evolutionary mechanisms have been proposed for homo-oligomerization. One of them is so-called domain swapping (Bennett et al. 1994), which was first observed back in the 1960s by Crestfield et al. (1962). Typically, the N- and C-termini are swapped, though in principle any part(s) of a protein can be exchanged (Liu and Eisenberg 2002; Gronenborn 2009). The length of the swapped region can vary considerably, from a few amino acids to the whole domain. There are two mechanisms of domain swapping that have been suggested based on computational studies. Mechanism 1 implies that the protein has to undergo unfolding of monomers, and such an unfolded state represents the free energy trap during the oligomerization process—competition between intra- and interchain interactions will lead to the formation of monomeric and oligomeric states, respectively (Yang et al. 2004). In this mechanism, the monomer topology, but not the specific elements of the sequence (e.g., proline residues in the hinge areas, which were suggested as warning signs for possible domain swapping (Cornish-Bowden and Koshland 1971; Jones and Thornton 1995)) determines the possible area of domain swapping. In mechanism 2 unfolding of monomers does not occur, and domain swapping is believed to be a continuous process with a relatively flat energy profile (Malevanets et al. 2008). The swapping is initiated at the N- or C-terminus of the folded protein and gradually progresses in the cooperative manner between two protein chains until the stable conformer is found (Hashimoto et al. 2011).

Clearly the experimental analysis of the oligomeric state is an essential step during the characterization of a novel protein in order to understand its function(s). In addition, oligomerization state analysis might be required during ordinary biochemical/biophysical routines, such as protein purification, protein reconstitution, crystallization, etc. Several techniques can be used to perform such analysis: gel filtration, macromolecular crystallography, analytical ultracentrifugation, mass spectroscopy and light scattering. Here, we will discuss in greater detail, a special subtype of the light scattering—multi-angle laser light scattering (MALLS)—and its implications for the characterization of protein samples.

6.2 Multi-angle light scattering

6.2.1 Mathematical apparatus

Multi-angle light scattering is a spectroscopic technique that allows determination of the molar mass of polymers in solution (Takagi 1990; Wyatt 1993; Mogridge 2004;

Folta-Stogniew 2006). This technique is particularly suitable for the determination of oligomerization states, especially in the case of difficult targets such as membrane protein–detergent complexes or post-translationally modified proteins. For the latter applications light scattering measurements are coupled to refractometry and absorbance measurements, also known as the three-detector method—a combination of ultraviolet absorbance (UV), light scattering (LS) and refractive index (RI) detectors and a size-exclusion chromatography (SEC) column (Wyatt 1993). In this setup the only prior knowledge required for determination of the oligomerization state is the protein sequence; if the nature of the modification (detergent) is known it is also possible to determine the amount of the latter (Kendrick et al. 2001; Wei et al. 2004).

The weight-average molar mass M_w of a studied protein can be derived via the following equations:

$$M_w = k_1 \frac{\Delta LS}{(dn/dc)\Delta RI} \quad (6.1)$$

and

$$\frac{dn}{dc} = k_2 A \frac{\Delta RI}{\Delta UV} \quad (6.2)$$

thus

$$M_w = \frac{k_1 \Delta LS \Delta UV}{k_2 A (\Delta RI)^2}. \quad (6.3)$$

ΔLS is the response from the light scattering detector, which can be expressed via classical Rayleigh relationship (Tanford 1961):

$$\Delta LS = \left(\frac{I_\theta}{I_0}\right)_{solution} - \left(\frac{I_\theta}{I_0}\right)_{buffer} = K\left(\frac{dn}{dc}\right)^2 M_w C. \quad (6.4)$$

In other words, the LS detector registers the change in the intensity between the incident and the scattered light at the angle θ. So ΔLS is the excess of the scattered light from the sample (a protein of the interest in a buffer) over the buffer without the protein added.

It must be noted that equation (6.4) is valid for ideal solutions. Thus the following restraints must be applied for real solutions: the protein concentration should not be greater than 0.1 mg ml^{-1} (Folta-Stogniew and Williams 1999) and the size of a protein should be smaller than the light wavelength used. This condition is easily met by the globular proteins up to several hundred kDa.

A is the extinction coefficient (typically A_{280}, the extinction coefficient of a protein at a wavelength of 280 nm; in other words the absorbance of the 1 mg ml^{-1} solution of a given protein at 280 nm). It can be estimated from the protein sequence via the molar extinction coefficient ε_{280} (M^{-1} cm^{-1}):

$$\varepsilon_{280} = (5500 N_{Trp}) + (1490 N_{Tyr}) + (125 N_{Cys}) \quad (6.5)$$

where N_{Trp}, N_{Tyr} and T_{Cys} are the number of tryptophan, tyrosine and cysteine residues in the given sequence. The extinction coefficient can also be determined experimentally using quantitative amino acid analysis.

A_{280} (ml mg^{-1} cm^{-1}) is obtained by dividing ε_{280} by the molecular weight calculated from the amino acid sequence (M_c):

$$A_{280} = \frac{\varepsilon_{280}}{M_c}. \tag{6.6}$$

A_{280} can be used to calculate the protein concentration C (mg ml^{-1}) if protein absorbance is measured at 280 nm (UV_{280}) and ΔUV_{280} [the difference in the detected absorbance between sample (protein and buffer) and blank (buffer alone)]:

$$C = \frac{\Delta UV_{280}}{A_{280}} \tag{6.7}$$

The term dn/dc is the increment in refractive index, relating the changes in the protein concentration to the changes in refractive index. In case of soluble proteins dn/dc is nearly constant (≈ 0.187 ml g^{-1}) though it depends on the composition of the buffer (Ball and Ramsden 1998; Wen and Arakawa 2000). The value of dn/dc can be determined experimentally by measuring ΔRI and ΔUV_{280}:

$$\frac{dn}{dc} = \frac{\Delta RI}{\Delta UV_{280}/A_{280}} = \frac{A_{280}(RI_{solution} - RI_{buffer})}{\Delta UV_{280}}. \tag{6.8}$$

The parameters k_1 and k_2 from equations (6.1) and (6.2) are physical and instrument constants that may also be determined experimentally using the three-detector method with protein standards of well-known molar mass. The constant K (equation 6.4) can be expressed as:

$$K = \frac{2\pi^2 n^2}{\lambda_0^4 N_A}\left(\frac{1+\cos^2\theta}{r^2}\right). \tag{6.9}$$

where n is the refractive index of the solution without a protein, λ_0 is the wavelength of light, N_A is the Avogadro number (6.022 × 10^{23}), θ is the angle between the incident and the scattered rays of light and r is the distance between the scattering molecule and a detector.

In case of globular soluble proteins, the molecular mass (and therefore the oligomerization state, N) can be estimated even if only two detectors are available: LS + UV or LS + RI:

$$M_w = \frac{\Delta LS}{K(dn/dc)^2(\Delta UV_{280}/A_{280})}$$

$$= \frac{A_{280}\Delta LS}{K(dn/dc)^2\Delta UV_{280}} \tag{6.10}$$

$$= \frac{\Delta LS}{K(dn/dc)\Delta RI}$$

and
$$N = \frac{M_w}{M_c} \quad (6.11)$$

where M_c is the molecular weight calculated from the sequence. However, the value of dn/dc must be known or measured experimentally.

In case of membrane proteins we also need to estimate the amount of detergent bound to a protein, thus the value of M_w for the protein–detergent micelle ($M_{w,complex}$) contains the additional contribution δ from the detergent:

$$M_{w,complex} = (1 + \delta) M_{w,protein}. \quad (6.12)$$

Similarly:
$$C_{complex} = (1 + \delta) C_{protein}. \quad (6.13)$$

Thus in this case equations (6.7) and (6.8) will become more complex:

$$C_{complex} = \frac{\Delta UV_{280}}{A_{280,complex}} = \frac{\Delta UV_{280}}{[1/(1+\delta)]A_{280,protein} + [\delta/(1+\delta)]A_{280,detergent}} \quad (6.14)$$

$$\left(\frac{dn}{dc}\right)_{complex} = \left(\frac{1}{1+\delta}\right)\left(\frac{dn}{dc}\right)_{protein} + \left(\frac{\delta}{1+\delta}\right)\left(\frac{dn}{dc}\right)_{detergent}$$

$$= \frac{\Delta RI}{\Delta UV_{280}}\left[\left(\frac{1}{1+\delta}\right)A_{280,protein} + \left(\frac{\delta}{1+\delta}\right)A_{280,detergent}\right] \quad (6.15)$$

Fortunately, many commonly used detergents do not absorb at 280 nm, thus the $A_{280,detergent}$ term may be nullified. Still, the value of δ is usually unknown, and it is therefore not possible to determine the molecular weight of the protein–detergent complex directly. However, to determine the oligomerization state of the studied protein, we need to estimate only the molecular weight of protein fraction in the protein–detergent complex and we can modify equation (6.10):

$$M_{w,protein} = \frac{\Delta LS}{K(dn/dc)^2_{apparent} C_{protein}} \quad (6.16)$$

where
$$\left(\frac{dn}{dc}\right)_{apparent} = \left(\frac{dn}{dc}\right)_{protein} + \delta\left(\frac{dn}{dc}\right)_{detergent} \quad (6.17)$$

and with the assumption that a detergent does not absorb at 280 nm:

$$\left(\frac{dn}{dc}\right)_{apparent} = \frac{\Delta RI}{C_{protein}} = \frac{A_{280,protein}\Delta RI}{\Delta UV_{280}} \quad (6.18)$$

and (equation 6.3) will look like:

$$M_w = \frac{\Delta LS \Delta UV_{280}}{K A_{280,protein}(\Delta RI)^2}. \quad (6.19)$$

As seen from equation (6.19), input from all three detectors is required to determine the molecular weight of the protein fraction.

6.3 Applications

6.3.1 Determination of detergent micelle size

Detergents are widely used to extract membrane proteins from lipid bilayers and to keep them stable in the absence of a membrane environment for further investigations. Detergents are amphipathic molecules, which are able to self-aggregate in aqueous solutions with the formation of micelles under the condition that the critical micellar concentration (CMC) is achieved (le Maire et al. 2000). Micelles of different detergents have different sizes and aggregation numbers. To determine the actual size of a micelle different methods might be applied, such as fluorescence quenching (Tummino and Gafni 1993), neutron scattering (Clifton et al. 2013) or analytical centrifugation (Ebel 2011). Due to the fact that micelles are large and scatter considerable amounts of light, their size might be measured using multi-angle laser light scattering (MALLS). Slotboom et al. (2008) studied the behavior of four commonly used maltoside detergents, DDM, DM, UDM and TDM, with MALLS and determined the respective sizes for their micelles (Fig. 6.3). These data extend previously reported values for other detergents ($C_{12}E_9$, FOS-14, LDAO) also obtained with MALLS (Strop and Brunger 2005).

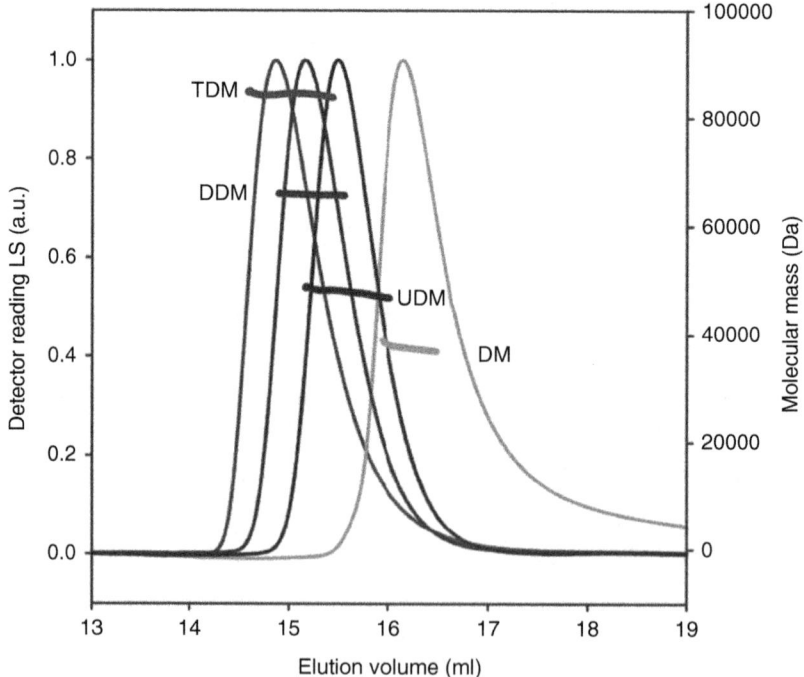

Fig. 6.3 MALLS analysis of detergent micelles (adapted after Slotboom et al. 2008).

6.3.2 Determination of the oligomeric state of membrane proteins

The determination of the oligomeric state of a membrane protein with SEC-MALLS is becoming more common because it is not easy to unambiguously deduce the oligomeric state using the elution volume of the protein in size-exclusion chromatography analysis.

For example, the exact stoichiometry of protein subunits in glutamate transporters had been controversial for a long time until the structure of the archeal glutamate transporter homologue was solved (Yernool et al. 2004). Before the crystal structure of the glutamate transporter homologue Glt$_{Ph}$ was published, the trimeric oligomerization state for two bacterial glutamate transporters had been determined by MALLS (Yernool et al. 2003). Furthermore, Yernool et al. (2003) tested the performance of MALLS by studying the oligomerization state for two additional membrane proteins (the α-HL heptamer and maltoporin LamB) for which the stoichiometry was already well established. The determined molecular weights were within 5–10% of the previously obtained values.

Another more recent example is a fluoride ion channel, which has not yet been resolved with X-ray crystallography but most probably has an unusual dual topology (Stockbridge et al. 2013). The dimeric state of this protein has been revealed with MALLS, and confirmed by photobleaching experiments.

MALLS was successfully used to study the mechanism of assembly of heteromeric kainate receptors (Kumar et al. 2011). The formation of heterodimers between GluR6 and KA2 subunits upon mixing was registered by the increased amplitude of the dimer peak with a corresponding decrease in the monomer peak (when separated GluR6 and KA2 exist as dimeric and monomeric species, respectively).

MALLS is also a very useful technique for testing the homogeneity of a protein sample. For example, the truncated version of the magnesium-specific divalent cation transporter CorA exists in solution as a pentamer and a dimer mixture with ratio 1:1.5 (unpublished result). However, for successful crystallization in most cases the sample should be as homogeneous as possible. Thus we screened a wide range of Mg^{2+} (the native substrate for this protein) concentrations with MALLS to evaluate the substrate stabilization effect, if any. Only the addition of 100 mM Mg^{2+} caused the dimeric species to completely vanish and the sample yielded high-quality diffraction crystals.

MALLS allows relatively fast determination of the amount of detergent and lipid bound to the membrane protein of interest (equation 6.19). The only prerequisite is that the value of dn/dc for the detergent and lipid should either be obtained from literature or measured experimentally. The latter is done by plotting the value of Δn (obtained with the RI detector) versus concentration; the slope of this linear plot gives dn/dc (Strop and Brunger 2005). An example of an analysis of the bound fraction of detergent is the study of the YiiP metal transporter by Wei et al. (2004). Wei et al. investigated the oligomeric state of this transporter and the associated amounts of detergent/lipids in four maltoside detergents. The dn/dc values for detergents were determined experimentally offline, yielding values of 0.138 to 0.142 for the different detergents. Also a dn/dc value of 0.138 for a lipid mixture was determined. Taking an average value of 0.140, authors simplified the equation (6.17) to:

$$\left(\frac{dn}{dc}\right)_{apparent} = 0.187 + 0.140\delta \tag{6.20}$$

where δ is the combined weight ratio of lipids and detergents bound to the YiiP micellar complex. The calculated mass of bound detergent/lipids constituted around 70% of the micelle mass (Wei et al. 2004).

6.4 Outlook

Multi-angle light scattering coupled with size exclusion chromatography is a powerful technique allowing fast and reliable determination of the oligomeric state and homogeneity of a protein sample and the quantities of molecules (detergents, lipids, oligosaccharides) bound to a protein of interest. The main factor preventing the ubiquitous use of MALLS in academic research is probably the cost of the equipment, but with cheaper models and collaborative research centers this problem will hopefully vanish in the near future.

References

Ahern, T.J. et al., 1987. Control of oligomeric enzyme thermostability by protein engineering. *Proceedings of the National Academy of Sciences of the United States of America*, 84(3), 675–679.

Ali, M.H. and Imperiali, B., 2005. Protein oligomerization: How and why. *Bioorganic and Medicinal Chemistry*, 13(17), 5013–5020.

Bahadur, R.P. et al., 2003. Dissecting subunit interfaces in homodimeric proteins. *Proteins*, 53(3), 708–719.

Ball, V. and Ramsden, J.J., 1998. Buffer dependence of refractive index increments of protein solutions. *Biopolymers*, 46(7), 489–492.

Bennett, M.J., Choe, S. and Eisenberg, D., 1994. Domain swapping: entangling alliances between proteins. *Proceedings of the National Academy of Sciences of the United States of America*, 91(8), 3127–3131.

Borchert, T.V. et al., 1994. Design, creation, and characterization of a stable, monomeric triosephosphate isomerase. *Proceedings of the National Academy of Sciences of the United States of America*, 91(4), 1515–1518.

Brooijmans, N., Sharp, K.A. and Kuntz, I.D., 2002. Stability of macromolecular complexes. *Proteins*, 48(4), 645–653.

Clackson, T. and Wells, J.A., 1995. A hot-spot of binding-energy in a hormone–receptor interface. *Science*, 267(5196), 383–386.

Clifton, L.A., Neylon, C. and Lakey, J.H., 2013. Examining protein–lipid complexes using neutron scattering. *Methods in Molecular Biology*, 974, 119–150.

Cornish-Bowden, A.J. and Koshland, D., Jr, 1971. The quaternary structure of proteins composed of identical subunits. *Journal of Biological Chemistry*, 246(10), 3092–3102.

Creighton, T.E., 1997. *Protein Function*. Oxford University Press, Oxford.

Crestfield, A.M., Stein, W.H. and Moore, S., 1962. On the aggregation of bovine pancreatic ribonuclease. *Archives of Biochemistry and Biophysics*, Suppl. 1, 217–222.

Dey, S. et al., 2010. The subunit interfaces of weakly associated homodimeric proteins. *Journal of Molecular Biology*, 398(1), 146–160.

Dobson, C.M., 2002. Protein-misfolding diseases: Getting out of shape. *Nature*, 418(6899), 729–730.

Ebel, C., 2011. Sedimentation velocity to characterize surfactants and solubilized membrane proteins. *Methods*, 54(1), 56–66.

Folta-Stogniew, E., 2006. Oligomeric states of proteins determined by size-exclusion chromatography coupled with light scattering, absorbance, and refractive index detectors. *Methods in Molecular Biology*, 328, 97–112.

Folta-Stogniew, E. and Williams, K.R., 1999. Determination of molecular masses of proteins in solution: Implementation of an HPLC size exclusion chromatography and laser light scattering service in a core laboratory. *Journal of Biomolecular Techniques*, 10(2), 51–63.

Franke, I. et al., 1998. Genetic engineering, production and characterisation of monomeric variants of the dimeric *Serratia marcescens* endonuclease. *FEBS Letters*, 425(3), 517–522.

Gerk, L.P., Leven, O. and Muller-Hill, B., 2000. Strengthening the dimerisation interface of Lac repressor increases its thermostability by 40 deg. C. *Journal of Molecular Biology*, 299(3), 805–812.

Goodsell, D.S. and Olson, A.J., 2000. Structural symmetry and protein function. *Annual Review of Biophysics and Biomolecular Structure*, 29(1), 105–153.

Griffin, M.D.W. and Gerrard, J.A., 2012. The relationship between oligomeric state and protein function. In: *Protein Dimerization and Oligomerization in Biology*. Springer, New York, pp. 74–90.

Gronenborn, A.M., 2009. Protein acrobatics in pairs—dimerization via domain swapping. *Current Opinion in Structural Biology*, 19(1), 39–49.

Hammer, N.D. et al., 2008. Amyloids: Friend or foe? *Journal of Alzheimers Disease*, 13(4), 407–419.

Hashimoto, K. et al., 2011. Caught in self-interaction: evolutionary and functional mechanisms of protein homooligomerization. *Physical Biology*, 8(3), 035007.

Jones, S. and Thornton, J.M., 1995. Protein–protein interactions—a review of protein dimer structures. *Progress in Biophysics and Molecular Biology*, 63(1), 31–65.

Jones, S. and Thornton, J.M., 1996. Principles of protein–protein interactions. *Proceedings of the National Academy of Sciences of the United States of America*, 93(1), 13–20.

Jones, S. and Thornton, J.M., 1997. Analysis of protein–protein interaction sites using surface patches. *Journal of Molecular Biology*, 272(1), 121–132.

Kendrick, B.S. et al., 2001. Online size-exclusion high-performance liquid chromatography light scattering and differential refractometry methods to determine degree of polymer conjugation to proteins and protein-protein or protein-ligand association states. *Analytical Biochemistry*, 299(2), 136–146.

Klotz, I.M., Langerma, N. and Darnall, D.W., 1970. Quaternary structure of proteins. *Annual Review of Biochemistry*, 39, 25–62.

Kumar, J., Schuck, P. and Mayer, M.L., 2011. Structure and assembly mechanism for heteromeric kainate receptors. *Neuron*, 71(2), 319–331.

Liu, Y. and Eisenberg, D., 2002. 3D domain swapping: As domains continue to swap. *Protein Science*, 11(6), 1285–1299.

Lo Conte, L., Chothia, C. and Janin, J., 1999. The atomic structure of protein–protein recognition sites. *Journal of Molecular Biology*, 285(5), 2177–2198.

le Maire, M., Champeil, P. and Moller, J.V., 2000. Interaction of membrane proteins and lipids with solubilizing detergents. *Biochimica et Biophysica Acta-Biomembranes*, 1508(1–2), 86–111.

Maji, S.K. et al., 2009. Functional amyloids as natural storage of peptide hormones in pituitary secretory granules. *Science*, 325(5938), 328–332.

Malevanets, A., Sirota, F.L. and Wodak, S.J., 2008. Mechanism and energy landscape of domain swapping in the B1 domain of protein G. *Journal of Molecular Biology*, 382(1), 223–235.

Marianayagam, N.J., Sunde, M. and Matthews, J.M., 2004. The power of two: protein dimerization in biology. *Trends in Biochemical Sciences*, 29(11), 618–625.

Mazurek, S. et al., 2005. Pyruvate kinase type M2 and its role in tumor growth and spreading. *Seminars in Cancer Biology*, 15(4), 300–308.

Miller, S. et al., 1987. The accessible surface area and stability of oligomeric proteins. *Nature*, 328(6133), 834–836.

Mogridge, J., 2004. Using light scattering to determine the stoichiometry of protein complexes. *Methods in Molecular Biology*, 261, 113–118.

Monod, J., Wyman, J. and Changeux, J.P., 1965. On the nature of allosteric transitions: a plausible model. *Journal of Molecular Biology*, 12, 88–118.

Mossing, M.C. and Sauer, R.T., 1990. Stable, monomeric variants of lambda Cro obtained by insertion of a designed beta-hairpin sequence. *Science*, 250(4988), 1712–1715.

Nishi, H. et al., 2013. Evolutionary, physicochemical, and functional mechanisms of protein homooligomerization. *Progress in Molecular Biology and Translational Science*, 117, 3–24.

Nooren, I.M.A. and Thornton, J.M., 2003a. Diversity of protein–protein interactions. *EMBO Journal*, 22(14), 3486–3492.

Nooren, I.M.A. and Thornton, J.M., 2003b. Structural characterisation and functional significance of transient protein-protein interactions. *Journal of Molecular Biology*, 325(5), pp.991–1018.

Ofran, Y. and Rost, B., 2003. Analysing six types of protein-protein interfaces. *Journal of Molecular Biology*, 325(2), 377–387.

Slotboom, D.J. et al., 2008. Static light scattering to characterize membrane proteins in detergent solution. *Methods*, 46(2), 73–82.

Stockbridge, R.B. et al., 2013. A family of fluoride-specific ion channels with dual-topology architecture. *eLIFE*, 2(0), e01084.

Strop, P. and Brunger, A.T., 2005. Refractive index-based determination of detergent concentration and its application to the study of membrane proteins. *Protein Science*, 14(8), 2207–2211.

Sun, J. et al., 2014. Crystal structure of the plant dual-affinity nitrate transporter NRT1.1. *Nature*, 507(7490), 73–77.

Takagi, T., 1990. Application of low-angle laser-light scattering detection in the field of biochemistry—review of recent progress. *Journal of Chromatography*, 506, 409–416.

Tanford, C., 1961. *Physical Chemistry of Macromolecules*. John Wiley and Sons, New York.

Traut, T.W., 1994. Dissociation of enzyme oligomers: a mechanism for allosteric regulation. *Critical Reviews in Biochemistry and Molecular Biology*, 29(2), 125–163.

Tummino, P.J. and Gafni, A., 1993. Determination of the aggregation number of detergent micelles using steady-state fluorescence quenching. *Biophysical Journal*, 64(5), 1580–1587.

Uversky, V.N., 2013. A decade and a half of protein intrinsic disorder: Biology still waits for physics. *Protein Science*, 22(6), 693–724.

Wei, Y.N., Li, H.L. and Fu, D., 2004. Oligomeric state of the *Escherichia coli* metal transporter YiiP. *Journal of Biological Chemistry*, 279(38), 39251–39259.

Wen, J. and Arakawa, T., 2000. Refractive index of proteins in aqueous sodium chloride. *Analytical Biochemistry*, 280(2), 327–329.

Wyatt, P.J., 1993. Light-scattering and the absolute characterization of macromolecules. *Analytica Chimica Acta*, 272(1), 1–40.

Yang, S.C. et al., 2004. Domain swapping is a consequence of minimal frustration. *Proceedings of the National Academy of Sciences of the United States of America*, 101(38), 13786–13791.

Yernool, D. et al., 2003. Trimeric subunit stoichiometry of the glutamate transporters from *Bacillus caldotenax* and *Bacillus stearothermophilus*. *Biochemistry*, 42(44), 12981–12988.

Yernool, D. et al., 2004. Structure of a glutamate transporter homologue from *Pyrococcus horikoshii*. *Nature*, 431(7010), 811–818.

Zhanhua, C. et al., 2005. Protein subunit interfaces: heterodimers versus homodimers. *Bioinformation*, 1(2), 28–39.

Part 3

Plant development: from genes to growth

7
Mechanisms controlling time measurement in plants and their significance in natural populations

George COUPLAND

Max Planck Institute for Plant Breeding Research, Cologne, Germany

Abstract

Plants, in common with many other organisms, measure time across different scales. Such time measurement is essential for the adaptation of plants to a wide range of environments. This chapter describes several of the mechanisms by which plants measure the time of day and the time of year. The circadian clock and its significance in regulating thousands of genes that control daily rhythms and contribute to a wide range of processes are described. How plants measure day length or the duration of winter cold to synchronize developmental decisions with the changing seasons is also discussed. Many of these processes have been deciphered in controlled environments in the laboratory, and the final section considers how predictive these results have been for the variation that is found in nature to allow plants to adapt to different environments.

Keywords

Biological timing, *Arabidopsis thaliana*, circadian rhythms, vernalization, seasonal responses, seasonal timing of gene expression, adaptation, quantitative genetics, association genetics

Chapter Contents

7 Mechanisms controlling time measurement in plants and their significance in natural populations   187
    George COUPLAND

7.1 Introduction	189
7.2 The plant circadian clock	189
7.2.1 Structure of the oscillator	190
7.2.2 Regulation of output pathways	192
7.2.3 Entrainment	193
7.3 Seasonal timing	194
7.3.1 Measurement of day length	194
7.3.2 Vernalization	196
7.4 Timing in natural populations	198
7.5 Conclusion	201
References	202

7.1 Introduction

Plants are sedentary and must therefore adapt to the environment in which they live. In temperate climates the ambient environment fluctuates considerably during the day or with the changing seasons. Plants have evolved mechanisms for detecting and responding to such changes in a wide range of environmental parameters such as the intensity, wavelength or direction of light, high or low temperatures and water availability. In common with all other organisms on earth, plants evolved under a regime of light and dark that varies with the seasons. They have therefore evolved mechanisms to utilize this information to measure time and thus anticipate regular changes in their environment (Sung and Amasino, 2005; de Montaigu et al., 2010; Andres and Coupland, 2012). By exploiting such mechanisms, plants can anticipate the onset of light in the morning, the advent of darkness in the evening and the progression of the seasons during the annual cycle. This chapter describes the molecular mechanisms used by plants to measure time during the daily cycle and to synchronize flowering with the changing seasons. These processes have been extensively studied in controlled conditions in the laboratory, and the extent to which these results can be extended to natural populations is also discussed (Alonso-Blanco et al., 2009; Weigel, 2012).

7.2 The plant circadian clock

Plants can assess the time of day using a timing mechanism called the circadian clock. This clock generates circadian rhythms with a periodicity of approximately 24 h (the word circadian deriving from the Latin *circa*, "approximately", and *dies*, "day"). The growth of clock mutants under daily cycles of different durations has elegantly demonstrated the significance of clock regulation to plant metabolism, growth and development (Dodd et al., 2005). Conceptually this clock is usually considered in three parts: input pathways that use environmental parameters to synchronize the timing mechanism with the daily cycle; the central oscillator that generates the 24-h timekeeping mechanism; and output pathways that control biochemical or physiological processes in a 24-h rhythm (Harmer, 2009). Rhythms controlled by the circadian clock can be distinguished from those driven directly by daily environmental conditions such as light/dark transitions by transferring plants from the daily cycle to continuous conditions (Harmer, 2009). Under such continuous conditions circadian rhythms continue to oscillate whereas cycles driven by the environment are lost (Fig. 7.1).

The circadian clock pervades almost all aspects of plant biology. The prevalence of this mode of gene regulation became clear with the advent of genome-wide transcriptomic approaches. Based on such methods it has been estimated that around 30% of plant genes are regulated by the circadian clock, so that the abundance of their transcripts cycles in a circadian-clock regulated fashion (Harmer et al., 2000; Covington et al., 2008).

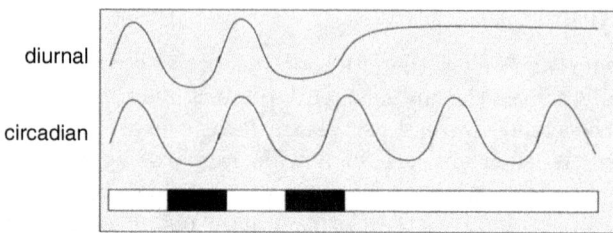

Fig. 7.1 Diurnal and circadian rhythms. Both rhythms are observed under daily cycles of light and dark. However, after transferring plants to continuous light diurnal rhythms are no longer observed while circadian rhythms continue.

7.2.1 Structure of the oscillator

Central oscillators that generate the 24-h time-keeping mechanism central to the circadian clock have been elucidated in many systems and are based on negative feedback loops. A simple example of such a loop is shown in Fig. 7.2 (Dunlap, 1999).

In plants the first clock proteins to be identified were the closely related MYB transcription factors LATE ELONGATED HYPOCOTYL (LHY) and CIRCADIAN CLOCK ASSOCIATED 1 (CCA1) as well as the PSEUDO RESPONSE REGULATOR (PRR) protein TIMING OF CAB1 (TOC1) (Schaffer et al., 1998; Wang and Tobin, 1998; Strayer et al., 2000). The mRNAs of *LHY* and *CCA1* were found to peak in expression in the morning, while *TOC1* mRNA peaked in the evening. Furthermore, mutations in *LHY* and *CCA1* affected expression of *TOC1*, while TOC1 affected

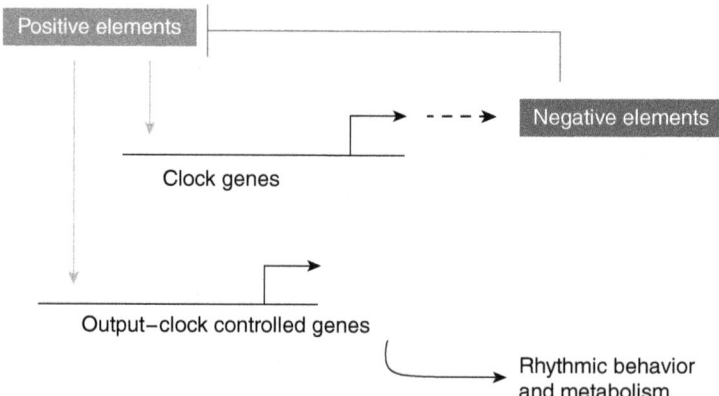

Fig. 7.2 Representation of a simple feedback loop. Positive elements, typically transcription factors, activate transcription of clock genes that encode proteins that feed back to repress the activity of the positively acting transcription factors. Delays incorporated into the system ensure that one cycle lasts for 24 h. The positive elements also activate genes in output pathways generating circadian rhythms in genes not encoding clock proteins.

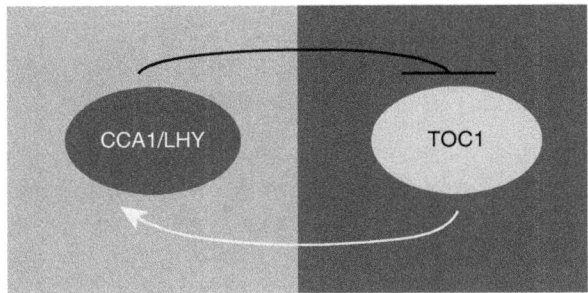

Fig. 7.3 The first plant circadian feedback loop to be discovered. CCA1 and LHY are MYB transcription factors, expressed in the morning, that repress TOC1 transcription. TOC1 is a PSEUDO RESPONSE REGULATOR, now known to bind DNA, proposed to activate transcription of *LHY* and *CCA1*.

the amounts of *LHY* and *CCA1* mRNA (Alabadi et al., 2001). Finally LHY and CCA1 bound to the promoter of *TOC1*; much later TOC1 was shown to bind to the promoter of *CCA1* (Huang et al., 2012). These observations led to the formulation of a simple feedback loop, shown in Fig. 7.3.

This model was later extended. The loop including CCA1/LHY and TOC1 remained central, but interlocked morning and evening loops were added that included other PSEUDO RESPONSE REGULATORS (Locke et al., 2005). The existence of these three loops could explain why the *lhy cca1* double mutant retained low-amplitude rhythms for several cycles after transfer to continuous conditions. Mathematical modeling of this three-loop model could explain the rhythms observed in wild-type plants and in the *lhy cca1* double mutants and was supported by further genetic work (Ding et al., 2007).

Later, deeper molecular analysis of the function of TOC1 raised difficulties with the interactions essential for the central loop (Huang et al., 2012). Experiments based on identifying genome-wide binding sites of TOC1 and assessing its effect on the gene expression of these targets showed that TOC1 acts as a transcriptional repressor, whereas all of the previous models had assumed that it is a transcriptional activator. Therefore TOC1 binds to the promoter of and represses *CCA1* and *LHY* (Huang et al., 2012). This important modification of the central loop model meant that there was no activator component within the system. This difficulty was resolved recently by showing that homologues of LHY and CCA1, called REVEILLE (RVE), are clock-regulated proteins that activate transcription of evening-expressed genes such as *TOC1* (Hsu et al., 2013). In particular RVE8 was shown to have a central role in activating clock genes in the evening and controlling the transcription of a range of output genes (Hsu et al., 2013). These results suggest a more complex structure for the oscillator mechanism than previously envisaged, with mutual repression between morning elements encoded by LHY/CCA1 and evening elements encoded by PRR proteins, with the latter activated by RVE (Hsu and Harmer, 2014).

7.2.2 Regulation of output pathways

Description and analysis of clock-regulated output pathways provide an indication of the adaptive significance of the circadian clock. Such studies have been greatly facilitated by genome-wide transcriptome analyses that allow a description of all circadian clock-controlled genes in the genome and a detailed determination of the time of day at which they are expressed (Harmer et al., 2000; Covington et al., 2008). Results from these studies have provided some key examples that demonstrate the significance of circadian clock control.

Circadian clock control enables plant gene expression to anticipate changes in the daily environment rather than responding directly to environmental cues. In this way the plant can be prepared for dramatic changes in its environment. This effect is compellingly demonstrated by circadian clock regulation of the phenylpropenoid biosynthetic pathway. Phenylpropenoids are secondary metabolites that protect plants from the effect of UV light. Twenty-three genes encoding enzymes in the phenylpropenoid pathway are coordinately regulated, peaking in expression before dawn and allowing biosynthesis of protective phenylpropenoids prior to the onset of light exposure (Harmer et al., 2000). This example illustrates the roles of the circadian clock in coordination of gene expression and anticipation of environmental conditions.

The circadian clock also modulates the response to environmental cues by influencing the temporal activity of downstream signaling pathways (Hotta et al., 2007). In this way response to environmental cues can be restricted to particular times of day. For example, transcription of the *Arabidopsis CAB* gene, which encodes chlorophyll binding protein, is induced by light. Under diurnal conditions this gene shows a strong peak in transcription soon after dawn that is conferred by both clock and light regulation. However it responds to light much less effectively at other times of day, suggesting that the clock antagonizes light activation of *CAB* transcription for most of the day, effectively restricting its responsiveness to light to the morning (Millar and Kay, 1996). Similar effects were observed with responsiveness to low temperature. Transcription of genes encoding CBF transcription factors is induced by exposure to low temperatures of 4°C. However, their induction only occurs if they are exposed to low temperatures around 4 h after dawn (Fowler et al., 2005). At 16 h after dawn low-temperature exposure does not induce *CBF* expression. This effect was shown to be due to the LHY/CCA1 MYB transcription factors that participate in the central oscillator described in Section 7.2.1 binding to the promoters of *CBF* genes in the morning and enhancing their transcription (Dong et al., 2011).

A final example of the importance of regulation of clock-regulated output pathways is in the modulation of plant metabolism. Plants fix carbon dioxide by photosynthesis during the day when they are exposed to light. During this time excess fixed carbon is stored as starch. During the night, when photosynthesis is not possible, starch is broken down and used as an energy source. The rate of starch breakdown during the night is precisely modulated by the circadian clock to ensure that it is available until dawn and not exhausted prematurely (Scialdone et al., 2013) (Fig. 7.4). This process requires the circadian clock because mutants that disrupt the clock mechanism cannot modulate the rate of starch breakdown so effectively. Although the mechanisms controlling this

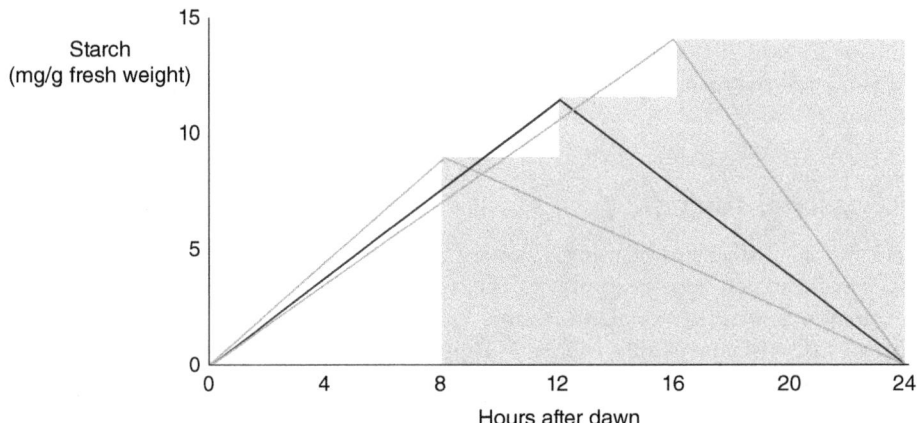

Fig. 7.4 The rate of starch breakdown is modulated to ensure that it is not exhausted until dawn. The rate is higher when the night is short than when the night is long (Scialdone et al., 2013).

phenomenon are not understood, it has been modeled by assuming interaction of a timer function that allows the duration of night to be anticipated along with a function that links the rate of starch breakdown to the amount of starch available (Scialdone et al., 2013).

7.2.3 Entrainment

The circadian clock is synchronized to the diurnal cycle by environmental cues in a process termed entrainment. Classical entraining signals are transitions from dark to light at dawn, from light to dark at dusk or steps up or down in temperature at dusk and dawn, respectively (Harmer, 2009). In plants light signals influence the activity of the oscillator components in several ways. Classical photoreceptors, including cryptochromes and phytochromes, entrain the oscillator and activate transcription of clock genes expressed around dawn, such as *CCA1* and *LHY* (Devlin and Kay, 2000; Martinez-Garcia et al., 2000). In addition, light can act at the post-translational level to regulate the stability of clock proteins. Strikingly, plants contain a small family of proteins that both absorb light and act as ubiquitin ligases targeting substrate proteins for degradation by the proteasome. These proteins are called FLAVIN F-BOX KELCH REPEATS (FKF1), ZEITLUPE (ZTL) and LOV KELCH PROTEIN 2 (LKP2) (Nelson et al., 2000; Somers et al., 2000). All these proteins bind a flavin chromophore enabling the absorption of light, encode an F-box domain that is characteristic of a particular class of ubiquitin ligase and contain kelch repeats to interact with substrates (Somers and Fujiwara, 2009). At the genetic level these genes act redundantly in controlling circadian rhythms (Baudry et al., 2010). ZTL has been shown to interact with clock components of the PRR family. In particular PRR5 is degraded in the dark but less efficiently in the light; this difference is not observed in the *ztl* mutant, in which PRR5 is more stable (Kiba et al., 2007). Similar data are available for TOC1 (Mas et al., 2003). Therefore these results suggest that one mechanism by which the clock is entrained by light is by modulation of the activity of ZTL and related proteins that target PRR clock proteins for ubiquitination and degradation.

7.3 Seasonal timing

In Section 7.2 we discussed timing of gene expression and plant responses during the day. However, plants also respond to the changing seasons, synchronizing developmental decisions with seasonal conditions to ensure that mature structures or organs develop at the optimal time to ensure reproduction or survival. Here we focus on the effect of day length (or photoperiod) and winter temperatures (vernalization) on the transition of plants from vegetative growth to flowering, the first step in plant reproduction.

7.3.1 Measurement of day length

Plants detect seasonal changes in day length and use this information to regulate their development (Andres and Coupland, 2012). Characteristic decisions controlled by day length are the transition to flowering, the onset of bud dormancy in trees in autumn and the initiation of tuberization in potatoes. Flowering of *Arabidopsis thaliana* is triggered by exposure to spring or summer long days of 16-h light and is repressed under winter short days of 10-h light. This system has been widely used as a model to decipher the mechanisms by which plants detect and respond to day length. These mechanisms have been reviewed extensively (Turck et al., 2008; Andres and Coupland, 2012) and are only briefly summarized here.

Intensive genetic analysis identified a pathway that confers early flowering of *A. thaliana* in response to long days (Koornneef et al., 1991). Isolation of the affected genes, together with epistasis analysis, construction of transgenic plants overexpressing the genes and analysis of their expression in different mutant backgrounds, placed the genes in a regulatory hierarchy (Turck et al., 2008). The CONSTANS (CO) transcription factor occupies a central position within the pathway. CO has two B-box zinc fingers and a DNA-binding domain closely related to that of the PRR clock components (Putterill et al., 1995; Robson et al., 2001; Khanna et al., 2009). The role of CO within this pathway is to activate transcription of the downstream genes *FLOWERING LOCUS T (FT)* and *TWIN SISTER OF FT (TSF)*, which are closely related paralogues encoding proteins related to the lipid-binding proteins of animals (Kardailsky et al., 1999; Kobayashi et al., 1999). CO activates transcription of *FT* and *TSF* only under long days, not under short days (Kardailsky et al., 1999; Kobayashi et al., 1999; Samach et al., 2000; Jang et al., 2009). The first indication of how this day length-dependent regulation of CO activity occurs came from the discovery that *CO* mRNA is controlled by the circadian clock and accumulates around 12–20 h after dawn (Suarez-Lopez et al., 2001). The timing of this peak causes *CO* mRNA to be present when plants are exposed to light under long days, whereas under short days it is expressed only in the dark. This temporal control of expression regulates the

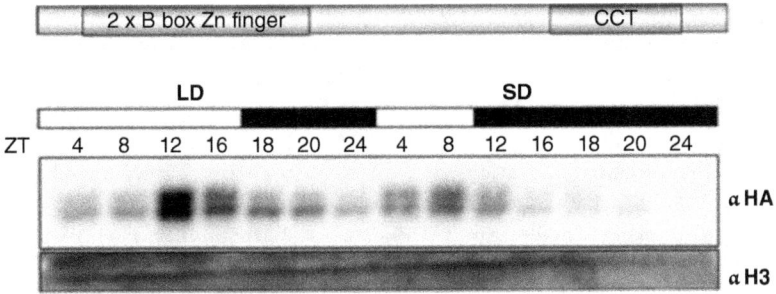

Fig. 7.5 The CONSTANS (CO) transcription factor plays a central role in promoting flowering in response to long days. CO contains two N-terminal B-box zinc fingers and a C-terminal DNA-binding domain of the CO COL TOC1 (CCT) class. Under long days (16-h light, 8-h dark) CO protein accumulates soon after dawn and at the end of the day. Under short days (10-h light, 14-h dark) CO protein does not accumulate to the levels required to promote flowering.

activity of CO because the protein is rapidly degraded in the dark but accumulates in the light (Valverde et al., 2004) (Fig. 7.5). Thus CO promotes transcription of *FT* and *TSF* under long days because its mRNA expression coincides with the exposure of plants to light, which stabilizes the protein, whereas under short days it is only expressed in the dark when the protein is rapidly degraded.

How is this light-dependent stabilization of CO protein achieved? In the dark CO protein is a substrate for a ubiquitin ligase complex that includes CONSTITUTIVE PHOTOMORPHOGENIC 1 (COP1), which is a RING finger protein, and the related protein SUPPRESSOR OF PHYTOCHROME A 1 (SPA1) (Laubinger et al., 2006; Jang et al., 2008). CO protein is stable in the dark in mutants in which activity of this complex is impaired, and both SPA1 and COP1 physically interact with CO. The COP1–SPA1 complex is more active in the dark than in the light, and therefore targets substrate proteins such as CO for degradation more effectively in the dark than in the light (Hoecker, 2005). This occurs because photoreceptors, particularly blue light photoreceptors called CRYPTOCHROMES (CRY) and red/far-red photoreceptors called PHYTOCHROMES (PHY), inactivate the COP1–SPA1 complex in the light. The C-terminal domain of CRY photoreceptors interacts directly with COP1 to inactivate the ubiquitin ligase (Yang et al., 2000; Wang et al., 2001). However, there is no evidence that this interaction is light dependent. By contrast CRY does interact with SPA1 in a light-dependent manner (Zuo et al., 2011), and this is assumed to inactivate the ubiquitin ligase complex allowing its substrates, such as CO, to accumulate specifically in the light.

CO transcription occurs in the vascular tissues of leaves, where CO protein binds directly to the *FT* promoter and activates its transcription (An et al., 2004; Tiwari et al., 2010). FT protein, which has a size of only 19.8 kDa, can move from this position in the leaves through the vascular tissue to the shoot apical meristem where it causes transcriptional reprogramming leading to the developmental transition from vegetative development to flowering and ultimately to the initiation of floral development (Schmid et al., 2003; Corbesier et al., 2007; Mathieu et al., 2007; Torti et al., 2012). Although FT is structurally similar to lipid-binding proteins in animals, this class of proteins in plants is implicated in transcriptional regulation by direct interaction with several classes of transcription factors (Abe et al., 2005; Wigge et al., 2005; Niwa et al., 2013).

This photoperiodic flowering pathway of Arabidopsis confers a seasonal response to lengthening days, ensuring that flowers develop in spring. However, the core of the pathway has proven highly conserved and has been used by a variety of plant species to control flowering or other developmental decisions that occur in response to seasonal changes in day length. These include bud dormancy in trees in autumn (Bohlenius et al., 2006), tuberization in potatoes (Navarro et al., 2011) and flowering of distantly related species such as rice (Hayama et al., 2003).

7.3.2 Vernalization

Many plants only flower if they have been exposed to winter cold for an extended period. Such species have evolved mechanisms for detecting and measuring the duration of cold. In the model species *A. thaliana* this response requires the MADS box transcription factor FLOWERING LOCUS C (FLC) (Michaels and Amasino, 1999; Sheldon et al., 1999). FLC is expressed prior to winter, when it acts a floral repressor, preventing flowering until the plant is exposed to low temperatures for an extended period. When the plant is exposed to cold for several weeks, *FLC* mRNA level falls progressively during the cold treatment. When *A. thaliana* is returned from cold to warm, *FLC* expression remains repressed even in tissues that grow and develop in the warm, and thus this provides a mitotic memory of cold exposure (Fig. 7.6)

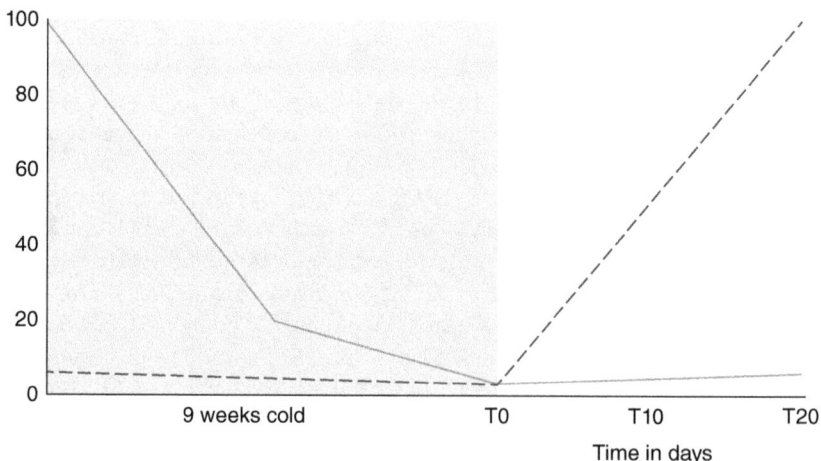

Fig. 7.6 The Lov-1 variety of *Arabidopsis thaliana* must be exposed to vernalization to flower. On exposure to cold for 9 weeks *FLC* mRNA falls progressively (gray full line). After return to the warm (T0) the *FLC* mRNA level remains low and *FT* mRNA (broken dark gray line) starts to rise to promote flowering (Coustham et al., 2012).

(Song et al., 2012). After return to the warm, *FT* is expressed and flowering occurs. Thus the key to vernalization is the reduction in abundance of *FLC* mRNA, which occurs slowly in the cold to ensure that the plant does not flower prematurely if exposed to low temperatures for just a short period of a few days, and the stable repression of FLC after return to warmer temperatures.

The mechanism by which repression of *FLC* occurs was elucidated by identifying a series of mutants in which repression of *FLC* in the cold was impaired (Sung and Amasino, 2005). Two of the genes defined by these mutations, *VERNALIZATION 2* (*VRN2*) (Gendall et al., 2001) and *VERNALIZATION INSENSITIVE 3* (*VIN3*) (Sung and Amasino, 2004), were particularly informative. *VRN2* encodes a component of the POLYCOMB REPRESSIVE 2 (PRC2) complex. This protein complex is highly conserved in metazoans and plants (Kohler and Villar, 2008). Its enzymatic activity is to modify histone 3, which is found as a component of nucleosomes around which DNA is wound in nuclei. PRC2 causes post-translational modification of a lysine residue 27 amino acids from the amino terminus of histone 3 (H3K27). The activity of PRC2 attaches three methyl groups to this amino acid, a modification termed H3K27 on histone 3. This H3K27 mark represses gene expression. Identification of mutations in this complex preventing vernalization suggested that histone modification plays a significant part in repression of *FLC* in the cold. However, analysis of *FLC* expression in the *vrn2* mutant showed that expression of *FLC* mRNA still falls in the cold, but that after return to the warm *FLC* expression rises again in the *vrn2* mutant but not in the wild type (Gendall et al., 2001). Thus the H3K27 mark seems more associated with mitotic memory of transcriptional repression than the immediate response to the cold. *VIN3* also encodes a protein implicated in histone modification (Sung and Amasino, 2004). This protein contains a PLANT HOMEODOMAIN (PHD), which is often found associated with chromatin, and a VERNALIZATION5/VIN3-LIKE (VEL) domain found in a small family of related proteins (Sung and Amasino, 2004; Greb et al., 2007). Mutations in *VIN3* prevent *FLC* downregulation in the cold (Sung and Amasino, 2004). Also, unlike *VRN2*, transcription of *VIN3* is induced during vernalization. The VIN3 protein is required for H3K27 modification of *FLC* and interacts directly with PRC2. These results suggest that transcriptional induction of *VIN3* in the cold could allow it to interact with PRC2, changing the activity of the complex, so that it more efficiently modifies histones at the *FLC* gene (Song et al., 2012).

A critical aspect of *FLC* repression is that it should be timed appropriately to correlate with the duration of winter at locations where *A. thaliana* is found in nature. Strikingly, different varieties of *A. thaliana* vary in the duration of cold treatment required for *FLC* repression and flowering to occur (Shindo et al., 2006). For example, some accessions require only 4 weeks' vernalization to flower whereas others require 9 weeks'. Genetic analysis of two varieties showing these different responses demonstrated that this difference in the length of vernalization required to induce flowering was conferred by allelic variation at the *FLC* locus (Shindo et al., 2006). The LOV-1 variety from northern Sweden requires longer vernalization than the Col accession from central Europe; this is due to sequence variation at *FLC*. Analysis of H3K27 modification of the two alleles showed that histone 3 on the LOV-1 allele is less heavily modified than on the Col allele prior to vernalization, and that the mark accumulates on the LOV-1 allele more slowly during vernalization (Coustham et al., 2012). Recombinant versions of *FLC* were then constructed swapping parts of the Col and LOV-1 alleles. This analysis identified four polymorphisms near the transcriptional start site of *FLC* that contribute a large part of the difference in the rate of modification of histones at *FLC* (Coustham et al., 2012). The region containing these polymorphisms was previously identified as a domain called the nucleation region, where H3K27 first accumulates on the *FLC* gene prior to spreading across the whole locus (Angel et al., 2011). These experiments indicate that the vernalization pathway can adapt to repress *FLC* at different rates that correlate with the duration of winter.

An additional intriguing aspect of vernalization is that in some species it is strictly related to the age of the plant. In such species, which include relatives of *A. thaliana*, vernalization does not induce flowering of young plants, but rather they must achieve a certain age before becoming responsive to vernalization. In these species repression of *FLC* is not age-related (Wang et al., 2011), demonstrating that the age-related process must occur in parallel to allow flowering to occur when *FLC* is repressed. This age-related process was recently shown to be caused by downregulation of a microRNA called miRNA15, which falls progressively with age in several species (Bergonzi et al., 2013; Zhou et al., 2013). This miRNA binds to the mRNA of a class of transcription factors called SQUAMOSA BINDING PROTEIN LIKE (SPL), repressing expression of the SPLs (Chuck et al., 2007; Wang et al., 2009; Wu et al., 2009). In relatives of *A. thaliana* that show age-related vernalization, miRNA156 must reach trough levels and some of its target SPL transcription factors must rise in expression before flowering can occur. This is an additional time measurement phenomenon associated with vernalization, which ensures that plants do not flower if they are exposed to winter temperatures when extremely young.

7.4 Timing in natural populations

All the above analyses of time measurement were carried out in carefully controlled artificial growth conditions, but are assumed to be relevant in natural environments. To find out if this is indeed the case several experimental approaches have been used to address the significance of these pathways under normal environmental conditions.

One approach is to study plant varieties (or accessions) that are adapted to different local environments and to determine whether the timing pathways described in Section 7.3 are modified among these varieties in a way that could explain their local adaptation (Bohlenius et al., 2006). One of the most striking examples of adaptation is latitudinal clines, in which plants vary in behavior gradually and consistently across a latitudinal gradient (Savolainen et al., 2013). For example, trees vary in the precise time that they terminate growth in the autumn, and this correlates with the latitude at which they are found in nature (Fig. 7.7).

The initiation of growth termination is controlled by day length, and the correlation of day length with latitude underlies this cline. So the growth of poplar trees at high latitude stops when the day length is longer than for poplar trees

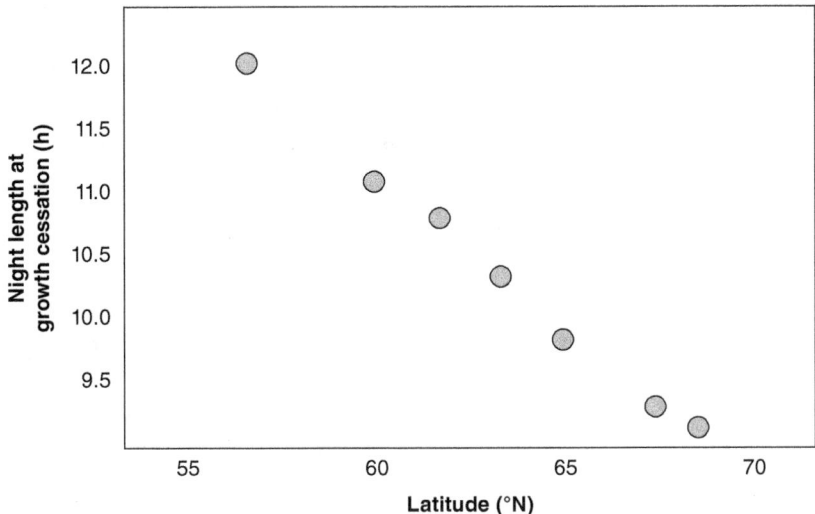

Fig. 7.7 A cline in growth cessation in the tree *Betula pendula*. Plants growing at the latitude shown were transferred to the greenhouse and the day length at which growth stops was measured. The day length at which growth stops progressively increases (or night length falls as shown) with the latitude at which they are found (Savolainen et al., 2013).

growing at lower latitude. This response ensures that those trees growing at high latitude stop growing sooner in the autumn and are therefore prepared for the earlier onset of a harsher winter. Variation in growth termination along the cline was shown to be conferred by the same pathway that controls photoperiodic flowering in *A. thaliana* (Bohlenius et al., 2006). The poplar orthologue of *FT* promotes growth, and its expression must be reduced to allow growth termination. A latitudinal cline that controls differences in the timing of growth termination was shown to correlate with the timing of expression of the poplar orthologue of *CO* during the day. Plants from higher latitudes expressed *CO* mRNA later in the day, so that when the day length dropped below about 21 h *CO* was already expressed in the dark and therefore growth stopped. In plants growing at lower latitudes *CO* was expressed in the light even if the day length dropped to about 16 h. This elegant set of experiments demonstrated that genetic variation in the photoperiodic pathway defined in *A. thaliana* contributes to ecologically important variation along a latitudinal cline (Bohlenius et al., 2006).

Another approach for testing the significance of timing pathways defined in nature is to isolate genes that contribute to ecologically significant variation in natural populations. In this case, the genes identified in controlled conditions are not used as candidates; rather genetic approaches are used to isolate the genes that contribute variation in natural populations. Here the methods of quantitative genetics are used, such as quantitative trait loci (QTL) mapping, association genetics or nested association mapping (Nordborg and Weigel, 2008; McMullen et al., 2009; Atwell et al., 2010;

Bergelson and Roux, 2010). These methods allow the isolation of genes that contribute variation among natural accessions.

One of the earliest examples of QTL mapping in *A. thaliana* studied variation in flowering time between varieties collected in the Cape Verde Islands and in central Europe (El-Assal et al., 2001). This analysis studied populations made by crossing these varieties in controlled conditions, and convincingly showed that allelic variation

Table 7.1 *Arabidopsis thaliana* genes for which natural genetic variation has been found to modify the timing of flowering. (Modified from Alonso-Blanco et al., 2009, and Weigel, 2012.)

GENE	PATHWAY AFFECTED	COMMENTS	REFERENCES
FRIGIDA	Vernalization	Many independent alleles	Lempe et al. (2005), Shindo et al. (2005)
FLOWERING LOCUS C	Vernalization	Many independent alleles	Michaels et al. (2003), Lempe et al. (2005), Shindo et al. (2005)
FRIGIDA-LIKE 1	Vernalization		Schlappi (2006)
FRIGIDA-LIKE 2	Vernalization		Schlappi (2006)
FLM	General repressor		Werner et al. (2005)
MAF2	General repressor		Rosloski et al. (2010)
FT	Photoperiod		Schwartz et al. (2009)
TWIN SISTER OF FT	Photoperiod	Candidate from nested association mapping in nature	Brachi et al. (2010)
PHYTOCHROME C	Photoperiod		Balasubramanian et al. (2006)
CRYPTOCHROME 2	Photoperiod	A case study of how to isolate and confirm a QTL	El-Assal et al. (2001)
SHORT VEGETATIVE PHASE	General repressor		Mendez-Vigo et al. (2013)

at the *CRYPTOCHROME 2* gene contributed to differences in flowering time between these natural varieties. This gene contributes to the photoperiodic flowering pathway to regulate the stability of CO, as already described, so this example again supported the idea that genetic variation found in nature is closely related to that found by mutagenesis in laboratory varieties. Many such QTL studies have now been carried out and the results of some of these are summarized in Table 7.1. Generally these find variation at genes that were known from mutagenesis studies in the laboratory. However, mutations in *FRIGIDA* (*FRI*) and *FLC* are only found if appropriate accessions are used for mutagenesis (Michaels and Amasino, 1999), because FRI is required to activate the expression of *FLC* and many laboratory accessions carry weak or null alleles at *FLC* or *FRI* (Michaels et al., 2003; Lempe et al., 2005; Shindo et al., 2005). Also the *FRI-LIKE* genes were easily identified by natural variation (Schlappi, 2006), although *FRI-LIKE 1* was recovered by mutagenesis of a late-flowering vernalization-requiring variety (Michaels et al., 2004). The data summarized in Table 7.1 also illustrate that variation in the vernalization response pathway seems much more prevalent than variation in the photoperiodic pathway. Perhaps, the major components of this pathway, *FRI* and *FLC*, are more dispensable than the photoperiodic pathway, whose components might have other roles in light-regulated development that make them more strongly selected for.

A final approach that has been used to study the significance of timing in natural populations is to perform association genetics or nested association mapping directly on populations growing in natural conditions. This method contrasts with previous examples that studied genetic variation between plants isolated in nature but scored the phenotypic variation in controlled conditions. Mapping directly in natural conditions has not been carried out very often, but one experiment scored large populations of natural varieties and mapping populations in a garden in Lille (Brachi et al., 2010). This experiment identified candidate loci, particularly *TSF*, as causing more pronounced differences in flowering time than expected from what was known from work in controlled environments.

7.5 Conclusion

We have now a broad outline of the pathways used by plants to measure time during the daily cycle and across the changing seasons. Many of these processes involve plant-specific proteins, although the concepts are transferable to similar systems found in animals (Wood and Loudon, 2014). The genes identified by intensive genetics in controlled environments do seem to underlie natural variation during adaptation in nature, although some pathways and components are much more variable than others. The vernalization pathway is more variable than the photoperiodic pathway. Also the variation found in the photoperiodic pathway seems to be mainly concentrated in photoreceptor-encoding genes and *FT*, but not in other central components such as *CO*. Such patterns could not be immediately predicted from the analysis of model varieties in controlled conditions, justifying the study of such adaptive traits in the full complexity of natural environments.

References

Abe, M., Kobayashi, Y., Yamamoto, S., Daimon, Y., Yamaguchi, A., Ikeda, Y., Ichinoki, H., Notaguchi, M., Goto, K. and Araki, T. (2005). FD, a bZIP protein mediating signals from the floral pathway integrator FT at the shoot apex. *Science* **309**, 1052–1056.

Alabadi, D., Oyama, T., Yanovsky, M.J., Harmon, F.G., Mas, P. and Kay, S.A. (2001). Reciprocal regulation between TOC1 and LHY/CCA1 within the Arabidopsis circadian clock. *Science* **293**, 880–883.

Alonso-Blanco, C., Aarts, M.G., Bentsink, L., Keurentjes, J.J., Reymond, M., Vreugdenhil, D. and Koornneef, M. (2009). What has natural variation taught us about plant development, physiology, and adaptation? *Plant Cell* **21**, 1877–1896.

An, H., Roussot, C., Suarez-Lopez, P., Corbesier, L., Vincent, C., Pineiro, M., Hepworth, S., Mouradov, A., Justin, S., Turnbull, C.G.N. and Coupland, G. (2004). CONSTANS acts in the phloem to regulate a systemic signal that induces photoperiodic flowering of Arabidopsis. *Development* **131**, 3615–3626.

Andres, F. and Coupland, G. (2012). The genetic basis of flowering responses to seasonal cues. *Nature Reviews Genetics* **13**, 627–639.

Angel, A., Song, J., Dean, C. and Howard, M. (2011). A Polycomb-based switch underlying quantitative epigenetic memory. *Nature* **476**, 105–108.

Atwell, S., Huang, Y.S., Vilhjalmsson, B.J., Willems, G., Horton, M., Li, Y., Meng, D., Platt, A., Tarone, A.M., Hu, T.T., Jiang, R., Muliyati, N.W., Zhang, X., Amer, M.A., Baxter, I., Brachi, B., Chory, J., Dean, C., Debieu, M., de Meaux, J., Ecker, J.R., Faure, N., Kniskern, J.M., Jones, J.D., Michael, T., Nemri, A., Roux, F., Salt, D.E., Tang, C., Todesco, M., Traw, M.B., Weigel, D., Marjoram, P., Borevitz, J.O., Bergelson, J. and Nordborg, M. (2010). Genome-wide association study of 107 phenotypes in Arabidopsis thaliana inbred lines. *Nature* **465**, 627–631.

Balasubramanian, S., Sureshkumar, S., Agrawal, M., Michael, T.P., Wessinger, C., Maloof, J.N., Clark, R., Warthmann, N., Chory, J. and Weigel, D. (2006). The PHYTOCHROME C photoreceptor gene mediates natural variation in flowering and growth responses of *Arabidopsis thaliana*. *Nature Genetics* **38**, 711–715.

Baudry, A., Ito, S., Song, Y.H., Strait, A.A., Kiba, T., Lu, S., Henriques, R., Pruneda-Paz, J.L., Chua, N.H., Tobin, E.M., Kay, S.A. and Imaizumi, T. (2010). F-box proteins FKF1 and LKP2 act in concert with ZEITLUPE to control Arabidopsis clock progression. *Plant Cell* **22**, 606–622.

Bergelson, J. and Roux, F. (2010). Towards identifying genes underlying ecologically relevant traits in *Arabidopsis thaliana*. *Nature Reviews Genetics* **11**, 867–879.

Bergonzi, S., Albani, M.C., Ver Loren van Themaat, E., Nordstrom, K.J., Wang, R., Schneeberger, K., Moerland, P.D. and Coupland, G. (2013). Mechanisms of age-dependent response to winter temperature in perennial flowering of *Arabis alpina*. *Science* **340**, 1094–1097.

Bohlenius, H., Huang, T., Charbonnel-Campaa, L., Brunner, A.M., Jansson, S., Strauss, S.H. and Nilsson, O. (2006). CO/FT regulatory module controls timing of flowering and seasonal growth cessation in trees. *Science* **312**, 1040–1043.

Brachi, B., Faure, N., Horton, M., Flahauw, E., Vazquez, A., Nordborg, M., Bergelson, J., Cuguen, J. and Roux, F. (2010). Linkage and association mapping of Arabidopsis thaliana flowering time in nature. *PLoS Genetics* **6**, e1000940.

Chuck, G., Cigan, A.M., Saeteurn, K. and Hake, S. (2007). The heterochronic maize mutant Corngrass1 results from overexpression of a tandem microRNA. *Nature Genetics* **39**, 544–549.

Corbesier, L., Vincent, C., Jang, S., Fornara, F., Fan, Q., Searle, I., Giakountis, A., Farrona, S., Gissot, L., Turnbull, C. and Coupland, G. (2007). FT protein movement contributes to long-distance signaling in floral induction of Arabidopsis. *Science* **316**, 1030–1033.

Coustham, V., Li, P., Strange, A., Lister, C., Song, J. and Dean, C. (2012). Quantitative modulation of polycomb silencing underlies natural variation in vernalization. *Science* **337**, 584–587.

Covington, M.F., Maloof, J.N., Straume, M., Kay, S.A. and Harmer, S.L. (2008). Global transcriptome analysis reveals circadian regulation of key pathways in plant growth and development. *Genome Biology* **9**, R130.

Devlin, P.F. and Kay, S.A. (2000). Cryptochromes are required for phytochrome signaling to the circadian clock but not for rhythmicity. *Plant Cell* **12**, 2499–2510.

Ding, Z., Doyle, M.R., Amasino, R.M. and Davis, S.J. (2007). A complex genetic interaction between *Arabidopsis thaliana* TOC1 and CCA1/LHY in driving the circadian clock and in output regulation. *Genetics* **176**, 1501–1510.

Dodd, A.N., Salathia, N., Hall, A., Kevei, E., Toth, R., Nagy, F., Hibberd, J.M., Millar, A.J. and Webb, A.A. (2005). Plant circadian clocks increase photosynthesis, growth, survival, and competitive advantage. *Science* **309**, 630–633.

Dong, M.A., Farre, E.M. and Thomashow, M.F. (2011). Circadian clock-associated 1 and late elongated hypocotyl regulate expression of the C-repeat binding factor (CBF) pathway in Arabidopsis. *Proceedings of the National Academy of Sciences of the United States of America* **108**, 7241–7246.

Dunlap, J.C. (1999). Molecular bases for circadian clocks. *Cell* **96**, 271–290.

El-Assal, S.E.D., Alonso-Blanco, C., Peeters, A.J.M., Raz, V. and Koornneef, M. (2001). A QTL for flowering time in *Arabidopsis* reveals a novel allele of *CRY2 Nature Genetics* **29**, 435–440.

Fowler, S.G., Cook, D. and Thomashow, M.F. (2005). Low temperature induction of Arabidopsis CBF1, 2, and 3 is gated by the circadian clock. *Plant Physiology* **137**, 961–968.

Gendall, A.R., Levy, Y.Y., Wilson, A. and Dean, C. (2001). The VERNALIZATION 2 gene mediates the epigenetic regulation of vernalization in Arabidopsis. *Cell* **107**, 525–535.

Greb, T., Mylne, J.S., Crevillen, P., Geraldo, N., An, H., Gendall, A.R. and Dean, C. (2007). The PHD finger protein VRN5 functions in the epigenetic silencing of Arabidopsis FLC. *Current Biology* **17**, 73–78.

Harmer, S.L. (2009). The circadian system in higher plants. *Annual Review of Plant Biology* **60**, 357–377.

Harmer, S.L., Hogenesch, J.B., Straume, M., Chang, H.S., Han, B., Zhu, T., Wang, X., Kreps, J.A. and Kay, S.A. (2000). Orchestrated transcription of key pathways in Arabidopsis by the circadian clock. *Science* **290**, 2110–2113.

Hayama, R., Yokoi, S., Tamaki, S., Yano, M. and Shimamoto, K. (2003). Adaptation of photoperiodic control pathways produces short-day flowering in rice. *Nature* **422**, 719–722.

Hoecker, U. (2005). Regulated proteolysis in light signaling. *Current Opinion in Plant Biology* **8**, 469–476.

Hotta, C.T., Gardner, M.J., Hubbard, K.E., Baek, S.J., Dalchau, N., Suhita, D., Dodd, A.N. and Webb, A.A. (2007). Modulation of environmental responses of plants by circadian clocks. *Plant, Cell and Environment* **30**, 333–349.

Hsu, P.Y. and Harmer, S.L. (2014). Wheels within wheels: the plant circadian system. *Trends in Plant Science* **19**, 240–249.

Hsu, P.Y., Devisetty, U.K. and Harmer, S.L. (2013). Accurate timekeeping is controlled by a cycling activator in Arabidopsis. *eLIFE* **2**, e00473.

Huang, W., Perez-Garcia, P., Pokhilko, A., Millar, A.J., Antoshechkin, I., Riechmann, J.L. and Mas, P. (2012). Mapping the core of the Arabidopsis circadian clock defines the network structure of the oscillator. *Science* **336**, 75–79.

Jang, S., Torti, S. and Coupland, G. (2009). Genetic and spatial interactions between FT, TSF and SVP during the early stages of floral induction in Arabidopsis. *The Plant Journal* **60**, 614–625.

Jang, S., Marchal, V., Panigrahi, K.C., Wenkel, S., Soppe, W., Deng, X.W., Valverde, F. and Coupland, G. (2008). Arabidopsis COP1 shapes the temporal pattern of CO accumulation conferring a photoperiodic flowering response. *EMBO Journal* **27**, 1277–1288.

Kardailsky, I., Shukla, V.K., Ahn, J.H., Dagenais, N., Christensen, S.K., Nguyen, J.T., Chory, J., Harrison, M.J. and Weigel, D. (1999). Activation tagging of the floral inducer FT. *Science* **286**, 1962–1965.

Khanna, R., Kronmiller, B., Maszle, D.R., Coupland, G., Holm, M., Mizuno, T. and Wu, S.H. (2009). The Arabidopsis B-box zinc finger family. *Plant Cell* **21**, 3416–3420.

Kiba, T., Henriques, R., Sakakibara, H. and Chua, N.H. (2007). Targeted degradation of PSEUDO-RESPONSE REGULATOR5 by an SCFZTL complex regulates clock function and photomorphogenesis in *Arabidopsis thaliana*. *Plant Cell* **19**, 2516–2530.

Kobayashi, Y., Kaya, H., Goto, K., Iwabuchi, M. and Araki, T. (1999). A pair of related genes with antagonistic roles in mediating flowering signals. *Science* **286**, 1960–1962.

Kohler, C. and Villar, C.B. (2008). Programming of gene expression by Polycomb group proteins. *Trends in Cell Biology* **18**, 236–243.

Koornneef, M., Hanhart, C.J. and Van Der Veen, J.H. (1991). A genetic and physiological analysis of late flowering mutants in *Arabidopsis thaliana*. *Molecular and General Genetics* **229**, 57–66.

Laubinger, S., Marchal, V., Le Gourrierec, J., Wenkel, S., Adrian, J., Jang, S., Kulajta, C., Braun, H., Coupland, G. and Hoecker, U. (2006). Arabidopsis SPA proteins regulate photoperiodic flowering and interact with the floral inducer CONSTANS to regulate its stability. *Development* **133**, 3213–3222.

Lempe, J., Balasubramanian, S., Sureshkumar, S., Singh, A., Schmid, M. and Weigel, D. (2005). Diversity of flowering responses in wild *Arabidopsis thaliana* strains. *PLoS Genetics* **1**, 109–118.

Locke, J.C., Southern, M.M., Kozma-Bognar, L., Hibberd, V., Brown, P.E., Turner, M.S. and Millar, A.J. (2005). Extension of a genetic network model by iterative experimentation and mathematical analysis. *Molecular Systems Biology* **1**, 2005 0013.

McMullen, M.D., Kresovich, S., Villeda, H.S., Bradbury, P., Li, H., Sun, Q., Flint-Garcia, S., Thornsberry, J., Acharya, C., Bottoms, C., Brown, P., Browne, C., Eller, M., Guill, K., Harjes, C., Kroon, D., Lepak, N., Mitchell, S.E., Peterson, B., Pressoir, G., Romero, S., Oropeza Rosas, M., Salvo, S., Yates, H., Hanson, M., Jones, E., Smith, S., Glaubitz, J.C., Goodman, M., Ware, D., Holland, J.B. and Buckler, E.S. (2009). Genetic properties of the maize nested association mapping population. *Science* **325**, 737–740.

Martinez-Garcia, J.F., Huq, E. and Quail, P.H. (2000). Direct targeting of light signals to a promoter element-bound transcription factor. *Science* **288**, 859–863.

Mas, P., Kim, W.Y., Somers, D.E. and Kay, S.A. (2003). Targeted degradation of TOC1 by ZTL modulates circadian function in *Arabidopsis thaliana*. *Nature* **426**, 567–570.

Mathieu, J., Warthmann, N., Kuttner, F. and Schmid, M. (2007). Export of FT protein from phloem companion cells is sufficient for floral induction in Arabidopsis. *Current Biology* **17**, 1055–1060.

Mendez-Vigo, B., Martinez-Zapater, J.M. and Alonso-Blanco, C. (2013). The flowering repressor SVP underlies a novel Arabidopsis thaliana QTL interacting with the genetic background. *PLoS Genetics* **9**, e1003289.

Michaels, S.D. and Amasino, R.M. (1999). *FLOWERING LOCUS C* encodes a novel MADS domain protein that acts as a repressor of flowering. *Plant Cell* **11**, 949–956.

Michaels, S.D., He, Y., Scortecci, K.C. and Amasino, R.M. (2003). Attenuation of FLOWERING LOCUS C activity as a mechanism for the evolution of summer-annual flowering behavior in Arabidopsis. *Proceedings of the National Academy of Sciences of the United States of America* **100**, 10102–10107.

Michaels, S.D., Bezerra, I.C. and Amasino, R.M. (2004). FRIGIDA-related genes are required for the winter-annual habit in Arabidopsis. *Proceedings of the National Academy of Sciences of the United States of America* **101**, 3281–3285.

Millar, A.J. and Kay, S.A. (1996). Integration of circadian and phototransduction pathways in the network controlling CAB gene transcription in Arabidopsis. *Proceedings of the National Academy of Sciences of the United States of America* **93**, 15491–15496.

de Montaigu, A., Toth, R. and Coupland, G. (2010). Plant development goes like clockwork. *Trends in Genetics* **26**, 296–306.

Navarro, C., Abelenda, J.A., Cruz-Oro, E., Cuellar, C.A., Tamaki, S., Silva, J., Shimamoto, K. and Prat, S. (2011). Control of flowering and storage organ formation in potato by FLOWERING LOCUS T. *Nature* **478**, 119–122.

Nelson, D.C., Lasswell, J., Rogg, L.E., Cohen, M.A. and Bartel, B. (2000). FKF1, a clock-controlled gene that regulates the transition to flowering in Arabidopsis. *Cell* **101**, 331–340.

Niwa, M., Daimon, Y., Kurotani, K., Higo, A., Pruneda-Paz, J.L., Breton, G., Mitsuda, N., Kay, S.A., Ohme-Takagi, M., Endo, M. and Araki, T. (2013). BRANCHED1 interacts with FLOWERING LOCUS T to repress the floral transition of the axillary meristems in Arabidopsis. *Plant Cell* **25**, 1228–1242.

Nordborg, M. and Weigel, D. (2008). Next-generation genetics in plants. *Nature* **456**, 720–723.

Putterill, J., Robson, F., Lee, K., Simon, R. and Coupland, G. (1995). The CONSTANS gene of Arabidopsis promotes flowering and encodes a protein showing similarities to zinc finger transcription factors. *Cell* **80**, 847–857.

Robson, F., Costa, M.M., Hepworth, S.R., Vizir, I., Pineiro, M., Reeves, P.H., Putterill, J. and Coupland, G. (2001). Functional importance of conserved domains in the flowering-time gene CONSTANS demonstrated by analysis of mutant alleles and transgenic plants. *Plant Journal* **28**, 619–631.

Rosloski, S.M., Jali, S.S., Balasubramanian, S., Weigel, D. and Grbic, V. (2010). Natural diversity in flowering responses of *Arabidopsis thaliana* caused by variation in a tandem gene array. *Genetics* **186**, 263–276.

Samach, A., Onouchi, H., Gold, S.E., Ditta, G.S., Schwarz-Sommer, Z., Yanofsky, M.F. and Coupland, G. (2000). Distinct roles of CONSTANS target genes in reproductive development of Arabidopsis. *Science* **288**, 1613–1616.

Savolainen, O., Lascoux, M. and Merila, J. (2013). Ecological genomics of local adaptation. *Nature Reviews Genetics* **14**, 807–820.

Schaffer, R., Ramsay, N., Samach, A., Corden, S., Putterill, J., Carre, I.A. and Coupland, G. (1998). The late elongated hypocotyl mutation of Arabidopsis disrupts circadian rhythms and the photoperiodic control of flowering. *Cell* **93**, 1219–1229.

Schlappi, M.R. (2006). FRIGIDA LIKE 2 is a functional allele in Landsberg erecta and compensates for a nonsense allele of FRIGIDA LIKE 1. *Plant Physiology* **142**, 1728–1738.

Schmid, M., Uhlenhaut, N.H., Godard, F., Demar, M., Bressan, R., Weigel, D. and Lohmann, J.U. (2003). Dissection of floral induction pathways using global expression analysis. *Development* **130**, 6001–6012.

Schwartz, C., Balasubramanian, S., Warthmann, N., Michael, T.P., Lempe, J., Sureshkumar, S., Kobayashi, Y., Maloof, J.N., Borevitz, J.O., Chory, J. and Weigel, D. (2009). Cis-regulatory changes at FLOWERING LOCUS T mediate natural variation in flowering responses of *Arabidopsis thaliana*. *Genetics* **183**, 723–732, 721SI–727SI.

Scialdone, A., Mugford, S.T., Feike, D., Skeffington, A., Borrill, P., Graf, A., Smith, A.M. and Howard, M. (2013). Arabidopsis plants perform arithmetic division to prevent starvation at night. *eLIFE* **2**, e00669.

Sheldon, C.C., Burn, J.E., Perez, P.P., Metzger, J., Edwards, J.A., Peacock, W.J. and Dennis, E.S. (1999). The FLF MADS box gene: A repressor of flowering in Arabidopsis regulated by vernalization and methylation. *Plant Cell* **11**, 445–458.

Shindo, C., Aranzana, M.J., Lister, C., Baxter, C., Nicholls, C., Nordborg, M. and Dean, C. (2005). Role of FRIGIDA and FLOWERING LOCUS C in determining variation in flowering time of Arabidopsis. *Plant Physiology* **138**, 1163–1173.

Shindo, C., Lister, C., Crevillen, P., Nordborg, M. and Dean, C. (2006). Variation in the epigenetic silencing of FLC contributes to natural variation in Arabidopsis vernalization response. *Genes and Development* **20**, 3079–3083.

Somers, D.E. and Fujiwara, S. (2009). Thinking outside the F-box: novel ligands for novel receptors. *Trends in Plant Science* **14**, 206–213.

Somers, D.E., Schultz, T.F., Milnamow, M. and Kay, S.A. (2000). ZEITLUPE encodes a novel clock-associated PAS protein from Arabidopsis. *Cell* **101**, 319–329.

Song, J., Angel, A., Howard, M. and Dean, C. (2012). Vernalization - a cold-induced epigenetic switch. *Journal of Cell Science* **125**, 3723–3731.

Strayer, C., Oyama, T., Schultz, T.F., Raman, R., Somers, D.E., Mas, P., Panda, S., Kreps, J.A. and Kay, S.A. (2000). Cloning of the Arabidopsis clock gene TOC1, an autoregulatory response regulator homolog. *Science* **289**, 768–771.

Suarez-Lopez, P., Wheatley, K., Robson, F., Onouchi, H., Valverde, F. and Coupland, G. (2001). *CONSTANS* mediates between the circadian clock and the control of flowering in *Arabidopsis*. *Nature* **410**, 1116–1120.

Sung, S.B. and Amasino, R.M. (2004). Vernalization in Arabidopsis thaliana is mediated by the PHD finger protein VIN3. *Nature* **427**, 159–164.

Sung, S. and Amasino, R.M. (2005). Remembering winter: toward a molecular understanding of vernalization. *Annual Review of Plant Biology* **56**, 491–508.

Tiwari, S.B., Shen, Y., Chang, H.C., Hou, Y., Harris, A., Ma, S.F., McPartland, M., Hymus, G.J., Adam, L., Marion, C., Belachew, A., Repetti, P.P., Reuber, T.L. and Ratcliffe, O.J. (2010). The flowering time regulator CONSTANS is recruited to the FLOWERING LOCUS T promoter via a unique cis-element. *New Phytologist* **187**, 57–66.

Torti, S., Fornara, F., Vincent, C., Andrés, F., Nordström, K., Göbel, U., Knoll, D., Schoof, H. and Coupland, G. (2012). Analysis of the Arabidopsis shoot meristem transcriptome during floral transition identifies distinct regulatory patterns and a leucine-rich repeat protein that promotes flowering. *Plant Cell* **24**, 444–462.

Turck, F., Fornara, F. and Coupland, G. (2008). Regulation and identity of florigen: FLOWERING LOCUS T moves center stage. *Annual Review of Plant Biology* **59**, 573–594.

Valverde, F., Mouradov, A., Soppe, W., Ravenscroft, D., Samach, A. and Coupland, G. (2004). Photoreceptor regulation of CONSTANS protein and the mechanism of photoperiodic flowering. *Science* **303**, 1003–1006.

Wang, Z.Y. and Tobin, E.M. (1998). Constitutive expression of the CIRCADIAN CLOCK ASSOCIATED 1 (CCA1) gene disrupts circadian rhythms and suppresses its own expression. *Cell* **93**, 1207–1217.

Wang, H., Ma, L.G., Li, J.M., Zhao, H.Y. and Deng, X.W. (2001). Direct interaction of Arabidopsis cryptochromes with COP1 in light control development. *Science* **294**, 154–158.

Wang, J.W., Czech, B. and Weigel, D. (2009). miR156-Regulated SPL transcription factors define an endogenous flowering pathway in *Arabidopsis thaliana*. *Cell* **138**, 738–749.

Wang, R., Albani, M.C., Vincent, C., Bergonzi, S., Luan, M., Bai, Y., Kiefer, C., Castillo, R. and Coupland, G. (2011). Aa TFL1 confers an age-dependent response to vernalization in perennial Arabis alpina. *Plant Cell* **23**, 1307–1321.

Weigel, D. (2012). Natural variation in Arabidopsis: from molecular genetics to ecological genomics. *Plant Physiology* **158**, 2–22.

Werner, J.D., Borevitz, J.O., Warthmann, N., Trainer, G.T., Ecker, J.R., Chory, J. and Weigel, D. (2005). Quantitative trait locus mapping and DNA array hybridization identify an FLM deletion as a cause for natural flowering-time variation. *Proceedings of the National Academy of Sciences of the United States of America* **102**, 2460–2465.

Wigge, P.A., Kim, M.C., Jaeger, K.E., Busch, W., Schmid, M., Lohmann, J.U. and Weigel, D. (2005). Integration of spatial and temporal information during floral induction in Arabidopsis. *Science* **309**, 1056–1059.

Wood, S. and Loudon, A. (2014). Clocks for all seasons: unwinding the roles and mechanisms of circadian and interval timers in the hypothalamus and pituitary. *Journal of Endocrinology* **222**, R39–R59.

Wu, G., Park, M.Y., Conway, S.R., Wang, J.W., Weigel, D. and Poethig, R.S. (2009). The sequential action of miR156 and miR172 regulates developmental timing in Arabidopsis. *Cell* **138**, 750–759.

Yang, H.Q., Wu, Y.J., Tang, R.H., Liu, D., Liu, Y. and Cashmore, A.R. (2000). The C termini of Arabidopsis cryptochromes mediate a constitutive light response. *Cell* **103**, 815–827.

Zhou, C.M., Zhang, T.Q., Wang, X., Yu, S., Lian, H., Tang, H., Feng, Z.Y., Zozomova-Lihova, J. and Wang, J.W. (2013). Molecular basis of age-dependent vernalization in *Cardamine flexuosa*. *Science* **340**, 1097–1100.

Zuo, Z., Liu, H., Liu, B., Liu, X. and Lin, C. (2011). Blue light-dependent interaction of CRY2 with SPA1 regulates COP1 activity and floral initiation in Arabidopsis. *Current Biology* **21**, 841–847.

8
Forces in plant development

Olivier HAMANT

Plant Reproduction and Development Lab, ENS Lyon, France

Abstract

The development of living organisms relies on complex molecular regulations. Because organisms also are physical objects, the mechanics of growth need to be incorporated to fully understand how changes in shape are controlled. This chapter explores this molecular and mechanical nexus, mainly using examples from the plant world obtained from the recent literature. The chapter is divided in three main sections. First, some basic biomechanical concepts are explained, with a focus on the relationship between force and shape. In the second section we investigate how the mechanical properties of cells help to build a tissue and, in turn, how tissue shape can prescribe a stress pattern that controls cell behavior. The third section explores mechanotransduction at the subcellular scale, with its implications for cell polarity. The overall picture is one in which the robustness of development relies on coordinated feedback loops that are both molecular and mechanical in essence.

Keywords

Mechanics, stiffness, stress, strain, mechanotransduction, patterning, growth, microtubules, meristem, plant

Chapter Contents

8 Forces in plant development — 209
Olivier HAMANT

- **8.1** Overview — 211
  - 8.1.1 Definitions — 211
  - 8.1.2 Basics — 212
  - 8.1.3 The structure behind the architecture — 215
  - 8.1.4 Dynamics: forces and movements in plants — 219
- **8.2** Understanding the role of forces at the tissue scale — 221
  - 8.2.1 From material properties to shape — 221
  - 8.2.2 Shape-derived stress — 223
  - 8.2.3 Growth-derived stress — 226
  - 8.2.4 Mechanical signals in animals — 228
- **8.3** Understanding the role of forces at the cell scale — 236
  - 8.3.1 Subcellular stress — 236
  - 8.3.2 Cell polarity and membrane tension — 240
  - 8.3.3 Mechanotransduction — 241
- **8.4** Conclusion — 242

  References — 243

8.1 Overview

8.1.1 Definitions

A full grasp on biomechanics requires a proper understanding of the terms used, and thus correct definitions. The definitions below are adapted from those proposed by Bruno Moulia in a recent review (Moulia 2013). These concepts will be discussed in a wider context in the rest of the chapter, while also integrating the contribution from the molecular effectors of the cell.

8.1.1.1 Growth

Growth is an irreversible increase in volume. In all organisms it depends on the ability of the cell to produce new material (and thus largely on the S phase and the production and activity of ribosomes). In plants, growth involves the deformation (strain) of existing polymeric cell walls, so measuring the intensity and spatial distribution of growth requires kinematic methods.

8.1.1.2 Kinematics

Kinematics is the quantitative study of the motion, deformation and flow of bodies, without analyzing their causes.

8.1.1.3 Mechanics

Mechanics is the science that deals with the movement (displacements and deformations) of bodies under the influence of forces.

8.1.1.4 Rheology

Rheology is the quantitative and qualitative study of the deformation and flow of matter. Rheological studies lead to the definition of phenomenological models that relate stresses to strains and strain rates, and potentially to other physical variables (e.g., temperature).

8.1.1.5 Stiffness

Stiffness is the extent to which an object resists deformation without any damage (not to be confused with strength). The stiffer the object, the more force has to be applied for the same deformation. The stiffness of a material can be characterized as the amount of stress per unit strain. It may display isotropy or anisotropy. Stiffness can also be characterized at the level of a structure (e.g., a beam or a shell). For example, flexural rigidity (or bending stiffness) is the amount of torque needed to produce a unit change in curvature. The opposite of stiffness is compliance.

8.1.1.6 Strain

Strain is the degree to which the length of a small virtual cube of material changes in a particular direction relative to its original length (normal strain), or to which the cube's angle changes (shear strain). Deformation is defined using strain. When subjected to the action of forces, a deformable body will become strained. The link between strains, strain rates and stresses is analyzed in rheology.

8.1.1.7 Strength

Stress is the limit of the capacity of a material or structure to withstand stresses or strains before it weakens or breaks. In materials, the strength is described by failure criteria, the simplest being the maximum stress or strain at which failure starts under uniaxial stress. But more complex failure criteria may be necessary to describe multiaxial stresses and anisotropic rheologies.

8.1.1.8 Stress

Stress is the density of reaction forces per unit area acting on an oriented surface and measured in pascal (Pa) or $N\ m^{-2}$. The surface may be a physical interface (e.g., between neighboring cells or at the boundary of an organ) or a virtual slice through a block of tissue or cell wall. A *normal stress* is one that acts perpendicularly to the surface. A *shear stress* acts tangentially to the surface. *Pressure* is therefore a normal stress and a positive pressure is considered as compression. In general, a surface will be subject to both normal and shear stresses. For a small virtual cube of material (e.g., within a cell wall or within a fluid inside or outside a cell), normal and shear stresses act on each of the cube's six faces. The stress exerted on the cube by a uniform external pressure is isotropic (the pressure has the same magnitude on each face), as is usually the case in fluids. In many solids occurring in plants, however, the shear and normal stresses acting on the different faces of the cube will have different magnitudes, making the local stress distribution *anisotropic*. However, if the cube is at rest (or at least not accelerating rapidly), then the net forces acting on the surfaces of the cube must balance. Stress cannot be measured directly (it is not observable and can only be estimated using a model).

8.1.1.9 Thigmomorphogenesis

Thigmomorphogenesis (from the Greek word *thigma* meaning "touch") the physiological response of a plant to external mechanical stimuli, which in natural conditions come mostly from drag by wind or currents, raindrops and contact with and rubbing by neighboring objects, plants or passing animals.

8.1.2 Basics

8.1.2.1 Deformations

There are three different ways to deform a solid material: pulling (i.e., applying tension/tensile stress, leading to an increase in length), pushing (i.e., applying a compression/compressive stress, leading to a decrease in length) and twisting (i.e., applying torsion/shear stress, leading to an angular deformation) (Fig. 8.1). In this chapter we will mainly focus on tension, as the presence of turgor pressure in plants (5–10 bar;

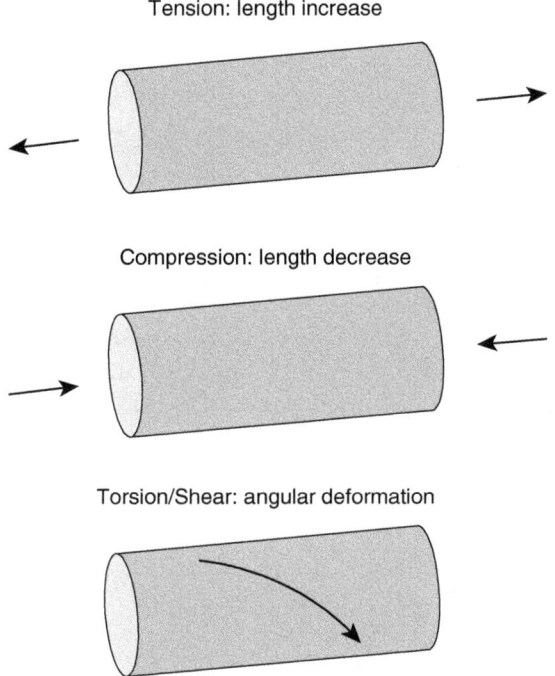

Fig. 8.1 Three modes of deformation. From J. Dumais, *Lectures on Biomechanics*.

1 bar = 10^5 Pa) puts the wall, and sometimes the tissues under high tension. This in turn may affect the behavior of intracellular effectors.

8.1.2.2 Stress versus force

In biomechanics, we often use the term "mechanical stress" instead of "mechanical force". What is the difference? As explained in Section 8.1.1, mechanical stress represents a mechanical force that is normalized by the geometrical properties of the material. For example, for a given force on a beam, the amount of stress in the beam is going to decrease if the section of the beam increases: a larger beam will exhibit more resistance to the force, i.e., a reduced mechanical stress (Fig. 8.2).

If stress (like force) is in essence invisible, its effect on material is made visible by the amount of deformation, which is also called strain. By definition, strain is the rate of deformation, and is thus dimensionless.

Stress and strain are related, as force and deformation are. For a spring with a stiffness k, applying a given force F will extend the spring of certain distance ($L - L_0$, where L_0 is the resting length of the spring). This rule, $F = k(L - L_0)$, also called Hooke's law, is valid for a spring because it is approximated as a one-dimensional (1D) system. For 2D and 3D objects, the geometry of the material will affect the deformation, and now stress and strain must be used. Similar to Hooke's law, the

214 Forces in plant development

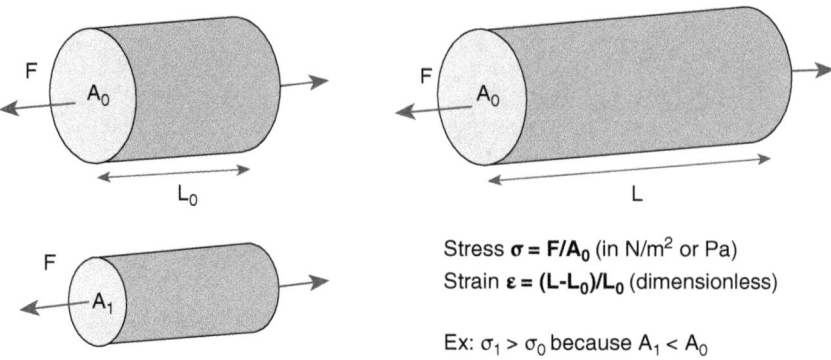

Fig. 8.2 Definition of stress and strain. From J. Dumais, *Lectures on Biomechanics*.

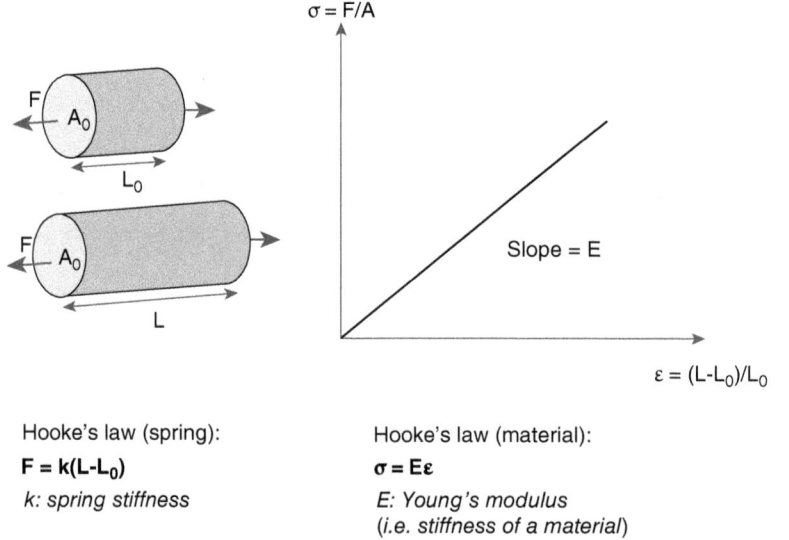

Fig. 8.3 The relationship between stress and strain: definition of Young's modulus.

relation between stress (usually denoted σ) and strain (usually denoted ε) is $\sigma = E\varepsilon$, where E (also called Young's modulus) is the equivalent of the spring stiffness, and describes the material's stiffness (Fig. 8.3).

8.1.2.3 Stiff versus brittle versus tough versus compliant

On the basis of its behavior under stress, the properties of a material can be defined as stiff (high stress leads to little strain), brittle (the material breaks without any warning/a phase of rapid deformation), tough (it does not break easily) or

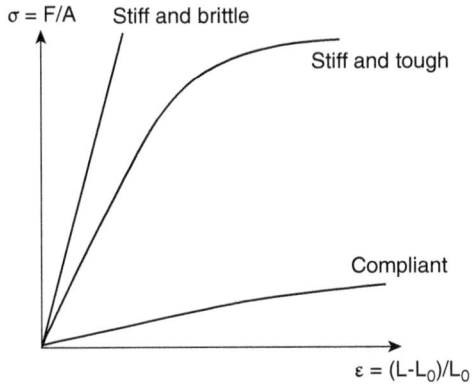

Fig. 8.4 Definition of stiff, brittle, tough and compliant. From J. Dumais, *Lectures on Biomechanics*.

Material	Young's modulus (Mpa)
Rubber	7
Human cartilage	24
Hair	ca. 6,000
Plywood	14,000
Cellulose (crystalline)	25,000
Glass	70,000
Diamond	1,200,000

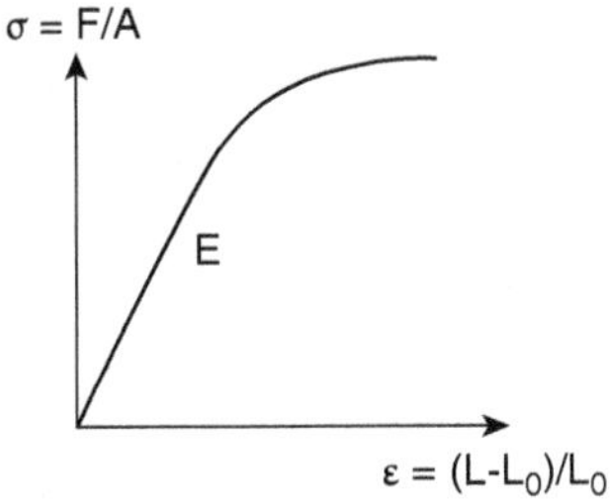

Fig. 8.5 Examples of material properties. From J. Dumais, *Lectures on Biomechanics*.

compliant (high strain for little stress) (Fig. 8.4). Some examples of Young's moduli and mechanical behaviors are shown in Fig. 8.5.

8.1.3 The structure behind the architecture

8.1.3.1 Analogies between architecture and biology

Architecture can help us understand the mechanics of living organisms. In particular, in a building one can easily observe that walls and columns are usually vertical. From this we can deduce that the building is resisting a directional force, and since Newton we know that this force is gravity ($F = mg$). This also means that, in turn, for a given architecture one can deduce the pattern of forces. This will be important in the rest of the chapter, as it will serve as an important clue for calculating stress patterns, and also identify those regions where stress is more important.

The fact that a building does not collapse implies that it is under a balance of forces. In fact, such a balance of forces can explain most shapes (Fig. 8.6). For instance, the

216 Forces in plant development

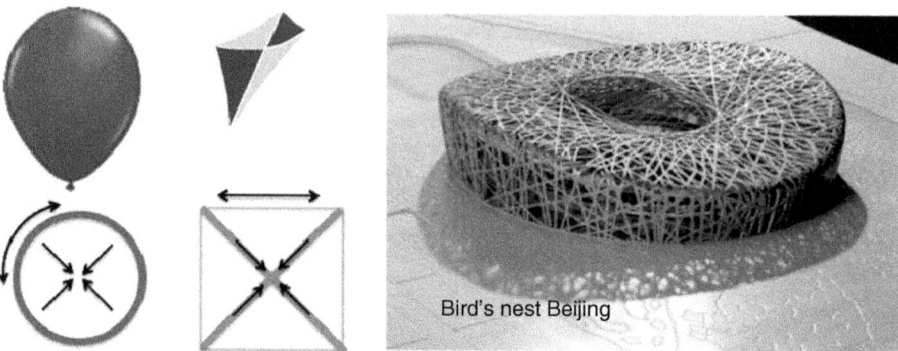

Fig. 8.6 Shape: a balance of forces. (See also discussions around the tensegrity concept in Ingber, 1998.) From left to right: a sphere (balloon), a flat sheet (kite) and more complex shapes (the Bird's Nest in Beijing).

spherical shape of a balloon results from the balance between a gas that is under a compressive force and an envelope that resists this pressure and is under tension. Interestingly, in earthquake-prone regions, buildings often exhibit reinforcements in directions other than the vertical; this is to resist to shear stress as the building needs to withstand an episode of swaying. Similarly, the stable shape of a kite can be explained by the fact that a membrane under tension maintains two rods (arranged as a cross) under compression. In modern architecture, this balance of forces is often explored to the limit of the possibilities offered by new materials. For instance, the Bird's Nest in Beijing is a now classic example of a very complex network of beams in many orientations, the combination of which (some under compression, others under tension) maintains the shape of the building (i.e., it does not collapse).

Such a balance of forces is also observed in living organisms. In the nineteenth century, the German anatomist and surgeon Julius Wolff observed that the internal architecture of bones (trabeculae) can match the pattern of stress to which the bone is exposed. In fact, recent modeling has shown that, depending on the posture of the animal, a local response to stress can explain the observed diversity in internal bone architecture. This has consequences for anthropology or paleontology for instance, as the internal architecture of a primate femur can reflect whether the animal stood upright or not (Fig. 8.7).

The rest of this section will show how plant shapes depend on mechanical stress, using a multiscale approach. For more details see Hamant (2013).

8.1.3.2 Resistance to wind

Plants, like all other organisms, respond to forces. Maybe the most obvious example of such behavior is seen by looking at the shape of a tree that is often exposed to high winds (e.g., along the coastline or when isolated at high elevation). Its "flag" shape means that the tree has developed its branches in a dominant orientation. This orientation reduces exposure to the wind and thus the amount of stress perceived by the

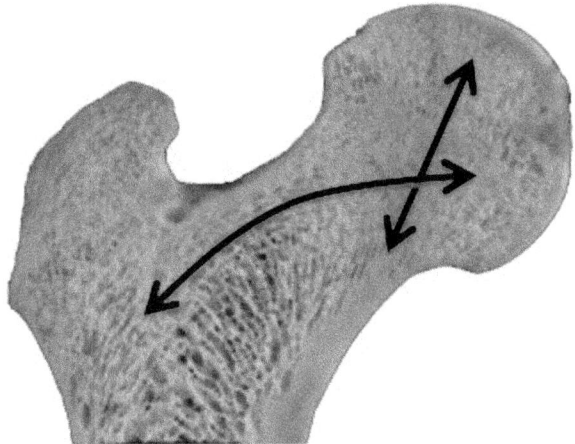

Fig. 8.7 Wolff's law: the internal architecture of a femur matches the pattern of forces to which it is exposed (mainly caused by the weight of the upper body on the femur).

plant. Other responses include the thickening of the trunk and the biochemical stiffening of the cell walls. This response is not anecdotal, as touching plants can induce similar responses, but to a much lesser extent. In particular, contact with neighboring plants was shown to induce stem shortening and wall stiffening in Arabidopsis (Braam 2005). The intrinsic architecture of a tree seems to depend largely on wind. In particular, while its maximal height is classically related to the ability to conduct water up to 100 m from the roots, without inducing cavitation, wind is at least as important in explaining why trees are very rarely taller than 100 m. More recently, it was proposed that the entire aerial architecture of a tree can be explained by the movements of branches exposed to wind.

8.1.3.3 Reaction wood

At the scale of plant organs it is well known that concentric rings (which correspond to secondary xylem, or wood) can be found in sections of the trunks of trees. Such rings can also be found in branches, but in branches these rings are usually not symmetrical. It is now well established that trees reinforce their branches on one side to resist to gravity, in particular if the branch carries fruits or snow is deposited on long branches. For this reason this type of wood is called "reaction wood" (Fig. 8.8). In angiosperms, reaction wood is enriched in cellulose, while in gymnosperms it is usually enriched in lignin. While this represents two different biochemical strategies, in the end both cellulose and lignin stiffen the tissue.

8.1.3.4 A balance of forces in tissues

Although this is rarely mentioned in textbooks, the balance of forces is also present in tissues. For instance, the convoluted shape of the intestine in mammals results from

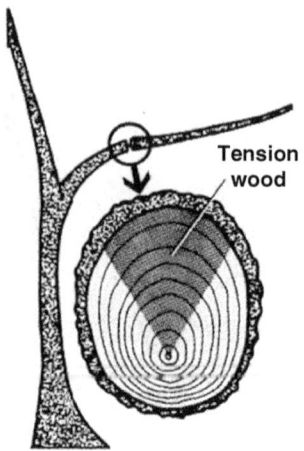

Fig. 8.8 Reaction wood.

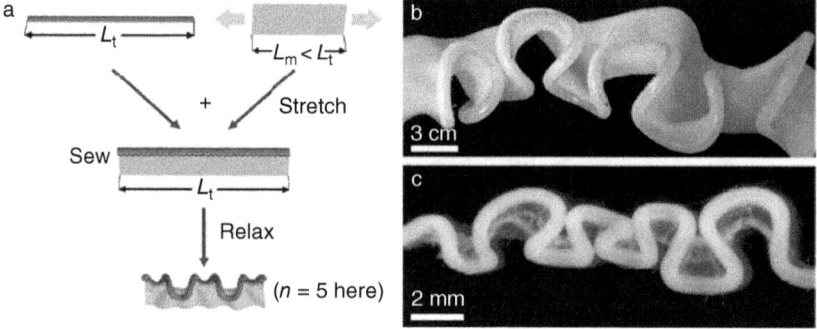

Fig. 8.9 An analogy (a, b) to represent the balance of force in the intestine (c). From Savin et al. (2011). Reprinted by permission from Macmillan Publishers Ltd.

the balance between the compression of the intestinal tube and the tension in the accompanying mesentery. Such a balance of forces can be represented by analogy with a model where a piece of rubber (representing the mesentery) is pulled and glued to another piece of rubber (Fig. 8.9).

Such a balance of forces is also present in plant tissues. For instance, when one cuts a radish it opens to show that the outer layers of the tissue were under tension, while the internal tissues were under compression. Similarly, when peeling the epidermis off a stem, it contracts (thus it was under tension) and the remaining medulla expands (thus it was under compression) (Fig. 8.10).

This balance of forces is also present at the cellular level. In particular, a plant cell can be represented as a balloon, with turgor pressure putting the cell wall under tension. Interestingly, the way a plant cell grows can be entirely explained from a

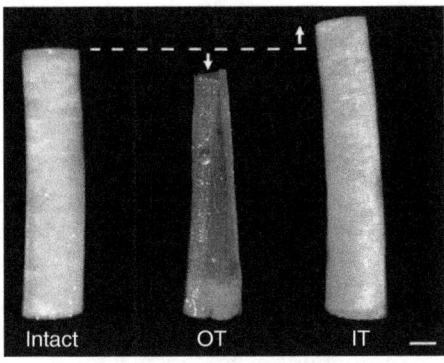

Fig. 8.10 Balance of forces in a plant stem: the epidermis (outer tissue, OT) is under tension while the inner tissues (IT) are under compression. From Kutschera and Niklas (2007), with permission from Elsevier.

mechanical point of view as a race to maintain a balance of forces: the presence of turgor pushes the wall, which in turn yields to the pressure. If nothing happened, the wall would become more and more extended, and thus thinner and thinner. In a real cell, this is compensated by wall synthesis to maintain a steady state wall thickness that modulates the growth rate. More fundamentally, the presence of cell walls in living organisms other than animals can be explained by the fact that the very first cell had to face a hypo-osmotic environment, and thus was exposed to tension in its outer membrane. Building a cell wall is one way to resist this tension. In animals, another strategy was developed: instead of building a stiff cage, animal cells developed a contractile acto-myosin-based cortical cytoskeleton.

8.1.4 Dynamics: forces and movements in plants

Beyond the general role of mechanics in shaping plant architecture, we shall now give three examples to illustrate how mechanical forces can also contribute to more specific, yet crucial, plant functions. Many other examples exist, and this list should certainly not be considered exhaustive.

8.1.4.1 Seed dispersal and ballistics

The role of mechanical forces in plant biology is probably most spectacular when considering seed and spore dispersal. Plants possess an array of mechanics-based tools in which mechanical energy is stored to generate a powerful catapult. Recently, the case of spore dispersal in ferns has been studied in more detail (Noblin et al. 2012). In ferns, spores are stored in organs called sporangia that exhibit a mechanically reinforced outer layer (the annulus) consisting of 12 to 13 cells with thick walls. Importantly, the outer wall is much thinner and weaker than the other walls. As the sporangium dries out the thick walls maintain their length while the thinner wall contracts. This leads to the slow opening of the capsule: the catapult is armed. Water tension then reaches a threshold beyond which cavitation occurs: air bubbles form in the annulus cells, which thus inflate, opposing the preceding motion. Because cavitation is much faster, this leads to immediate (10 ms) valve unfolding and spore dispersal.

8.1.4.2 Bud packing and lobed leaves

What is determining the compound shape of leaves (Couturier et al. 2009)? Classically, the formation of leaflets is thought to be driven by local peaks of auxin, through PIN-dependent auxin transport, and to rely on boundary genes such as CUC (Blein et al. 2008; Bilsborough et al. 2011). This mechanism is homologous to what is observed in the shoot apical meristem for the initiation of primordia. Mechanical stress may also contribute to shaping compound leaves. In fact, in trees, compound leaves are generated in buds and this means that they have to be produced in a 3D confined environment. The bud walls thus act as a barrier to the development of the leaf. This means that growth may be restricted as the leaf grows and folds inside the bud. Quantitative analysis of compound leaf shapes and modeling have demonstrated that this may be sufficient to explain the diversity of compound leaf shapes observed in nature. In fact, the shape of an unfolded young leaf in a bud is relatively similar to the shape of a fully grown leaf, suggesting that most of the shaping occurs during growth inside the bud. This proposition has been further supported by microsurgical experiments. Therefore, if the formation of compound leaves involves auxin-based patterning, the presence of compound leaves may simply be the consequence of how a young leaf is packed inside the bud, i.e., on how growth is arrested by the physical limit of the bud (Couturier et al. 2009, 2011, 2012).

8.1.4.3 Tropism and proprioception

Gravitropism is a well-studied process involving the motion of heavy amylopasts (called statoliths) which serve as cue to the plant to indicate the direction of the gravity vector. This leads the root to grow downwards (gravitropism) and allows it to fulfill its function, mainly plant nutrition and anchoring. While this in essence represents a typical plant response to a mechanical signal, gravitropism is not sufficient to explain the deformation of the plant after gravistimulation. For example, when an Arabidopsis plant with a fully developed inflorescence is suddenly positioned horizontally (in the dark to avoid the contribution of phototropism), the stem progressively bends upwards to generate a vertical stem at the tip. However, a modeling approach indicates that gravisensing is not sufficient to explain this response. In fact, if this response was only due to a response to gravity, the stem would overshoot left and right and would keep oscillating around the vertical vector before stabilizing in a vertical orientation. In real plants no such oscillation is observed. It has been proposed, and validated *in silico*, that adding proprioception, i.e., perception by the plant of its own shape (here curvature of the stem) slows down the oscillation and allows the stem to rapidly become vertical, as observed in real plants (Bastien et al. 2013). Shape perception may involve biochemical pathways as well as mechanosensing, and this represents a stimulating area of research in plant development for the future.

8.2 Understanding the role of forces at the tissue scale

8.2.1 From material properties to shape

8.2.1.1 Revealing the existence of stress in tissues

The easiest way to reveal the presence of mechanical stress in an object is to cut it and record the way it deforms afterwards. As shown in Section 8.1.3.4, cuts in plants and animal tissues lead to deformations, from which the balance of forces (and sometimes a stress pattern) can be deduced (e.g., Dumais and Steele 2000; Kutschera and Niklas 2007; Savin et al. 2011). Although this method can be quite approximate, it is very valuable for validating predictions from models.

8.2.1.2 Calculating a stress pattern

While stress patterns can be predicted based on intuition (e.g., this is relatively easy for a rectangular building that is under gravity), it can be much more complicated when the shape and/or the pattern of force is heterogeneous. Let us take the example of the plant epidermis. Experimental data show that the epidermis is under tension. Using this knowledge, one can use an analogy with a balloon to calculate a stress pattern. In particular, it is well known that the pattern of stress in cylindrical pressure vessels is anisotropic: to resist to the higher stress in the circumferential direction than in the axial direction, reinforcement hoops are often visible. Now, to calculate the exact stress ratio between axial and circumferential stress, one simply needs to know the definition of stress: a force divided by the surface area of the section on which the force is normal (i.e., perpendicular to). With this one can measure the surface area across a longitudinal section and across a transverse section. Upon doing the maths, one finds that the ratio between circumferential stress and axial stress (i.e., the ratio between the surface area of a longitudinal section and a transverse section) is 2.

8.2.1.3 Models

While the balloon represents a good analogy (and also the basis for a continuous model), it cannot properly represent the cellular nature of the tissue and the relevant discontinuities. To address this, two main types of mechanical models have been built. In the simplest one, each cell wall (or membrane portion) is represented as a spring, and the force balance is achieved when all the springs are connected. This model has the advantage of being mechanically simple (governed by Hooke's law only) and it can include some degree of regulation: for instance, to represent the fact that the cell is resisting to stress, the stiffness of certain springs can be increased. This model was used to represent the microtubule-dependent reinforcement of a cell wall under stress (Hamant et al., 2008). A major disadvantage of this type of model is that it can only describe 2D surfaces. To represent 3D objects, a finite element method is most often developed. In this type of model each cell wall (or sub-cell wall region) is represented as an elastic sheet. A cell is then represented as a cage with several elastic sheets at its periphery, and the behavior of several contiguous cells can be modeled. As in the spring model, because mechanical elements are used (elastic membranes), their stiffness and mechanical anisotropy can be modulated in the model, and conversely the amount of stress in relation to shape can be deduced (see Box 8.1).

> **Box 8.1** What is a cell-based computational model?
>
> A model can be defined as a comprehensive description of an object or concept. The main goal of modeling in a developmental context is to establish a link of causality between (sub)cellular features and global morphogenesis. Because this involves a huge number of parameters and interactions, the use of computers is absolutely necessary to accurately simulate the behavior of populations of cells. Computer modeling is thus an excellent tool for predicting the global emergent properties of a tissue from local cellular rules. How does one generate a cell-based computer model? First, and most importantly, the definition of the local rules requires a careful examination of cell behavior. Quantitative imaging is thus developed to precisely measure cellular parameters, ranging from cellular strain rate to subcellular localization of markers, like PIN1 or the cytoskeleton. These quantified data support the formalization of cellular hypotheses (e.g., the polarity of PIN1 is oriented toward the cell with the highest auxin content). These hypotheses are subsequently translated into algorithms, and the behavior of populations of cells can be simulated in virtual tissues. The morphogenesis and behavior of these virtual tissues are then compared with real templates. If the real and virtual tissues behave similarly, this demonstrates that the tested local hypotheses are plausible. This, however, does not prove that these hypotheses are the only valid ones.
>
> From Uyttewaal et al. (2010).

8.2.1.4 *Atomic force microscopy*

While stress cannot be visualized (hence the need for models), its impact on changes in shape (strain) can be used to measure the force, assuming the mechanical properties of the material are known. This is typically what a weighing scale is doing: an object is put on scale that deforms, and as the mechanical properties of the material that is deformed are known, one can deduce the force, which in this case is the weight P. As $P = mg$, and g on earth is about 10, one can deduce the mass. Conversely, weighing scale can be used to measure the force that a spring can apply to an object: if the shortening by half of a spring that one pushes against a weighing scale leads to a mass of 1 kg, this means that this spring exerts a force of 10 N when it is similarly deformed on a given material.

If one knows how much force is applied and how much deformation results, then one can calculate Young's modulus as the ratio between stress and strain. Many methods exists to calculate Young's modulus, but the challenge in biology is to do this at a very local scale (that of a cell or even a molecule). Atomic force microscopy is an indentation method that allows one to obtain estimates of Young's modulus. In this

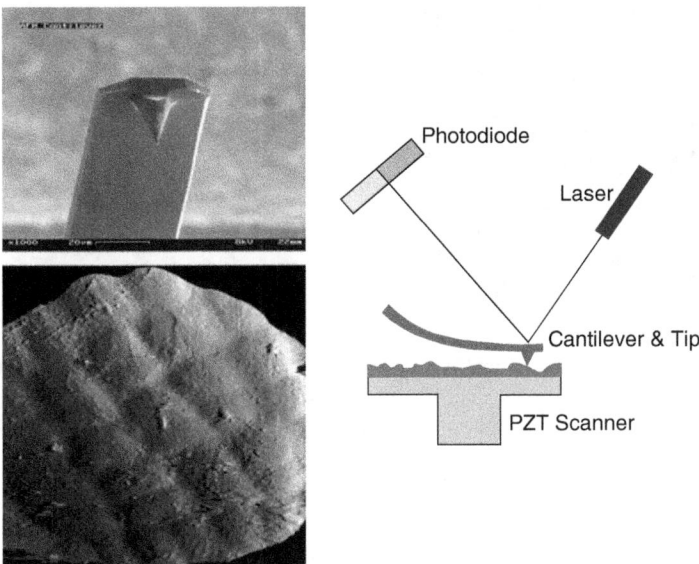

Fig. 8.11 By applying a cantilever at the surface of the meristem and by recording its deflection as it contacts the surface, the elastic modulus of the substrate (i.e., the meristem surface in our case) can be obtained. PZT, piezoelectric tube.

method a microcantilever is applied, with a known force, and the deformation of the cantilever is measured with a laser that is reflected from the arm of the cantilever (Fig. 8.11). If calibrated, this method can provide the stiffness of the sample, which relates to its Young's modulus. For a full review of nanoindentation methods see Milani et al. (2013).

Sections 8.2.2 and 8.2.3 will use examples from our own research to illustrate how one can integrate shape-derived and growth-derived stresses in the molecular control of shape changes in plants.

8.2.2 Shape-derived stress

This section draws on the work of Hamant et al. (2008) and others. Cell wall stiffness relies heavily on the mechanical properties of cellulose microfibrils (which exhibit an elastic modulus comparable to that of steel). Interestingly, cellulose microfibrils are often deposited in a parallel orientation, thus providing mechanical anisotropy to the cell wall (Baskin 2005) (Fig. 8.12).

It is now well established that this orientation is regulated by cortical microtubules (CMTs) (Lloyd 2011). In addition to collinearities between the CMTs and cellulose in fixed tissues, live imaging of cellulose synthase (CESA) complexes (CSC) revealed that the trajectories of the CSC do indeed follow the microtubule tracks (Paredez et al. 2006). The recently identified protein POM2/CSI1 further confirms this model, as this protein is able to bind both microtubules and CESA, and as in the *pom2/csi1* mutant,

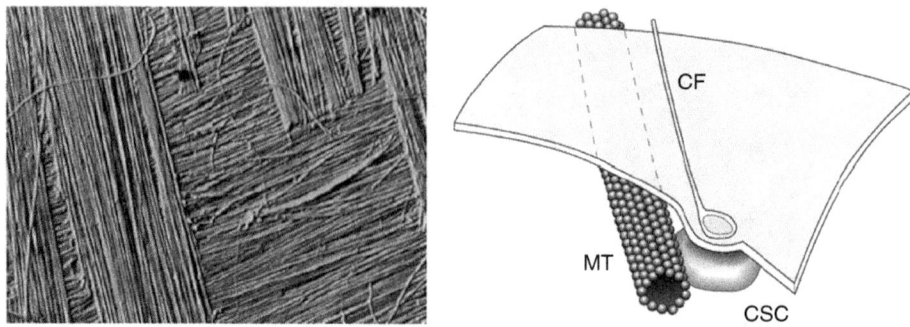

Fig. 8.12 Left: Cellulose microfibrils imaged on the inner face of a cell wall. Right: a model showing the cortical microtubules (MT) guiding the cellulose synthase complex (CSC) in the plasma membrane which extrudes cellulose microfibrils (CF) into the extracellular space. From Burgert and Fratzl (2009), by permission of Oxford University Press.

the movement of CESA becomes largely uncoupled from that of the microtubules (Bringmann et al. 2012). In a collaborative project with Jan Traas, Yves Couder, Elliot Meyerowitz and Henrik Jönsson, we investigated whether CMT orientation is controlled by mechanical stress, building on previous theoretical (Thompson 1917; Castle 1938; Green and King 1966; Williamson 1990) and experimental (Wymer et al. 1996) work.

The patterns and the dynamics of the CMTs in the shoot apical meristem both fitted this hypothesis. In particular, in the boundary domain, where mechanical stress is believed to be highly anisotropic, the CMT orientation was stabilized and parallel to the maximal stress direction, independent of cell shape (Fig. 8.13, left). A mechanical model, in which each cell orients its CMTs according to the local stress pattern, was able to reproduce the supracellular alignment of the CMTs in the boundary

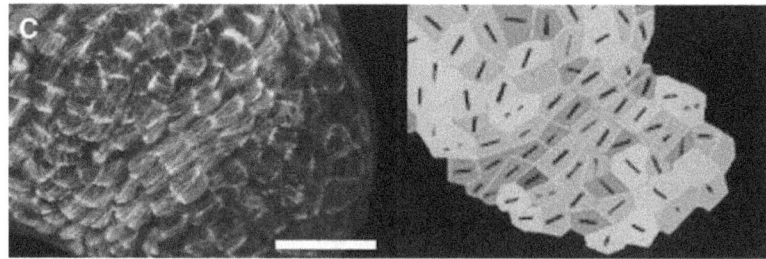

Fig. 8.13 Left: p35S::GFP-MBD marking cortical microtubules and the cell shape at the surface of a meristem generating a young primordium (P). Scale bar = 20 μm. Right: Microtubule orientation (bars) in the 2D stress-feedback model on cells extracted from the confocal data, notably reproducing the original microtubule orientations in the boundary. From Hamant et al. (2008). Reprinted with permission from AAAS.

(Fig. 8.13, right), thus providing mathematical proof supporting the plausibility of our hypothesis.

To test this model experimentally, we changed the stress pattern and followed the response of the CMTs. Using ablation (inducing a circumferential stress pattern), compression (inducing a global tensional stress pattern perpendicular to the compression) and cell wall thinning using drugs (increasing the stress intensity), we showed that the CMTs did indeed reorient according to the new stress pattern. We also confirmed the contribution of CMTs to meristem shape. Depolymerization of microtubules using oryzalin induced major defects: cell growth became isotropic and the crease in the boundary of the meristem was abolished (Fig. 8.14).

To conclude, this work demonstrates a feedback loop (Fig. 8.15), in which CMTs control anisotropic growth via the deposition of cellulose and the shape of the meristem. This modifies the global pattern of stress, which in turn affects the orientation of the CMTs (Hamant et al. 2008).

One of the implications of this work is that plant cells respond to mechanical stress to control changes in shape, and in a way this response recalls Wolff's law in which

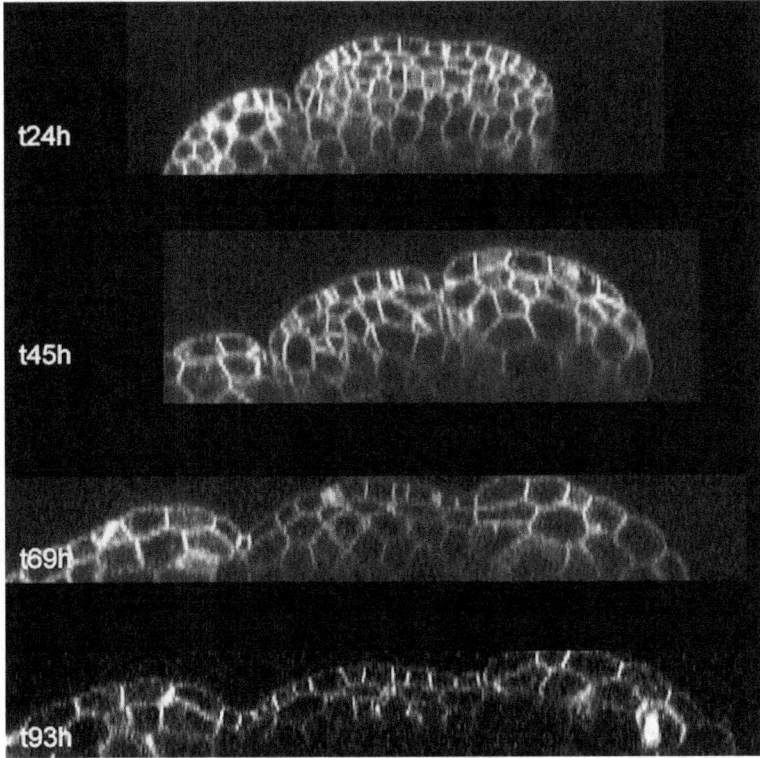

Fig. 8.14 Longitudinal optical sections of an existing crease being flattened over time after depolymerization of microtubules with oryzalin. Scale bar = 50 μm. From Hamant et al. (2008). Reprinted with permission from AAAS.

226 Forces in plant development

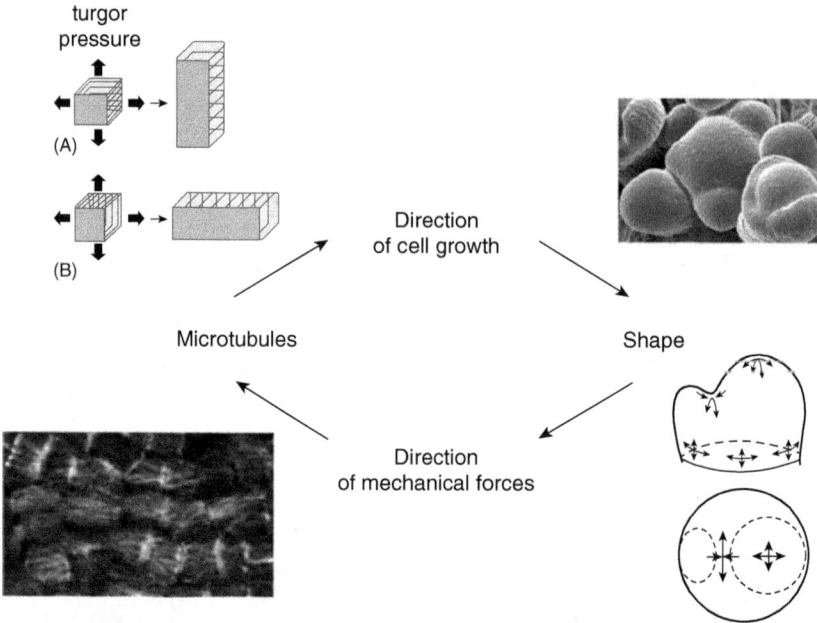

Fig. 8.15 A mechanical feedback loop at the shoot apical meristem.

bones reinforce their architecture in the direction parallel to maximal stress (Wolff 1892; Thompson 1917). This rule also seems to be followed other cellular contexts. For instance, motile animal cells usually reinforce their cytoskeleton and adhesion to the substrate, and thus become polarized, in order to resist the maximal directional stress (e.g., Dalous et al. 2008).

8.2.3 Growth-derived stress

This section draws on the work in Uyttewaal et al. (2012). It was hypothesized that cells could regulate their ability to respond to mechanical stress by modulating the dynamics of their microtubules. Several proteins have been shown to affect microtubule dynamics, and we chose to focus our attention on katanin because of its well-known biochemical function (microtubule severing) and documented impact on microtubule self-organization. In particular, by cutting microtubules, katanin provides a large number of free microtubules, thus increasing the number of encounters and zippering events. Consistently, overexpression of katanin leads to a transient increase in the number of microtubule bundles, while the absence of katanin leads to disorganized, less bundled microtubule arrays throughout the cell (Bichet et al. 2001; Stoppin-Mellet et al. 2006). Microtubules in the katanin mutant (*atktn1*) are less dynamic than in the wild type (WT). We thus investigated the impact of the absence of katanin on the microtubule response. As expected, CMTs were less well organized in *atktn1* than in WT, with a decreased number of bundles, and a weaker supracellular organization (Fig. 8.16).

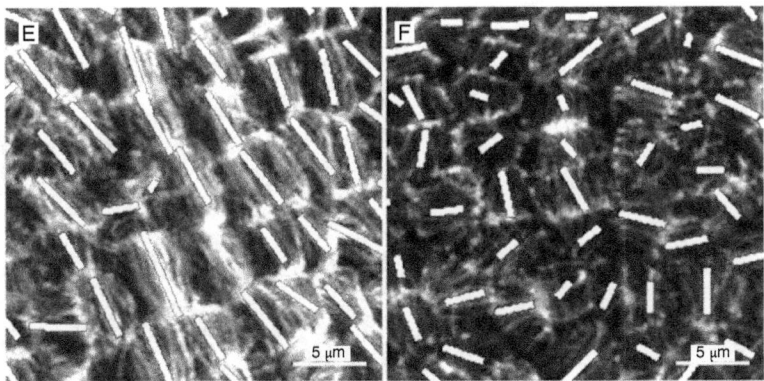

Fig. 8.16 p35S::GFP-MBD expression in the peripheral zone in the wild type (left) and *atktn1* (right). CMT orientation and anisotropy were determined with an ImageJ macro: the orientation and length of the white line in each cell indicate the average CMT orientation and anisotropy of the array, respectively. From Uyttewaal et al. (2012).

To test whether this could be related to a decreased response to mechanical stress, we next performed compression, ablation and isoxaben treatments, and observed that the CMTs were slower to respond to the new stress pattern under these conditions, consistent with a role of CMT dynamics in the ability of the cell to respond to mechanical stress. This is actually consistent with other data showing that CMTs are slower to reorient in response to other signals in *atktn1* (Fig. 8.17).

Using the *atktn1* mutant background, we next addressed a basic question: what is the impact on growth and shape of a reduced ability to respond to stress? In particular, if one cell is growing faster than its neighbors, this should induce tensions (residual stress) between adjacent cells. Two scenarios are then possible: either the cells accommodate the stress and suppress growth heterogeneity or they use and amplify the differences, promoting growth heterogeneity. Karen Alim and Arezki Boudaoud developed a model in which the impact of the response of cells to mechanical stress is computed and local growth heterogeneity is measured as an output. The model predicts that if the mechanical stress response is strong enough, growth heterogeneity always increases. In other words, cells adjoining a faster growing cell should try to resist this differential growth by aligning their microtubules circumferentially around the fast-growing cell; however, by overreacting to the stress generated by the rapidly growing cell, the adjacent cells modify their growth parameters so much that growth heterogeneity is actually amplified (Figs 8.18 and 8.19).

We next measured growth rates in WT and *atktn1* meristems, and this analysis revealed that growth heterogeneity is reduced in *atktn1*, consistent with the predictions of the model (Fig. 8.20). Growth heterogeneity was notably higher in the boundary domains of WT meristems, i.e., in sites of predicted higher stress, and growth was more homogeneous in that domain in *atktn1* (Fig. 8.21).

In *atktn1* the shape of the shoot apical meristem was altered, with ill-defined boundaries. The shoot apical meristem had a crater-like shape. This can be related to

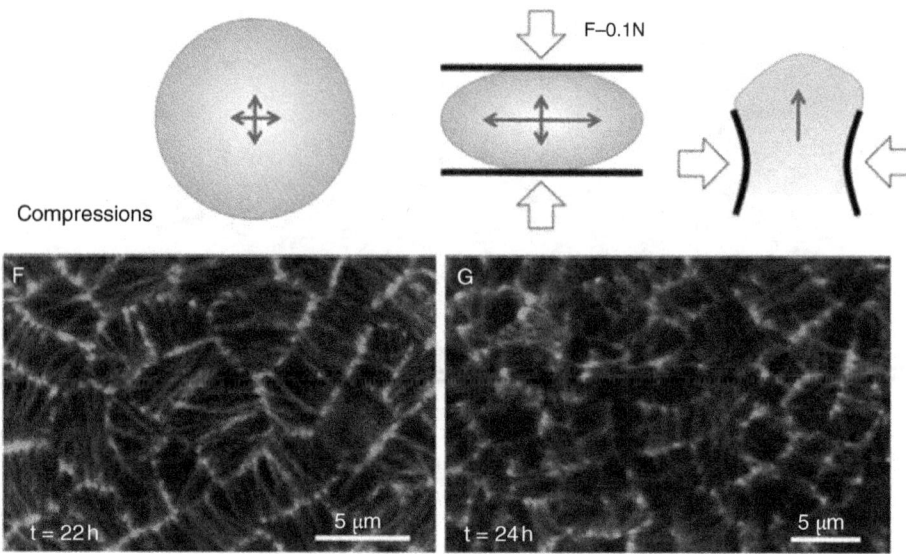

Fig. 8.17 After compression, CMTs form parallel bundles (strongly anisotropic arrays) in the wild type (left) and bundles with multiple orientations (nearly isotropic) in *atktn1* (right). From Uyttewaal et al. (2012).

the decreased growth anisotropy in the peripheral zone of *atktn1*. Interestingly, *pin1* meristems treated with auxin also exhibit a ring-shaped shoot apical meristem. This may suggest that the shoot apical meristem exhibits a crater-like shape by default, and that the balance between the growth rate and growth anisotropy is tightly regulated to prevent organogenesis and maintain a dome shape (Fig. 8.22). To some extent, this finding echoes the work of Nath et al. (2003), which suggested that plant leaves are ruffled by default and that growth parameters must be tightly regulated to maintain a flat shape.

In summary, and to quote a commentary on this work, "in both plants and animals, the interplay between mechanical force generation and mechanical sensing plays a stabilizing role in many developmental processes. This work demonstrates that cells in the Arabidopsis shoot apical meristem respond to local mechanical stresses by reorienting their growth, thereby guiding morphogenesis. Notably, the mechanism underlying such guidance is amplification—not suppression—of growth-rate heterogeneity" (Gibson and Gibson 2012) (Fig. 8.23).

8.2.4 Mechanical signals in animals

While much detail is known about the microtubule example in Section 8.2.3, it is not an exception—there is accumulating evidence to suggest that mechanical forces can act as signals for many other processes in all living organisms. This section discusses some examples (again following a multiscale approach), and the list is of course far from exhaustive.

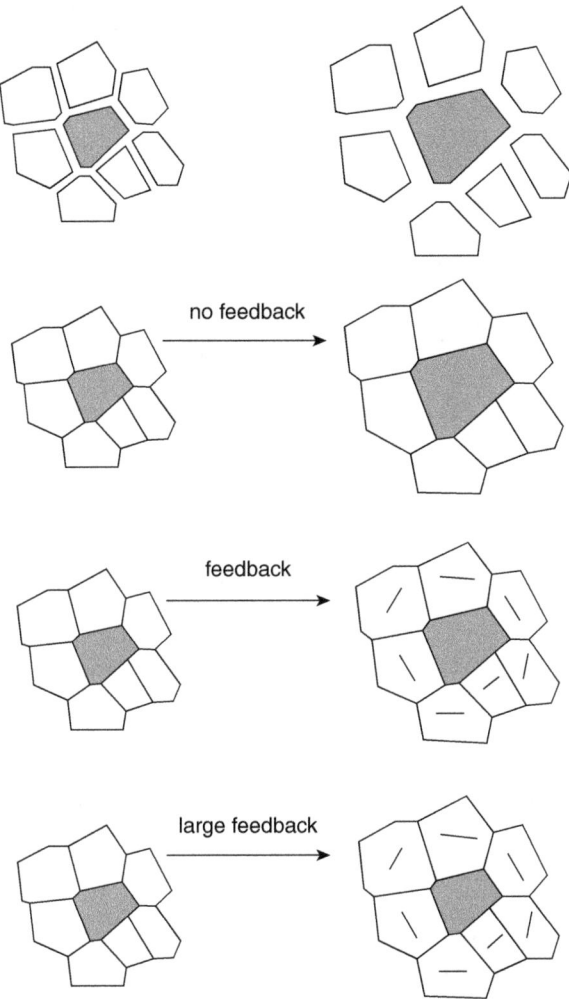

Fig. 8.18 Diagrammatic representation of the hypothesis implemented in the model. Each cell is assigned a random target growth rate that it would achieve if isolated from its neighbors (here the gray cell has a large target growth rate). When feedback of stress on growth is implemented, cells restrain growth in the direction of the light gray lines (i.e., the direction of the CMTs), thus reducing the actual growth of the gray cell. From Uyttewaal et al. (2012).

8.2.4.1 Skeletogenesis and thigmomorphogenesis

Although the comparison is a bit far fetched, there are some analogies between the way plants respond to wind and the way mammalian embryos build their skeleton. In mammals, movement of the limbs *in utero* is necessary for the proper differentiation of the joints between the bones. In fact, in a paralyzed embryo, joints become

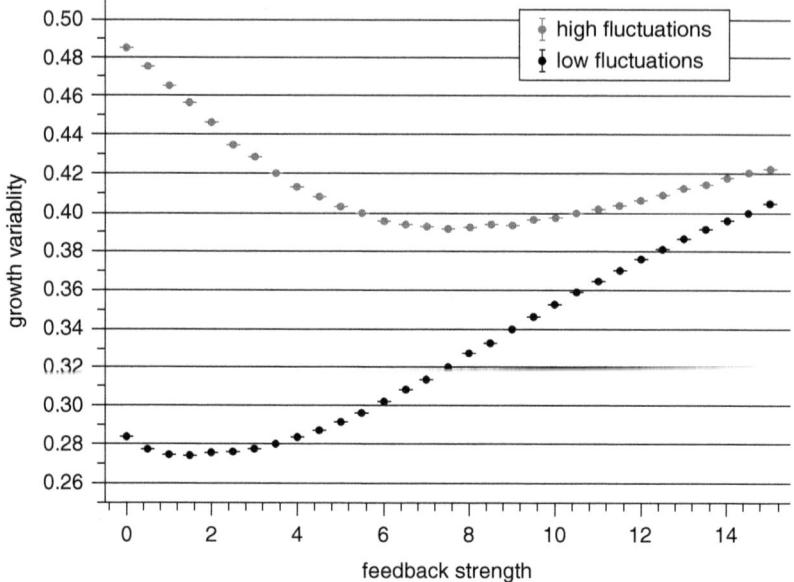

Fig. 8.19 Growth variability measures the difference in growth rate between a cell and its neighbors and takes a value of zero when all cells grow at the same rate. If target growth experiences high fluctuations, the response to mechanical stress is first predicted to render growth among neighboring cells more homogenous, and as stress increases, growth variability then increases. If target growth is more stable (low fluctuations), the response to mechanical stress is predicted to increase the variability of growth among neighboring cells. From Uyttewaal et al. (2012).

stiff and the bones fuse. In other words, elastic deformation of the limbs leads to their plastic differentiation. In plants, as mentioned earlier, wind generates elastic deformations that lead to the differentiation of shorter, stiffer stems. In both cases, elastic deformations have had a dramatic impact on the developmental program.

8.2.4.2 Patterning

Mechanical forces are involved in determining cell identity. For instance, left–right asymmetry is established in mammalian embryos via a leftward flow of extraembryonic fluid that is generated by motile cilia (Nonaka et al. 2002.). Similarly, the separation between anterior and posterior cell populations in the imaginal disk of the *Drosophila* wing is a site of increased tension that could prevent mixing of cell populations (Landsberg et al. 2009; Aliee et al. 2012). Last, the large polarity field that is present in the developing *Drosophila* wing has been shown to be caused by the contraction of the hinge domain of the wing, that then pulls on the wing blade, thus polarizing the entire wing towards the hinge (Aigouy et al. 2010) (Fig. 8.24).

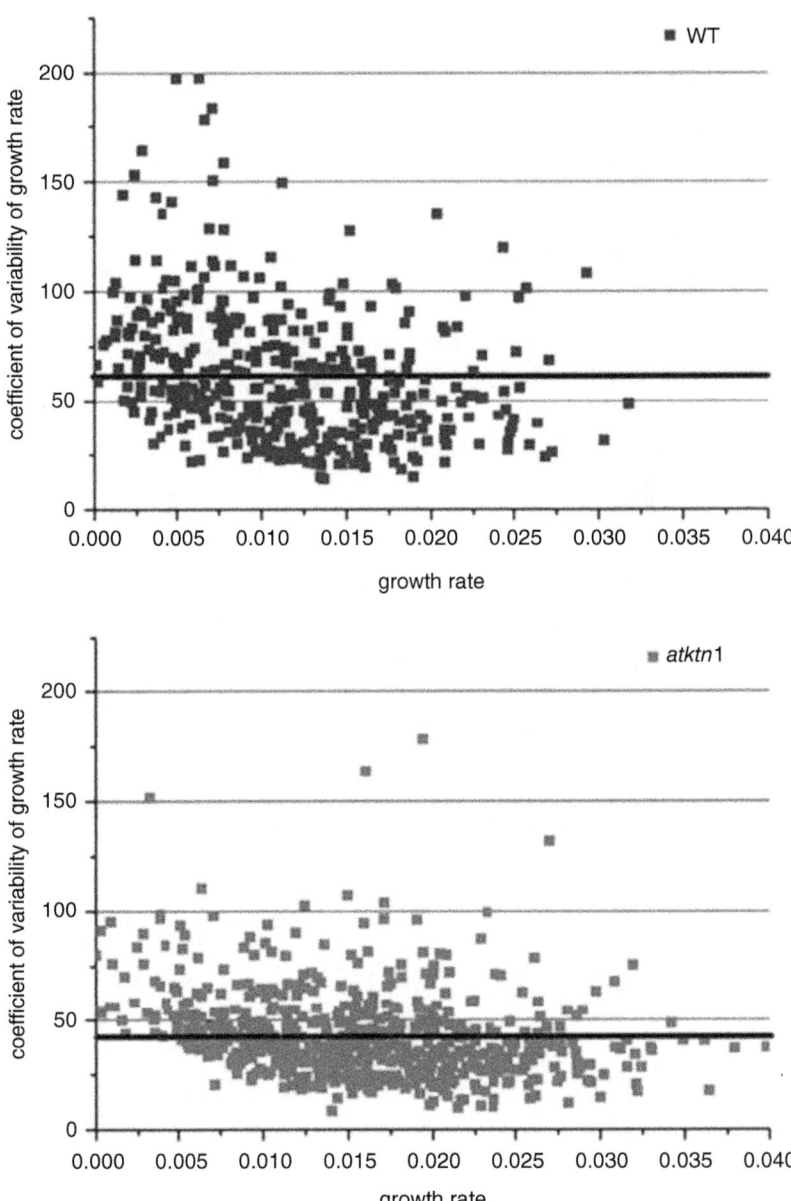

Fig. 8.20 Coefficient of variability (local variability representation) of growth rates plotted against the growth rate of the individual shoot apical meristem cells (excluding boundaries and primordia) of the wild type (WT; dark gray, top) and *atktn1* (light gray, bottom). The coefficient of variability was computed for a cell and all its adjoining neighbors and assigned to this cell, for a total of $n = 410$ WT cells (six meristems) of and $n = 607$ *atktn1* cells (six meristems). The dark horizontal lines indicate the mean coefficient of variability. From Uyttewaal et al. (2012).

232 *Forces in plant development*

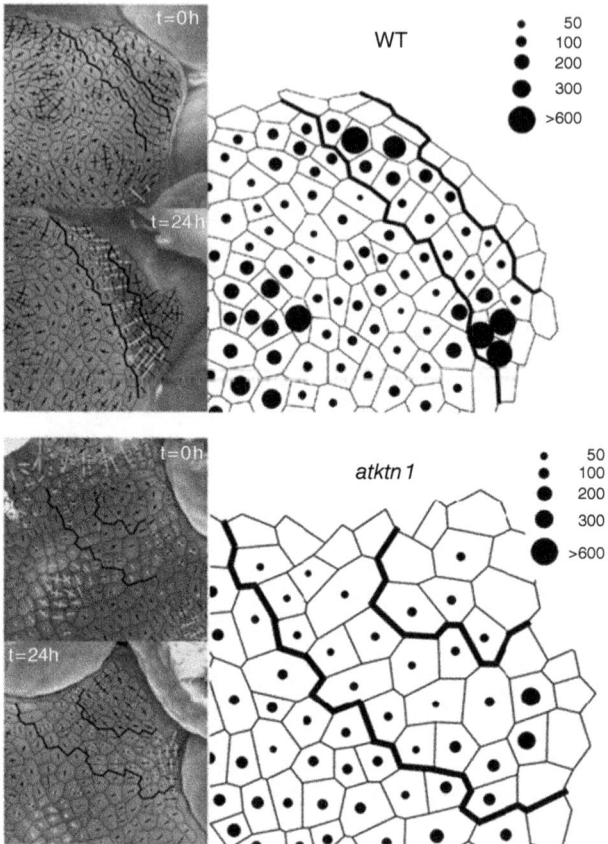

Fig. 8.21 Spatial distribution of coefficient of variability (in percent) of growth rates in the boundaries of primordia exhibiting similar plastochron ages in the wild type (WT; top) and *atktn1* (bottom). The size of the symbol increases with the value of the coefficient. Scanning electron micrographs with overlaid principal curvature directions for both genotypes are included. Regions that will give rise to boundaries (creases with a negative curvature in the meridional direction) during the next 24 h are delineated with thick lines. From Uyttewaal et al. (2012).

8.2.4.3 Cell division

The idea that mechanical stress contributes to the control of cell division is not new. In fact, Errera's rule, which states that new division planes follow the shortest path, is based on the observation that cells somehow resemble soap bubbles and that surface tension plays a key role in the position of new division planes (for a review see, e.g., Louveaux and Hamant 2013). More recently, modeling and micromechanical approaches have supported the idea that mechanical signals play an instrumental role in defining the rate of cell division and the orientation of the plane of cell division. For

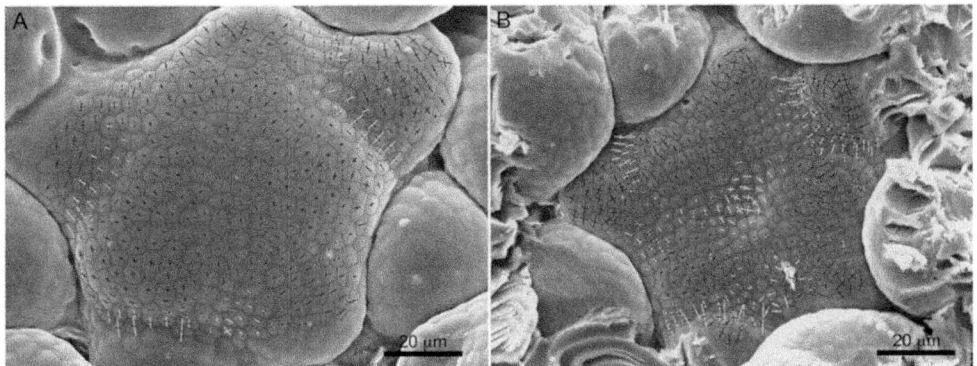

Fig. 8.22 Scanning electron micrographs with overlaid principal curvature directions for wild-type (A) and *atktn1* (B) meristems. The indicators appear in white if in this direction the surface is concave (negative curvature) and in black when the surface is convex (positive curvature). From Uyttewaal et al. (2012).

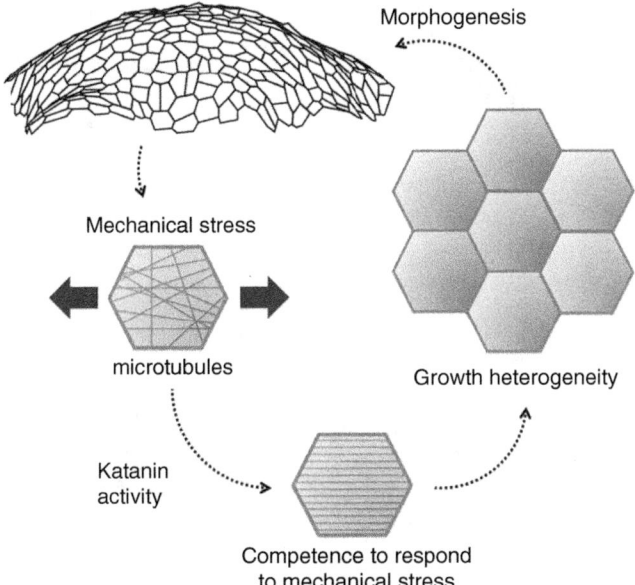

Fig. 8.23 A refined model of the mechanical feedback loop operating at the shoot apical meristem.

instance, in the imaginal disk of the *Drosophila* wing, cell division arrests synchronously at a certain time point. Because a biochemical gradient would rather lead to a progressive arrest of cell division, mechanical forces have been proposed to trigger this arrest (and explain its synchronicity: mechanical signals travel at the speed of sound in a homogeneous material) (Fig. 8.25).

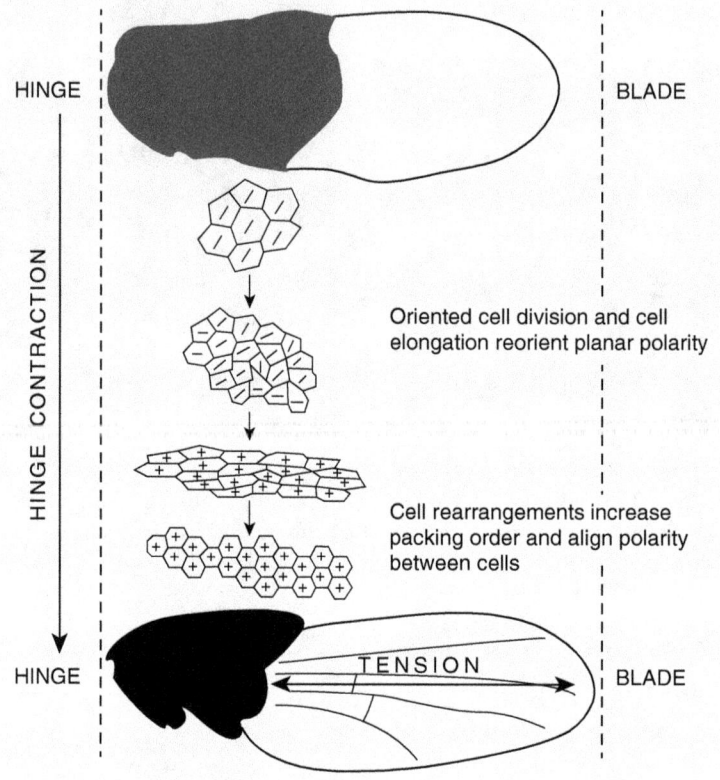

Fig. 8.24 Polarity is reoriented in *Drosophila* wings (pupal stage) by epithelial remodeling and cell flow: hinge contraction induces remodeling by increasing epithelial tension. Reprinted from Aigouy *et al.* (2010), with permission from Elsevier.

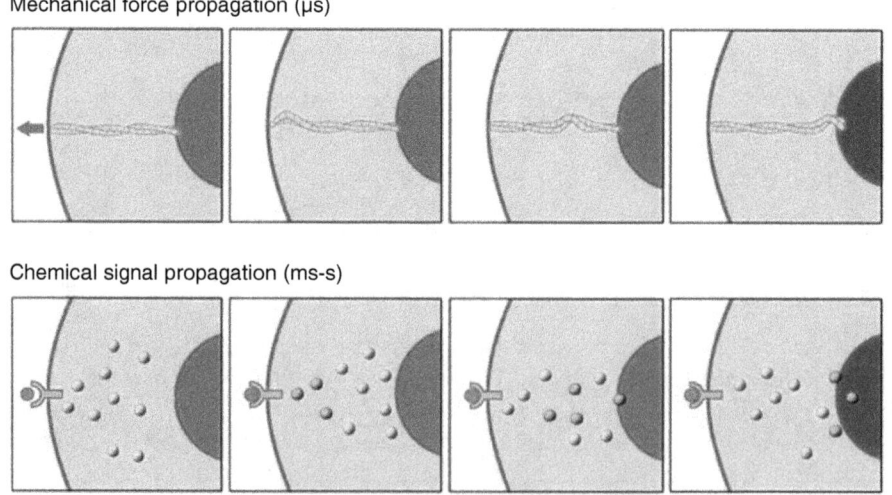

Fig. 8.25 The propagation of mechanical signals is faster than chemical signals. From mechanobio.info.

While all this remains a matter of debate, experimental evidence support at least a contribution from mechanical forces to this process. Notably, as the wing disk grows, the internal cells would be compressed, and beyond a certain threshold cell division would stop (Shraiman 2005; Hufnagel et al. 2007). In HeLa cells, it was also shown that the orientation of the mitotic spindle depends on the pattern of adhesion of the cell to its matrix, and thus on the internal pattern of stress (Thery et al. 2007).

8.2.4.4 Cell identity

The impact of mechanical signals on cell identity is clearly shown in the case of mesenchymal stem cells grown on matrices exhibiting different stiffnesses: when grown on soft matrices, these cells differentiate into neuroblasts, while on stiffer matrices they differentiate into osteoblasts (Engler et al. 2006) (Fig. 8.26).

Interestingly, these effects have recently been shown to involve nuclear lamins, suggesting that cells perceive and transduce their mechanical environment down to their nuclear envelope (Swift et al. 2013). This somehow echoes the tensegrity concept developed by Donald Ingber, where mechanical forces may be perceived inside the cell though a continuum of stiff elements, themselves being arranged to reflect a balance of forces (Ingber 1998).

In parallel to these *in vitro* data, it has also been shown that mechanical forces can channel gene expression in developing *Drosophila* embryos *in vivo*. The expression of the *TWIST* gene occurs during gastrulation, i.e., a time when large-scale tissue movements occur, and in a domain under strong compression (Farge 2003). Mechanical perturbations with a microvice, magnetic beads or via cell ablation further confirmed the induction of *TWIST* expression when the local pattern of stress was modified (Farge 2003; Desprat et al. 2008). Interestingly, the *NOTAIL* gene in zebrafish is also induced by mechanical forces during epiboly, suggesting that differentiation of the mesoderm in vertebrates and invertebrates is triggered by a mechanical signal (Brunet et al. 2013).

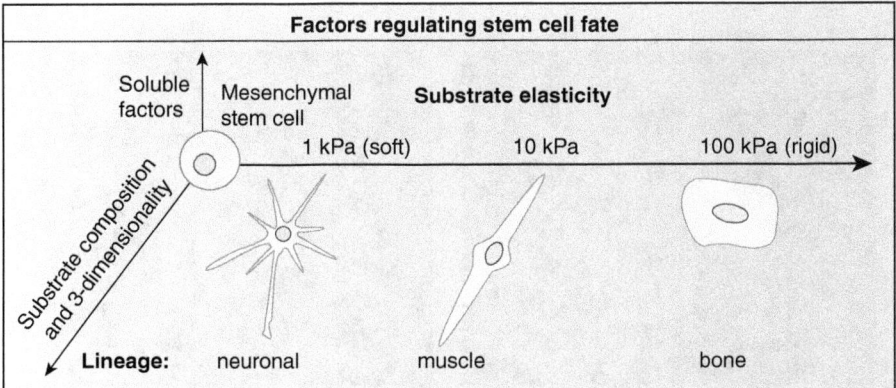

Fig. 8.26 The contribution of substrate elasticity to stem cell fate. From Engler et al. (2006), with permission from Elsevier.

Although this has only just started to be addressed in plants, gene expression is also affected by mechanical perturbations in this kingdom. A long list of touch-induced genes is available (Braam 2005) and it has been shown that mechanical stress rapidly induces the expression of *ZFP2* after bending a stem. Interestingly, the accumulation of *ZFP2* mRNA is linearly correlated to the mechanical strain, strongly suggesting a fine-tuned and supracellular regulation of this gene in response to mechanical forces (Coutand et al. 2009).

8.3 Understanding the role of forces at the cell scale

8.3.1 Subcellular stress

8.3.1.1 The case of pavement cells

This subsection draws on the work of Sampathkumar et al. (2014). In plant leaves, the epidermis often exhibits cells shaped like jigsaw-puzzle pieces, called pavement cells. This shape has been associated with the presence of bundled microtubules in the cell neck domains, where growth is restricted. It is well established that the pattern of microtubules in these cells depends on complex molecular regulation, relying on small Rho GTPase proteins (Fu et al. 2005). Mechanical stress could also contribute to orienting the microtubules in these cells. To test this possibility, a finite element model was used to calculate the stress pattern at the cell scale; it revealed that microtubules orient along the maximal tension lines, as prescribed on the outer wall of the cell (Fig. 8.27).

After ablation, compression or pharmacological treatment, microtubules can re-orient along the new stress pattern in those cells, thus showing that the mechanical feedback that we observed in the shoot apical meristem is also true at the single-cell scale. Interestingly, mechanical perturbations also lead to an increased rate of microtubule severing, thus suggesting that mechanical stress not only channels the orientation of microtubules but also promotes their competence to self-organize, in a positive feedback loop (Fig. 8.28).

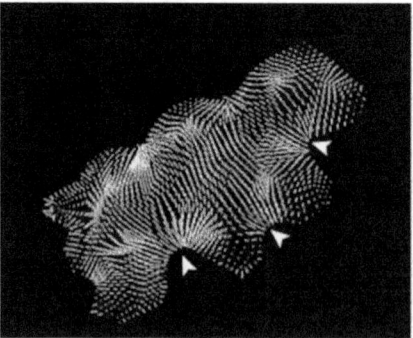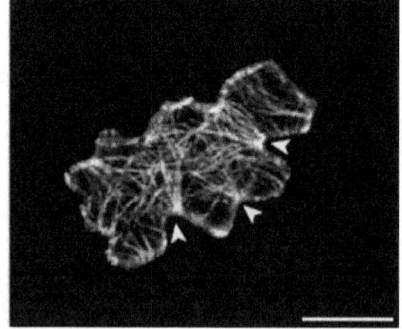

Fig. 8.27 Maximal tensile stress (left) and microtubule (right) patterns in a pavement cell. Adapted from Sampathkumar et al. (2014).

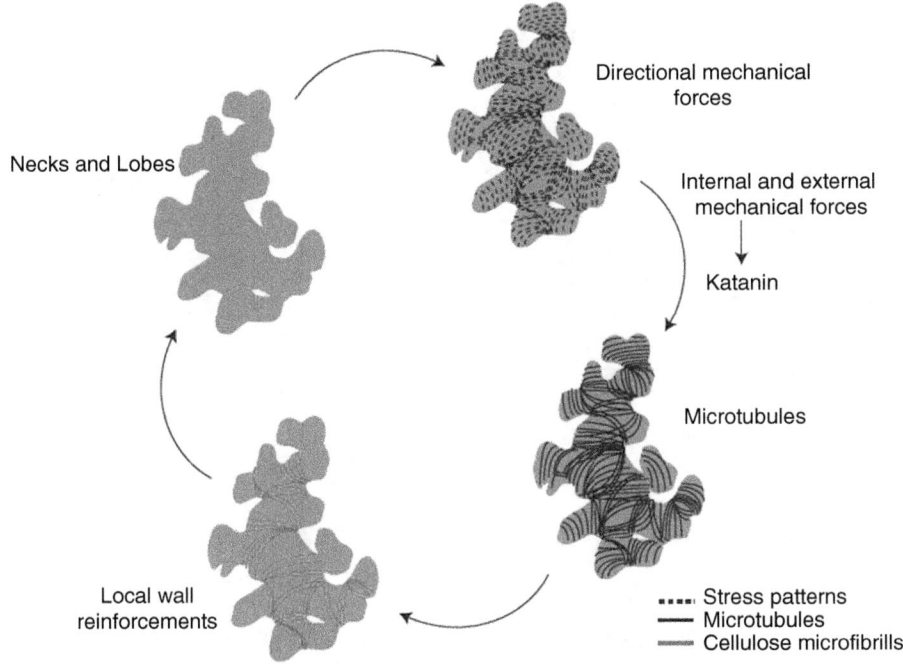

Fig. 8.28 Mechanical forces regulate pavement cell shape by controlling microtubule organization and cellulose deposition. Adapted from Sampathkumar et al. (2014).

8.3.1.2 PIN1 and membrane tension

The polar localized auxin efflux carrier PIN1 is concentrated on membranes that usually are parallel to CMTs in cells, as viewed from the top (Heisler et al. 2010) (Fig. 8.29).

Several scenarios could explain this observation. One is that PIN1 is preferentially recruited on membranes that are less rich in CMTs. However, as the depolymerization of microtubules does not dramatically affect PIN1 dynamics at the shoot apical meristem this scenario seems unlikely. This observation is consistent with previous reports showing that actin, and not microtubules, is the main driver of PIN1 polarity on plant cells (Geldner et al. 2001). We thus tested the hypothesis that a common upstream signal controls both microtubule orientation and PIN1 polarity, focusing on mechanical stress. After ablation, PIN1 reorient circumferentially around a wound, i.e., parallel to microtubule orientation and to stress patterns. Isoxaben is a well-known inhibitor of cellulose synthesis. Thus, after isoxaben treatment, walls are predicted to become thinner, and thus mechanical stress (i.e., the tension in the wall generated by turgor pressure divided by the cross section of the wall) increases (Figs 8.30 and 8.31).

Clearly, CMTs exhibited a circumferential pattern in the shoot apical meristem, with thicker bundles and a clearer microtubule anisotropy, better related to the global predicted stress pattern. This is consistent with CMTs orienting according

238 *Forces in plant development*

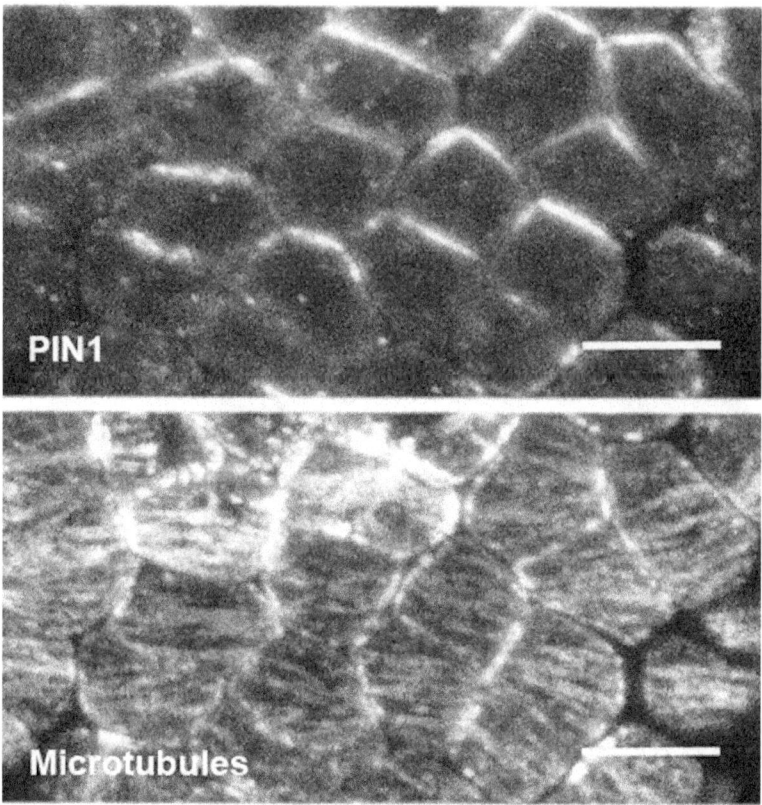

Fig. 8.29 Immunolocalization of PIN1 (top) and α-tubulin microtubules (bottom) in the boundary domain of the shoot apical meristem: PIN1 and microtubule patterns are correlated. Scale bars = 5 μm. From Heisler et al. (2010).

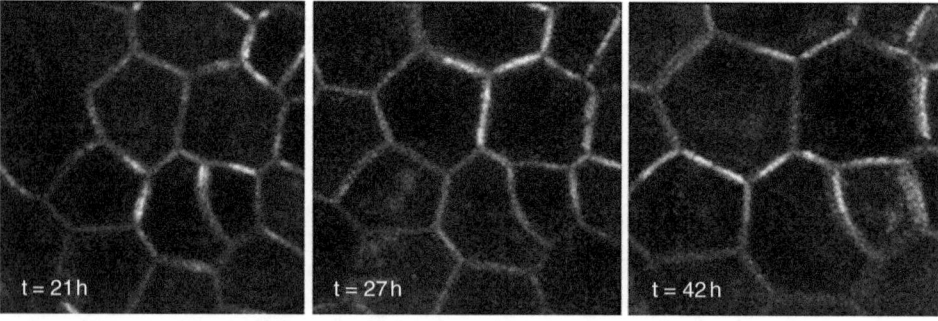

Fig. 8.30 Kinetics of PIN1–GFP reorientation at the surface of an oryzalin-treated meristem (from left to right 21, 27 and 42 h after oryzalin treatment). The GFP signal switches from one side of the cell to another, showing that PIN1 retains the ability to reorient in the absence of microtubules. From Heisler et al. (2010).

The role of forces at the cell scale 239

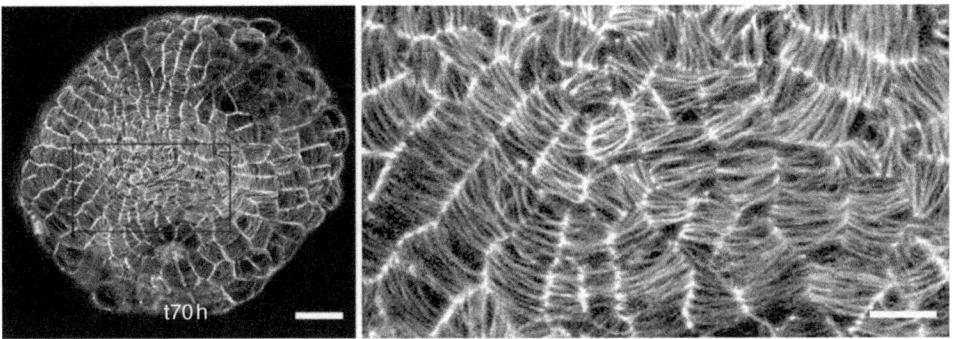

Fig. 8.31 Left: Surface of a GFP-MBD meristem 70 h after isoxaben treatment. Note the presence of circumferential bundles of microtubules. Scale bar = 20 μm. Right: Close-up showing that the microtubules become circumferential even at the tip of the meristem, and this can be correlated to the dome shape of the meristem in the wild-type background. Scale bar = 10 μm. From Heisler et al. (2010).

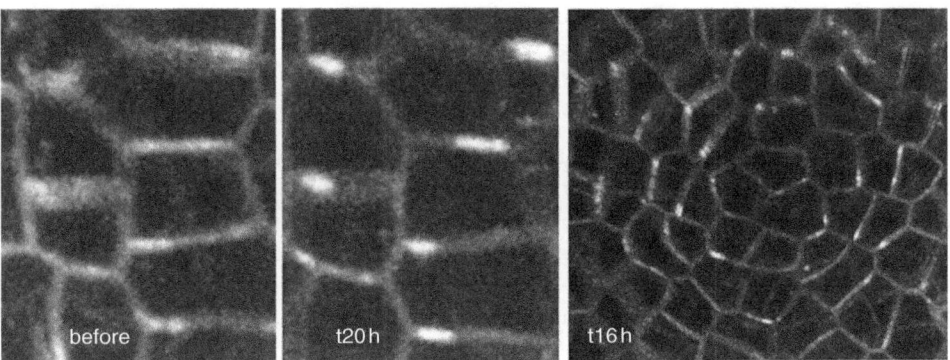

Fig. 8.32 Surface of a meristem expressing PIN1–GFP before (left) and 20 h after (center) isoxaben treatment. The GFP signal becomes localized to a subdomain of the plasma membrane. Right: Surface of a meristem expressing PIN1–GFP 16 h after isoxaben treatment, showing a preferential localization of PIN1 on the circumferential membranes. From Heisler et al. (2010).

to the maximal stress direction. Interestingly, the PIN1–GFP signal was also altered after isoxaben treatment. PIN1 was mainly localized on the circumferential membranes, again following the correlation with the microtubules. Interestingly, PIN1 was concentrated at cell vertices, which are predicted to be local stress maxima (Fig. 8.32).

Modeling data from Pawel Krupinski further supported these data: in particular, it was shown *in silico* that if PIN1 concentrates on the membrane that experiences the highest tensional stress, PIN1 relocates in circumferential membranes around an ablation. This model was also consistent with other models in which PIN1 polarity depended on the auxin concentration in neighboring cells. These models were sufficient

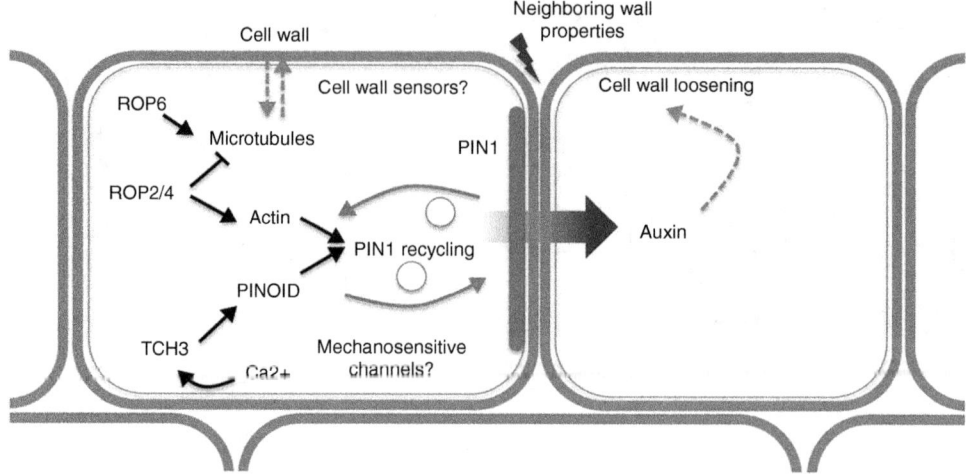

Fig. 8.33 A model integrating molecular effectors of PIN1 recycling (black lettering and arrows) and mechanical cues (gray lettering) to drive PIN1 polarity (thin gray arrows in the left-hand cell) and auxin flux (thick arrow from left to right). From Hamant et al. (2010), with permission from Elsevier.

to explain large-scale auxin distributions in the shoot apical meristem, driving the sequence of organ initiation or phyllotaxis. As auxin leads to softer wall, a neighboring cell with more auxin can induce localized tension between adjacent cells, and thus our mechanical hypothesis provides a plausible model to explain how a cell could "measure" the auxin content in adjacent cells. In particular, mechanical tension could inhibit endocytosis locally, thus trapping the PIN1-containing vesicle in the tensest membranes (Heisler *et al.* 2010) (Fig. 8.33).

8.3.2 Cell polarity and membrane tension

The idea that membrane tension plays a major role in cell polarity is not new (for a review see, e.g., Asnacios and Hamant 2012). There are many examples showing that cell polarity is defined by the cell cortex. In fact, cells without a nucleus can become polarized: fragments of fish epidermal keratocytes were shown to polarize their cytoskeleton and move, as a whole cell, like a fibroblast (actin at the front and myosin at the back; Verkhovsky et al. 1999). Interestingly, in those fragments a slight push with a needle was sufficient to trigger polarity from an initially apolar and isotropic cell, thus suggesting that mechanical cues can serve as cues to break symmetry. Similarly, motile cells that are placed in a hydrodynamic flow polarize their cytoskeleton, and their motility is driven toward resisting the flow (Dalous et al. 2008). This is another illustration of a general rule in life: cells do what they can to resist mechanical stress.

Interestingly, by definition, membrane tension not only depends on the cortical cytoskeleton but also on the cell's osmotic pressure. While this has been widely studied in plants, the contribution of osmotic pressure to cell polarity in animals is a rather recent discovery. For instance, when placed in hyperosmotic conditions, the membrane tension of *Caenorhabidtis elegans* sperm cells is decreased, actin filaments are disorganized and this leads to a reduction in cell motility. Conversely, a hypoosmotic environment has the opposite effect (Batchelder et al. 2011). Similarly, in neutrophils, the impact of suction on the cell (which increases tension) can be mimicked by placing the cell in a hypotonic solution. Interestingly, these effects are still observed in the presence of blebbistatin (which inhibits myosin activity), thus suggesting that membrane tension, rather than the cortical cytoskeleton, is driving cell polarity (Houk et al. 2012).

8.3.3 Mechanotransduction

Transduction pathways of any kind largely rely on changes in protein conformation, either when a ligand binds its receptor, a channel opens or a kinase phosphorylates its substrate. Mechanotransduction is no different: mechanical stress has been shown to induce changes in protein conformation, leading to a cascade of intracellular events, as observed in other transduction pathways. The main difference is that the first conformational change is induced by a mechanical strain of the protein, rather than by ligand–receptor interaction or post-translational modification (Fig. 8.34).

Such mechanically induced conformational changes can be visualized by fluorescent resonance energy transfer (FRET). For instance, p130Cas, a Src target, was used to reveal src activation after pulling on human umbilical vein cells with a laser-tweezer on

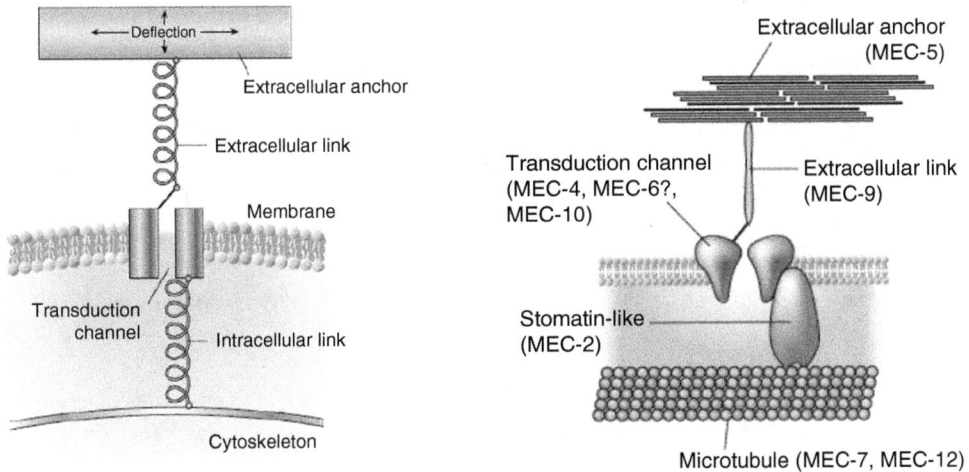

Fig. 8.34 Mechanotransduction: general principles (left) and an example (touch receptor in *C. elegans*, right). From Gillespie and Walker (2001). Reprinted with permission from Macmillan Publishers Ltd.

242 Forces in plant development

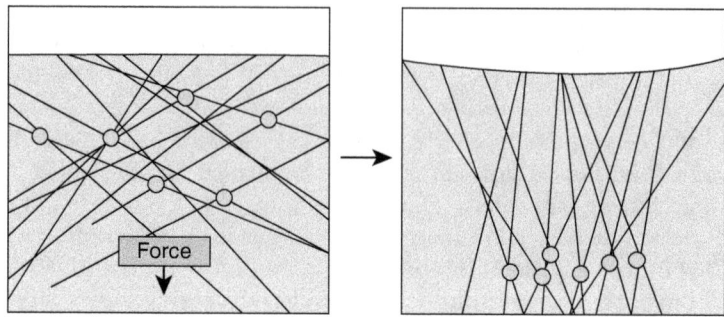

Fig. 8.35 Forces can bring molecules together. From mechanobio.info.

fibronectin-coated beads adhering to the cells (Wang et al. 2005). Similarly, conformational changes occur at adhesion sites: talin and α-catenin have been shown to change their conformation upon mechanical stimulation, leading to the recruitment of vinculin and reinforcement of focal adhesion sites (e.g., Yonemura et al. 2010). These conformational changes propagate, and can again be monitored by FRET. For instance, the conformational changes of vinculin have been used to design a FRET-based tension sensor (Grashoff et al. 2010).

The membrane context can also affect protein conformation. For instance, the thickness of the lipid bilayer could better match the open or closed state of a channel, and if membrane tension affects membrane thickness, then the channel becomes at least in part mechanosensitive (see, e.g., Hamill and Martinac 2001). Note that the boundary between mechanically based and biochemically based transduction can be blurred, notably because of the fact that strains lead to biochemical modifications but also because all the pathways are interconnected. For instance, membrane tension, by preventing endocytosis, can also affect transduction of many pathways, notably by impacting the turnover of transmembrane receptors. In addition, by reorganizing the cytoskeleton internally, mechanical stress can also cluster/concentrate components inside the cell, and thus modify their activity (Fig. 8.35).

8.4 Conclusion

We can conclude this chapter with some words from D'Arcy Thompson (1917) (Fig. 8.36):

It behoves us always to remember that in physics it has taken great men to discover simple things. They are very great names indeed which we couple with the explanation of the path of a stone, the droop of a chain, the tints of a bubble, the shadows in a cup. It is but the slightest adumbration of a dynamical morphology that we can hope to have until the physicist and the mathematician shall have made these problems of ours their own.

D'Arcy Thompson, *On Growth and Form* (1917)

Fig. 8.36 D'Arcy Thompson.

References

Aigouy, B., R. Farhadifar, D. B. Staple, A. Sagner, J. C. Roper et al., 2010 Cell flow reorients the axis of planar polarity in the wing epithelium of Drosophila. *Cell* **142**: 773–786.

Aliee, M., J. C. Roper, K. P. Landsberg, C. Pentzold, T. J. Widmann et al., 2012 Physical mechanisms shaping the Drosophila dorsoventral compartment boundary. *Curr Biol* **22**: 967–976.

Asnacios, A. and O. Hamant, 2012 The mechanics behind cell polarity. *Trends Cell Biol* **22**: 584–591.

Baskin, T. I., 2005 Anisotropic expansion of the plant cell wall. *Annu Rev Cell Dev Biol* **21**: 203–222.

Bastien, R., T. Bohr, B.Moulia and S. Douady, 2013 Unifying model of shoot gravitropism reveals proprioception as a central feature of posture control in plants. *Proc Natl Acad Sci USA* **110**: 755–760.

Batchelder, E. L., G. Hollopeter, C.Campillo, X.Mezanges, E. M. Jorgensen et al., 2011 Membrane tension regulates motility by controlling lamellipodium organization. *Proc Natl Acad Sci USA* **108**: 11429–11434.

Bichet, A., T. Desnos, S. Turner, O. Grandjean and H. Hofte, 2001 BOTERO1 is required for normal orientation of cortical microtubules and anisotropic cell expansion in Arabidopsis. *Plant J* **25**: 137–148.

Bilsborough, G. D., A. Runions, M. Barkoulas, H. W. Jenkins, A. Hasson et al., 2011 Model for the regulation of *Arabidopsis thaliana* leaf margin development. *Proc Natl Acad Sci USA* **108**: 3424–3429.

Blein, T., A. Pulido, A. Vialette-Guiraud, K. Nikovics, H. Morin et al., 2008. A conserved molecular framework for compound leaf development. *Science* **322**: 1835–1839.

Braam, J., 2005 In touch: plant responses to mechanical stimuli. *New Phytol* **165**: 373–389.

Bringmann, M., E. Li, A. Sampathkumar, T. Kocabek, M. T. Hauser et al., 2012 POM-POM2/CELLULOSE SYNTHASE INTERACTING1 is essential for the functional association of cellulose synthase and microtubules in Arabidopsis. *Plant Cell* **24**: 163–177.

Brunet, T., A. Bouclet, P. Ahmadi, D. Mitrossilis, B. Driquez et al., 2013 Evolutionary conservation of early mesoderm specification by mechanotransduction in Bilateria. *Nat Commun* **4**: 2821.

Burgert, I. and P. Fratzl, 2009 The plant cell wall acts as a sophisticated mechanical device [Abstract]. *Integr Comp Biol* **49**(Suppl. 1): e23.

Castle, E. S., 1938 Membrane tension and orientation of structure in the plant cell wall. *J Cell Comp Physiol* **10**: 113–121

Coutand, C., L. Martin, N. Leblanc-Fournier, M. Decourteix, J. L. Julien et al., 2009 Strain mechanosensing quantitatively controls diameter growth and PtaZFP2 gene expression in poplar. *Plant Physiol* **151**: 223–232.

Couturier, E., S. Courrech du Pont and S. Douady, 2009 A global regulation inducing the shape of growing folded leaves. *PLoS ONE* **4**: e7968.

Couturier, E., S. C. du Pont and S. Douady, 2011 The filling law: a general framework for leaf folding and its consequences on leaf shape diversity. *J Theor Biol* **289**: 47–64.

Couturier, E., N. Brunel, S.Douady and N. Nakayama, 2012 Abaxial growth and steric constraints guide leaf folding and shape in *Acer pseudoplatanus*. *Am J Bot* **99**: 1289–1299.

Dalous, J., E. Burghardt, A.Muller-Taubenberger, F.Bruckert, G. Gerisch et al., 2008 Reversal of cell polarity and actin-myosin cytoskeleton reorganization under mechanical and chemical stimulation. *Biophys J* **94**: 1063–1074.

Desprat, N., W. Supatto, P. A. Pouille, E. Beaurepaire and E. Farge, 2008 Tissue deformation modulates twist expression to determine anterior midgut differentiation in *Drosophila* embryos. *Dev Cell* **15**: 470–477.

Dumais, J. and C. R. Steele, 2000 New evidence for the role of mechanical forces in the shoot apical meristem. *J Plant Growth Regul* **19**: 7–18.

Engler, A. J., S. Sen, H. L. Sweeney and D. E. Discher, 2006 Matrix elasticity directs stem cell lineage specification. *Cell* **126**: 677–689.

Farge, E., 2003 Mechanical induction of Twist in the *Drosophila* foregut/stomodeal primordium. *Curr Biol* **13**: 1365–1377.

Fu, Y., Y. Gu, Z. Zheng, G. Wasteneys and Z. Yang, 2005 Arabidopsis interdigitating cell growth requires two antagonistic pathways with opposing action on cell morphogenesis. *Cell* **120**: 687–700.

Geldner, N., J. Friml, Y. D. Stierhof, G. Jurgens and K. Palme, 2001 Auxin transport inhibitors block PIN1 cycling and vesicle trafficking. *Nature* **413**: 425–428.

Gibson, W. T. and M. C. Gibson, 2012 Growing cells push back under pressure. *Cell* **149**: 259–261.

Gillespie, P. G. and R. G. Walker, 2001 Molecular basis of mechanosensory transduction. *Nature* **413**: 194–202.

Grashoff, C., B. D. Hoffman, M.D. Brenner, R.Zhou, M. Parsons et al., 2010 Measuring mechanical tension across vinculin reveals regulation of focal adhesion dynamics. *Nature* **466**: 263–266.

Green, P. B. and A. King, 1966 A mechanism for the origin of specifically oriented textures in development with special reference to *Nitella* wall structure. *Aust J Biol Sci* **19**: 421–437.

Hamant, O., 2013 Widespread mechanosensing controls the structure behind the architecture in plants. *Curr Opin Plant Biol* **16**: 654–660.

Hamant, O., M. G. Heisler, H. Jonsson, P. Krupinski, M. Uyttewaal et al., 2008 Developmental patterning by mechanical signals in Arabidopsis. *Science* **322**: 2008 1650–1655.

Hamant, O., J. Traas and A. Boudaoud, 2010 Regulation of shape and patterning in plant development. *Curr Opin Genet Dev* **20**: 454–459.

Hamill, O. P. and B. Martinac, 2001 Molecular basis of mechanotransduction in living cells. *Physiol Rev* **81**: 685–740.

Heisler, M. G., O. Hamant, P.Krupinski, M.Uyttewaal, C. Ohno et al., 2010 Alignment between PIN1 polarity and microtubule orientation in the shoot apical meristem reveals a tight coupling between morphogenesis and auxin transport. *PLoS Biol* **8**: e1000516.

Houk, A. R., A. Jilkine, C. O. Mejean, R. Boltyanskiy, E. R. Dufresne et al., 2012 Membrane tension maintains cell polarity by confining signals to the leading edge during neutrophil migration. *Cell* **148**: 175–188.

Hufnagel, L., A. A. Teleman, H. Rouault, S. M. Cohen and B. I. Shraiman, 2007 On the mechanism of wing size determination in fly development. *Proc Natl Acad Sci USA* **104**: 3835–3840.

Ingber, D. E., 1998 The architecture of life. *Sci Am* **278**: 48–57.

Kutschera, U. and K. J. Niklas, 2007 The epidermal-growth-control theory of stem elongation: an old and a new perspective. *J Plant Physiol* **164**: 1395–1409.

Landsberg, K. P., R. Farhadifar, J. Ranft, D. Umetsu, T. J. Widmann et al., 2009 Increased cell bond tension governs cell sorting at the *Drosophila* anteroposterior compartment boundary. *Curr Biol* **19**: 1950–1955.

Lloyd, C., 2011 Dynamic microtubules and the texture of plant cell walls. *Int Rev Cell Mol Biol* **287**: 287–329.

Louveaux, M. and O. Hamant, 2013 The mechanics behind cell division. *Curr Opin Plant Biol* **16**: 774–779.

Milani, P., S. A. Braybrook and A. Boudaoud, 2013 Shrinking the hammer: micromechanical approaches to morphogenesis. *J Exp Bot* **64**: 4651–4662.

Moulia, B., 2013 Plant biomechanics and mechanobiology are convergent paths to flourishing interdisciplinary research. *J Exp Bot* **64**: 4617–4633.

Nath, U., B. C. Crawford, R. Carpenter and E. Coen, 2003 Genetic control of surface curvature. *Science* **299**: 1404–1407.

Noblin, X., N. O. Rojas, J. Westbrook, C. Llorens, M.Argentina et al., 2012 The fern sporangium: a unique catapult. *Science* **335**: 1322.

Nonaka, S., Shiratori, H., Saijoh, Y. and H. Hamada, 2002. Determination of left–right patterning of the mouse embryo by artificial nodal flow. *Nature* **418**: 96–99.

Paredez, A. R., C. R. Somerville and D. W. Ehrhardt, 2006 Visualization of cellulose synthase demonstrates functional association with microtubules. *Science* **312**: 1491–1495.

Sampathkumar, A., P. Krupinski, R. Wightman, P. Milani, A. Berquand et al., 2014 Subcellular and supracellular mechanical stress prescribes cytoskeleton behavior in Arabidopsis cotyledon pavement cells. *eLIFE* **3**: e01967.

Savin, T., N. A. Kurpios, A. E. Shyer, P.Florescu, H. Liang et al., 2011 On the growth and form of the gut. *Nature* **476**: 57–62.

Shraiman, B. I., 2005 Mechanical feedback as a possible regulator of tissue growth. *Proc Natl Acad Sci USA* **102**: 3318–3323.

Stoppin-Mellet, V., J. Gaillard and M. Vantard, 2006 Katanin's severing activity favors bundling of cortical microtubules in plants. *Plant J* **46**: 1009–1017.

Swift, J., I. L. Ivanovska, A. Buxboim, T. Harada, P. C. Dingal et al., 2013 Nuclear lamin-A scales with tissue stiffness and enhances matrix-directed differentiation. *Science* **341**: 1240104.

Thery, M., A. Jimenez-Dalmaroni, V. Racine, M. Bornens and F. Julicher, 2007 Experimental and theoretical study of mitotic spindle orientation. *Nature* **447**: 493–496.

Thompson, D'A. W., 1917 *On Growth and Form Cambridge*. Cambridge University Press, Cambridge, UK.

Uyttewaal, M., J. Traas and O. Hamant, 2010 Integrating physical stress, growth, and development. *Curr Opin Plant Biol* **13**: 46–52.

Uyttewaal, M., A. Burian, K. Alim, B. Landrein, D. Borowska-Wykret et al., 2012 Mechanical stress acts via katanin to amplify differences in growth rate between adjacent cells in Arabidopsis. *Cell* **149**: 439–451.

Verkhovsky, A. B., T. M. Svitkina and G. G. Borisy, 1999 Self-polarization and directional motility of cytoplasm. *Curr Biol* **9**: 11–20.

Wang, Y., E. L. Botvinick, Y. Zhao, M. W. Berns, S. Usami et al., 2005 Visualizing the mechanical activation of Src. *Nature* **434**: 1040–1045.

Williamson, R. E., 1990 Alignment of cortical microtubules by anisotropic wall stresses. *Aust J Plant Physiol* **17**: 601–613.

Wolff, J., 1892 *The Law of Bone Remodeling*. (Reprinted 1986, Springer, Berlin.)

Wymer, C. L., S. A. Wymer, D. J. Cosgrove and R. J. Cyr, 1996 Plant cell growth responds to external forces and the response requires intact microtubules. *Plant Physiol* **110**: 425–430.

Yonemura, S., Y. Wada, T. Watanabe, A. Nagafuchi and M. Shibata, 2010 Alpha-catenin as a tension transducer that induces adherens junction development. *Nat Cell Biol* **12**: 533–542.

9
An introduction to modeling the initiation of the floral primordium

Christophe GODIN[1], Eugenio AZPEITIA[2], and Etienne FARCOT[3]

[1]Inria, Virtual Plants Inria-Cirad-Inra Team, Montpellier, France.
[2]Inria, Virtual Plants Inria-Cirad-Inra Team, Montpellier, France.
[3]School of Mathematical Sciences, University of Nottingham, Nottingham, UK.

Abstract

This chapter presents models of the processes involved in floral initiation and development. It begins by briefy presenting models of hormonal transport. The focus is on two key aspects of floral development, namely floral initiation, due to the periodic local accumulation of auxin (a plant hormone) near the plant apex, and the genetic regulation of floral development. The main assumptions about auxin transport that have been proposed and tested in the literature are described, and it is shown how the use of models makes it possible to test assumptions expressed in terms of local cell–cell interaction rules and to check if they lead to patterning in the growing tissue that is consistent with observation. The gene regulatory networks (GRNs) that control the initial steps of floral development and differentiation are investigated. In a simplified form, this network contains dozens of components that interact with each other in space and time. The understanding of such a complex system also requires a modeling approach in order to quantify these interactions and analyze their properties. There are two main formalisms that are used to model GRN: the Boolean and the ODE formalisms. Both these are illustrated on a submodule of the floral GRN and their main advantages and drawbacks are discussed. It is shown how manipulations of the network models can be used to make predictions corresponding to possible biological manipulations of the GRN (e.g., loss-of-function mutants). Throughout, specific mathematical

From Molecules to Living Organisms: An Interplay Between Biology and Physics. First Edition. Eva Pebay-Peyroula et al. © Oxford University Press 2016.
Published in 2016 by Oxford University Press.

topics of particular interest for the development of these ideas are detailed in separate boxes which can be read relatively independently of the main text.

Keywords

Developmental models, auxin transport, gene regulatory network, floral initiation, dynamical systems, Boolean networks, ordinary differential equations, meristem

Chapter Contents

**9 An introduction to modeling
the initiation of the floral primordium** 247
Christophe GODIN, Eugenio
AZPEITIA, and Etienne FARCOT

    9.1 Introduction 250
    9.2 Specifying growth points on meristem domes 251
        9.2.1 Auxin import and export in cells 252
        9.2.2 Auxin flows in the meristem 253
        9.2.3 Allocating PIN1 to cell membranes 256
        9.2.4 Accumulation of auxin at the meristem surface 262
    9.3 Modeling the regulation of floral initiation 265
        9.3.1 Expression patterns in the floral meristem 266
        9.3.2 Modules controlling development 267
        9.3.3 Modeling the genetic regulation of bud fate 269
    9.4 Conclusion 277

    References 278

9.1 Introduction

A major current challenge in developmental biology is to link gene regulation to shape development. In the last two decades, spectacular progresses in molecular biology and live imaging have made it possible to observe animal or plant development *in vivo* with unprecedented resolution in both space and time. As a consequence, a wealth of new and key data are now available on the various facets of growth: gene expression patterns, tissue geometry at cellular resolution, physical properties of tissues, hormone concentrations, etc.

In this chapter we will focus on plant development, and in particular on the organ that produces all the other plant organs: the apical meristem. Apical meristems consist of small domes of cells located at the tip of each axis, containing undifferentiated cells. Two main types of meristem are usually distinguished, the shoot and the root apical meristems, depending on whether they are at the tip of a shoot axis or root axis. In this chapter we focus our attention on shoot apical meristems (SAMs).

Stem cells at the summit of the SAM slowly divide and provide cells to the surrounding peripheral region, where they proliferate rapidly and differentiate into leaf or flower primordia. Some of these cells give rise to other meristems in a recursive manner, which makes it possible for the plant to build up branching systems (Fig. 9.1; see also Chapter 3 which provides a detailed biological account of flower development).

Our aim in this chapter is twofold. We first want to introduce some of the key processes that occur in the functioning and growth of the SAM and how they are

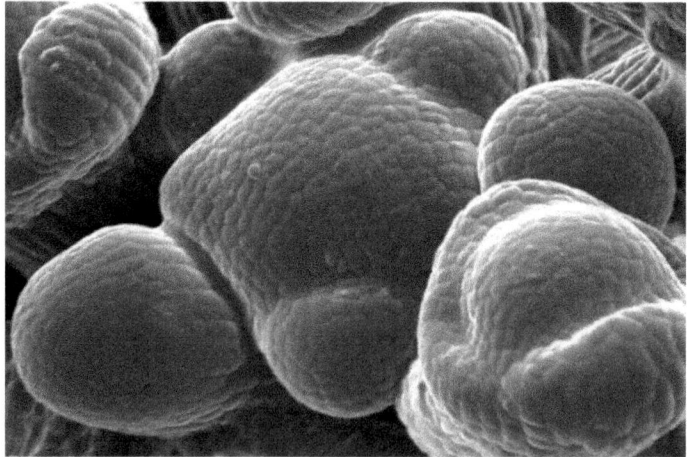

Fig. 9.1 Shoot apical meristem of *Arabidopsis thaliana*. At the tip of the shoot apical meristem (SAM), a dome-like population of stem cells proliferates. Organs (here flower primordia) are initiated one after the other at the periphery of this central region in a spiral pattern, where each organ forms an angle of approximately 137° with the previous one. At the bottom right of the picture, a young flower bud is developing and itself starts to produce new lateral organs (here the sepals). (Photo from Jan Traas.)

quantified and modeled. We have chosen to focus on the very moment when an organ is created and a form emerges at the flank of the meristem dome. We shall investigate this morphogenetic event upstream of primordium initiation, where hormonal signals specify the location of the future primordium, and downstream of organ initiation, where the molecular cascades regulating flower development are triggered. Importantly, our second aim is to introduce researchers who are not familiar with the modeling background used to describe these processes to the mathematical and computational language in a very progressive though precise way. Overall, our aim is to show how models can be used to better to decipher the complexity of the observed phenomena and how they make it possible to deduce facts that can be tested against actual observations.

9.2 Specifying growth points on meristem domes

After initiation at the SAM, the young flower primordium develops as a growing dome. This process obviously involves mechanical forces that locally deform the tissue based on variations in the stresses in cell walls. The question then arises as to what biophysical or biochemical processes locally trigger the profound physical changes upstream of these deformations?

In recent decades, researchers have identified potential candidates for these factors. In particular, at the SAM, one of the earliest events—maybe the earliest one—that precedes the initiation of a primordium has been shown to be local accumulation of the plant hormone auxin [60]. This accumulation is thought to induce a cascade of biochemical and biophysical events that locally lead to the outgrowth of a primordium. Therefore two major questions need to be answered from a morphogenetic point of view: (1) what drives the accumulation of auxin in meristems at precise positions and (2) how is the chemical auxin signal translated into mechanical instructions that lead to outgrowth of the primordium?

The latter question is the subject of active research and debate at the moment [28, 52, 10, 64, 9, 4]. The general idea is that auxin decreases the rigidity of the cell walls where it accumulates, in turn creating a bulge in the tissue at the corresponding position, as discussed in Section 9.1. We will not describe this molecular mechanism in any more detail here.

The former question arises at a wider scale in the plant. Auxin is known to be transported over long distances, and the mechanisms that drive this transport have been studied for decades in different tissues since the pioneering work of Sachs [66]. Interestingly, to achieve differential accumulation of the hormone in different places, this transport process needs be very dynamic and thoroughly coordinated at level of the whole cell population. As the cells can only interact locally with their immediate neighbors in the tissue, this coordination must therefore essentially be an emergent property of this complex cell–cell interaction network. To better understand this phenomenon, researchers have attempted to model the emerging organization of auxin transport through the cell network on the basis of different hypothetical cell–cell interaction rules. We will now describe the principles of these models.

9.2.1 Auxin import and export in cells

The term auxin refers to a group of small molecules, for example indole acetic acid (IAA), that can be transported through cells in plant tissues. This transport is mainly due to the presence in cells of membrane efflux or influx carriers that make it possible for auxins to pass through cell membranes (see [35, 63] for general presentations of models of auxin transport in plants).

Let us first take a closer look at the general mechanism behind the transport of auxin molecules through cell membranes and walls. The principle of this transport was proposed in the 1970s in [62, 55], and is known as the chemiosmotic theory (Fig. 9.2). In brief, the idea is that auxin exists in two forms in plant tissues, a neutral protonated form, IAAH, and an anionic form, IAA$^-$. The protonated form can diffuse freely across the cell wall while the anionic form cannot. The relative proportion of these two forms depends on the pH of the medium (e.g., [34]). In the cytoplasm, which has a neutral pH (7.0), auxin is mostly in its anionic form IAA$^-$, but in the acidic intercellular compartment (the apoplast; pH 5.5) the neutral form exists in much higher proportions. Therefore, in the absence of other processes, auxin can passively diffuse from the apoplastic compartment inside the cells but then becomes trapped in

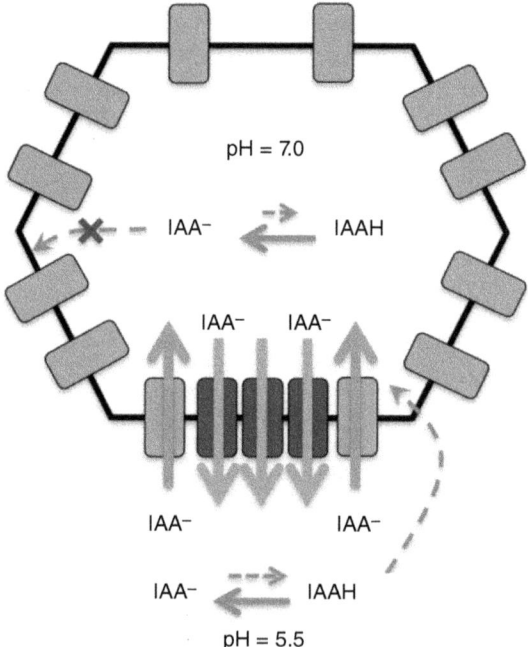

Fig. 9.2 Auxin transport across cell plasma membranes. In the cytoplasm (neutral pH), the anionic form largely dominates and can exit the cell only via efflux carriers (e.g., PIN1 proteins, dark gray with grey arrows). In the apoplastic compartment, the neutral form can diffuse freely to the cytoplasm. Additional transport of the anionic form from the apoplast to the cytoplasm requires an influx carrier (e.g., AUX proteins, lighter gray).

its anionic form in the cytoplasm. To get out of the cell into the apoplast, the anionic form requires specific molecular carriers.

This is done using membrane transporter proteins of the PIN1-FORMED (PIN1) family. PIN1 proteins are putative efflux carriers, i.e., they actively contribute to the transport of auxin from the cell to the intercellular compartment (the apoplast). PIN1 proteins are polarly located in the cell membranes (i.e., at a given moment in time, they are located on a particular side of the cell). Another family of transporters, the AUX/LAX proteins, contribute to the transport of auxin from the apoplast into the cell cytoplasm (the AUX/LAX proteins are active influx carriers). They are not polarized in the cell and usually uniformly cover the surfaces of the cell membranes.

9.2.2 Auxin flows in the meristem

Now let us consider the question of auxin transport at a larger scale, over an organ or particular tissue. Such coherent long-distance transport is due to the ability of the cells to coordinate their activities and consistently orient their PIN1 polarities. In this way, the cells create pathways through the tissue in which auxin is transported from place to place. In the SAM, this dynamic network of PIN1 pumps is coordinated in such a way that it periodically leads to the creation of small auxin patches in the L1 layer close to the tip [61].

To understand how such a system can work at the level of cell populations, it is useful to model quantitatively this dynamic transport process. Let us start by modeling auxin transport for a fixed configuration of PIN1 polarity within the cells. The description of the transport then consists of quantifying the conservation of auxin molecules at any point in the tissue. Note that for sake of simplicity we neglect here the action of influx transporters of the AUX/LAX family (however, their role may be important in maintaining a certain level of robustness in the patterning; see [34, 7]).

Let us formalize this. Consider a meristem tissue made up of a collection of cells i whose states are defined by an auxin concentration a_i (mol m^{-3}) and a volume v_i. If PIN1 proteins in cell i are polarized toward a neighboring cell n, then we consider that auxin can be actively exported from cell i to cell n. In this transport network, processes that may affect the auxin concentration a_i of each cell i are of three different types:

- Transport: auxin molecules may move within the network of cells. It is usually considered that this movement has two potential sources (chemiosmotic model):
  - passive transport: due to thermal noise, auxin molecules may move randomly from cell to cell.
  - active transport: auxin molecules can be exported outside the cell by membrane transporters.
- Synthesis: auxin may be synthesized locally in cells.
- Degradation: a percentage of the auxin molecules may be regularly degraded (or inactivated) within the cell itself.

The variation in auxin concentration in each cell can be modeled by quantifying these processes and combining their effect in an equation reflecting the conservation of auxin molecules in the network of cells through space and time. Fluxes in the tissue will make it possible to quantify the net auxin movement between cells (see Box 9.1). The rate of change of a_i is therefore defined by:

$$\frac{\partial a_i}{\partial t} = -\frac{1}{v_i} \sum_{n \in N_i} s_{i,n} \phi_{i \to n} - \delta_a a_i + \sigma_a, \qquad (9.1)$$

where $s_{i,n}\phi_{i \to n}$ is the net flux of auxin from cell i to cell n through the interface surface $s_{i,n}$, σ_a (mol m^{-3} s^{-1}) is a constant that describes the rate at which auxin is produced in cells and δ_a (s^{-1}) defines the rate of auxin degradation. Both constants are considered here independently of the cell state (i.e., no subscript i).

Box 9.1 Relation between fluxes and concentrations of molecules in cells

To compute the net number of molecules that enter the cell during a certain time, one needs to consider the *net flux* of molecules through the different faces of the cell.

Let us consider two compartments A and B separated by a membrane of surface $s_{A,B}$ (m^2). We define the flux density $\phi_{A \to B}$ (mol m^{-2} s^{-1}) through the surface $s_{A,B}$ as the net number of molecules that cross the membrane from compartment A to compartment B per unit surface and per unit time (note that the direction A to B is meaningful here). A positive flux density from A to B means that molecules are transferred from A to B. Reciprocally, a negative flux means that molecules are transferred from B to A. The total flux through surface $s_{A,B}$ separating compartments A and B is then defined by $s_{A,B}\phi_{A \to B}$.

Based on fluxes, it is easy to compute the net flux Φ_i of molecules that enter a cell i per unit time. In a tissue, consider the membrane that separates a cell i from one of its neighboring cells n, crossed by a flux density $\phi_{n \to i}$. If $s_{i,n}$ denotes the surface of this membrane, then the net flux that crosses the surface $s_{i,n}$ per unit time is (in mol s^{-1}):

$$\Phi_{n \to i} = s_{i,n}\phi_{n \to i}, \qquad (9.1)$$

and by summing up on all the interfaces of cell i with other cells n, we get the total net number of molecules that entered cell i during unit time:

$$\Phi_i = \sum_{n \in N_i} s_{i,n}\phi_{n \to i}$$

$$= -\sum_{n \in N_i} s_{i,n}\phi_{i \to n} = -\sum_{n \in N_i} \Phi_{i \to n}. \qquad (9.3)$$

Consequently, if we assume that molecules can only remain or move in or out of cells to their neighbors, the concentration a_i of a particular molecule in cell i of volume v_i will change at a rate:

$$\frac{\partial a_i}{\partial t} = \frac{\Phi_i}{v_i}$$

$$= -\frac{1}{v_i} \sum_{n \in N_i} s_{i,n} \phi_{i \to n}. \quad (9.4)$$

By decomposing the net total flux $\Phi_{i \to n}$ through the membrane between i and n in a net flux due to diffusion $\Phi^{D}_{i \to n}$ and a net flux due to active transport $\Phi^{A}_{i \to n}$, we get:

$$\phi_{i \to n} = \phi^{D}_{i \to n} + \phi^{A}_{i \to n}, \quad (9.5)$$

and by replacing $\phi_{i \to n}$ in equation (9.1), it becomes:

$$\frac{\partial a_i}{\partial t} = -\frac{1}{v_i} \sum_{n \in N_i} s_{i,n} \phi^{D}_{i \to n} - \gamma_a \sum_{n \in N_i} s_{i,n} \phi^{A}_{i \to n} - \delta_a a_i + \sigma_a. \quad (9.6)$$

Let us now make the diffusion and the active transport terms in this equation explicit. The density of flux due to diffusion between two compartments is classically assumed to be proportional to the difference in concentration between these compartments (a particular case of Fick's first law):

$$\phi^{D}_{i \to n} = \gamma_D (a_i - a_n), \quad (9.7)$$

where γ_D is the constant of permeability reflecting the capability of auxin to move across the membrane (in m s^{-1}).

The net density of flux due to active transport depends on the distribution of carrier proteins that can be found in membranes on both sides of a given wall between two cells i and n. A model of this type was originally proposed by Mitchison [48, 49], making it possible to express the density of flux through the wall $s_{i,j}$ as a function of the auxin concentration in the neighboring cells:

$$\phi^{A}_{i \to n} = \gamma_A (a_i p_{i,n} - a_n p_{n,i}) \quad (9.8)$$

where γ_A (m^3 mol^{-1} s^{-1}) characterizes the transport efficiency of the PIN1 proteins and $p_{i,n}$ (mol m^{-2}) is the surface concentration of PIN1 proteins in the membrane of cell i facing cell n facilitating transport from cell i to cell n:

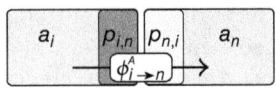

The equation can be interpreted as if each single PIN1 carrier protein in the wall transports $\gamma_A a_i$ molecules of auxin. Since there are $p_{i,n}$ PIN1 molecules per unit of surface in the wall, each unit of surface exports $\gamma_A a_i p_{i,n}$ molecules [25]. Equation (9.1) thus becomes:

$$\frac{\partial a_i}{\partial t} = -\frac{\gamma_D}{v_i} \sum_{n \in N_i} s_{i,n} (a_i - a_n) - \frac{\gamma_A}{v_i} \sum_{n \in N_i} s_{i,n} (a_i p_{i,n} - a_n p_{n,i}) - \delta_a a_i + \sigma_a, \quad (9.9)$$

with the terms on the right-hand side, from left to right, being a diffusion term, an active transport term, an auxin decay term and an auxin synthesis term.

Different variants of this model have been used in the literature to simulate auxin flows in the meristem. A first application was to estimate auxin distribution in meristems based on observed and precise maps of the polar PIN1 transporters in cells [5]. Indeed, while auxin concentrations are extremely difficult to measure at cellular resolution (in a cell or a small group of cells), the presence of PIN1 transporters in cell membranes can be observed with good accuracy using confocal microscopy. The idea was then to estimate the auxin distributions in the observed meristem at a given time from the observed distribution of PIN1 proteins in confocal images by simulating the transport of auxin in the observed networks of PIN1 "efflux pumps". This made it possible to confirm that auxin was indeed accumulating and to predict that, according to the observed distribution of transporters, auxin should accumulate in the central zone of the meristem [5]. This prediction was confirmed experimentally several years later with the use of a new auxin sensor [75].

Other applications have used variants of this transport model to assess the plausibility of different assumptions concerning the origin of the observed PIN1 polarity. Some of these variants integrate a more detailed description of the chemiosmotic model. For example, an apoplast compartment has been added to the previous model and protonated and non-protonated versions of auxin are distinguished (e.g. [73]). Other refinements include the introduction of AUX/LAX influx carriers (e.g. [36]), a representation of the trafficking inside the cell between PIN1 molecules embedded in the plasma membrane and a pool of "free" molecules available in the cytoplasm (e.g., [31, 80]).

In a growing tissue the PIN1 distributions are very dynamic. Their allocation to different membranes of a particular cell can change in an hour, thus leading to a very plastic and dynamic configuration of the auxin transport network in the meristem. Here again the processes that drive this dynamic reconfiguration of the network are poorly understood. However, several hypotheses have been proposed to explain the various self-organized patterns of the PIN1 molecules observed in different plant tissues.

9.2.3 Allocating PIN1 to cell membranes

While the regulation of PIN1 polarization is a key determinant in the formation of primordia, little is known about the underlying biochemical or physical processes. Research on the details of plausible molecular mechanisms is currently in progress (e.g. [29, 80]). However, at a coarser scale, different hypotheses have been proposed to interpret the dynamics of PIN1 carriers in different tissues and their self-organizing ability. All these mechanisms rely on a positive feedback loop between the auxin concentration or the auxin flux and the dynamic PIN1 allocation process.

9.2.3.1 Concentration-based hypothesis

It has been observed in the meristem that auxin accumulates in small spots close to the tip. When PIN1 carriers were labeled with green fluorescent protein, a closer inspection of the distribution of the PIN1 polarities in the corresponding cells revealed that, in these cells, PIN1 tends to accumulate on the cell's membrane facing the center of the spot. As the auxin signal appears to be maximal at the center of the spot, it has been suggested that the PIN1 carriers could actually be allocated to the cell membranes in proportion to the amount of auxin contained in the corresponding adjacent cells [69, 31] (Fig. 9.3A). This behavior can be captured by the following polarity law [69]:

$$p_{i,n} = \frac{s_{i,j} b^{a_j}}{Z} p_i, \qquad (9.10)$$

where $p_{i,n}$ is the concentration of PIN1 molecules in cell i embedded in the membrane facing cell n, p_i is the total concentration of PIN1 molecules in cell i, $s_{i,j}$ is the surface separating cells i and j, b is a parameter controlling the intensity of the influence of cell j on cell i and a_j is the auxin concentration of cell j. Z is a normalizing coefficient, corresponding to the fact that, in a given cell, all the $p_{i,n}$ must sum to p_i:

$$Z = \sum_j s_{i,j} b^{a_j}, \qquad (9.11)$$

In the concentration-based hypothesis, auxin peaks (if any) are reinforced by the neighboring cells that orient their PIN1 pumps towards the peak, against the auxin gradient. Similarly to Turing reaction–diffusion systems, this system is able to spontaneously generate spatial patterns [69, 31, 73] (Fig. 9.3B). A tissue with an initial uniform auxin distribution and subject to local random fluctuations will break this spatial symmetry and evolve toward a stable state where the system self-organizes with spatial motifs corresponding to periodic accumulations of auxin concentration throughout the tissue (see Fig. 9.3B).

9.2.3.2 Flux-based hypothesis

The concentration-based hypothesis was introduced rather recently as a means to explain the auxin patterns in the SAM. However, in the 1970s, another feedback mechanism was proposed by Sachs to explain the development of vascular tissues during leaf development [65, 67]. Sachs suggested that positive feedback in the tissue reinforces and stabilizes existing auxin fluxes. This mechanism would be able to reconfigure the transport network dynamically in case of obstacles or wounds and would be responsible for the establishment of the venation patterns in leaves. In 1980, Mitchison used quantitative models and computer simulations to show that if the feedback of the flux on the cell export in the direction of the flux is strong enough, the system can indeed create canals [48, 49]. This so-called "canalization" system is now recognized as a fairly plausible mechanism for vein formation [58, 72, 11], Figure 9.3.C. This canalization

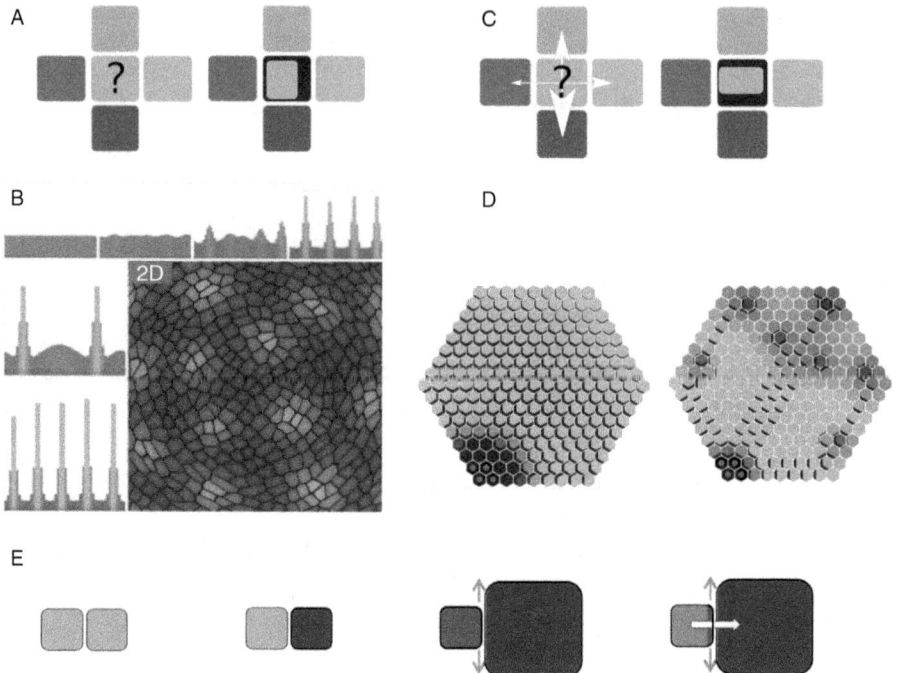

Fig. 9.3 Different hypotheses for the dynamic allocation of PIN1 molecules to membranes. (A) Concentration-based hypothesis. Cells neighboring cell i are assumed to have different concentrations of auxin (left: lighter shades are high concentrations, darker shades are low concentrations). The concentration-based rule indicates that PIN1s should be allocated in proportion to the auxin contained in each neighbor (right). (B) Turing-like patterns emerging spontaneously from the concentration-based model: the emergence of auxin peaks due to noise perturbation of an initially homogeneous concentration. Different wavelengths can be achieved depending on the model parameters (bottom left panels). Spatial periodicity generated by the system in 2D (bottom right panel). (C) Flux-based hypothesis. White arrows indicate the intensity of the auxin fluxes through the different walls (bigger arrows indicate higher flux). The flux-based rule allocates more PIN1s to membranes crossed by a higher flux (right). (D) Patterns resulting from the flux-based rule: a sink (four dark cells at the bottom left) attracts auxin. Two regimes can appear depending on the strength of the feedback function F: if the function is linear no canals are formed (left); if it is superlinear canals appear (right). (E) Stress-based hypothesis. Starting from two cells having a similar amount of auxin (left), and assuming that one of the two cells attains a higher auxin level (right cell in darker gray in the center left panel), the right cell is thus instructed to grow faster (center right panel), which puts the cell wall of the left cell in tension and triggers greater allocation of PIN1s to this wall according to the stress-based rule (right panel). Note that this mechanism amplifies the concentration differences and thus implements a concentration-based rule. Parts A, C and D reproduced from Stoma et al., Plos Comput. Biol. 2008, 4(10): e1000207. Part B reproduced from R. S. Smith et al., PNAS 2006, 103: 1301–1306.

system was later revisited by Rolland-Lagan [59], who suggested that the feedback could be mediated by the PIN1 proteins in plant tissues, and by Feugier et al. [25] who further explored the mathematical properties of this dynamic system. Box 9.2 illustrates the type of mathematical analysis that can be made of such a model and the insights into the system that may be gained by mathematical derivation.

Formally, it is assumed that the concentration of PIN1 proteins $p_{i,n}$ in cell i transporting auxin to cell n is changed due to (1) insertion of PIN1 proteins in the membrane induced by the flux, (2) a background insertion and (3) removal of PIN1 from the membrane. The balance between these processes can be captured by the following equation:

$$\frac{\partial p_{i,n}}{\partial t} = F(\phi_{i \to n}) + \alpha_p - \beta_p p_{i,j}, \qquad (9.12)$$

where

$$\begin{aligned} F(\phi_{i \to n}) &> 0 \quad \text{if} \quad \phi_{i \to n} > 0, \\ &= 0 \quad \text{if} \quad \phi_{i \to n} \leq 0. \end{aligned} \qquad (9.13)$$

F is a function that expresses how the flux of auxin through a particular membrane feeds back on the allocation of PIN1 to that membrane. Feugier et al. [25] showed that only superlinear functions (i.e., functions that grow faster than the identity function) induce canals. For example, a quadratic feedback function ($F(x) = \gamma_p x^2$) would lead to:

$$\frac{\partial p_{i,n}}{\partial t} = \gamma_p \phi_{i \to n}^2 + \alpha_p - \beta_p p_{i,j},$$

The feedback generated by superlinear functions is such that when a cell receives enough auxin it switches to a mode where a large efflux through just one wall is preferred to smaller effluxes through many walls simultaneously (see [22]). This corresponds to a bifurcation in the dynamical system that switches the system from a non-canalized to a canalized mode when auxin fluxes are sufficient (see Box 9.4) (Fig. 9.3D). At the tissue level, superlinear functions thus induce the formation of canals that are believed to underlie the origin of vascular tissues in plants (Fig. 9.3D).

Box 9.2 Flux-based transport: local properties

To make the analysis easier and gain insight into the model, the system can be simplified and studied locally. For this, a single cell i is considered and the auxin and PIN1 in its immediate neighbors are assumed to be negligible ($\forall n \in N_i, a_n = 0$ and $p_{n,i} = 0$), though we consider the interface with these neighbors and the

(*continued*)

Box 9.2 *Continued*

distribution of PIN1 towards them. In this "isolated" context the variations of auxin (equation (9.9)) takes the simpler form:

$$\frac{\partial a_i}{\partial t} = -\left(\frac{\gamma_D}{v_i} \sum_{n \in N_i} s_{i,n}\right) a_i - \frac{\gamma_A}{v_i} \sum_{n \in N_i} s_{i,n} a_i p_{i,n} - \delta_a a_i + \sigma_a,$$

which makes it possible to compute the steady-state level of auxin ($\frac{\partial a_i}{\partial t} = 0$) in terms of the concentration of PIN1 at the different neighbor interfaces, using the short-hand notation $\delta = \delta_a + \frac{\gamma_D}{v_i} \sum_{n \in N_i} s_{i,n}$:

$$a_i = \frac{\sigma_a}{\delta + \frac{\gamma_A}{v_i} \sum_{n \in N_i} s_{i,n} p_{i,n}}.$$

To determine a_i one therefore needs to estimate the values of $p_{i,n}$ for each $n \in N_i$ at steady state. In the same single-cell context, equation (9.12) for the PIN1 variables gives, for each $n \in N_i$:

$$p_{i,n} = \frac{1}{\beta_p} F(p_{i,n} a_i) + \frac{\alpha_p}{\beta_p} = \frac{1}{\beta_p} F\left(\frac{\sigma_a p_{i,n}}{\delta + \frac{\gamma_A}{v_i} \sum_{l \in N_i} s_{i,l} p_{i,l}}\right) + \frac{\alpha_p}{\beta_p}.$$

Since a_i is entirely determined by $p_{i,n}$ (see the above expression for a_i at equilibrium), the "isolated" steady state is given by solving the equation above for each $n \in N_i$. Interestingly, one particular solution, in which all $p_{i,n}$ are equal ($p_{i,n} = p$ for every $n \in N_i$), corresponds to a non-polarized cell. This solution is given by a fixed point equation:

$$p = G(p)$$

with

$$G(p) = \frac{1}{\beta_p} F\left(\frac{\sigma_a p}{\delta + p \frac{\gamma_A}{v_i} \sum_{l \in N_i} s_{i,l}}\right) + \frac{\alpha_p}{\beta_p}.$$

This analysis makes it possible to discuss symmetry issues, which are key to the study of cell polarization. For example, depending on the exact form of the function F, there can be other solutions in which the $p_{i,n}$ are not all identical. These solutions are less symmetric than the former and correspond to cell polarization. In general, both symmetric and asymmetric solutions co-exist, but only one of them might be stable (see Box 9.4), depending on parameter values; i.e., there exist critical parameter values at which a bifurcation occurs.

The precise conditions in which this breaking of symmetry occurs, i.e., the form of the function F and the parameter values at which it occurs, are not fully known and would be a useful research topic for mathematical modelers.

Another issue related to the symmetry of the problem is due to the fact that since all $p_{i,n}$ are given by exactly the same equation, any "polar" solution where PIN1s are not all identical is geometrically not completely determinate. For instance, it is not possible to distinguish between a solution where $p_{i,1}$ is low on a wall and $p_{i,2}$ high on another wall and vice versa. This ambiguity is in fact resolved by considering the problem at a larger scale, as discussed in Box 9.3.

9.2.3.3 Stress-based hypothesis

A strong correlation between the positioning of PIN1 molecules and microtubule orientation in cell membranes has been observed in apical meristems [29]. That study showed that microtubules are preferentially oriented parallel to the positioning of PIN1 in cell membranes. However, the orientation of the microtubules themselves appears to be correlated with mechanical stresses in cell walls [28]. This therefore suggests that the polarization of PIN1 in cells could be a consequence of local stress variations in cell walls. To interpret these observations, a model coupling the allocation of PIN1 to mechanical stresses has been introduced [29]. In a cell i, this model relates the amount of PIN1 allocated to a membrane facing a neighboring cell n to the amount of stress, $\sigma_{i,n}$, existing in the wall between i and n in a proportional manner:

$$p_{i,n} = \lambda_1 \frac{\sigma_{i,n}^m}{Z} p_i, \tag{9.14}$$

where p_i is the total concentration of PIN1 molecules in cell i, m is a constant parameter controlling the relative intensity of the stress feedback in each wall on the PIN1 allocated to the walls, a_i is the auxin concentration in cell i, and λ_1 is a constant parameter. Z is a normalizing coefficient, corresponding to the fact that, in a given cell, all the $p_{i,n}$ must sum to p_i:

$$Z = 1 + \lambda_1 \sum_n \sigma_{i,n}^m, \tag{9.15}$$

where the 1 on the right-hand side is included to avoid the cancellation of the denominator when the cell has no stress in its walls.

The model then assumes that the amount of auxin present in a cell modulates the elastic properties of the cell walls: the more auxin, the less rigid the cell walls. This rule is formalized by the following equation relating the elastic modulus E in the walls of cell i to the auxin concentration in the cell, a_i:

$$E(a_i) = E_{\min} + \frac{\lambda_2(E_{\min} - E_{\max})}{\lambda_2 + a_i^r}, \tag{9.16}$$

where $E_{\min}$ and $E_{\max}$ are the minimum and maximum values, respectively, of the elastic modulus of the cell, λ_2 is a constant parameter and r is a parameter controlling the strength of the dependence between the elasticity of the wall and the auxin concentration in the cell.

Hence, if a cell locally has more auxin than its neighbors (Fig. 9.3E) the rigidity of its wall is decreased and the cell walls tend to elongate. As the walls between cells are rigidly connected, this extension induces and augments the stress in the walls of neighboring cells. Viewed from the neighboring cell, this corresponds to a differential augmentation of stress in its wall, which in turn leads the neighboring cell to allocate more PIN1 molecules to this wall according to equation 9.14. Interestingly, one can observe that this mechanism is a special case of the concentration-based hypothesis discussed earlier, where additional hypotheses are made about the mechanism by which a cell allocates more auxin on the side where the neighbouring cells contain higher auxin concentration.

9.2.4 Accumulation of auxin at the meristem surface

These different hypotheses have been used to model the dynamic accumulation of auxin at the tip of SAMs upstream of organ formation. Based on additional assumptions, they could all be used successfully to reproduce the spiral phyllotaxis patterns observed in real Arabidopsis meristems. Two teams independently showed that the concentration-based hypothesis was a plausible explanation for phyllotaxis [69, 31]. In these works, auxin transport was modeled in the L1 layer of cells (the outermost layer of cells in the meristem) by assuming that auxin is produced and degraded in every cell and is transported up the auxin gradient by PIN1 molecules, according to concentration-based rules similar to that of equation 9.10. Both models were implemented and simulated on an artificial growing tissue, including basic rules for cell growth and division. Simulations showed that phyllotactic patterns of different types, both spiral and whorls, could be recapitulated from the dynamic models based on purely local cell–cell interaction rules (here concentration-based rules for the allocation of PIN1 to membranes, and auxin transport rules similar to equation 9.9).

However, the success of these models in reproducing the dynamics of phyllotactic patterns in the growing meristem raised an important new conundrum: although it appears that a concentration-based process is at work to polarize PIN1 proteins in the L1 layer, it is seems likely that veins in plant organs are initiated by canalization according to the flux-based hypothesis. This latter view is supported by a considerable number of experimental observations and models [65, 48, 59, 58, 24, 72, 80]. The picture concerning the dynamic allocation of PIN1 to cell membranes would then be that a concentration-based process prevails in the L1 layer to explain the periodic accumulation of auxin peaks, while a flux-based process is involved in other parts of the tissue where veins are observed. This idea of a combination of different models in different tissues was developed further by Bayer et al. [13] who tried to assemble both processes in a single model. For this, they assumed that both mechanisms (flux- and concentration-based) can be triggered by a cell. However, the relative importance of each mechanism is controlled by the level of auxin in the cell: at low auxin concentrations, the PIN1 allocation process is dominated by the concentration-based mechanism, while at high auxin concentrations it is dominated by the flux-based mechanism. In the L1 layer, where levels of auxin are low (except in the newly formed primordia), the concentration-based rule therefore dominates the transport process, leading to regular accumulation of auxin at the periphery of the central zone. In the incipient primordium, however, the accumulation of auxin is maximal and triggers the flux-based mechanism while weakening accordingly the action of the concentration-based one (the combined model hypothesis). Consequently, a flux of auxin is initiated in the initium downward, towards the inner tissues, and the flux-based mechanism builds up a canal from the auxin source (the initium) to the auxin sink (the inner parts of the tissue). This canal is assumed to initiate the provascular tissues beneath the new organ. Interestingly, this model predicted that at the moment of initiation, just before switching on of the flux-based process, a transient polarization of PIN1 pointing upwards in the inner tissues should be observed. Despite the fact that such transient phenomena are difficult to capture, this prediction was verified on confocal images of tomato and Arabidopsis meristems, [13].

An alternative approach to this combined model was developed by Stoma et al. [70], based on the idea that a unique model could possibly be active at both the surface of the meristem and in the inner layers. These authors then tested the possibility that a flux-based mechanism could explain the accumulation of auxin in the L1 layers. They showed that when primordia are considered to be imperfect sinks for auxin, flux-based systems can transport auxin against the auxin gradient. They then showed that such a transport hypothesis can lead to the correct phyllotactic patterns and therefore constitute a plausible transport mechanism in the L1 layer as well. Based on this result, they showed that a flux-based system could be used in both the L1 layer and the inner tissues. However, they could not find a unique region in the parameter space where both L1 fluxes against the gradient and canalization in inner tissues would work simultaneously. This defect in the parsimony of the approach was later corrected by [79] who showed that parameters of the flux-based model common to the L1 layer (sink-driven patterning) and the inner layers (source-driven patterning) can be selected in order to reproduce both the diffused patterns of PIN1 polarization in the L1 layer and the strand patterns in the inner layers.

There are advantages and disadvantages to all the models considered so far, and they fit the biological facts to varying degrees of accuracy. Based on our current knowledge of the transport processes, it is not possible to come out in favor of any one particular model. In recent years, other related models or variants of previous models have been introduced (e.g., [80, 2]) and these rely on a refined understanding of the molecular processes underlying the transport systems. There is no doubt that a better biological understanding of the molecular and biophysical processes involved in hormone transport at the SAM will eventually make it possible to make progress in the modeling of hormone transport at the SAM and to eliminate progressively unrealistic hypotheses. However, these models are complex because their global behavior emerges from local rules of interactions between cells and feedback loops between biochemical or physical variables. Progress with these models will certainly also require a systematic analysis and better understanding of their mathematical and formal properties (steady states, symmetry breaking, bifurcations, network properties, etc.) (e.g., [25, 53, 79, 74, 26, 23]; see also Box 9.3).

Box 9.3 Flux-based transport: global constraints

The distributions of auxin and PIN1 at steady state are not only determined by the local, intracellular dynamics (see Box 9.2), but also by how the fluxes of auxin combine at the multicellular level.

Mathematically, the spatial domain on which auxin transport is studied is a graph $G = (V, E)$, whose node set V represents cells and edges $E \subset V \times V$ represent areas of cell membrane that are at the interface between neighboring cells.

One important property comes from the fact that

$$\phi_{i \to j} = -\phi_{j \to i},$$

as follows immediately from the definition $\phi_{i \to j} = \gamma_D (a_i - a_j) + \gamma_A (a_i p_{i,j} - a_j p_{j,i})$. This property entails a global constraint: the sum of all fluxes in the tissue must be zero.

Also, from the relation above and the fact that in equation (9.12) the function F is zero for non-positive flux, it follows that at steady state it is always true that either

$$p_{i,j} = \frac{\alpha_P}{\beta_P} \quad \text{or} \quad p_{j,i} = \frac{\alpha_P}{\beta_P}.$$

Indeed, the steady-state equations are, for any $(i, j) \in E$

$$p_{i,j} = \frac{1}{\beta_P} (F(\phi_{i \to j}) + \alpha_P) \quad \text{and} \quad p_{j,i} = \frac{1}{\beta_P} (F(\phi_{j \to i}) + \alpha_P),$$

where one of $\phi_{i \to j}$, $\phi_{j \to i}$ is non-positive. In other words, at least one of the two variables $p_{i,j}$, $p_{j,i}$ must be fixed to the value $\frac{\alpha_P}{\beta_P}$ while the other one is either at the same value or at a higher value (because the function F is non-negative).

This property forces some PIN1 variables to the low concentration $\frac{\alpha_P}{\beta_P}$. In Box 9.2 we saw that within each cell, the solution of steady-state equations can provide us with information about "how many" PIN1 variables take a given value, but not which ones exactly. Now that we are considering the whole tissue, we can see that additional constraints will actually partially resolve this ambiguity.

One can loosely interpret the situation as follows: each edge (i, j) of the graph G can be oriented in the direction of the globally constrained flux, which forces one of the two variables $p_{i,j}$, $p_{j,i}$ to a low value. In addition, the internal flux within each cell determines how many "high PIN1 walls" are possible. Every configuration that simultaneously matches the two constraints, one global and one local, is a steady state.

9.3 Modeling the regulation of floral initiation

As we discussed in Section 9.2, floral primordia are initiated in the SAM due the self-organized accumulation of auxin at the periphery of the central zone of the SAM. This accumulation locally triggers floral development [60] and a cascade of genetic regulation. In this section, we will describe how such regulatory networks can be modeled and the main mathematical formalisms for achieving models of dynamical systems (see Box 9.4).

Box 9.4 Dynamical systems: definitions

A dynamical system is a system that evolves in time. This notion can be formalized mathematically in an abstract way, encompassing all the more specific forms that are used in this chapter.

To describe how a system evolves in time, one needs to have a representation of time, and a representation of the state of the system at a given instant. One also needs a transformation rule to describe the effect of time on the state of the system. Mathematically, a dynamical system is nothing more than these three elements.

A dynamical system is a triple (S, T, Φ) where:

- S is a set, very often endowed with a notion of neighborhood or distance, called *state space*.
- T is a set describing time, usually endowed with a notion of order, an addition rule $+$, and an origin 0 which is neutral for addition ($t + 0 = t$).
- Φ is a map $S \times T \to S$ which, given a state $x \in S$ and a time $t \in T$, returns the new state $x' = \Phi(x, t)$ of the system after a duration t if it is in state x at $t = 0$. This *evolution operator*, also called *flow*, must satisfy the following intuitive axioms:
  - $\Phi(x, 0) = x$ for any state $x \in S$.
  - $\Phi(\Phi(x, t), t') = \Phi(x, t + t')$ for any $x \in S$ and $t, t' \in T$.

Although the notion of flow covers all examples of dynamical systems that will be seen in this book, it is not always used explicitly (see Boxes 9.6 and 9.7). However, it is still a very useful concept as it allows us to define important notions in a unified way. Such notions include:

- The *trajectory* (sometimes also called the *orbit*) of a state x is the set of all past and future states $o(x) = \{\Phi(x, t) : t \in T\}$.
- A *steady state* is a state that does not change in time: $\Phi(x, t) = x$ for all t. Its trajectory is a single point: $o(x) = \{x\}$.
- A *periodic trajectory* is a trajectory $o(x)$ such that there exists a time $\tau \in T$, called the period, such that $\Phi(y, \tau) = y$ for any point $y \in o(x)$ (including $y = x$). Geometrically, $o(x)$ is a closed loop in state space.
- A steady state x (or periodic trajectory $o(x)$) is called *stable* if for every state y close enough to x, the future states $\Phi(y, t)$ become arbitrarily close to $o(x)$ when $t \to \infty$. It is called *unstable* otherwise, i.e., if one can find a y arbitrarily close to x such that $o(x)$ and $\Phi(y, t)$ do not get close in the limit $t \to \infty$.

9.3.1 Expression patterns in the floral meristem

During its development, the floral meristem is divided in regions having differential gene expression patterns. There are many genes correlated with the different regions and developmental phases of the floral meristem. However, here we will only cover the main ones (for a more comprehensive review see Chapter 3 and [15, 1, 43]). Inflorescence meristems produce floral primordia at the axils of cryptic bracts, at points of high auxin concentration [82, 15]. The auxin peaks are generated in a phyllotactic pattern created by auxin flux, as already described. Then, *LEAFY* (*LFY*) and *APETALA1* (*AP1*) expression is established over the whole floral meristem [56]. *LFY* and *AP1* are important genes for floral development and the establishment of floral meristem identity, since their mutants have several floral defects ranging from late flowering, floral reversion and floral organ defects, among others [47, 78]. Later, *WUSCHEL* (*WUS*) and its counterpart *CLAVATA3* (*CLV3*) are upregulated in the central zone and the stem cell layers of the floral meristem, respectively. Both genes have distinctive, central expression patterns and interact in a negative feedback loop which specifies and maintains a pool of stem cells in the meristem. Mutations in *WUS* produce a premature differentiation of the stem cells, while *CLV3* mutants expand the stem cell zone [50, 71, 81]. Finally, the floral meristem will differentiate into four annular, concentric zones known as whorls, where the floral organs (sepals, petals, stamens and carpels) will develop. The control of these whorls is regulated by three families of genes, called A, B and C, as described in the so-called ABC model [16]. The A genes, *AP1* and *APETALA* (*AP2*) define the sepals. A plus B genes [i.e., *APETALA3* (*AP3*) and *PISTILLATA* (*PI*)] characterize the petals. The C gene, *AGAMOUS* (*AG*), together with B genes, define the stamens while C alone defines the carpels (Fig. 9.4 [16, 12]. Hence, the presence of different sets of genes, differentially characterizes the floral meristem spatially and temporally.

So, how do genes dynamically obtain their spatial and temporal expression patterns? Experimental research has demonstrated that genes interact with each other at many different levels, including chromatin modifications, physical interaction and transcriptional and post-transcriptional regulation. All these interactions are indispensable for regulating their spatio-temporal expression patterns.

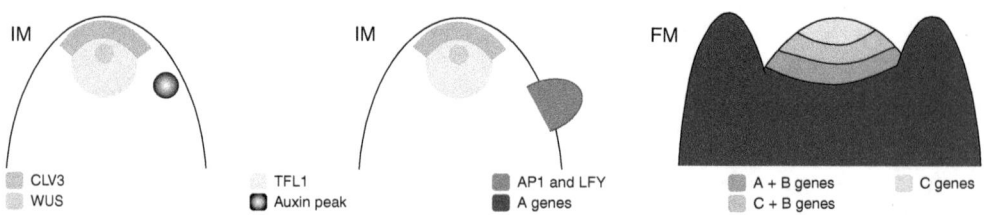

Fig. 9.4 Gene expression patterns during floral transition and development (IM, inflorescence meristem; FM, floral meristem.

9.3.2 Modules controlling development

Interestingly, at the beginning of floral development, auxin directly promotes *LFY* expression thought its signaling pathway [82]. However, it has been clearly demonstrated that *LFY* expression does not just depend on auxin concentration. *LFY* and *AP1* are inhibited by and inhibit *TFL1*, a negative regulator of flowering time. On the other hand, *AP1* and *LFY* positively upregulate each other [56, 57, 40, 37, 77]. In agreement with this, expression of *TFL1* is complementary to expression of *AP1* and *LFY*. The expression domains of the latter two genes greatly overlap within the floral meristem, while *TFL1* is exclusively expressed in the inflorescence meristem [57]. Moreover, *AGL*24 and *SOC*1, two flower-inducing genes, positively regulate *AP1* and *LFY* [27, 39], and together with *AP1* they act to repress *TFL1* [42]. Interestingly, while *SOC*1 and *AGL*24 regulate each other positively [39], creating another positive feedback loop, *AP1* represses them [44]. Consequently, *AGL*24 and *SOC*1 are downregulated once *AP1* expression is established. Downregulation of *AGL*24 and *SOC*1 is important for floral development to proceed, given that *AGL*24 and *SOC*1 inhibit the SEPALLATA (SEP) genes, which are important for upregulation of ABC genes [54, 19]. Once *AGL*24 and *SOC*1 have been downregulated, SEP genes repress *TFL1* [42].

As floral development continues, although still in the early floral developmental phases, expression of *WUS* and *CLV3* is upregulated in the floral meristem. *WUS* and *CLV3* form a loop, where *WUS* (non-cell autonomously) upregulates *CLV3* while *CLV3* inhibits *WUS* expression. These regulations create a negative feedback loop controling the size of the stem cell niche and the number of stem cells within it [50, 71, 30]. In the floral meristem, *WUS* expression is not only important for stem cell maintenance but also for inducing *AG* in conjunction with *LFY* [41]. During floral organ development, *AG* and *AP1* delimit two main regions, since they repress each other. The expression of B genes, which share a part of their expression domain with the A genes and another with the C genes, also depends on the expression of a different set of genes, like *UFO* [17]. Close to the end of floral patterning, *AG* turns off *WUS*. This is an important step in completing and ending floral development (Fig.9.5) [38].

Finally, it is important to note that genes and hormones also communicate. *LFY* modifies the metabolism, transport and signaling of auxin [45, 77]. *AP1* modifies gibberellin metabolism and response [37]. Gibberellins also act as an inductor of flowering by promoting genes like *SOC*1 [51] and LFY expression [8]. And the *CLV3*/*WUS* regulatory module interacts with the cytokinin metabolism and signaling pathway [14], indicating that gene and hormonal regulation are not independent processes.

Due to high gene interconnectivity, the non-linearity of the gene interactions and the complexity of the molecular regulation, computational and mathematical models are needed for in-depth integrated study. Many different approaches can be used to model molecular regulation. Here we will focus on two of the most common ones, Boolean and ordinary differential equation (ODE) gene regulatory network (GRN) models (for a more comprehensive review of GRN modeling formalism see, e.g., [18])

268 *Modeling the initiation of the floral primordium*

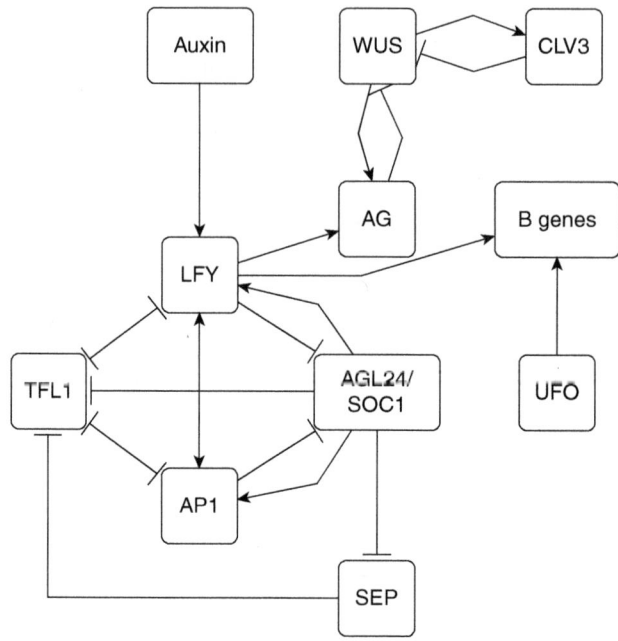

Fig. 9.5 Graph of the gene regulatory network that regulates floral transition and development. Nodes represent the genes and the edges the interactions among them. Bar head and arrowhead edges stand for negative and positive interactions, respectively.

(Box 9.5). Both Boolean and ODE models provide meaningful information about system dynamics. Consequently, they have already been employed for a wide variety of systems, including floral development. For example, a flowering transition GRN has been modeled recently [32] while the GRN for floral organ development has been modeled over the years using different approaches [46, 20, 76].

Box 9.5 Mathematical modeling of regulatory networks

Mathematically, a network of n genes interacting with each other is represented by a list (or *vector*) of n numbers:

$$X = (X_1, \ldots, X_n),$$

where each X_i represents the level of expression, or activity, or the gene i.

The entries X_i are sometimes continuous and are typically interpreted as the concentration of the mRNA or protein coded by gene i. This leads to a continuous state space (see Box 9.4) of the form $\mathbb{R}^n$.

The entries X_i can also be discrete, representing the gene i being transcribed "Not at all (0), A little (1), A little more (2)..." with a finite number of levels. The most common case, called Boolean, is when X_i takes only the values 0 (not transcribed) or 1 (transcribed), with a state space $\{0, 1\}^n$.

In both cases, referred to as "continuous" and "discrete" hereafter, the evolution of the network is defined as a dynamical system (see Box 9.4).

In the discrete case, one thus needs a successor map

$$F : \{0, 1\}^n \to \{0, 1\}^n.$$

In the context of gene network modeling, one refers to the flow induced by this map as *synchronous*, because this updates all the genes simultaneously at each time step. An alternative approach, where it is deemed more realistic to update at most one gene at a given time, is called *asynchronous*: only one coordinate F_i of the map F is applied at each time step, all other variables being left unchanged. A rule has to be introduced to determine which variables are updated (it can be deterministic or stochastic). It is easy to prove that steady states are identical for synchronous and asynchronous updates, and are the fixed points of the successor map F.

In the continuous case, the differential equation describing the system's dynamics is usually derived from the laws of chemical kinetics. For simplicity, these laws are often simplified using "quasi-steady state assumptions", whereby some variables are supposed to evolve so fast that they reach steady state instantly. A standard form obtained from this assumption is the so-called Hill function:

$$h(X_i) = \frac{X_i^p}{X_i^p + \theta_i^p},$$

which rapidly increases near $X_i \approx \theta_i$ from a plateau $h \approx 0$ (when $X_i < \theta_i$) to a plateau $h \approx 1$ (when $X_i < \theta_i$). Intuitively, one can see here a connection with Boolean models. This connection can be made more precise, although this is not discussed in detail in this chapter. The whole system is then described by the system of ODEs

$$\frac{dX}{dt} = F(X) = (F_1(X), \ldots, F_n(X)),$$

where the jth coordinate F_j is a sum of products of terms of the form $h(X_i)$.

9.3.3 Modeling the genetic regulation of bud fate

As stated by[33] in the Boolean context, the steady states of a GRN correspond to the different gene configuration states observed in a biological system. For example, in the case of the floral meristem, they could correspond to the gene configuration of the whorl observed experimentally. This hypothesis has been independently corroborated by different groups (e.g., [3, 20, 32]).

9.3.3.1 GRN formalisms

We can build a network using the experimental information about gene interactions and gene expression patterns presented previously. Let us take as an example the expression patterns and gene interactions of *AP1*, *LFY* and *TFL1*, which are at the core of the transition from inflorescence meristem to floral meristem. According to the above description, *TFL1* is a negative regulator of *AP1* and *LFY*. *LFY* and *AP1* are negative regulators of *TFL1* and positive regulators of each other (Fig. 9.5). This situation can be formalized in either discrete or continuous time (see Boxes 9.6 and 9.7).

Box 9.6 Dynamical systems: discrete time

The archetype of discrete time is given by integer numbers, i.e. $T = \mathbb{N}$ or $\mathbb{Z}$. In that case, one can use the flow to define a notion of **successor** map, denoted $F : S \to S$. Given a state s, this map returns the new state of the system after one unit of time:

$$F(x) = \Phi(x, 1).$$

Very often, discrete time dynamical systems will be presented directly in terms of their successor map instead of the flow Φ. If needed, the flow can always be retrieved by repeatedly applying the successor map: $\Phi(x, t) = \underbrace{F \circ F \circ \cdots F}_{t \text{ times}}(x)$.

Example 1: Lindenmayer algae
This system represents the shape of an alga using a word written with two symbols A and B. The state space is thus the set of all words that can be written with these two symbols, denoted $\{A, B\}^*$. The successor map is defined in two steps. First, one introduces the two rules:

$$A \longrightarrow AB$$
$$B \longrightarrow A$$

The successor map is the transformation which, given a word $w \in \{A, B\}^*$ returns the new word obtained by replacing every occurrence of A by AB, and then every occurrence of B by A.

This system is usually considered only with the initial state A and not the whole state space (i.e., one considers a single trajectory and in this case the notion of stability is not relevant), and defining the flow would introduce unnecessary complications. ◇

Example 1 has a discrete state space. One can also have a discrete time and a continuous state space.

Example 2: logistic map
The logistic map $F(x) = \lambda x(1-x)$ is a famous dynamical system, representing the evolution of a population (of size x) in a medium with limited resources (represented by the term $(1-x)$).

This system has two steady states, 0 and $x_\lambda = \frac{\lambda-1}{\lambda}$, which are the solutions of the fixed point equation $F(x) = x$. One can prove that 0 is stable for $0 < \lambda \leq 1$ and unstable otherwise. Also, x_λ is stable for $1 < \lambda \leq 3$ and unstable for $\lambda > 3$. In the latter case, the trajectories of this system can become very complicated, despite its simple formulation. ◇

Example 2 illustrates an important concept: a *bifurcation* occurs when the number or stability of steady states in a system change for a particular value of some parameter. A parameter is a number that appears in the formulae defining a dynamical system and which does not evolve in time (i.e., it is not a variable).

Box 9.7 Dynamical systems: continuous time
___

The archetype of continuous time is the set of real numbers, i.e., $T = \mathbb{R}$. In this case, one cannot define a successor, because for any positive time t one can find a shorter time t' that would be a nearer successor, and the limit $t \to 0$ gives the current time, and hence is not a successor. However, one knows from calculus that the following limit makes sense, as the derivative

$$\frac{\partial \Phi(x,t)}{\partial t} = \lim_{dt \to 0} \frac{\Phi(x, t+dt) - \Phi(x,t)}{dt}.$$

Continuous time dynamical systems are very often presented in terms of this derivative, i.e., the rate of change of the state. A function F is used as an evolution rule, providing the rate of change for each state x of the system. This provides us with a *differential equation* of the form

$$\frac{\partial x}{\partial t} = F(x).$$

Then, the notion of flow is identical to the concept of solution of the differential equation. Some standard terminology is associated with differential equations: the state x is usually a vector $x = (x_1, x_2, \ldots, x_n)$ of real numbers which depends

(*continued*)

Box 9.7 *Continued*

on time and space, i.e., it is a function $x(X,Y,Z,t)$ where X,Y,Z are spatial coordinates. If the right-hand side $F(x)$ involves partial derivatives other than $\partial/\partial t$ (e.g., $\partial/\partial X$, $\partial^2/\partial X^2$,...) the model is a system of *partial differential equations* (PDEs). Otherwise, one uses the the term *ordinary differential equations* (ODEs), and the notation is slightly altered:

$$\frac{dx}{dt} = F(x).$$

Example 1: reaction–diffusion.
The diffusion in space of a substance of concentration $c(X,Y,Z,t)$ is modelled by the Laplacian operator (in Cartesian coordinates)

$$\Delta c = \frac{\partial^2 c}{\partial X^2} + \frac{\partial^2 c}{\partial Y^2} + \frac{\partial^2 c}{\partial Z^2}.$$

Alternatively, one may consider a discretized spatial domain, with a finite number of sites $i \in \{1,\ldots,N\}$ (e.g., cells in a tissue) with concentration c_i in site i. In this framework each site i has a set $\mathcal{N}_i$ of neighbors and diffusion is modeled by

$$\Delta c = \sum_{j \in \mathcal{N}_i} (c_j - c_i)$$

The notation Δ is ambiguous here but would always be clear in a given context. When two substances of concentration c_A and c_B interact chemically and diffuse in space, their evolution is described by a system of the form:

$$\begin{cases} \dfrac{\partial c_A}{\partial t} = F(c_A, c_B) + D_A \nabla c_A \\ \dfrac{\partial c_B}{\partial t} = G(c_A, c_B) + D_B \Delta c_B \end{cases}$$

where D_A and D_B are the diffusion rates of the two substances and the functions F, G represent the kinetics of their interactions. This class of model, first proposed by Alan Turing, is capable of generating a large variety of spatial patterns, modelled as steady states of these two equations. The ratio between the two diffusion rates is often used as a parameter for studying bifurcations, which in this context are interpreted as changes in the spatial patterns of substances A and B.

Note that with a discrete spatial domain, this system is in fact a system of ODEs and the notation d/dt should be used. ◇

Remark: with differential equations, steady states can be found by solving a system of algebraic equations. Indeed, by definition a steady state is such that its rate of change is zero, i.e., a solution of $F(x) = 0$. These solutions can sometimes be computed by hand.

In discrete time, Boolean networks are often used as they are the simplest discrete modeling formalism for studying GRNs. The successor map is usually defined gene by gene using logical statements on how these genes are regulated. A possible set of Boolean functions describing the *AP1*, *LFY*, *TFL1* network is the following:

$$TFL1(t+1) = !AP1(t) \& !LFY(t)$$
$$AP1(t+1) = !TFL1(t) \,|\, LFY(t)$$
$$LFY(t+1) = !TFL1(t) \,|\, AP1(t)$$

where !, & and | stand for the logical NOT, AND and OR, respectively. The state space of this system made up of three Boolean variables has eight elements in total ($2^3 = 8$). For each state one can calculate a successor using the rules given above—the resulting successor map is represented graphically in Fig. 9.6. In general, an analogue of this figure can be produced that recapitulates the complete dynamics of a Boolean GRN model, but in practice the state space can be too large to perform the full calculation or represent it in a readable way. It is usually possible, though, to find the system's steady states. In our example, there are two steady states, 011 and 100, which correspond to the 'inflorescence meristem' and 'floral meristem' identities (Fig.9.6).

One can see in Fig. 9.6 that the Boolean representation does not allow for gradual changes of the variables. In particular, for the chosen rules the "inflorescence meristem" state 100 has no predecessor (a situation sometimes referred to as the "garden of Eden"), so that any perturbation of this state will immediately enter the "basin of attraction" of the other steady state. In fact, during plant development, we will transit from an "inflorescence meristem" to a "floral meristem" identity. However, this transition only happens when a new meristem is produced under specific conditions [i.e., a high concentration of FLOWERING LOCUS T (FT) and an auxin peak; see Section 9.3.2 and Chapter 3] and not under any type of perturbation. While there are many different way to overcome this and other limitations of the Boolean modeling approach, one of the best solutions, which gives a more gradual representation of gene

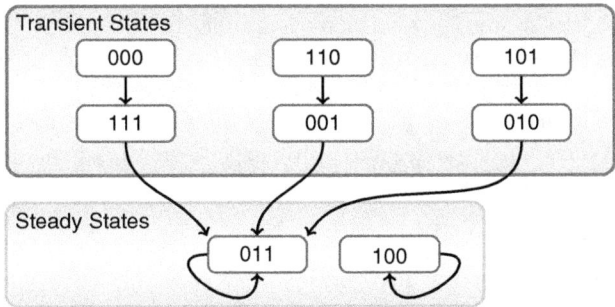

Fig. 9.6 Gene transitions and steady states for the Boolean GRN example. Each node represents one state of variables *TFL1*, *AP1* and *LFY* (in this order), and arrows represent the action of the successor map. The two steady states 011 and 100 are respectively interpreted as "floral meristem" and "inflorescence meristem".

dynamics and is able to overcome many of the Boolean limitations, is the use of an ODE model. In this case, the variables are the concentrations $[TFL1]$, $[AP1]$ and $[LFY]$. Using Hill functions, a set of equations analogous to the Boolean model can be written in the following form:

$$\frac{d[TFL1]}{dt} = \frac{\theta_{AP1}^p}{[AP1]^p + \theta_{AP1}^p} \frac{\theta_{LFY}^p}{[LFY]^p + \theta_{LFY}^p} - \delta_{TFL1}[TFL1],$$

$$\frac{d[AP1]}{dt} = \frac{1}{2}\left(\frac{[LFY]^p}{[LFY]^p + \theta_{LFY}^p} + \frac{\theta_{TFL1}^p}{[TFL1]^p + \theta_{TFL1}^p}\right) - \delta_{AP1}[AP1],$$

$$\frac{d[LFY]}{dt} = \frac{1}{2}\left(\frac{[AP1]^p}{[AP1]^p + \theta_{AP1}^p} + \frac{\theta_{TFL1}^p}{[TFL1]^p + \theta_{TFL1}^p}\right) - \delta_{LFY}[LFY],$$

where δ_X accounts for the degradation of each of the genes in the network. The values of the parameters need to be obtained from experiments or estimated. There are different methods for parameter estimation (e.g., [68]). For this example, the θ_X values, which represent the substrate concentration at which the reaction rate is half its maximal, was set to 1, all decays $\delta_X = 0.3$ and the Hill coefficient $p = 3$. Solving the system above cannot be achieved by hand any more, but numerical simulations give steady states which are analogous to the Boolean case (Fig.9.8). Importantly, with the ODE model, we can observe the expected transitions from an inflorescence to a floral meristem identity. However, in the continuous case, the transition can be simulated in a more realistic way. For example, the transition can be achieved by gradually modifying the initial $[TFL1]$, as observed in Fig.9.8. At a particular threshold, a transition is made from a steady state with low $[TFL1]$ to a steady state with a high $[TFL1]$. Hence, we can see here that the use of an ODE model makes it possible to solve two problems: (1) to identify perturbations that are able to modify the steady state of the system and (2) to provide plausible mechanisms that reproduce developmental transitions in a realistic way.

Regulatory networks usually have complex structures and behaviors and cannot readily be analyzed in a simple qualitative manner. Instead, it is usually necessary to quantify GRNs in order to analyze and understand their properties. It is not obvious, for example, that a network such as the one presented in Fig. 9.5, built from different sources of information, would eventually be self-consistent. The use of models allows us to study such consistencies and analyze the emerging properties of the GNR. In Section 9.3.3.2 we will show how such analyses can be performed with the aid of a network modeling approach.

9.3.3.2 Analysis of network properties

As observed in previous examples, even if we are dealing with a simple network motif, some interesting properties, like the steady states, can be studied in a systematic way. Likewise, from the models other network properties can be analyzed in order to validate, predict or understand the molecular mechanisms. For example, a posterior analysis that could be performed in any biological GRN is a mutant analysis.

In both continuous and discrete networks, fixing the node values to 0 simulates loss-of-function mutants, while the overexpression of a gene is simulated by fixing its value to 1 in the Boolean case and to a high value in the continuous case. For example, we could simulate *TFL1* overexpression. According to experimental research, overexpression of *TFL1* delays upregulation of *LFY* and *AP1*, but does not forbid it [56], suggesting that the positive feedback loop between *LFY* and *AP1* is stronger than *TFL1* repression. When we perform this simulation using the Boolean or ODE models, *TFL1* overexpression does not forbid upregulation of *AP1* and *LFY*. However, in the Boolean case we obtain a biologically meaningless cyclic steady state (Fig. 9.7), which appears due to their time limitation and synchronous gene updating. In contrast, with the ODE modeling approximation we do not obtain any unrealistic cyclic steady state. In fact, many mutant effects are easier to represent in a continuous framework, and can be used to determine plausible ranges for the parameters. The list of properties that we can analyze using GRN models is large, and includes structural (e.g., connectivity, clustering, network motifs) and dynamic (evolvability, robustness, stability) properties.

In both the continuous and discrete cases, we find in the *TFL1* overexpression simulation that the steady states with *TFL1* and that with all three genes upregulated are congruent with the experimental information. Both steady states and mutant simulations are important for model validation. Incongruences between the network results and the experimental data suggest that the model is incorrect and needs to be modified.

It is necessary to use modeling approaches to understand the molecular mechanisms behind developmental processes. As observed from our examples, even with a simple network motif there are some non-trivial behaviors, such as the appearance of different steady states and the effect of mutants. We only used a toy model, but many research groups have tried to gain a more comprehensive understanding of the floral transition and other developmental processes [46, 3, 20, 76, 14, 32]. The knowledge generated from these works has provided many insights and detailed information.

Importantly, with both Boolean and continuous modeling approaches we can sometimes obtain equivalent information. However, discrete formalisms, including the Boolean one, are by nature qualitative and time is not represented explicitly. Moreover, in general the update of the network state is synchronous for all the nodes. These limitations can generate artifacts in the network dynamics, like the unrealistic

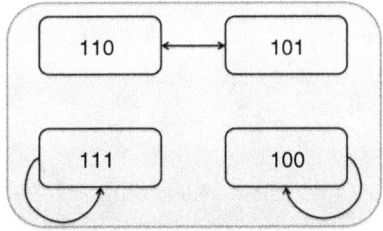

Fig. 9.7 Steady states obtained for overexpression of *TFL1*.

transitory states already described. We observed, for example, that if we start from the state $AP1 = 1$, $LFY = 0$ and $TFL1 = 1$ in the Boolean version of the model, we will pass a transitory state with $AP1 = 0$, $LFY = 1$ and $TFL1 = 0$ before reaching the steady-state configuration (Fig.9.6). Downregulation of $AP1$ in the transitory state has never been observed experimentally during floral transition, and in fact it is not observed in the continuous version of the model (Fig.9.8), indicating that it is an artefact of the Boolean formalism.

Thus, even when there are a number of possibilities for overcoming discrete modeling limitations, such as translating the Boolean model into a qualitative continuous one and using an asynchronous update of the gene's values, ODE models directly override many of the limitations of discrete formalism at the cost of many parameters which are usually unavailable and need to be estimated. This is why the modeling formalism must be selected according to the available data and the research aim. Networks can be used to study gene dynamics, development robustness, molecular pathway architecture and to predict missing, wrong or incoherent information, among many other things. They provide an extremely useful tool for the study of any molecular process.

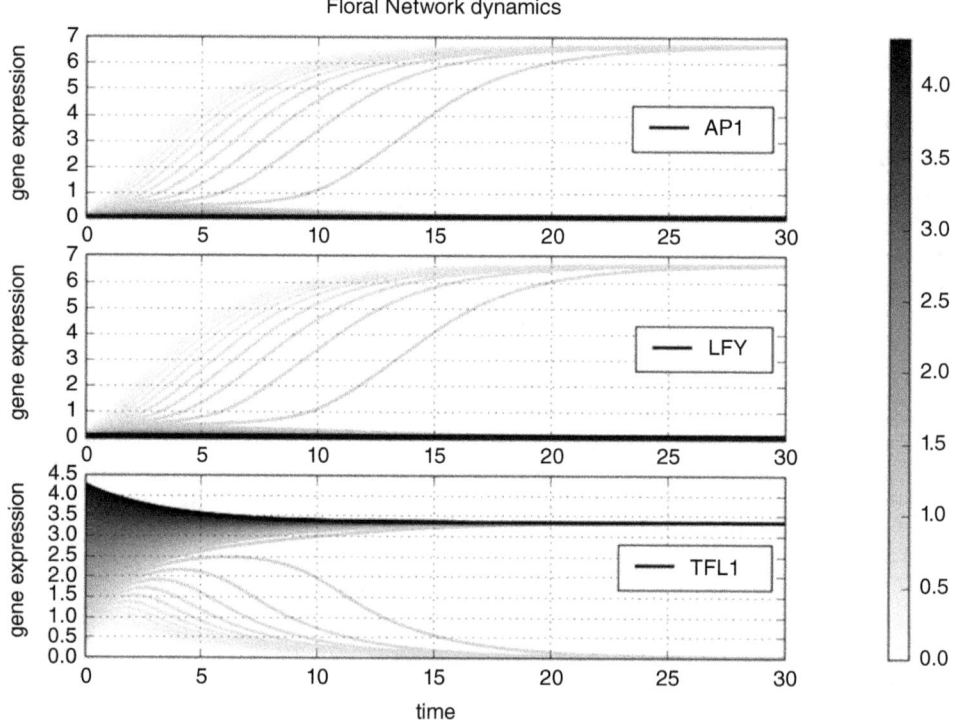

Fig. 9.8 Time course for each of the three genes, using the ODE model. Parameters are as given in the text. The bar on the right-hand side indicate changes in the initial concentration of $TFL1$, which is gradually increased from 0 to 4. Low values of $TFL1$ lead to a steady state analogous to the Boolean 011, while high values lead to a state analogous to 100.

9.4 Conclusion

In this chapter we have discussed how two key processes in the formation of floral primordia can be described and modeled in a quantitative way. In Section 9.2, we briefly reviewed the models that make it possible to explain the dynamic accumulation of auxin at the SAM. We described the main assumptions about auxin transport that have been proposed and tested in the literature (concentration-based and flux-based hypotheses). Models make it possible to check whether these assumptions expressed in terms of local cell–cell interaction rules lead to consistent patterning in the growing tissue as an emerging property. When compared with data, each model has its own strengths and weaknesses. However, in our current state of knowledge it is difficult to identify clearly whether the assumptions underlying a particular transport model are more realistic than those of others. To make further progress two main avenues must be explored: (1) we first need to gain deeper knowledge about the biological processes underlying the transport between cells; (2) we also need to gain a better mathematical insight into these complex models that mix spatial structure, growth, biochemical reactions and feedback loops between these elements.

Based on the assumption that a flower is initiated at the SAM downstream of spatial accumulation of auxin, we then investigated the GRN that controls the initial steps of floral development and differentiation. In a simplified form, this network contains dozens of actors interacting with each other in space and time. The understanding of such a complex system here also requires a modeling approach in order to quantify these interactions and analyze their properties. We briefly presented the two main formalisms used to model GRNs: the Boolean and ODE formalisms. We illustrated on a submodule of the floral GRN both types of models and discussed their main advantages and drawbacks. We showed how manipulations of the network models can be used to make predictions corresponding to possible biological manipulations of the GRN (e.g., loss-of-function mutants).

An integrated view of floral initiation would require us to address other important biological processes. For example, a mechanical model would be required to study how the floral dome bulges out from the meristem dome (and how the young organs such as sepals and petals physically emerge from the floral primordium). Such models are currently being developed (e.g., [28, 9, 4]) and provide a rigorous basis for further quantifying the effect of the forces that shape organs. These models will also need to be coupled, for example by connecting auxin transport models to GRNs on realistic three-dimensional structures obtained from confocal microscopy (e.g., [21, 6]). This will make it possible to test the consistency between several modeling modules in the context of a more integrated spatio-temporal understanding of floral development. Of course, to do this simplifications of the detailed models will have to be made and adaptation between their typical time and spatial scales will have to be considered. As a result, it will be possible to compare and assess the output of model simulations with observations of development in a precise quantitative manner, leading to the possibility of further developing these models as powerful tools for the study of development.

References

[1] E R Alvarez-Buylla, M Benítez, A Corvera-Poiré, A Chaos Cador, S de Folter, A Gamboa de Buen, A Garay-Arroyo, B García-Ponce, F Jaimes-Miranda, R V Pérez-Ruiz, A Piñeyro-Nelson, Y E Sánchez-Corrales. Flower development. *The Arabidopsis Book*, 8:e0127, 2010. doi: 10.1199/tab.0127.

[2] K Abley, P Barbier de Reuille, D Strutt, A Bangham, P Prusinkiewicz, A F M Maree, V A Grieneisen and E Coen. An intracellular partitioning-based framework for tissue cell polarity in plants and animals. *Development*, 140(10):2061–2074, 2013.

[3] R Albert and H G Othmer. The topology of the regulatory interactions predicts the expression pattern of the segment polarity genes in *Drosophila melanogaster*. *Journal of Theoretical Biology*, 223(1):1–18, 2003.

[4] F Boudon, J Chopard, O Ali, B Gilles, O Hamant, A Boudaoud, J Traas and C Godin. A computational framework for 3D mechanical modeling of plant morphogenesis with cellular resolution. *PLoS Computational Biology*, 11(1):e1003950, 2015.

[5] P Barbier de Reuille, I Bohn-Courseau, K Ljung, H Morin, N Carraro, C Godin and J Traas. Computer simulations reveal properties of the cell-cell signaling network at the shoot apex in Arabidopsis. *Proceedings of the National Academy of Sciences USA*, 103(5):1627–1632, 2006.

[6] P Barbier de Reuille, A-L Routier-Kierzkowska, D Kierzkowski, G W Bassel, T Schüpbach, G Tauriello, N Bajpai, S Strauss, A Weber, A Kiss, A Burian, H Hofhuis, A Sapala, M Lipowczan, M B Heimlicher, S Robinson, E M Bayer, K Basler, P Koumoutsakos, A Hk Roeder, T Aegerter-Wilmsen, N Nakayama, M Tsiantis, A Hay, D Kwiatkowska, I Xenarios, C Kuhlemeier and R S Smith. MorphoGraphX: A platform for quantifying morphogenesis in 4D. *eLIFE*, 4:05864, 2015.

[7] K Bainbridge, S Guyomarc'h, E Bayer, R Swarup, M Bennett, T Mandel and C Kuhlemeier. Auxin influx carriers stabilize phyllotactic patterning. *Genes & Development*, 22(6):810–823, 2008.

[8] M A Blázquez, R Green, O Nilsson, M R Sussman and D Weigel. Gibberellins promote flowering of Arabidopsis by activating the LEAFY promoter. *The Plant Cell*, 10(5):791–800, 1998.

[9] B Bozorg, P Krupinski and H Jönsson. Stress and strain provide positional and directional cues in development. *PLoS Computational Biology*, 10(1):e1003410, 2014.

[10] S A Braybrook and A Peaucelle. Mechano-chemical aspects of organ formation in *Arabidopsis thaliana*: the relationship between auxin and pectin. *PloS ONE*, 8(3):e57813, 2013.

[11] G D Bilsborough, A Runions, M Barkoulas, H W Jenkins, A Hasson, C Galinha, P Laufs, A Hay, P Prusinkiewicz and M Tsiantis Model for the regulation of *Arabidopsis thaliana* leaf margin development. *Proceedings of the National Academy of Sciences USA*, 108(8):3424–3429, 2011.

[12] J L Bowman, D R Smyth and E M Meyerowitz. Genetic interactions among floral homeotic genes of Arabidopsis. *Development*, 112(1):1–20, 1991.

[13] E M Bayer, R S Smith, T Mandel, N Nakayama, M Sauer, P Prusinkiewicz and C Kuhlemeier. Integration of transport-based models for phyllotaxis and midvein formation. *Genes & Development*, 23(3):373–384, 2009.

[14] V S Chickarmane, S P Gordon, P T Tarr, M G Heisler and E M Meyerowitz. Cytokinin signaling as a positional cue for patterning the apical–basal axis of the growing Arabidopsis shoot meristem. *Proceedings of the National Academy of Sciences USA*, 109(10):4002–4007, 2012.

[15] J W Chandler. Floral meristem initiation and emergence in plants. *Cellular and Molecular Life Sciences*, 69(22):3807–3818, 2012.

[16] E Coen and E Meyerowitz. The war of the whorls: genetic interactions controlling flower development. *Nature*, 353(6339):31–37, 1991.

[17] E Chae, QK Tan, T A Hill and V F Irish. An Arabidopsis F-box protein acts as a transcriptional co-factor to regulate floral development. *Development*, 135(7):1235–1245, 2008.

[18] H De Jong. Modeling and simulation of genetic regulatory systems: a literature review. *Journal of Computational Biology*, 9(1):67–103, 2002.

[19] G S Ditta, A Pinyopich, P Robles, S Pelaz and M F Yanofsky. The SEP4 gene of *Arabidopsis thaliana* functions in floral organ and meristem identity. *Current Biology*, 14(21):1935–1940, 2004.

[20] C Espinosa-Soto, P Padilla-Longoria and E R Alvarez-Buylla. A gene regulatory network model for cell-fate determination during Arabidopsis thaliana flower development that is robust and recovers experimental gene expression profiles. *The Plant Cell Online*, 16(11):2923–2939, 2004.

[21] R Fernandez, P Das, V Mirabet, E Moscardi, J Traas, J-L Verdeil, G Malandain and C Godin. Imaging plant growth in 4D: robust tissue reconstruction and lineaging at cell resolution. *Nature Methods*, 7(7):547–553, 2010.

[22] F Feugier. Models of vascular pattern formation in leaves. *PhD Thesis*, University of Paris 6, 2006.

[23] C Feller, E Farcot and C Mazza. Self-organization of plant vascular systems: claims and counter-claims about the flux-based auxin transport model. *PloS ONE*, 10(3):e0118238, 2015.

[24] F G Feugier and Y Iwasa. How canalization can make loops: a new model of reticulated leaf vascular pattern formation. *Journal of Theoretical Biology*, 243(2):235–244, 2006.

[25] F G Feugier, A Mochizuki and Y Iwasa. Self-organization of the vascular system in plant leaves: inter-dependent dynamics of auxin flux and carrier proteins. *Journal of Theoretical Biology*, 236(4):366–375, 2005.

[26] E Farcot and Y Yuan. Homogeneous auxin steady states and spontaneous oscillations in flux-based auxin transport models. *SIAM Journal on Applied Dynamical Systems*, 12(3):1330–1353, 2013.

[27] V Grandi, V Gregis and M M Kater. Uncovering genetic and molecular interactions among floral meristem identity genes in *Arabidopsis thaliana*. *Plant Journal*, 69(5):881–893, 2012.

[28] O Hamant, M G Heisler, Henrik Jönsson, P Krupinski, M Uyttewaal, P Bokov, F Corson, P Sahlin, A Boudaoud, E M Meyerowitz, Y Couder and J Traas. Developmental patterning by mechanical signals in Arabidopsis. *Science*, 322(5908):1650–1655, 2008.

[29] M G Heisler, O Hamant, P Krupinski, M Uyttewaal, C Ohno, H Jönsson, J Traas and E M Meyerowitz. Alignment between PIN1 polarity and microtubule orientation in the shoot apical meristem reveals a tight coupling between morphogenesis and auxin transport. *PLoS Biology*, 8(10):e1000516, 2010.

[30] M Ikeda, N Mitsuda and M Ohme-Takagi. Arabidopsis WUSCHEL is a bifunctional transcription factor that acts as a repressor in stem cell regulation and as an activator in floral patterning. *The Plant Cell*, 21(11):3493–3505, 2009.

[31] H Jönsson, M G Heisler, B E Shapiro, E M Meyerowitz and E Mjolsness. An auxin-driven polarized transport model for phyllotaxis. *Proceedings of the National Academy of Sciences USA*, 103(5):1633–1638, 2006.

[32] K E Jaeger, N Pullen, S Lamzin, R J Morris and P A Wigge. Interlocking feedback loops govern the dynamic behavior of the floral transition in Arabidopsis. *The Plant Cell*, 25(3):820–833, 2013.

[33] S Kauffman. Homeostasis and differentiation in random genetic control networks. *Nature*, 224(5215):177–178, 1969.

[34] E M Kramer and M J Bennett. Auxin transport: a field in flux. *Trends in Plant Science*, 11(8):382–386, 2006.

[35] P Krupinski and H Jönsson. Modeling auxin-regulated development. *Cold Spring Harbor Perspectives in Biology*, 2(2):a001560, 2010.

[36] E M Kramer. PIN1 and AUX/LAX proteins: their role in auxin accumulation. *Trends in Plant Science*, 9(12):578–582, 2004.

[37] K Kaufmann, F Wellmer, J M Muiño, T Ferrier, S E Wuest, V Kumar, A Serrano-Mislata, F Madueño, P Krajewski, E M Meyerowitz, G C Angenent and J L Riechmann. Orchestration of floral initiation by APETALA1. *Science*, 328(5974):85–89, 2010.

[38] M Lenhard, A Bohnert, G Jürgens and T Laux. Termination of stem cell maintenance in Arabidopsis floral meristems by interactions between WUSCHEL and AGAMOUS. *Cell*, 105(6):805–814, 2001.

[39] C Liu, H Chen, H L Er, H M Soo, P P Kumar, J H Han, Y C Liou and H Yu. Direct interaction of AGL24 and SOC1 integrates flowering signals in Arabidopsis. *Development*, 135(8):1481–1491, 2008.

[40] S J Liljegren, C Gustafson-Brown, A Pinyopich, G S Ditta and M F Yanofsky. Interactions among APETALA1, LEAFY, and TERMINAL FLOWER1 specify meristem fate. *The Plant Cell*, 11(6):1007–1018, 1999.

[41] J U Lohmann, R L Hong, M Hobe, M A Busch, F Parcy, R Simon, and D Weigel. A molecular link between stem cell regulation and floral patterning in Arabidopsis. *Cell*, 105(6):793–803, 2001.

[42] C Liu, Z Wei N Teo, Y Bi, S Song, W Xi, X Yang, Z Yin and H Yu. A conserved genetic pathway determines inflorescence architecture in Arabidopsis and rice. *Developmental Cell*, 24(6):612–622, 2013.

[43] C Liu, Z Thong and H Yu. Coming into bloom: the specification of floral meristems. *Development*, 136(20):3379–3391, 2009.
[44] C Liu, J Zhou, K Bracha-Drori, S Yalovsky, T Ito and H Yu. Specification of Arabidopsis floral meristem identity by repression of flowering time genes. *Development*, 134(10):1901–1910, 2007.
[45] W Li, Y Zhou, X Liu, P Yu, J D Cohen and E M Meyerowitz. LEAFY controls auxin response pathways in floral primordium formation. *Science Signaling*, 6(270):ra23, 2013.
[46] L Mendoza and E R Alvarez-Buylla. Dynamics of the genetic regulatory network for Arabidopsis thaliana flower morphogenesis. *Journal of Theoretical Biology*, 193(2):307–319, 1998.
[47] M A Mandel, C Gustafson-Brown, B Savidge and M F Yanofsky. Molecular characterization of the Arabidopsis floral homeotic gene APETALA1. *Nature*, 360(6401):273–277, 1992.
[48] G J Mitchison. A model for vein formation in higher plants. *Philosophical Transactions of the Royal Society of London B: Biological Sciences*, 207:79–109, 1980.
[49] G J Mitchison. The polar transport of auxin and vein pattern in plants. *Philosophical Transactions of the Royal Society of London B: Biological Sciences*, 295:461–471, 1981.
[50] K F Mayer, H Schoof, A Haecker, M Lenhard, G Jürgens and T Laux. Role of WUSCHEL in regulating stem cell fate in the Arabidopsis shoot meristem. *Cell*, 95(6):805–815, 1998.
[51] J Moon, S S Suh, H Lee, K R Choi, C B Hong, N C Paek, S G Kim and I Lee. The SOC1 MADS-box gene integrates vernalization and gibberellin signals for flowering in Arabidopsis. *Plant Journal*, 35(5):613–623, 2003.
[52] A Peaucelle, S A Braybrook, L Le Guillou, E Bron, C Kuhlemeier and H Höfte. Pectin-induced changes in cell wall mechanics underlie organ initiation in Arabidopsis. *Current Biology*, 21(20):1720–1726, 2011.
[53] P Prusinkiewicz, S Crawford, R S Smith, K Ljung, T Bennett, V Ongaro and O Leyser. Control of bud activation by an auxin transport switch. *Proceedings of the National Academy of Sciences USA*, 106(41):17431–17436, 2009.
[54] S Pelaz, G S Ditta, E Baumann, E Wisman and M F Yanofsky. B and C floral organ identity functions require SEPALLATA MADS-box genes. *Nature*, 405(6783):200–203, 2000.
[55] J Raven. Transport of indoleacetic acid in plant cells in relation to pH and electrical gradients, and its significance for polar IAA transport. *New Phyotologist*, 74:163–172, 1975.
[56] O J Ratcliffe, I Amaya, C A Vincent, S Rothstein, R Carpenter, E S Coen and D J Bradley. A common mechanism controls the life cycle and architecture of plants. *Development*, 125(9):1609–1615, 1998.
[57] O J Ratcliffe, D J Bradley and E S Coen. Separation of shoot and floral identity in Arabidopsis. *Development*, 126(6):1109–1120, 1999.

[58] A Runions, M Fuhrer, B Lane, P Federl, A-G Rolland-Lagan and P Prusinkiewicz. Modeling and visualization of leaf venation patterns. *ACM Transactions on Graphics (TOG)*, 24(3):702–711, 2005.

[59] A G Rolland-Lagan and P Prusinkiewicz. Reviewing models of auxin canalization in the context of leaf vein pattern formation in Arabidopsis. *The Plant Journal*, 44(5):854–865, 2005.

[60] D Reinhardt, T Mandel and C Kuhlemeier. Auxin regulates the initiation and radial position of plant lateral organs. *The Plant Cell*, 12(4):507–518, 2000.

[61] D Reinhardt, E-R R Pesce, P Stieger, T Mandel, K Baltensperger, M Bennett, J Traas, J Friml and C Kuhlemeier. Regulation of phyllotaxis by polar auxin transport. *Nature*, 426(6964):255–260, 2003.

[62] P Rubery and A Sheldrake. Carrier-mediated auxin transport. *Planta*, 118(2):101–121, 1974.

[63] A Runions, R S Smith and P Prusinkiewicz. Computational models of auxin-driven development. In *Auxin and its Role in Plant Development*, pp. 315–357, Springer, Berlin, 2014.

[64] M Sassi, O Ali, F Boudon, G Cloarec, U Abad, C Cellier, X Chen, B Gilles, P Milani, J Friml, T Vernoux, C Godin, O Hamant and J Traas. An auxin-mediated shift toward growth isotropy promotes organ formation at the shoot meristem in Arabidopsis. *Current Biology*, 24(19):2335–2342, 2014.

[65] T Sachs. Polarity and the induction of organized vascular tissues. *Annals of Botany*, 33(2):263–275, 1969.

[66] T Sachs. A possible basis for apical organization in plants. *Journal of Theoretical Biology*, 37(2):353–361, 1972.

[67] T Sachs. Integrating cellular and organismic aspects of vascular differentiation. *Plant and Cell Physiology*, 41(6):649–656, 2000.

[68] J Sun, J M Garibaldi and C Hodgman. Parameter estimation using meta-heuristics in systems biology: a comprehensive review. *IEEE/ACM Transactions on Computational Biology and Bioinformatics*, 9(1):185–202, 2012.

[69] R S Smith, S Guyomarc'h, T Mandel, D Reinhardt, C Kuhlemeier and P Prusinkiewicz. A plausible model of phyllotaxis. *Proceedings of the National Academy of Sciences USA*, 103(5):1301–1306, 2006.

[70] S Stoma, M Lucas, J Chopard, M Schaedel, J Traas and C Godin. Flux-based transport enhancement as a plausible unifying mechanism for auxin transport in meristem development. *PLoS Computational Biology*, 4(10):e1000207, 2008.

[71] H Schoof, M Lenhard, A Haecker, K F Mayer, G Jürgens and T Laux. The stem cell population of Arabidopsis shoot meristems in maintained by a regulatory loop between the CLAVATA and WUSCHEL genes. *Cell*, 100(6):635–644, 2000.

[72] E Scarpella, D Marcos, J Friml and T Berleth. Control of leaf vascular patterning by polar auxin transport. *Genes & Development*, 20(8):1015–1027, 2006.

[73] P Sahlin, B Söderberg and H Jönsson. Regulated transport as a mechanism for pattern generation: capabilities for phyllotaxis and beyond. *Journal of Theoretical Biology*, 258(1):60–70, 2009.

[74] K van Berkel, R J de Boer, B Scheres and K ten Tusscher. Polar auxin transport: models and mechanisms. *Development*, 140(11):2253–2268, 2013.

[75] T Vernoux, G Brunoud, E Farcot, V Morin, H Van den Daele, J Legrand, M Oliva, P Das, A Larrieu, D Wells, Y Guédon, L Armitage, F Picard, S Guyomarc'h, C Cellier, G Parry, R Koumproglou, J H Doonan, M Estelle, C Godin, S Kepinski, M Bennett, L De Veylder and J Traas. The auxin signalling network translates dynamic input into robust patterning at the shoot apex. *Molecular Systems Biology*, 7(1):508, 2011.

[76] S van Mourik, A van Dijk, M de Gee, R Immink, K Kaufmann, G Angenent, R van Ham and J Molenaar. Continuous-time modeling of cell fate determination in Arabidopsis flowers. *BMC Systems Biology*, 4(1):101–101, 2010.

[77] C M Winter, RS Austin, S Blanvillain-Baufum e, M A Reback, M Monniaux, M F Wu, Y Sang, A Yamaguchi, N Yamaguchi, J E Parker, F Parcy, S T Jensen, H Li and D Wagner. LEAFY target genes reveal floral regulatory logic, cis motifs, and a link to biotic stimulus response. *Developmental Cell*, 20(4):430–443, 2011.

[78] D Weigel, J Alvarez, D R Smyth, M F Yanofsky and E M Meyerowitz. LEAFY controls floral meristem identity in Arabidopsis. *Cell*, 69(5):843–859, 1992.

[79] M Luke Walker, E Farcot, J Traas and C Godin. The flux-based PIN1 allocation mechanism can generate either canalyzed or diffuse distribution patterns depending on geometry and boundary conditions. *PloS ONE*, 8(1):e54802, 2013.

[80] K Wabnik, J Kleine-Vehn, J Balla, M Sauer, S Naramoto, V Reinöhl, R M H Merks, W Govaerts and J Friml. Emergence of tissue polarization from synergy of intracellular and extracellular auxin signaling. *Molecular Systems Biology*, 6:447, 2010.

[81] R K Yadav, M Tavakkoli and G V Reddy. WUSCHEL mediates stem cell homeostasis by regulating stem cell number and patterns of cell division and differentiation of stem cell progenitors. *Development*, 137(21):3581–3589, 2010.

[82] N Yamaguchi, M F Wu, C M Winter, M C Berns, S Nole-Wilson, A Yamaguchi, G Coupland, B A Krizek and D Wagner. A molecular framework for auxin-mediated initiation of flower primordia. *Developmental Cell*, 24(3):271–282, 2013.

Part 4

Forces in biology: reshaping membranes

10
Membrane remodeling: theoretical principles, structures of protein scaffolds and forces involved

Michael M. KOZLOV[1], Winfried WEISSENHORN[2,3,*], and Patricia BASSEREAU[4,5,6]

[1]Department of Physiology and Pharmacology, Sackler Faculty of Medicine, Tel Aviv University, Israel
[2]Université Grenoble Alpes, UVHCI, Grenoble, France
[3]CNRS, UVHCI, Grenoble, France
[4]Institut Curie, Centre de Recherche, Paris, France
[5]CNRS, PhysicoChimie Curie, UMR 168, Paris, France
[6]Université Pierre et Marie Curie, Paris, France
*Present address: Institut de Biologie Structurale (IBS) CEA-CNRS-Université Grenoble Alpes, Grenoble, France.

Abstract

Cellular membranes are dynamic structures that are constantly being remodeled to exert biological functions. This chapter is subdivided into four sections. After a general introduction we review the physical principles underlying the shaping of membranes into curved configurations expressed by the Helfrich model of membrane bending elasticity. We continue by presenting examples of protein scaffolds and their structures that either sense, induce or stabilize curved membranes, such as the BAR domain family of proteins and the ESCRT complexes that catalyze membrane fission. Then, based on specific examples of physical experiments with reconstituted membrane systems, we discuss the factors involved in membrane remodeling, which can lead to membrane constriction and fission. We further describe specific forces exerted on membranes by membrane-bending proteins, molecular motors and the cytoskeleton.

Keywords

Helfrich model, membrane curvature, elastic energy, budding, BAR domain, ESCRT-III, VPS4, line tension, nanotubes, membrane fission

Chapter Contents

10 Membrane remodeling: theoretical principles, structures of protein scaffolds and forces involved   287
Michael M. KOZLOV, Winfried WEISSENHORN, and Patricia BASSEREAU

10.1	Introduction	290
10.2	Theoretical principles of membrane remodeling	291
	10.2.1 Physics of membrane shaping	291
	10.2.2 The Helfrich model of membrane elasticity	291
10.3	Structural basis for membrane remodeling by a protein scaffold	304
	10.3.1 General parameters for protein–membrane interaction	304
	10.3.2 Parameters determining membrane interaction	304
	10.3.3 Structures of BAR domains and dynamin in membrane remodeling	305
	10.3.4 The ESCRT machinery in membrane remodeling processes	307
10.4	Forces involved in remodeling biological membranes	314
	10.4.1 Some physical methods for measuring forces and mechanics	314
	10.4.2 Membrane constriction due to line tension	320
	10.4.3 Bending membranes with proteins	323
	10.4.4 Forces produced by molecular motors and the cytoskeleton	327
	Acknowledgments	331
	References	331

10.1 Introduction

Cell membranes are nanofilms 4–5 nm thick consisting of phospholipids and proteins. In addition to serving as molecular exchangers and platforms for biochemical reactions between cells and their environment, membranes play a crucial role as an elastic shell determining the physical boundary of any cell or intracellular organelle.

In vivo, cell membranes exhibit a dual mechanical behavior. Firstly, the membranes acquire certain shapes, often characterized by large curvatures typical for specific types of cells and intracellular compartments. The most prominent examples are the membranes of the peripheral endoplasmic reticulum (ER), which consist of tubules 30–50 nm thick interconnected by three-way junctions into elaborate three-dimensional networks, and micron wide sheets with a thickness similar to that of the tubes [1,2]. The sheets can be stacked by peculiar helicoidal membrane connections [3], can have fenestrations [4] and their rims can be connected to the tubes. Another instance are the 10–20 nm thick and strongly fenestrated cisternae of the Golgi complex (GC) and the inner membranes of mitochondria that are compartmentalized into numerous cristae, the thin sheet-like structures similar in their geometrical parameters to the ER sheets and GC cisternae [5,6]. A common feature of all these structures is the large membrane curvature seen in their cross-sections. The radii of these curvatures, varying in the range between 10 and 30 nm, are only a few times larger than the membrane thicknesses (4–5 nm). Similarly large curvatures also characterize other intracellular membranes, such as endocytic vesicles [7–9] and caveolae [10,11].

Secondly, the shapes of membranes are not static but rather dynamic. They undergo continuous remodeling by fusion and fission. The processes of membrane remodeling include endocytosis (for a recent review see [12,13] and references therein) and exocytosis (for a review see [14] and references therein), mitochondrial fusion and fission (see for recent review [15]), breakage and reconnection of ER tubules as well as the redistribution of ER membranes between the sheets and tubules in the course of the cell cycle [2,16–18], constituent material exchange between ER and Golgi by budding, fusion and fission of spherical, tubular and pleomorphic traffic intermediates [19,20], entry of enveloped viruses into host cells and release of the newly assembled virions [21], cell-to-cell fusion [22,23] and cell division [24–26].

The mechanical properties of a cell membrane are provided by its basic structure—a phospholipid bilayer consisting of two lipid monolayers that is built and maintained by the powerful hydrophobic forces [27]. A symmetric lipid bilayer has a preferred flat conformation and resists deformation into curved shapes in general, and shapes with large curvatures in particular. Hence, to produce the peculiar shapes of the intracellular membranes, specialized mechanisms must exist within cells to generate forces and provide the energy necessary to bend membrane bilayers in the right places, at the right times and to the right extent. Further, the hydrophobic interactions between the lipid molecules guarantee structural integrity of the bilayer. Thus, generation of any kind of discontinuity in the bilayer, such as those accompanying, most probably, transient steps in membrane fusion and fission, consumes substantial energy, which again requires the action of intracellular machines performing the necessary work on the membranes.

Within the context of the cell, mechanical forces and the related energies can only come from proteins and protein machines. In this chapter we provide a general physical background for a qualitative and quantitative understanding of the mechanism underlying the action of such machines, discuss some experimental methods used to explore the structure of specific proteins involved in membrane shaping and remodeling, and address the experimental physico-chemical systems used to test the ability of these proteins to perform the required actions.

10.2 Theoretical principles of membrane remodeling

10.2.1 Physics of membrane shaping

To analyze the physics of membrane shaping we need a theory that will allow us to evaluate the elastic energy associated with the generation of every specific membrane shape and/or with a membrane remodeling event. There are two basic classes of approaches to considering membrane elastic energy: "microscopic" approaches that account for intermolecular interactions within the membrane with different degrees of accuracy, and "macroscopic" approaches that consider the membrane as an elastic continuum. Microscopic approaches require the use of numerical methods such as molecular dynamics and Monte Carlo simulations of the lipid molecular assemblies and the surrounding water molecules. Macroscopic approaches are based on the theory of elasticity.

Here we address only the macroscopic approaches. These can be subdivided into two types: the first type models a membrane as an elastic shell with zero thickness, i.e., as a two-dimensional (2D) elastic medium, and the second type considers a membrane as a layer of a finite thickness whose interior has the properties of a three-dimensional elastic continuum. The 2D model is used for the cases where the length scales of membrane deformations largely exceed the membrane thickness of a few nanometers, which is true for most intracellular membrane shaping processes. The finite-thickness membrane models (for recent reviews see [28,29]) are necessary to describe the processes of membrane deformation resulting from the generation of nanometer-size defects of the membrane structure, such as those developing during the intermediate stages of membrane fusion and fission or accompanying insertions into the membrane matrix of small protein domains. Here we describe and discuss in detail the 2D elastic model of membrane elasticity, which was developed by Wolfgang Helfrich more than 40 years ago [30] and has proved to be instrumental for understanding a broad class of membrane phenomena including membrane shaping and remodeling by proteins.

10.2.2 The Helfrich model of membrane elasticity

To formulate a continuous elastic theory of membrane elasticity one first needs to determine the appropriate physical variables characterizing the shape of the membrane and to postulate the relationship between these variables and the membrane's elastic energy.

10.2.2.1 Membrane curvature

The physical variables determining the shape of a membrane are the membrane *curvatures*. The first step in a rigorous introduction of this notion is presentation of the membrane as a surface with zero thickness. This is analogous to the classical description of interfaces between immiscible media introduced by Gibbs [31]. The background for such description is the choice within a membrane (or interface) of the so-called dividing surface, which must be parallel to the membrane–water boundary (Fig. 10.1A). Once a particular dividing surface has been chosen, its shape at every point represents the local membrane shape, and all the elastic characteristics of the membrane are attributed to it. Generally, the elastic model can be formulated for any arbitrarily selected dividing surface [32], and the physical variables depend on this choice. Therefore, there are several particular dividing surfaces, such as the neutral surface [33–35] for which some aspects of the system description are simplified. It is important to emphasize, however, that the overall elastic energy, which ultimately determines the physical behavior of the system, is independent of the position of the dividing surface. In the rest of this section we will identify the term "membrane" with that of the "dividing surface".

According to the fundamental geometry [36], the shape of the dividing surface at each point, and, hence the shape of the membrane, is determined by two independent

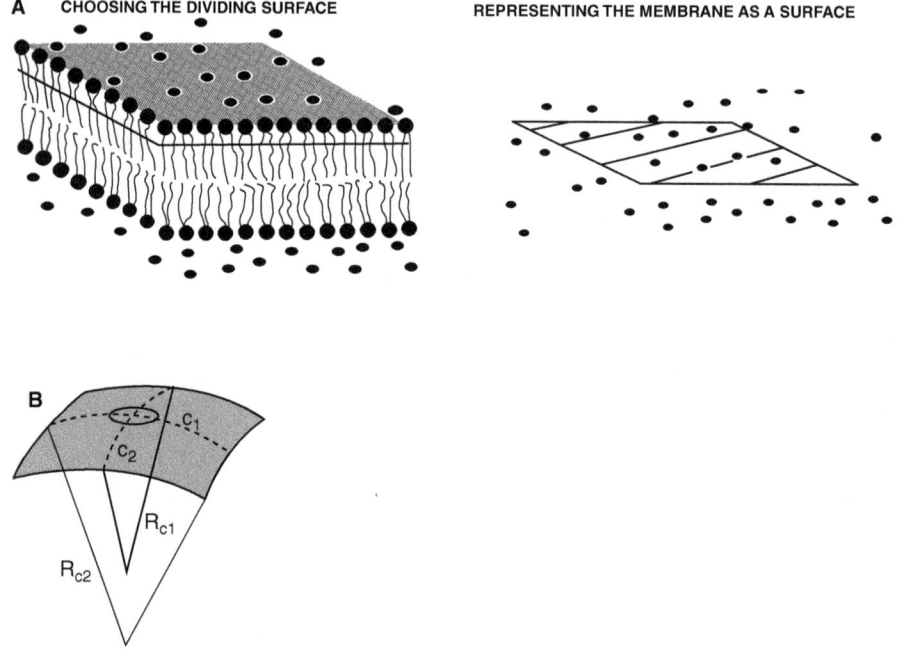

Fig. 10.1 Representation of a lipid bilayer by a surface. (A) Choosing the dividing surface to represent the membrane. (B) Principal curvatures of a membrane element.

geometrical values. It is most common and straightforward to use as these variables the two *principal curvatures*, c_1 and c_2 (Fig. 10.1B). The meaning of "principal curvatures" can be understood by the following thought experiment. Consider cross-sections of the dividing surface at a point under consideration by two planes, which are perpendicular to the dividing surface and orientated in two particular directions called the principal directions. We obtain two lines of intercepts between the planes and the surface, which have approximately circular shapes in close proximity to the point under consideration (Fig. 10.1B). The principal directions are determined by a requirement that the normal vector of the dividing surface must lie within the crossing planes when its origin moves along the lines of the intercepts. Geometric rules state that the principal directions are uniquely oriented and perpendicular to each other. The two circular fragments of the intercepting lines are characterized by their radii, R_1 and R_2 (Fig. 10.1B). The inverse values of these radii, $c_1 = 1/R_1$ and $c_2 = 1/R_2$, are referred to as the principal curvatures. Essentially, a curvature can be positive or negative. To define the sign of the curvature, one needs to distinguish between the two sides of the surface. For a lipid monolayer, one surface is covered by the phospholipid polar headgroups, whereas the other surface is hydrophobic. By convention, bending of a monolayer resulting in a bulging towards the hydrophilic surface produces a positive curvature, whereas bending in the opposite direction leads to negative curvature. For a closed bilayer membrane of a cell or an intracellular compartment, the difference between the membrane sides is determined by their orientation with respect to the volume that is enclosed by the membrane. Positive curvature is defined to describe a membrane bulging towards the external medium. However, it should be noted that the convention regarding the sign of the curvature is arbitrary. For example, in mathematical books, an agreement opposite to the one mentioned here is usually adopted.

While the meaning of principal curvatures is easy to illustrate, instead of c_1 and c_2 the Helfrich model uses their independent combinations called the *total curvature*, $J = c_1 + c_2$, and the *Gaussian curvature*, $K = c_1 \cdot c_2$. A deep physico-geometrical reason lies behind this choice of independent variables, as we shall now explain.

Most generally, the local shape of a surface is characterized by the curvature tensor b^α_β, a detailed understanding of which requires some knowledge of the differential geometry of surfaces [36]. In simple terms, the curvature tensor determines how the surface normal changes in the vicinity of the point being considered. The total and Gaussian curvatures are the orientation-invariant combinations (scalars) of the elements of the curvature tensor. The total curvature is linear in the elements of the curvature tensor, b^α_β, and represents the trace of the curvature tensor, $J = c_1 + c_2 = \text{Tr}(b^\alpha_\beta)$. The Gaussian curvature is quadratic in b^α_β and equals the determinant of the curvature tensor. The property of being orientation-invariant means that J and K are independent of the direction of the bending deformation with respect to any axis chosen in the membrane plane.

The total and Gaussian curvatures, J and K, must be the natural physical variables describing a lipid bilayer, for the following reason. At room temperature a homogeneous lipid bilayer of a common lipid composition has the properties of an incompressible 2D fluid, where the constituent lipid molecules can freely change their positions by

2D lateral diffusion in the plane of the monolayer [37]. Because of the 2D fluidity of the bilayer, its physical properties must be isotropic, i.e., orientationally invariant, in the bilayer plane. In particular, the bending energy of the bilayer must depend only on the orientation-invariant variables such as J and K.

There are four characteristic shapes corresponding to different combinations of the total and Gaussian curvature: a flat state for which the two curvatures vanish, $J = 0$ and $K = 0$; a cylindrical shape where $J \neq 0$ but $K = 0$; a saddle-like shape for which $J = 0$ but $K \neq 0$; and a spherical shape where $J \neq 0$ and $K \neq 0$.

For the following, it is useful to emphasize the relationship between the membrane curvatures and the deformations of the membrane interior. We will use the example of a bilayer, but the same reasoning is also correct for a monolayer. Consider a flat bilayer fragment (Fig. 10.2A). Naturally, for the reasons of symmetry, we choose the dividing surface to lie in the middle of the bilayer thickness (Fig. 10.2A). First, we bend the bilayer into a cylindrical shape, generating a positive total curvature, $J > 0$, and zero Gaussian curvature, $K = 0$. In the course of bending, we keep the area of the dividing surface constant. As illustrated in Fig. 10.2B, such deformation leads to stretching of the outer surface of the bilayer and compression of its inner surface with respect to the dividing surface. For a negative total curvature, $J < 0$, the outer surface becomes compressed while the inner one stretches (Fig. 10.2C). More generally, generation of the total curvature, J, leads to a relative stretching–compression of the bilayer surfaces located above and below the mid surface.

Now consider bending of the initial flat bilayer fragment into a saddle-like shape, corresponding to the generation of negative Gaussian curvature, $K < 0$, while keeping the value of the total curvature at zero, $J = 0$. In this case, the outer and inner bilayer surfaces retain equal areas, but both become compressed with respect to the dividing surface (Fig. 10.3). Hence, generally, generation of Gaussian curvature, K, leads to stretching–compression of the outer and inner membrane surfaces with respect to the

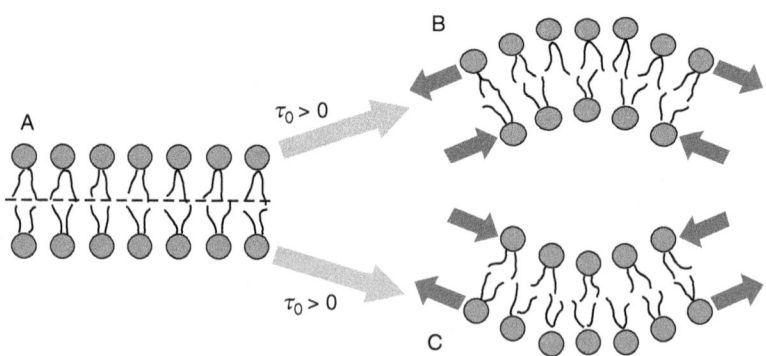

Fig. 10.2 Bending moment in the flat membrane state, τ_0. (A) The flat state. (B) Positive membrane curvature favored by a negative bending moment. (C) Negative membrane curvature favored by a positive bending moment.

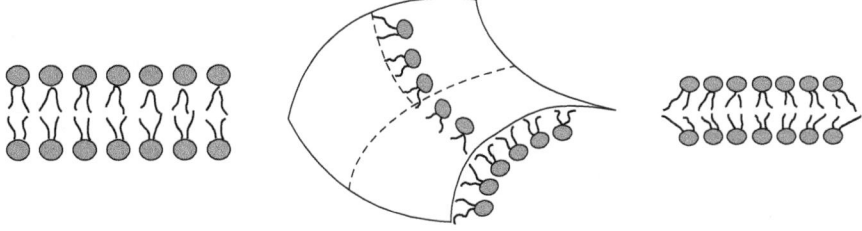

Fig. 10.3 The effect of Gaussian curvature on the intra-monolayer deformation of stretching–compression.

mid surface rather than relative to each other. A combined deformation changing both J and K, such as bending of a flat membrane into a spherical shape, results in stretching–compression of the outer and inner surfaces of the membrane with respect to each other and also to the mid surface.

10.2.2.2 Membrane elastic energy

The Helfrich model determines the energy of membrane bending with respect to a flat shape. The model only addresses weak bending deformations in the sense that the radii of principal curvature, R_1 and R_2, are much larger than the membrane thickness, d, so that $d/R_1 \ll 1$ and $d/R_2 \ll 1$. This means, in turn, that the principal curvatures, c_1 and c_2, are small fulfilling the conditions, $c_1 d \ll 1$ and $c_2 d \ll 1$.

The central formula of the Helfrich model [30] presents the curvature energy related to unit area of the membrane, f, taking into account the contributions up to quadratic order in the small parameters $c_1 d$ and $c_2 d$. In the language of the total, J, and Gaussian, K, curvatures, the energy f can be presented as an expansion in series up to the second power in J (since $J = c_1 + c_2$ is of the first order in c_1 and c_2) and the first power in K (since $K = c_1 \cdot c_2$ is of the second order in c_1 and c_2):

$$f = f_0 + \tau_0 \cdot J + \frac{1}{2}\kappa \cdot J^2 + \bar{\kappa} \cdot K \tag{10.1}$$

where f_0 is the energy in the flat state and τ_0, κ and $\bar{\kappa}$ are the elastic parameters independent of the deformation and accounting for the whole set of the membrane elastic properties in the flat state. These elastic parameters fully determine the behavior of the membrane near the flat state and are therefore very important.

The total elastic energy of a membrane, F, is given by integration of f over the membrane surface, A:

$$F = \oint f \mathrm{d}A. \tag{10.2}$$

Importantly, the elastic model expressed by equation (10.2) can be used to describe both a lipid monolayer and a bilayer. The values of the elastic parameters of a monolayer are different, of course, from those of a bilayer, but many of the general features and physical meanings are common for the monolayer and bilayer parameters. Therefore, the qualitative description presented in the following for the bilayer elasticity is also applicable for the elasticity of a lipid monolayer unless specifically stated.

Bending moment The physical meaning of the parameter τ_0 is the bending moment existing within the membrane in the flat state. More specifically, τ_0 determines the intrinsic tendency of every membrane element to bend away from the flat state. There are no general limitations on the sign of the bending moment, so that it can have either positive, $\tau_0 > 0$, or negative, $\tau_0 < 0$, values depending on the properties of the membrane. According to equation (10.1), if the bending moment is negative, $\tau_0 < 0$, the energy, f, decreases with development of a positive total curvature, $J > 0$, which means that the membrane tends to develop such a curvature, i.e., to bulge outwards (Fig. 10.2B). A positive bending moment, $\tau_0 > 0$, corresponds to the membrane's intrinsic tendency to develop a negative total curvature, $J < 0$, which means bulging inwards (Fig. 10.2C). Finally, if the bending moment vanishes, $\tau_0 = 0$, the membrane tends to remain flat. Obviously, the bending moment, τ_0, results from some asymmetry of the membrane structure with respect to the dividing surface. For the case of a lipid bilayer this means that, if the two monolayers have identical structures and elastic properties, for symmetry reasons the membrane must prefer to remain flat ($\tau_0 = 0$). The asymmetry can come from an intrinsic tendency of the two membrane monolayers to adopt either different curvatures or different areas, or from a difference in the lateral tensions in the monolayer. The dividing surfaces of the monolayers determining their curvatures and areas are chosen for convenience at the interface between the polar heads and the hydrocarbon tails where they have the properties of neutral surfaces [33–35]. The most common reason for different intrinsic curvatures of the monolayers is difference in their lipid compositions or protein content.

The effect of the area asymmetry of the monolayers depends on whether the monolayer areas are coupled by some physical or geometrical constraint, or can change independently of each other due to mutual slippage of the monolayers along the membrane plane. The regime of laterally coupled monolayers corresponds, for example, to the membrane of a closed vesicle. Because of the geometrical condition of the membrane closeness, the difference between the outer, A^{out}, and inner, A^{in}, areas of the monolayer, $\Delta A_m = A_{out} - A_{in}$, related to the area of the mid surface, A_m, is determined by the total curvature, J, averaged over the membrane area, $\bar{J}$, and multiplied by the membrane thickness (more exactly, the distance between the monolayer dividing surfaces), d, according to:

$$\Delta A_m / A_m = \bar{J} d. \tag{10.3}$$

Following from equation (10.3), a difference of the monolayer areas, ΔA_m, generates a total curvature, J, but does not determine the distribution of J along the membrane plane, only fixing its average value $\bar{J}$. In accordance with the relationship between the total curvature and the mutual stretching–compression of the upper and lower membrane surfaces mentioned in Section 10.2.2.1, equation (10.3) predicts that a positive $\bar{J} > 0$ develops if the area of the outer monolayer exceeds that of the inner one, $\Delta A_m > 0$, whereas the inverse relationship between the monolayer areas, $\Delta A_m < 0$, leads to a negative average total curvature, $\bar{J} < 0$. The coupled monolayer regime can also be realized within a limited membrane fragment, provided that the lipid exchange between the lipid monolayers of the fragment and those of the surrounding membrane is restricted by some physical barriers such as effective protein fences.

The situation where the areas of the monolayers can vary independently of each other through lipid exchange with external reservoirs is referred to as the uncoupled monolayer regime. In this regime, the monolayer area asymmetry, ΔA_m, is not a fixed parameter but rather an internal one (a degree of freedom) that the system can change to minimize its free energy. Therefore, for uncoupled monolayers, ΔA_m cannot serve as a primary factor generating membrane curvature. In this case, however, it is possible that the effective 2D pressures in the two monolayers determined by the lipid reservoirs are different, which could generate the curvature of the membrane fragment.

The uncoupled monolayer regime describes, for example, the evolution of a limited fragment of a large membrane involved in a process of membrane shaping, such as budding of an endocytic vesicle out of a large plasma membrane or of a transport intermediate out of the ER or Golgi complex. The two monolayers of such a membrane fragment can exchange lipid molecules with the monolayers of the surrounding membrane, the latter playing the role of material reservoirs.

Bending modulus The parameter κ is referred to as the *bending modulus*. Following equation (10.1), the bending modulus can be expressed as the second derivative of the energy with respect to the total curvature, $\kappa = \partial^2 f / \partial J^2$. Generally, according to the fundamentals of thermodynamics [38], the second derivative of the energy with respect to a thermodynamic variable has the meaning of a corresponding susceptibility, which must be positive to guarantee the thermodynamic stability of the system. Hence, in contrast to the bending moment, τ_0, the membrane bending modulus, κ, must be positive, $\kappa > 0$, and has the meaning of the membrane susceptibility with respect to the total curvature, J. Since the energy contribution determined by κ is quadratic in J, it is always positive no matter whether the generated total curvature is positive, $J > 0$, or negative, $J < 0$, which means that the bending modulus κ determines the resistance of a membrane to any bending of the initially flat state. Drawing a simplified analogy between a membrane and an elastic spring, the bending modulus, κ, is a equivalent to the elastic constant of the spring.

In his original article [30], Helfrich estimated the value of the bending modulus as $\kappa = 1\,\mathrm{eV} \approx 1.6 \times 10^{-12}$ erg $= 1.6 \times 10^{-19}$ J [30]. This estimation was based on the general idea that this modulus, which according to equation (10.1) has the units of energy, should be of the order of the energy of nearest-neighbor molecular interactions within the membrane, which in turn should be roughly equal to the molecular energy of condensation, of the order of 1 eV. A broad literature devoted to the experimental evaluation of the bending modulus by various assays has essentially confirmed this estimate. The values of κ reported for membranes of different compositions and obtained by different methods vary in a range between $10 k_\mathrm{B} T$ and $100 k_\mathrm{B} T$ (where $k_\mathrm{B} T \approx 4 \times 10^{-21}$ J is the characteristic energy of thermal fluctuations equal to the product of the Boltzmann constant, k_B, and the absolute temperature, T).

The typical value of the bending modulus of a lipid bilayer is about $\kappa \approx 20k_\mathrm{B}T = 8 \times 10^{-20}$ J (see, e.g., [39]).

It has to be emphasized that the basic value of the bending modulus is that of a lipid monolayer $\kappa_\mathrm{m} \approx 10k_\mathrm{B}T$. The value of the bilayer bending modulus depends on whether the monolayers are laterally coupled or decoupled. In the uncoupled regime, the bilayer bending modulus is just a sum of the monolayer ones, $\kappa = 2\kappa_\mathrm{m}$. If the monolayers are coupled, the bilayer bending modulus can reach $\kappa = 4\kappa_\mathrm{m}$ because of the mutual stretching–compression of the monolayers accompanying the bending. Finally, in the case of laterally coupled monolayers, the bilayer bending modulus can be decreased to the value of $\kappa = 2\kappa_\mathrm{m}$ by allowing flipping of the lipid molecules between the outer and inner monolayers. This process, referred to as flip-flop, can lead to relaxation of the extra stresses of the monolayer stretching–compression.

In the Helfrich presentation [30], the contributions to the energy from the total curvature, J, determined by the bending moment, τ_0, and the bending modulus, κ, are gathered in the form:

$$\tau_0 \cdot J + \frac{1}{2}\kappa \cdot J^2 = \frac{1}{2}\kappa \cdot J^2 - \kappa \cdot J_\mathrm{s} \cdot J + \kappa \cdot J_\mathrm{s}^2 = \frac{1}{2}\kappa(J - J_\mathrm{s})^2 \qquad (10.4)$$

where the value J_s is the spontaneous curvature of the membrane. The spontaneous curvature is directly related to the bending moment in the flat state by:

$$\tau_0 = -\kappa \cdot J_\mathrm{s} \qquad (10.5)$$

The term $\kappa \cdot J_\mathrm{s}^2$ added to the expression of the energy is a constant, enabling a compact presentation of the expression for the energy; however, it does not change the physics of the system since the energy is always determined up to a constant contribution.

The notion of the spontaneous curvature, J_s, which is basically a replacement for the bending moment in the flat state, τ_0, has been widely used in the membrane physics literature. Using the spontaneous curvature, one can express the variation of the membrane bending moment, τ, with the total curvature, J, as long as the latter remains small, $|J \cdot d| \ll 1$, and the higher-order contributions in $|J \cdot d|$ can be neglected. The bending moment, determined generally as $\tau = \partial f / \partial J$, is given in this approximation by:

$$\tau(J) = \tau_0 + \kappa \cdot J = \kappa \cdot (J - J_\mathrm{s}). \qquad (10.6)$$

In the literature J_s has often been interpreted as the total curvature the membrane adopts in a completely relaxed state corresponding to a minimum membrane energy. It has to be stressed, that, generally speaking, such an interpretation should be approximately correct only if the relaxed state of the membrane is close to the flat state so that $|J_\mathrm{s}d| \ll 1$. To understand this issue, one has to distinguish between the spontaneous curvature J_s determined through the bending moment in the flat state (equation 10.5) and the total curvature in the relaxed state we will refer to as the intrinsic curvature, J_i. Assume that the intrinsic curvature is relatively large so that $|J_\mathrm{i}d| \sim 1$. In a relaxed state, by definition, the bending moment must vanish:

$\tau(J_i) = 0$. Equation 10.6 for the bending moment is valid only for $|J \cdot d| \ll 1$. For larger curvatures, such as J_i, the terms of higher powers in the curvature have to be taken into account, so equation (10.5) has to be replaced by:

$$\tau(J) = \kappa \cdot (J - J_s) + \lambda_1 \cdot J^2 + \lambda_2 \cdot J^3 \ldots$$

where λ_1, λ_2 etc. are the higher-order elastic constants. Obviously, the curvature J_i satisfying the equation $\tau(J_i) = \kappa \cdot (J_i - J_s) + \lambda_1 \cdot J_i^2 + \lambda_2 \cdot J_i^3 + \ldots = 0$, is, generally, different from the spontaneous curvature, $J_i \neq J_s$. The intrinsic and spontaneous curvatures can only be equal for $|J_i d| \sim 1$ if, for some reason, the higher-order elastic constants vanish, $\lambda_1 = 0$, $\lambda_2 = 0$, etc.

This said, it has to be noted that, based on a series of extensive experimental investigations [40–42], strongly curved lipid monolayers of various compositions exhibit a surprisingly "linear" behavior described by equations (10.4) and (10.5) not only for small curvatures but also for large ones. Therefore, although not obvious a priori, identifying the spontaneous curvature of the monolayer, J_s^m, with the intrinsic curvature of the monolayer, J_i^m, seems to be a good approximation in practice.

The intrinsic curvatures of lipid monolayers, J_i^m, are accessible for experimental assessment. The systems used for this purpose are the so-called lipid mesophases, which are the 3D arrays resulting from lipid self-organization in aqueous solutions [43] (for review, see [44]). The formation of lipid mesophases is driven by the hydrophobic effect [27], namely, by the tendency of the hydrophobic moieties of the lipid molecules to be shielded from water by layers of polar heads. If the amount of water is limited or under special stabilizing conditions, the lipid mesophases have the properties of lyotropic liquid crystalline phases (see [43,45,46]). The building blocks of mesophases are lipid monolayers, which can have substantially different shapes depending on the type of lipid. The shape of the monolayer is used to characterize and classify lipid mesophases. In turn, lipids can be classified according to the shape of the monolayer and the corresponding type of mesophase they form in aqueous solutions (for review, see [47,48]). The most familiar type of lipid assembly is a planar bilayer consisting of two monolayers, which contact each other along their hydrophobic planes and whose hydrophilic surfaces cover the inner and outer surfaces of the bilayer. The lipid mesophase formed by planar lipid bilayers is called the lamellar (L) phase [43]. It consists of a stack of bilayers separated by a layers of water a few nanometers thick. Lipids such as phosphatidylcholines forming the lamellar phases are often referred to as "lamellar" lipids (Fig. 10.4A). Another common type of lipid mesophase, referred to as the inverted hexagonal (H_{II}) phase, consists of strongly curved lipid monolayers whose shapes can be seen to a good approximation as narrow cylinders with cross-sectional radii of a few nanometers (Fig. 10.4B). The lipid polar heads of such a cylindrical monolayer are oriented toward the internal space of the cylinder, which is filled with water, whereas the ends of the hydrocarbon chains form its external surface (Fig. 10.4B). Within the mesophase, the multiple lipid cylinders contact each other along their hydrophobic surfaces and are oriented such that their axes are parallel. This leads to a most compact packing of the monolayers so that the mesophase cross-section perpendicular to the axis of the cylinder reveals a hexagonal lattice of the individual cylinder cross-sections (Fig. 10.4B).

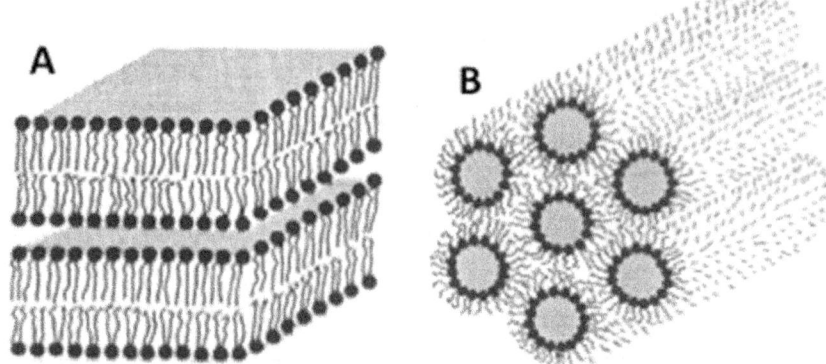

Fig. 10.4 Lipid mesophases: (A) lamellar (L) phase; (B) inverted hexagonal (H_{II}) phase.

Lipids such as dioleoylphosphatidylethanolamine (DOPE) (see [49]) that form the inverted hexagonal phase are referred to as "hexagonal" lipids. There are other mesophases that are less common but which should be briefly mentioned in order to illustrate the wide spectrum of possible lipid monolayer shapes and the corresponding diversity of lipid-phase behavior (for a review see [44] and references therein). Molecules of lysolipids having only one hydrocarbon tail tend to form monolayers that are strongly curved in the direction opposite to that of the "hexagonal" lipids. The resulting assemblies are represented by cylindrical or spherical micelles with internal volumes filled by the hydrocarbon chains. In a limited amount of water the cylindrical micelles pack into hexagonal phases, which in contrast to the inverted hexagonal phases, are denoted as H_I phases. Finally, there are lipids and lipid mixtures forming bicontinuous cubic (Q) phases whose monolayers are saddle-shaped and organized into periodic 3D surfaces. The mesophases self-assemble spontaneously in the absence of any external forces applied to the lipid monolayers. It is also commonly assumed that the inter-monolayer forces and geometrical constraints resulting from the mutual attachment of the monolayers along hydrophobic surfaces do not have a substantial effect on the shape of the monolayer. Hence the shapes of monolayers in mesophases must be close to the relaxed ones determined, practically solely, by the intra-monolayer interactions of lipid molecules. Therefore, the curvature of a monolayer within a mesophase is considered to represent its intrinsic curvature, J_i^m.

The notion of intrinsic curvature J_i^m has often been used to characterize individual lipid molecules [33–35,40,42]. Obviously, the intrinsic molecular curvature of a certain lipid must be understood in effective macroscopic rather than microscopic molecular terms as the intrinsic curvature of a monolayer within a mesophase built from lipid molecules of this particular type. It is even more common to ascribe a spontaneous molecular curvature, J_s^m to a certain type of lipid molecule [50]. The latter has to be understood as being derived according to equation (10.5) from the bending moment existing in a flat monolayer consisting of lipid of the given type [50].

Measurements of the intrinsic curvature of different lipids have been performed based on X-ray investigations of inverted hexagonal phases. For a detailed description of the method and the theoretical treatment of the results enabling evaluation of the intrinsic monolayer curvature, J_i^m, see [51]. In brief, the background system was a H_{II} phase formed by DOPE. Given a known number of DOPE molecules in the system, the chemical structure of DOPE and the water-to-lipid ratio within the cylindrical lipid monolayers in equilibrium with water vapor of a controllable pressure (the gravimetric approach), measurements of the Bragg spacing of the H_{II} phase enabled determination of the radius of the monolayer dividing surface, and, hence, of the intrinsic monolayer curvature attributed to the individual DOPE molecules. Mixing DOPE with relatively small amounts of some other lipid changed the intrinsic monolayer curvature in proportion to the amount of the second lipid added, enabling determination of the individual intrinsic curvature of the latter. Combination of this method with X-ray investigations of H_{II} phases within aqueous solutions of controllable osmotic pressure has allowed investigation of the elastic response of the monolayer to deformations around the strongly curved relaxed state and determination of the monolayer bending rigidities in this state. A table presenting the intrinsic curvatures of some lipids can be found in [48]. More data can be found on R.P.R. and https://www.brocku.ca/researchers/peter_rand/lipid/default.html.

Modulus of Gaussian curvature The elastic parameter $\bar{\kappa}$ determining the last term of equation (10.1) is referred to as the *modulus of Gaussian curvature* or the *saddle-splay modulus* [30]. This parameter, determined by $\bar{\kappa} = \partial f / \partial K$, can adopt either negative or positive values depending on the membrane structure. For a positive parameter, $\bar{\kappa} > 0$, generation of a negative Gaussian curvature, $K < 0$, provides a negative contribution to the energy f and hence is energetically favorable. Hence, a positive modulus of Gaussian curvature supports the formation of saddle-like shapes, which, as already mentioned, correspond to compression of the outer and inner membrane surfaces with respect to the dividing surface in the middle. If this modulus is negative, $\bar{\kappa} < 0$, the membrane tends to develop a positive Gaussian curvature, $K > 0$, corresponding to extension of its outer and inner surfaces with respect to the mid surface. This means that the modulus of Gaussian curvature, $\bar{\kappa}$, does not have a meaning of susceptibility but rather is another type of a bending moment related to generation of Gaussian curvature, K, in contrast to τ which is coupled to the total curvature, J.

The contribution of the Gaussian curvature, K, to the total elastic energy (equation 10.2), $F_K = \oint(\bar{\kappa} \cdot K) dA$, has an important geometrical property. In most cases, the modulus of Gaussian curvature is constant along the membrane so that this energy contribution can be presented as $F_K = \bar{\kappa} \oint K \, dA$. The integral of the Gaussian curvature, $\oint K \, dA$, over one or several closed surfaces obeys the Gauss–Bonnet theorem [36], according to which it does not depend on the surface shape(s) but rather only on the topological genus of the system, g, which reflects the connectivity of the membrane system:

$$\oint K \, dA = 4\pi(1 - g). \tag{10.7}$$

The genus for one closed surface, which is either spherical or can be obtained by deformations of a sphere without rupturing and/or reconnecting its surface, equals zero ($g = 0$). For two such surfaces $g = -1$; for three surfaces $g = -2$, etc. Generally, division of one closed surface into two corresponds to the change of the system genus by $\Delta g = -1$. Inversely, fusion of two closed surfaces into one generates the genus change of $\Delta g = 1$. Further, fusion of a closed surface with itself, resulting in generation of a toroidal hole in the surface shape referred to as a "handle", also generates a genus change $\Delta g = 1$. As a result a closed surface with one handle, which is equivalent, topologically, to a simple torus, has a genus $g = 1$, a double torus is characterized by $g = 2$, etc.

As a consequence of this geometrical property of Gaussian curvature, the energy of Gaussian curvature, F_K, changes only as a result of topological remodeling of the membrane. As a result, the sign and value of the modulus of Gaussian curvature, $\bar{\kappa}$, are relevant only for the processes of membrane fusion and fission and do not affect the energy of membrane bending deformations that maintain the membrane connectivity. This generates substantial obstacles to experimental determination of $\bar{\kappa}$. An early and, as far as we are aware, unique attempt to experimentally determine an apparent value of $\bar{\kappa}$ for a bilayer was undertaken in [52] by measuring the elastic torque acting on the edges of pierced vesicles. This was followed by estimations of bilayer $\bar{\kappa}$ based on an analysis of the mechanical behavior of cubic (Q_{II}) mesophases [44,53–55]. More recently, the modulus of Gaussian curvature of a lipid monolayer, $\bar{\kappa}_m$, was estimated based on observations and quantitative characterization of a temperature-driven phase transition between lamellar and cubic mesophases of a specific lipid composition [56].

An alternative way to evaluate the modulus of Gaussian curvature is by numerical modeling and simulations of the intra-membrane elastic stresses [57, 58].

To summarize these experimental and numerical studies: the moduli of Gaussian curvature of monolayers of common lipids are negative and vary in the range:

$$-1 < \bar{\kappa}_m/\kappa_m < 0 \tag{10.8}$$

where κ_m is the bending modulus of the monolayer. Specifically, for a monolayer of methylated DOPE and for mixture of DOPE/dioleoylphosphatidylcholine (DOPC) the evaluation gave $\bar{\kappa}_m/\kappa_m \approx -0.8$ [55, 56].

A positive contribution to the modulus of Gaussian curvature of a monolayer can be produced by molecular assemblies associated with the membrane surface. An example is a flexible polymer chain attached along its length to the surface of the monolayer and restricted to the plane of the monolayer while adopting different conformations [59]. An analysis showed that the conformational free energy of such a polymer chain depends on the Gaussian curvature of the membrane underneath and provides a positive contribution to the effective $\bar{\kappa}_m$ of the composed system.

Despite the negative values of $\bar{\kappa}_m$, the signs of the moduli of Gaussian curvature of bilayers can be different depending on the lipid composition of the monolayers. This is a consequence of the dependence of the modulus of Gaussian curvature of a bilayer, $\bar{\kappa}$, on the spontaneous curvatures of the monolayers, J_s^m, according to (see, e.g., [60, 61]):

$$\bar{\kappa} = 2\bar{\kappa}_m - 2\kappa_m \cdot J_s^m \cdot \kappa_m \cdot d \tag{10.9}$$

where κ_m is the monolayer bending modulus and d is the bilayer thickness (or, more precisely, the distance between the neutral surfaces of the two monolayers). The effect of J_s^m on $\bar{\kappa}$ can be understood, qualitatively, based on the above-discussed relationship between the sign of the Gaussian curvature, K, and the tendency of the membrane surfaces to undergo a stretching or compressing deformation with respect to the midplane. Assume the two membrane monolayers have a positive spontaneous curvature, $J_s^m > 0$. This means that each of the monolayers tends to bend in such a way that its outer (polar head) surface stretches with respect to the internal (hydrophobic) one. But the internal surfaces of the two monolayers coincide with the bilayer mid surface. Hence, for $J_s^m > 0$, the outer and inner bilayer surfaces tend to stretch with respect to the mid surface, which means, in turn, that the bilayer tends to adopt a positive Gaussian curvature, $K > 0$. Negative spontaneous curvatures of the two monolayers, $J_s^m < 0$, correspond to the tendency of the upper and lower surfaces of the bilayer to contract with respect to the bilayer mid surface and, hence, to the generation of a negative Gaussian curvature of the bilayer, $K < 0$. To summarize, the monolayer spontaneous curvature, J_s^m, influences the tendency of the bilayer to acquire Gaussian curvature of a certain sign, which means that it makes an effective contribution to the bilayer modulus of Gaussian curvature, $\bar{\kappa}$, and that, in accord with equation (10.9), this contribution must be proportional to $-J_s^m$.

According to equation (10.9), the presence in the two membrane monolayers of lipids with a sufficiently negative spontaneous curvature, $J_s^m < 0$, can provide a strong enough positive contribution to $\bar{\kappa}$ and guarantee a positive value for the bilayer modulus of Gaussian curvature, $\bar{\kappa}$.

Model summary The common form of the Helfrich formula for the membrane bending energy related to the unit area of the membrane surface is [30]:

$$f = f_0 + \frac{1}{2}\kappa(J - J_s)^2 + \bar{\kappa} \cdot K. \tag{10.10}$$

Generalization of this model to take into account the energy contributions of greater than second order in the principal curvatures can be found in [62,63]. Further generalization of the model accounting for tilting deformations of the lipid hydrocarbon chains has been overviewed recently in [29]. A model for monolayer bending deformation in the vicinity of a relaxed state characterized by large intrinsic curvature can be found in a series of works [33–35,41] overviewed and summarized in [51,64].

All these models have recently been used to understand the experimental results on membrane bending and remodeling by protein scaffolds described in subsequent sections.

10.3 Structural basis for membrane remodeling by a protein scaffold

10.3.1 General parameters for protein–membrane interaction

Glycerophospholipids make up about 70% of the total lipid content of a mammalian cell and glycosphingolipids, sphingomyelin and cholesterol account for the remaining 30%. Phosphatidylcholine (PC) is the most abundant glycerophospholipid (40–50%), followed by phosphatidylethanolamine (PE) (20–45%); the negatively charged phospholipids, phosphatidylserine (PS), phosphatidic acid (PA) and phosphatidylinositol (PI) are present in much smaller amounts [65]. However, the overall lipid composition varies further in different cell types and tissues. PS is the most abundant negatively charged phospholipid in eukaryotic membranes. PS is found mainly in the inner leaflet of the plasma membrane and in membranes of the endolysosomal system [66]. Most mammalian bilayers, including the plasma membrane, are asymmetric, and the asymmetric distribution of specific lipids is maintained by lipid transporters. Accordingly, PS can, for example, translocate to the outer leaflet of the plasma membrane during hemostasis [67] and apoptosis [68].

As well as affecting the physical properties of the bilayer, the lipid composition provides important platforms for interaction with peripheral membrane proteins, having roles in signaling, cellular movement and membrane remodeling processes, including budding and cytokinesis.

10.3.2 Parameters determining membrane interaction

Membrane interactions can be classified as either specific or non-specific. Specific interactions require the stereospecific recognition of phospholipid or lipid headgroups. Examples include the C1 domain that binds specifically to diacylglycerol and calcium-independent C2 domains (discoidin-like) that interact either with L-serine stereospecifically [69] or with phosphatidylinositols [70]. Furthermore, phosphatidylinositols are targets of a number of lipid-binding domains, which take advantage of the selective phosphorylation states of the inositol sugar thereby generating seven different forms of 3,4,5-phosphorylated phosphatidylinositols that are recognized by a large variety of small protein domains (PH, PX, FYVE, FERM, C2, GRAM, PDZ, PHD and PTB) [71]. Specific lipid head group recognition is sometimes complemented by hydrophobic interaction, such as immersion of hydrophobic amino acid side chains into the bilayer.

Non-specific membrane interactions require only a particular property of the membrane. They are often mediated in part or completely by electrostatic interactions between patch(es) of basic amino acids on the protein and acidic phospholipids in the membrane. A classic example is the calcium-dependent C2 domain that generate a basic patch upon calcium binding, which is transiently employed to interact with negatively charged phospholipids, mostly PS present in the inner leaflet of the plasma membrane [72].

Second, lipid packing can play an additional role in non-specific membrane recognition. The presence of certain lipids that do not follow the cylindrical shape of saturated phospholipids, such as the conical-shaped lipids PE and DAG or lysophosphatidic acid (LPA), may perturb lipid packing. In addition, the presence of long or short acyl chains in saturated or unsaturated form affects lipid geometry [73]. Amphiphatic helices such as the one present in the Sar1 GTPase, a protein resident in the ER, seem to prefer loose lipid packing or packing defects to insert into the bilayer [74].

A third parameter to consider as important for non-specific membrane interaction is the structure of the membrane itself. Many organelles or plasma membrane protrusions have highly curved membranes, which are either generated by membrane-deforming proteins or stabilized and/or recognized via their curved membrane shape. A major class of membrane curvature-sensing proteins are BAR domains [75]. Additional members of membrane curvature-sensing proteins are protein machines that assemble on curved membrane tubes to induce membrane fission, such as dynamin and endosomal sorting complexes required for transport (ESCRT), which are discussed in Section 10.3.4.

10.3.3 Structures of BAR domains and dynamin in membrane remodeling

The first BAR domain-containing protein that was associated with membrane curvature is endophilin-A1, which belongs to the superfamily of BAR (Bin-Amphiphysin-Rvs) domain-containing proteins that have been implicated in the generation of membrane curvature by deforming membranes into tubular structures [76,77]. BAR domains play an important role during endocytosis, underlined by the essential role of endophilin-A1 for the formation of synaptic vesicles at the plasma membrane [78].

The canonical BAR domain structure adopts a crescent-shaped dimeric conformation with (N-BAR) or without (BAR) an N-terminal amphiphatic helix that folds upon membrane contact [79–83]. A subset are connected to membrane-interacting PH or PX domains [83–86] and the superfamily extends to F-BAR [87,88] and I-BAR domains [89,90], which use similar structural conformations to bend membranes and/or recognize curved or flat membranes (Fig. 10.5). In vitro some BAR domains transform liposomes into narrow tubules and/or small vesicles. This depends largely on the protein:lipid ratio and on the presence of an amphiphatic helix at the N-terminus that can insert into a lipid bilayer and thus function as a wedge [78,91] that can eventually induce membrane fission [92]. The crescent-shaped structure of BAR domains was suggested to induce and/or stabilize membrane curvature by itself [78]. This latter function is supported by the helical polymerization of BAR domains upon membrane binding in vitro. This polymerization generates a membrane-bound protein scaffold with an intrinsic BAR domain-type curvature. F-BAR domains such as FBP17 form helical structures via tip to tip interactions as well as significant lateral interactions between filaments [93]. In contrast, N-BAR proteins such as endophilin-A1 form filaments without significant lateral interactions between endophilin dimers. These filaments are separated by the N-terminal amphiphatic helices that interact with each other perpendicular to the long axis of the BAR domain structure [94] (Fig. 10.5). Although BAR domains bend membranes or bind to curved membranes, direct interaction is mediated mostly by electrostatic interactions [93,94].

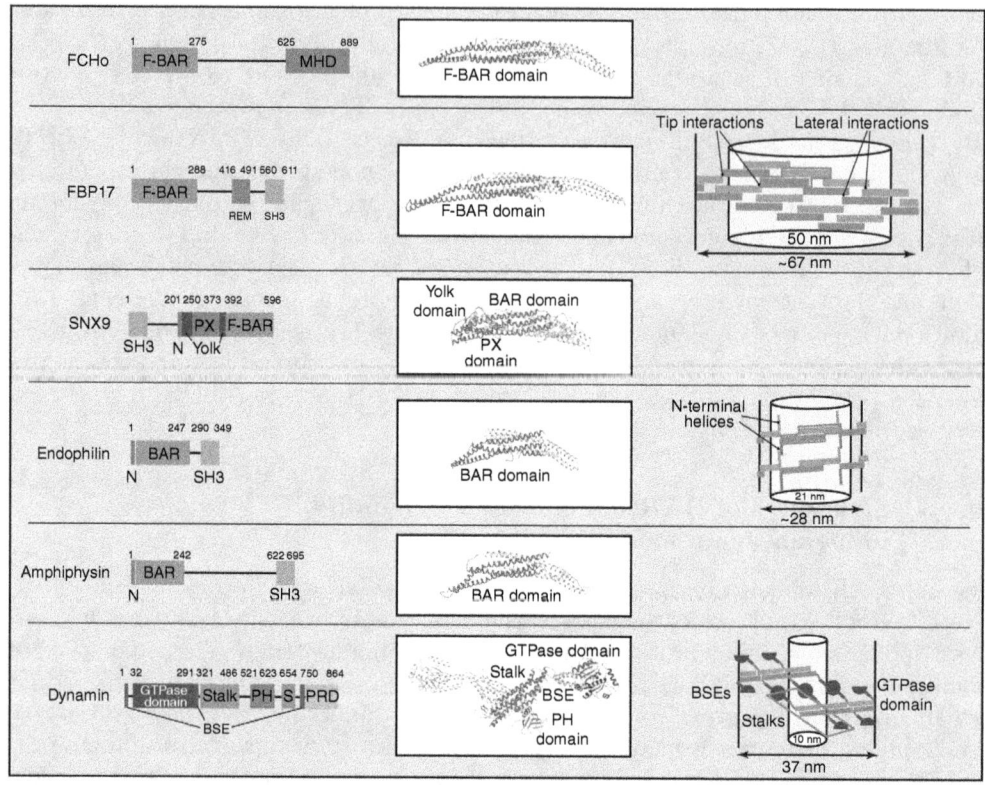

Fig. 10.5 Membrane-bending protein modules. Domain organization (left), structure (middle) and assembly (right) of the BAR domain proteins FCHo1, FBP17, SNX9, endophilin A1, amphiphysin-1 and dynamin-1. Subunits are shaded according to the domain organization. On the right the oligomerization modes of the F-BAR domain of FBP17 [93], endophilin [94] and of the constricted dynamin helix [95,96] are shown schematically. Dimeric polymer building blocks are boxed, cylinders indicate the approximate diameter of the lipid tubule and the outer diameter of the protein-coated oligomer is indicated. Figure adapted from [97].

Similar principles govern dynamin polymerization. Dynamin catalyzes the final membrane fission step during endocytosis and is composed of a GTPase domain, a helical stalk, a bundle signaling element and a membrane-binding PH domain (Fig. 10.5). It dimerizes via the helical stalk domain, forming flat tetramers in solution and curved tetramers when bound to membranes. The tetramers assemble into helical filaments with an external diameter of 50 nm *in vitro* alone or in the presence of membranes that are deformed into tubes having a final diameter of 20 nm and a 10 nm spacing between filaments. Looser helical coils with a spacing of 20 nm are assembled in the presence of BAR domains *in vitro* and on endocytotic vesicle necks *in vivo* [98]. Dimerization of GTPase domains from neighboring filaments induced by GTP binding (Fig. 10.5) is thought to constrict the membrane tube, and conformational changes associated with

GTP hydrolysis will induce membrane fission. Membrane fission is in addition largely dependent on membrane elasticity properties such as tension and rigidity [98].

10.3.4 The ESCRT machinery in membrane remodeling processes

Endosomal sorting complexes required for transport (ESCRT) function in a number of essential physiological and pathological membrane remodeling processes. These include plasma membrane receptor downregulation via multivesicular body (MVB) biogenesis [99], budding of some enveloped viruses from the plasma membrane [100], midbody abscission during cytokinesis [101,102] and repair of the plasma membrane [103]. Most of the approximately 32 proteins in the human ESCRT pathway form five different complexes called ESCRT-0, -I, -II, -III and the VPS4 complex that act during MVB biogenesis [99] (Fig. 10.6). However, only a subset of ESCRTs are recruited during budding of enveloped viruses and cytokinesis, including ESCRT-III and VPS4 that constitute the fission machinery [104,105].

10.3.4.1 The structure of ESCRT-III

Higher eukaryotes express 12 ESCRT-III proteins: the CHMP (CHarged Multivesicular body Protein) proteins and IST1 (Increased Sodium Tolerance 1 gene product) [106–109]. CHMP6 (Vps20), CHMP4 (A, B or C) (snf7), CHMP3 (Vps24) and CHMP2 (A or B) (Vps2) are recruited in this order and have been reported to form a core complex in yeast [110]; the remaining ESCRT-III proteins (CHMP1A, CHMP1B, CHMP5, CHMP7 and IST1) might exert regulatory functions. ESCRT-III proteins are small helical assemblies [106,111,112] and their autoinhibited conformation is controlled by the C-terminal region [106,111,113,114].

Removal of the C-terminus from ESCRT-III proteins leads to spontaneous membrane targeting *in vivo*, which employs a conserved N-terminal sequence motif for membrane insertion [128,129]. Direct ESCRT-III-driven membrane deformation *in vivo* has been observed upon CHMP4 [130] and CHMP2B over-expression [128]. CHMP2B membrane tubes contain a tight helical layer of CHMP2B filaments spaced by 30 Å [128] similar to the CHMP2A–CHMP3 helical tubes assembled *in vitro* [131]. Tube diameters in CHMP2A–CHMP3 range from 44 to 54 nm and their helical pitch is 30 Å or multiples (n) thereof forming n = 1-start to $n = 6$-start helical filament tubes [131] (Fig. 10.7). A medium-resolution electron microscopy (EM) reconstruction revealed an elongated 70-Å long asymmetric unit that is tilted by about 45° with respect to the tube axis, thus producing a tube width of 40 Å and an external dimension of the asymmetric unit of 40 Å, which is likely to fit two CHMP protomers, presumably CHMP3–CHMP2A. The outer surface of the tube structure is rather smooth [131], as would be expected for a membrane-interacting surface (Fig. 10.8). Such tubes have been shown to bind VPS4 on the inside which permits their disassembly *in vitro* [132].

In vitro CHMP4B forms loose helical arrays approximately 30 Å thick or ring-like structures [133,134] and CHMP2A polymerizes into circular coils of approximately the same thickness [131]. In addition, IST1 and CHMP1B tubes have large diameters (Fig. 10.7) and yeast Snf7-induced Vps24–Vps2 (CHMP3–CHMP2) helical tubes resemble their human counterparts [135]. However, it should be noted that yeast

Fig. 10.6 Hybrid structural models of ESCRT complexes. The models were assembled from crystallographic structures of core complexes and individual domains derived from homologous yeast or human proteins. The flexible connection of domains is indicated by dotted lines and their orientations are arbitrary. (a) The ESCRT-0 complex. The core of ESCRT-0 STAM (yeast Hse1) and HRS (yeast Vps27) have two domain-swapped GAT domains [115] with extended C-terminal regions of unknown structure. Preceding the GAT domains, both HRS and STAM contain ubiquitin-binding VHS domains [116]. In HRS, this is followed by a PI3P-specific FYVE domain, a double ubiquitin interacting domain (DUIM), a presumably unstructured region containing the PSAP motif responsible for the ESCRT-I–TSG101 interaction connecting to the GAT domain. In STAM, the VHS domain is linked to UIM and SH3 domains. Thus multiple ubiquitin recognition sites increase the avidity for K63-linked polyubiquitin chains potentially attached to cargo [116]. Notably, the STAM SH3 domain interacts with and activates the ubiquitin hydrolase AMSH (the N-terminal domain is indicated as a grey box and the structure of the C-terminal domain is shown in complex with di-ubiquitin) specific for K63-linked ubiquitin [117,118], thereby placing ubiquitin turnover at an early stage of the pathway. (b) The ESCRT-I complex. ESCRT-0 uses a PSAP motif to interact with TSG101 (Vps23), which is part of the elongated, approximately 27-nm long heterotetrameric ESCRT-I complex [Vps23 (TSG101); Vps28; Vps37; and Mvb12] [119]. The N-terminal ubiquitin E2 variant (UEV) domain of TSG101 binds to ESCRT-0 (HRS, PSAP), ALIX (PTAP and/or PSYP, present in the proline-rich domain, PRD), ubiquitin and viral L domains and is flexibly linked to the stalk region. Notably, Vps37 contains an N-terminal helix ($\alpha 0$) that enhances the membrane interaction of ESCRT-I. In humans, MVB12 exists as two and VPS37 as three different isoforms leading to the formation of distinct ESCRT-I complexes with yet unknown functions. In yeast, the C-terminal domain of Vps28 recruits ESCRT-II via an interaction with the N-terminal Npl4 zinc-finger (NZF) domain present in the Vps36-Glue domain of ESCRT-II [120]. (c) The ESCRT-II complex triggers ESCRT-III assembly. ESCRT-II is composed of VPS22, VPS36 and two copies of VPS25 that adopt a Y-shaped structure [121,122]. ESCRT-II is targeted to the endosomal membrane by lipid-binding activities of the VPS36 GLUE domain and the first helix of VPS22 ($\alpha 0$) [123]. The VPS36 GLUE domain also binds ubiquitin, and yeast Vps36 has two additional NZE zinc finger domains inserted [121]. VPS25 interacts with an N-terminal helical segment of CHMP6 (Vps20) [124], which triggers the assembly of the ESCRT-III complex in yeast [125]. Vps20 nucleates the assembly of an unknown number (n) of Snf7 (CHMP4) molecules which might be followed by a 1:1 stoichiometry of Vps2 (CHMP2) and Vps24 (CHMP3). In mammalian cells, the C-terminal regions of CHMP4 and CHMP3 recruit ALIX (via its BRO1 domain) and the deubiquitinating hydrolase AMSH (via its N-terminal domain), respectively [118]. All CHMP molecules contain MIT domain interacting motifs (MIMs) within their C-terminal regions (indicated by solid rectangles) that are targeted by the VPS4 MIT domain. VPS4 is modeled as a hexameric ring-structure based on electron microscopy single-particle analysis (the electron microscopy model of VPS4 is reproduced with permission from reference [126]). Note that this may not represent the VPS4 complex active in cells. ESCRT-II binds two Vps20 (CHMP6) molecules and both might trigger Snf7 polymer assembly [125], a process which likely requires a curved membrane for assembly initiation. Figure and figure legend adapted from [127].

Membrane remodeling by a protein scaffold 309

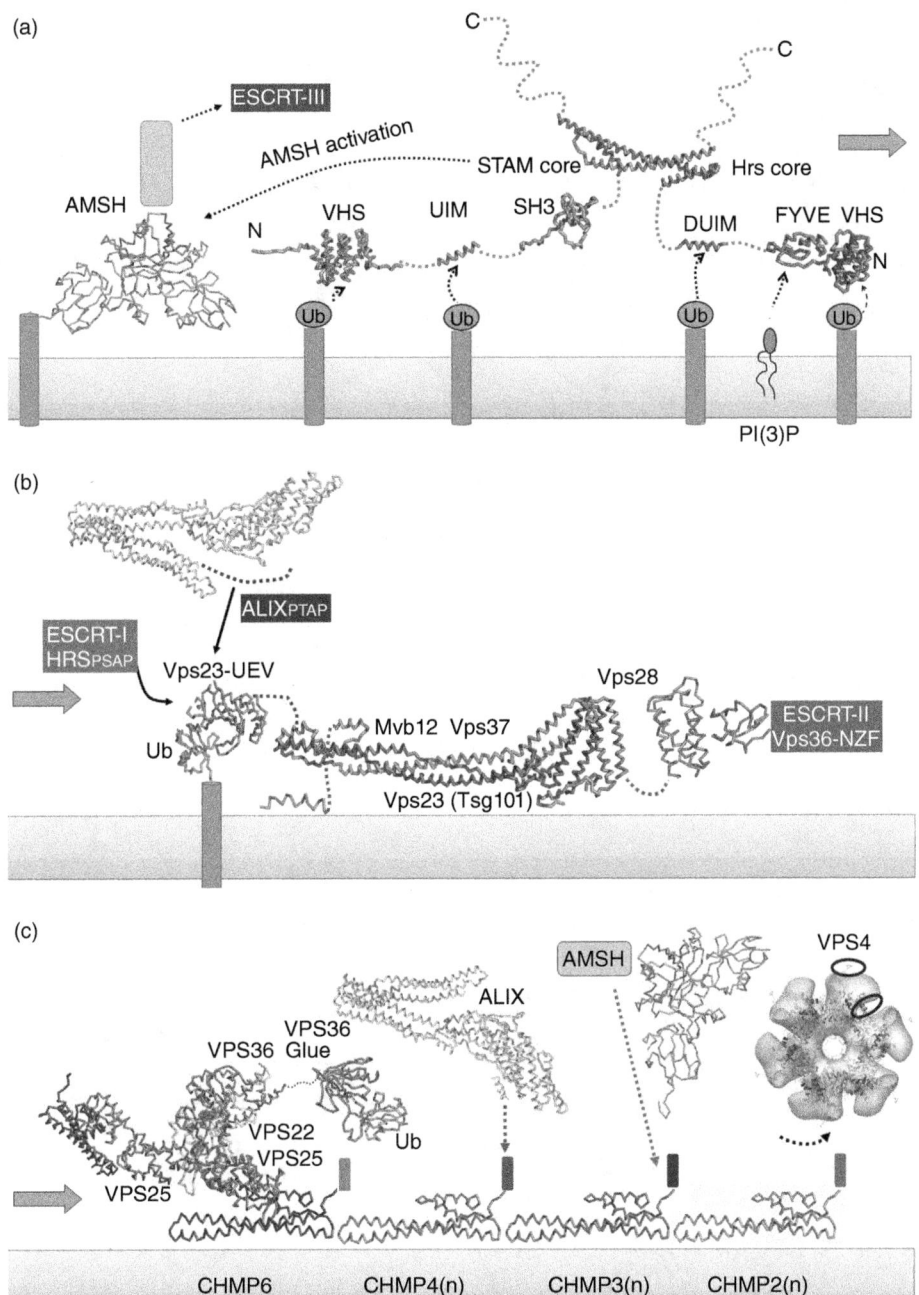

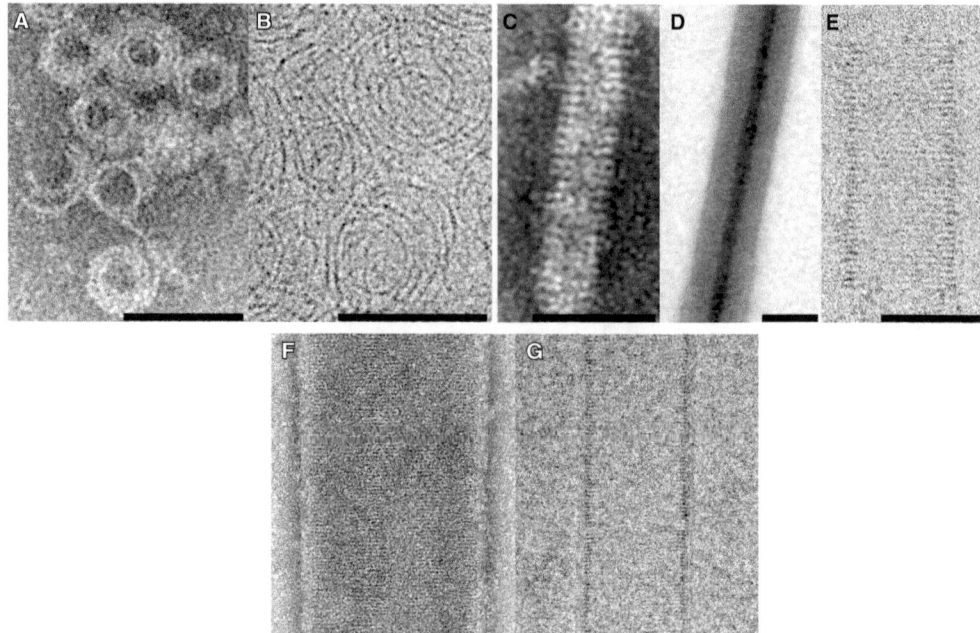

Fig. 10.7 Gallery of ESCRT-III polymers observed *in vitro* and *in vivo* by negative staining or cryoelectron microscopy (cryo-EM): (A) CHMP2A negative stain; (B) CHMP4 cryo-EM; (C) CHMP1B negative stain; (D) IST1 negative stain; (E) CHMP2A-3 cryo-EM; (F) CHMP2B (with membrane) negative stain; (G) CHMP2B (with membrane) cryo-EM. Scale bars are all 50 nm. Figure and figure legend adapted from [100].

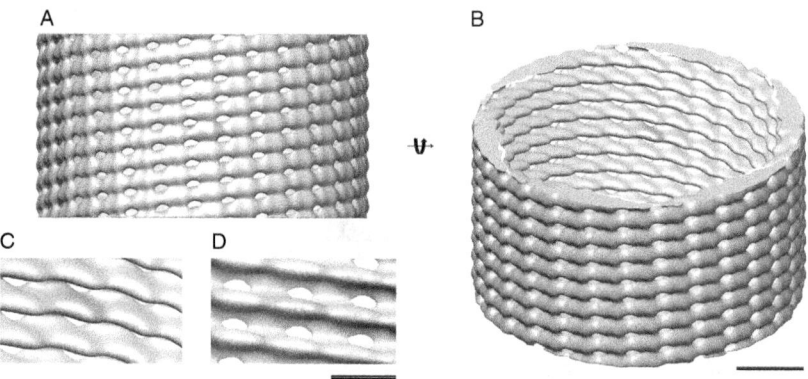

Fig. 10.8 3D reconstruction of CHMP2A–CHMP3 tubes (column pattern, $n = 6$). (A) Isosurface representation (blue) of the 3D reconstruction of a tube having an average diameter of 485 Å. The raise of one helical turn is indicated in pink and two helically related asymmetric units are colored green and red, respectively. (B) The same isosurface as in (A) rotated by 30° provides a view into the tube. Scale bar is 100 Å. (C), (D) Close up of the linear monomer–monomer interaction and the lateral interfilament contacts. C is the same view as B and D the same as A, but looking from the inside. Scale bar is 50 Å. Figure and figure legend adapted from [131].

Vps24 (CHMP3) can also assemble into two-stranded filaments [134]. In addition a 17-nm wide ESCRT-III spiral structure has been imaged at the midbody [136], which may contain multiple ESCRT-III filaments or represent a different structure.

10.3.4.2 ESCRTs and enveloped virus budding

Enveloped viruses such as HIV-1 enter the ESCRT machinery via short sequence motifs called late domains, present in viral structural proteins such as HIV-1 Gag. HIV-1 Gag contains two late domain motifs specific for the recruitment of ESCRT-I (via tsg101) and the ESCRT-associated protein Alix. Other viruses also employ late domains to interact with Hect domain-containing ubiquitin ligases [137–141]. Because the ESCRT access points seem to play no functional roles beyond their adaptor task, the main purpose of entering the ESCRT system at different stages is to obtain access to the ESCRT-III/VPS4 membrane fission machinery.

All ESCRT-III proteins are required for MVB biogenesis [99] and during cell division [142]. In contrast, HIV-1 release strictly requires only one CHMP4 and one CHMP2 family member that act sequentially [143]. Thus CHMP4 and CHMP2 isoforms may form the minimal ESCRT-III fission machinery. ESCRT-III members and VPS4 are transiently recruited to viral budding sites once Gag assembly has been completed [144]. CHMP4 is recruited first and serves as a platform for CHMP2A or CHMP2B interaction. The latter are essential for the fission reaction, and presumably CHMP4B accumulates within the neck structure of a budding virus in their absence [143]. Regarding the minimal fission machinery, it should be noted that CHMP1B is recruited during HIV-1 budding [144] and CHMP3 exerts an important synergistic effect on CHMP2A but not CHMP2B, which significantly augments budding efficiency [131]. This is also consistent with *in vitro* interaction data that revealed micromolar affinities of CHMP4B with CHMP2A and CHMP2B. However, the affinity of CHMP4B is about 10 times higher for CHMP3, indicating that CHMP3, if present, is preferentially recruited by CHMP4B [131].

Sequential recruitment of the ESCRT-III core to Gag-polymers was also recapitulated *in vitro* on giant unilamellar vesicle (GUV) membranes. Gag clusters on GUV membranes recruited ESCRT-I and Alix in a late domain-dependent manner. The presence of Alix allowed direct recruitment of CHMP4, while ESCRT-I in the absence of Alix required the presence of ESCRT-II and CHMP6 for CHMP4 recruitment. CHMP4 then recruited CHMP3 and CHMP2A synergistically, which is followed by CHMP1 [145]. Thus human ESCRT proteins follow the same order of recruitment as their yeast homologues [110,135,146,147].

The residence time of ESCRT-III and VPS4 has been estimated to be less than a few minutes [144,148]. Notably, release of VPS4 from the budding site precedes release of HIV-1 [148], consistent with its proposed active role in cytokinetic abscission [149] and during MVB formation [150]. Even though ESCRT-III-catalyzed membrane fission *in vitro* was observed without VPS4 ATPase activity [104,151], it is likely that VPS4 not only functions during ESCRT-III polymer disassembly and recycling from membranes [152] but its energy may also be directly converted into mechanical forces during the fission reaction.

10.3.4.3 Membrane fission models

Although circular and tubular structures of ESCRT-III polymers observed *in vitro* and *in vivo* can in principle assemble inside a bud neck, their large diameters, varying from 50 to over 200 nm, impede spontaneous membrane fission. Several models are currently under discussion in the literature. The first model is based on the observation that all ESCRT-III polymers potentially implicated in membrane fission (CHMP2A-CHMP3 or CHMP2A or CHMP2B) can form dome-like end-caps *in vitro* or *in vivo* [128,131,132] (Fig. 10.9). The first model thus predicts that CHMP4 polymers, which may induce a first narrowing of the neck, recruit CHMP2A–CHMP3 or CHMP2A or CHMP2B to build up a dome-like polymer. The successive narrowing of the helical filament and its affinity for membrane will "mold" the membrane [153] and induce constriction of the neck membrane. Due to the affinity of the ESCRT-III polymer for membranes, dome formation could theoretically constrict the neck up to a diameter of 6 nm, which would be energetically favorable for spontaneous fast fission *in silico* [154]. *In vivo* it is likely that VPS4 plays an active role in the fission reaction by targeting the final dome-like structure primed for fission. Disassembly of such an ESCRT-III structure could induce an additional stress on the membrane, and thus fission might occur concomitantly with disassembly of ESCRT-III (Fig. 10.10). The formation of

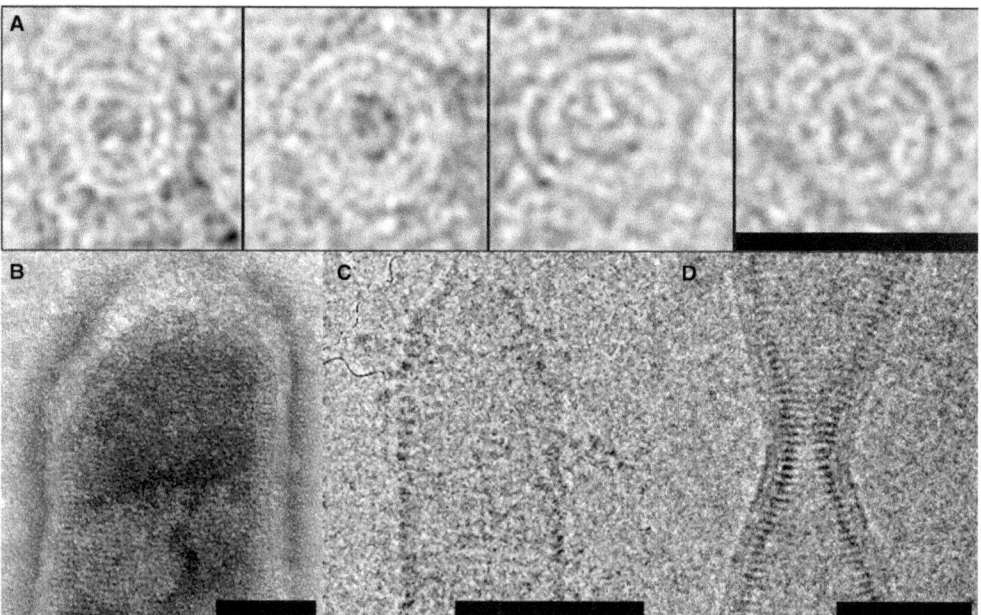

Fig. 10.9 Gallery of closed coil-like structures, dome-like structures and constricted polymers: (A) four CHMP2A spirals (negative stain); (B) CHMP2B dome (with membrane) (negative stain); (C) CHMP2A–CHMP3 dome (cryo-EM); (D) CHMP2B membranous bottleneck (cryo-EM). Scale bars are all 50 nm. Figure and figure legend adapted from [100].

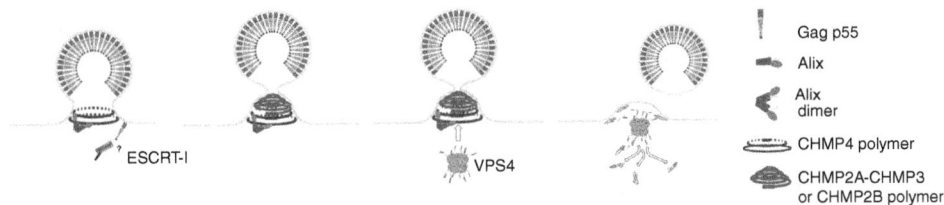

Fig. 10.10 Model of ESCRT-catalyzed HIV-1 budding. ESCRT-III CHMP4, CHMP2A–CHMP3 and CHMP2B were shown to form helical filaments *in vitro* or *in vivo* (CHMP2B) [128,131–134]. ESCRT-III is only recruited once Gag assembly has been completed and Tsg101 and Alix are already present at the budding site [144]. There are still open questions regarding the recruitment of ESCRT-III during HIV-1 budding. ESCRT-III CHMP4 can be recruited via CHMP6 and ESCRT-II or alternatively by Alix. However, ESCRT-II and Alix are dispensable for HIV-1 budding [143,163]. ESCRT-I VPS28 may therefore provide an alternative link to ESCRT-III via CHMP6 [164]. However, CHMP6 itself is also dispensable for HIV-1 budding, even in the absence of Alix [143,163]. The sequence of ESCRT-III recruitment is CHMP4B followed by CHMP2A or -B or by CHMP3–CHMP2A [143]. The model predicts that once Gag assembly is completed CHMP4B is recruited and polymerizes within the membrane neck structure (dark gray polymer), which may induce a first constriction of the neck. This first constriction may include a sliding mechanism supported by the action of VPS4. CHMP4B filaments then form the platform to recruit CHMP2B alone or CHMP3–CHMP2A filaments that can form dome-like structures [131]. The outside of the dome is predicted to have a high affinity for membranes, which could induce membrane neck constriction [154] together with the action of VPS4 fission. In this model it is likely that disassembly of ESCRT-III filaments and fission are concomitant processes. Figure adapted from [165].

dome-like or conical spiral ESCRT-III structures has been also observed *in vivo* in VPS4-depleted cells [155], indirectly supporting a physiological role for such structures. Furthermore ESCRT-III-induced lipid phase separation may play an additional role [156]. If ESCRT-III forms a protein scaffold on membranes its membrane-interacting properties could include and/or exclude certain lipids from the budding side. Such selective lipid sorting could contribute to constrictive line tension at the edge of the bud neck membrane, which could aid in fission [157]. Thus the principles of ESCRT-III-driven membrane fission may be similar to dynamin-catalyzed fission [98].

A variation of the dome model is the break and slide model. This suggests that wide spirals break and slide into lower-energy helices that narrow the intracellular bridge and promote the formation of a fission "dome" [135]. In the latter model VPS4 would play an important role during ESCRT-III remodeling to build up the dome-like end cap structure.

A significant variation of the dome model is the "whorl" model that involves ESCRT-I, -II and -III. This model predicts that ESCRT-III filaments may project into the bud neck pore in a whorl-like arrangement, ultimately serving as docking site for a dome-like ESCRT-III polymer. Alternatively, they might evolve into a dome-like polymer that executes fission [158,159]. These basic models are in principle in agreement with the role of ESCRT-III during cytokinesis [160] where tension release between daughter cells may trigger ESCRT-III assembly, leading to abscission [161].

A yet further modification of the dome model has been suggested based on super-resolution microscopy imaging of ESCRT-III and HIV-1 Gag. This study indicated that initial scaffolding of ESCRT subunits occurs within the interior of the viral bud and selective remodeling of ESCRTs leads to the formation of an internal dome structure that releases CHMP2A into the cytosol prior to fission [162]. This is in contrast to the other models that predict that the ESCRT-III scaffold is built up on the cytosolic side of the forming bud or virion.

10.4 Forces involved in remodeling biological membranes

10.4.1 Some physical methods for measuring forces and mechanics

10.4.1.1 Model membrane systems

In order to characterize lipid membrane mechanics and test theoretical models, physicists have had to develop minimal model systems. Prior to the 1980s biophysicists used simple cell systems such as red blood cells to study the physical properties of membranes [166]. Later, different techniques of preparation were established, and membranes with different geometries and controlled composition were gradually available, allowing for a direct comparison with theoretical models, which were developed in parallel. Currently, many model membrane systems are available for investigating membrane mechanics and to mimic *in vitro* biological processes that involve membranes [167,168]. The distinct geometries of the model systems correspond to the different types of technique for their study [169].

Vesicles Vesicles consist of a bilayer delimiting an internal aqueous compartment from the exterior (Fig. 10.11A). They usually have a spherical shape, but they can also adopt other shapes depending on asymmetry within lipid bilayers or on the surface to volume ratio, in agreement with the Helfrich model [170]. They are distinguished according to their size:

- Small unilamellar vesicles (SUVs) between 30 and 100 nm. They are obtained by sonication of a hydrated preparation of lipids [171]).
- Large unilamellar vesicles (LUVs) between about 100 nm and 1 µm. They are prepared either by extrusion through a filter of calibrated pore size [172] or by reverse-phase evaporation [173].
- Giant unilamellar vesicles (GUVs) with diameters ranging between a few micrometers and 100 µm.

GUVs are particularly suitable membrane model systems. Due to their size, single vesicles can be directly observed using optical microscopy methods and can be directly manipulated for mechanical characterization. Many protocols are currently available that allow their preparation, starting with the gentle hydration method, originally described by Reeves and Dowben [175], using electric fields at low salt concentration [176,177], originally introduced by Angelova et al. [176], or at biologically

Forces involved in membrane remodeling 315

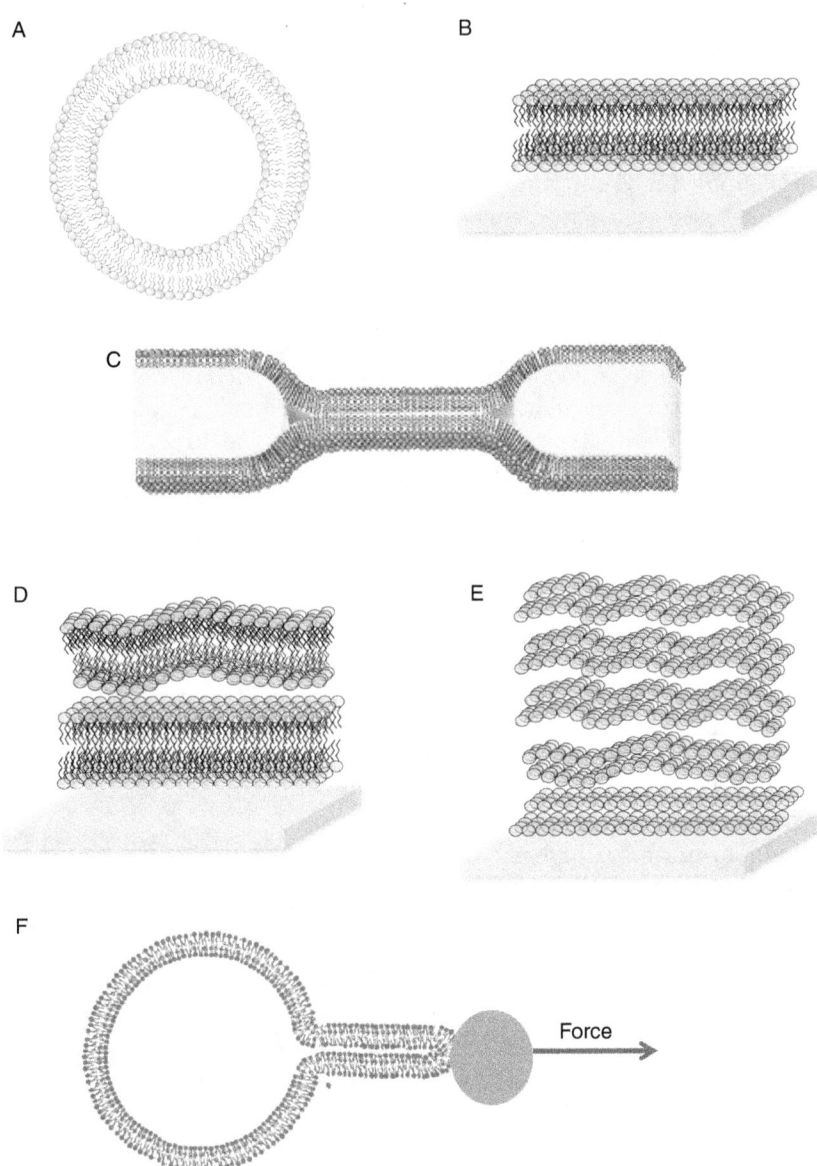

Fig. 10.11 Different types of model membranes: (A) vesicle, (B) supported bilayer, (C) suspended bilayer, (D) floating bilayer, (E) multilayer system, (F) membrane nanotube pulled by attaching a bead to a giant unilamellar vesicle (GUV) and pulling on it. The diameters of the GUV and of the nanotube are not to scale. Adapted from [174], with permission from Elsevier.

relevant salt concentrations [178] or gentle spontaneous swelling [175,179], improved when performed on a polymer gel [180].

GUVs can be prepared with a single type of lipid, but also with lipid mixtures that can phase separate [181–184], and phase diagrams can be established [168,185,186]. Transmembrane proteins can also be reconstituted if small proteoliposomes are used in the initial lipid film from which the GUVs are formed [187–192]. Finally, some recent methods based on inverted emulsions [193] have been used for the encapsulation of proteins inside GUVs [194,195], sometimes using microfluidics [196]. However, a drawback of this method is the presence of traces of oil in the membrane [197] that can perturb its mechanical properties. In order to study membranes with a composition similar to plasma membranes, GUVs can be manufactured using native membrane extract purified from cells deposited on a conductive surface and electroformation [178,198], but it is still unclear whether the asymmetry of the lipid between the two leaflets is preserved. Finally, a very promising method is the direct formation of GUVs by cell blebbing due to the detachment of the plasma membrane from the cortical cytoskeleton; thus the GUVs have the exactly the same composition and asymmetry in terms of lipids and membrane proteins as the native plasma membrane, but no interaction with actin filaments (for a review [199]). These blebs can be obtained (1) by chemical treatment forming giant plasma membrane vesicles (GPMV) [200], with the problem that membrane proteins can be cross-linked, or (2) by cell swelling, forming plasma membrane spheres (PMS), which preserves the distribution of proteins in the membrane [201]. Motivated by recent developments in synthetic biology, many efforts are currently being devoted to the design of artificial cells [202] with GUVs encapsulating gene expression system [203] that could possibly produce the proteins for their own replication and division.

Supported/suspended bilayers Planar bilayers can be formed over a solid substrate (supported bilayer) (Fig. 10.11B). They can be prepared either by fusion of SUVs on the solid substrate [204] or by the Langmuir–Blodgett technique [205]. Interactions with the substrate can be limited by the presence of a polymer "cushion" [206]. Formation of other types of supported bilayers is also possible by tethering the bilayer with anchors. Recently, new developments in the preparation of suspended films over holes based on the preparation of "black lipid films" have been reported (Fig. 10.11C); it is now possible to have a bilayer spanning a hole of a few hundred nanometers in diameter. Double bilayers with the top one floating on the other can be prepared by Langmuir–Schäffer deposition [207] (Fig. 10.11D); multilamellar oriented structures can be obtained by spinning lipid solutions on a solid substrate and further swelling them with a buffer [208] (Fig. 10.11E).

Membrane nanotubes Membrane nanotubes, which consist of a bilayer with a cylindrical cross-section, can be formed by applying a local force to a bilayer (Fig. 10.11E). Tube extrusion can be achieved by different methods: by locally adhering a GUV on a glass capillary or a pillar and applying a flow [209], by holding the GUV (with a micropipette, or by adhesion on the surface) and pulling on it (with another micropipette or a magnetic or optical tweezers) [210–212], by using gravity

[213] or a combination of electroporation and micromanipulation [214]. Their typical diameter ranges between a few tens of nanometers to 1 µm; it can be directly tuned by setting membrane tension (see Section 10.2.2) [215]. Tubes formed from GUVs have been used as probes to study mechanical interactions between proteins and membranes [216].

10.4.1.2 Some techniques for studying membrane mechanics

Many techniques have been designed to measure the mechanical properties of synthetic and natural membranes, which use Helfrich's mechanical description (see Section 10.2.2) for their interpretation. Here we will focus on only on three of them.

Micropipette aspiration This method was introduced by Evans and colleagues and used both for cells [217] and GUVs [218–221]. A floppy vesicle is aspirated in a micropipette connected to a water reservoir (Fig. 10.12A). The hydrostatic pressure ΔP sets the aspiration pressure that thus controls the membrane tension:

$$\sigma = \frac{D_p}{4(1 - D_p/D_v)} \Delta P$$

with D_p and D_v being the pipette and GUV diameter, respectively. The corresponding variation of the membrane area $A_{asp} - A_0$ when the membrane tension is changed from σ_0 to σ is measured optically by measuring the length of the part of the vesicle aspirated into the pipette (the "tongue"). The difference between the real area of the membrane at the microscopic scale (A) defined by the total number of lipids in the vesicle, and the observed area of the stretched membrane (A_p) is called the excess area, α [222]. Supposing the GUV volume to be constant, the tongue length L can be simply related to the excess area. L and the difference in the excess area $\Delta\alpha = \alpha_0 - \alpha$ when the tension is changed from σ_0 to σ, are related by the geometric relation [218]:

$$\Delta\alpha = \frac{(D_p/D_v)^2 - (D_p/D_v)^3}{D_p} L_p.$$

At low membrane tension (typically lower than 10^{-5} N/m), the excess area originates only from fluctuations. $\Delta\alpha$ is derived from Helfrich's expression [219]:

$$\Delta\alpha \cong \frac{k_B T}{8\pi\kappa} \ln\left(\frac{\sigma}{\sigma_0}\right). \tag{10.11}$$

This equation is the basis of the measurement of bending rigidity κ using micropipette aspiration.

Practically, the variation of the difference in excess area versus membrane tension is measured from the entropic regime dominated by the fluctuations (equation 10.11) to the enthalpic regime where the membrane is elastically stretched at high tension, and a crossover between these two regimes can be observed. The complete relation is [219,223]:

$$\Delta\alpha \cong \frac{k_B T}{8\pi\kappa} \ln\left(\frac{\sigma}{\sigma_0}\right) + c \times \frac{\sigma - \sigma_0}{K_a}$$

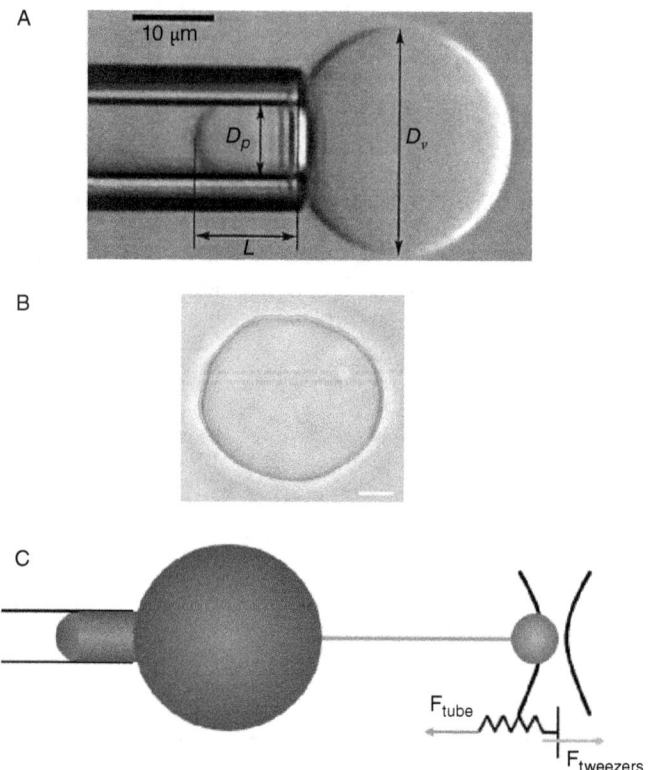

Fig. 10.12 Different methods for characterizing membrane mechanics. (A) Micropipette aspiration. A GUV is aspirated in a micropipette. From the length of the "tongue" of the GUV, the difference in excess area $\Delta \alpha = \alpha_0 - \alpha$ between the final state and a reference state is calculated. The membrane tension σ is deduced from the hydrostatic pressure and from the radius of the GUV and of the pipette. (B) Flickering spectroscopy. The contour of a freely fluctuating GUV is detected, corresponding to the darker gray line. Bar = 5 µm. (C) Membrane nanotube pulled from a GUV aspirated in a micropipette by a bead trapped by optical tweezers. The force exerted by the tube F_{tube} is measured with the optical tweezers, F_{tweezers}.

where K_a is the stretching modulus of the membrane and c is a constant of the order of 1.

Contour analysis This non-contact method involves optical detection of the thermal fluctuations of lipid membranes. The mean square amplitude of the undulation modes was calculated by Helfrich and Servuss [222] in a flat geometry, providing the fluctuation spectrum of a membrane:

$$\langle u_{q_\perp}^2 \rangle = \frac{k_B T}{\kappa q_\perp^4 + \sigma q_\perp^2} \tag{10.12}$$

where $u(\mathbf{r})$ is the local displacement in the normal direction to the membrane with respect to its mean position, and $\mathbf{q}_\infty$ is the wave vector corresponding to $\mathbf{r}$, the projection of the position of a point of the membrane on the mean plane. In principle, a mode analysis of the fluctuations can provide a measurement of the membrane bending rigidity κ and the tension σ, but it is complicated by the non-planar shape of GUVs. Over the years, technical improvements in computing capacity and video cameras have allowed us to detect vesicle contours from phase contrast video-microscopy experiments (Fig. 10.12B), with increasing space and time resolution [224–233]. Eventually, accounting for the integration time of the camera and the specific geometry of microscopy experiments that probe only a planar section of the GUV, the fluctuation spectrum could be fitted over a range of wave vectors for different lipid compositions [234] using:

$$\langle h_{q_x}^2 \rangle = \frac{kT}{2\sigma}\left[\frac{1}{q_x} - \frac{1}{\sqrt{\frac{\sigma}{\kappa}+q_x^2}}\right].$$

Recently, a more precise analysis has been proposed [235], in which the full probability distribution of the configurational fluctuations is taken into account.

Instead of the static fluctuation spectrum, fluctuation dynamics can be measured [236–238]. Additional sources of dissipation in the membrane, such as inter-monolayer friction, must be included in the analysis [239–242], showing the existence of hybrid mode coupling, bending and dilatational motion.

Nanotube pulling Measuring the force required to pull a nanotube from a GUV is another way to measure the mechanics of a membrane. In the case of homogeneous lipid tubes, the free energy of the membrane in the tube F_{tube} can be deduced from Helfrich's Hamiltonian:

$$F_{\text{tube}} = 2\pi RL\left(\frac{\kappa}{2R^2}+\sigma\right) - fL \tag{10.13}$$

where R and L are the tube radius and tube length, respectively, and f is the force applied to the membrane [215,240,243–246]. The equilibrium radius R_0 and the force required to hold the tube f_0, respectively are then given by:

$$R_0 = \sqrt{\frac{\kappa}{2\sigma}} \tag{10.14}$$

$$f_0 = 2\pi\sqrt{2\kappa\sigma}. \tag{10.15}$$

R_0 and f_0 thus depend only on the mechanical parameters of the membrane, κ and σ. These typically range from 10 nm to 1 µm, and from a few pN to 100 pN, respectively. Force can be thus measured using optical tweezers [212] (Fig. 10.12C) or the dragging force in a flow [209]. The dependence of the tube radius and of the force on membrane tension has been experimentally tested [210,212,247–250], allowing the bending rigidity of membranes to be measured.

10.4.1.3 Forces at the molecular scale

Methods have also been developed to measure forces between pairs of molecules such as ligands and receptors at the single-molecule level either *in vitro* by immobilizing proteins on surfaces and using very sensitive force detectors or, more recently, *in vivo*.

Bioforce probe (BFP) This method was initially introduced by E. Evans to measure the rupture force between pairs of ligands and receptors [251,252], in particular those involved in cell adhesion. In this method, the force sensor consists of a spherical red blood cell (RBC), the stiffness of which can be varied by micropipette aspiration. A bead coated with the ligand is bound to the RBC. Another micropipette holding a second bead coated with the receptor is brought into contact. After establishing contact, the second pipette is moved away at a controlled velocity and the elongation of the RBC measured (to deduce the force after calibration) until rupture occurs. The density of proteins on the bead is low enough to ensure that measurements correspond to adhesion/rupture between single pairs of proteins. Eventually, from the variation of the rupture force with the loading rate, the molecular characteristics of the ligand–receptor bond can be deduced using a model based on Kramer's theory (for an extensive review, see [253]).

Atomic force microscopy (AFM) AFM has been extensively used to measure the mechanical properties of single molecules, starting with DNA. It can also be used to measure rupture forces between ligands and receptors (which can be similar in the case of homotypic pairs). The tip of the AFM cantilever is functionalized with the ligand (often using streptavidin–biotin as a "glue" to achieve protein binding) while the substrate is coated with the immobilized receptor. As in the BFP experiments, the ligand is brought into contact with the receptor and the AFM tip is moved up at constant velocity while recording the force from the cantilever deflection until rupture is detected. An example of the technique applied to the E-cadherin pair can be found in [254].

FRET sensors Methods based on FRET (Förster/fluorescence resonance energy transfer) have been developed in recent years in order to directly measure forces in cells [255]. FRET is based on the design of a molecular module (the FRET sensor) that is inserted into the protein to be tested *in vivo*. In FRET, a fluorescent molecule is excited and its emitted fluorescence excites the fluorescence of the complementary molecule of the pair. Since the FRET efficiency depends on the distance between these two fluorescent molecules, measurement of the FRET intensity can be related to the distance between the two molecules and the force exerted on them if a calibrated elastic linker in added between them. This was used, for instance, to measure the tensile force exerted between two cells through E-cadherin junctions [256]. The force exerted on a single integrin in focal adhesion has measured using this very promising method [257].

10.4.2 Membrane constriction due to line tension

10.4.2.1 Lipid membrane domains produce membrane deformation

Membranes that contain lipid mixtures can phase separate depending on temperature or the lipid composition [168,181–186]. Interestingly, the presence of lipid domains can produce membrane bending in the absence of any protein on the membrane [184,258–261]. Buds made of the minority phase can bulge from the surface of the membrane. The associated force is the line tension γ at domain interfaces, which results from curvature or height mismatches that cause hydrophobic acyl chain segments to be exposed to bulk water molecules or to hydrophilic head groups from neighboring lipids [262]. Similarly to the surface tension between immiscible liquids that leads to a reduction of the area of contact between the liquids, line tension between lipid domains produces a constriction reducing the perimeter of the domain boundary and, thereby, the energy penalty. For a spherical bud with a curvature $C = 1/R$, an area S and a rim perimeter L (see Fig. 10.13A), the free energy is given by [263,264]:

$$F = \gamma 2\pi L + \frac{1}{2}\kappa S(C - C_0)^2 + \sigma(S - \pi L^2) \qquad (10.16)$$

We see here that line tension constriction (first term on the right-hand side of equation 10.16) is counterbalanced by the bending energy of the lipid domain (second term) and the membrane tension (third term) that tends to flatten the membrane. Thus, for a given lipid composition, bulging is controlled by membrane tension [264] (Fig. 10.13A). This has been used experimentally to measure line tension in phase-separated vesicles [267]. Depending on lipid composition and temperature, in particular on the distance to a critical point, line tension can vary between 0.1 (close to a critical point) and 10 pN in *in vitro* systems [267–269], and thus might become significant. The presence of proteins that induce a spontaneous curvature C_0 can facilitate bending, and thus budding; moreover, if proteins cluster on domains or in the absence of lipid domains, they can also contribute to the line tension.

10.4.2.2 Membrane scission due to line tension

If membrane tension is strongly reduced, for instance by adjusting the bulk osmotic pressure around GUVs, line tension can be sufficient to allow the membrane domain to bud and detach from the limiting membrane (see Fig. 10.13A, right) [265,270]. Scission also occurs in membrane nanotubes due to phase separation [271] along the domain edges [266,272] (Fig. 10.13B). It was proposed that the Gaussian curvature contribution allowed tube scission to occur at a finite tube radius [266]. Scission of a bud or a nanotube requires an energy barrier to be overcome. In the case of a membrane nanotube, the energy barrier is independent of the tension of the membrane and depends only on the bending moduli and the Gaussian curvature moduli of both phases. Nevertheless, the time scale of the scission event is strongly dependent on the initial radius of the tube; thus, tube scission is facilitated when the tension is high and the tube radius is small [266].

Line tension-induced scission has been proposed to be relevant for some biological processes related to membrane trafficking, in particular for the endocytosis of Shiga toxin [273]. Shiga toxin is able to produce its own endocytic tubules when binding to its lipid receptor, the glycolipid Gb3 [274]. It first clusters at the surface of the membrane and, if tension permits, forms tubular invaginations. As the toxin is enriched in

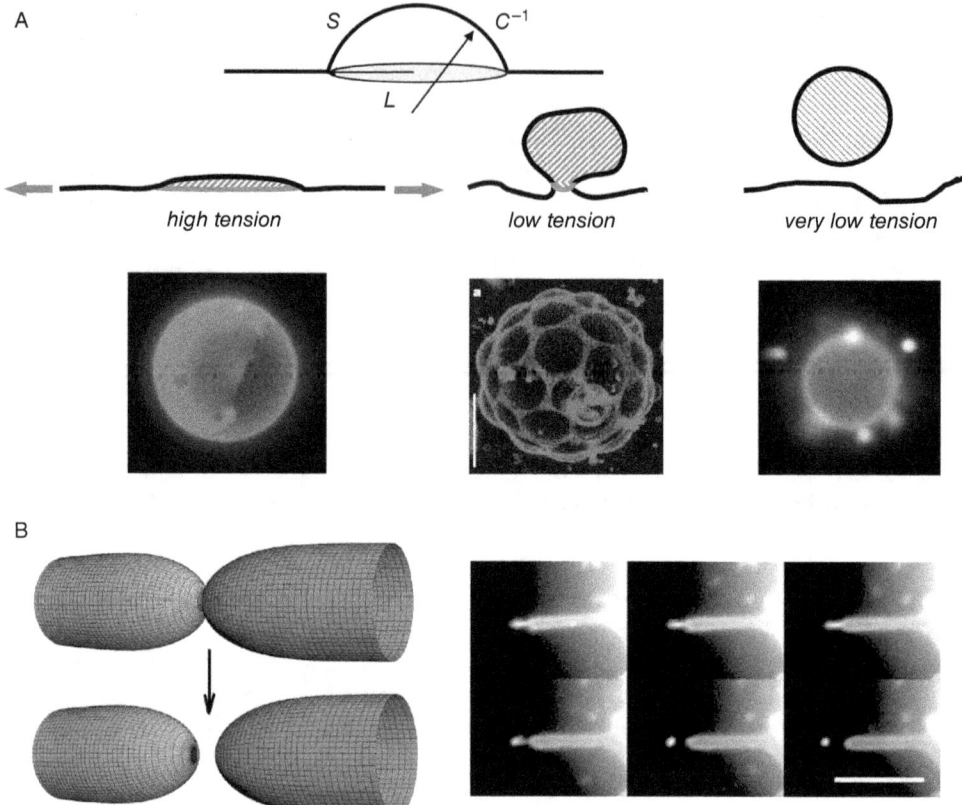

Fig. 10.13 The effect of line tension on membrane budding and scission. (A) Membrane tension and line tension compete for bud formation: at high tension, the membrane is flat and at low tension, line tension induces a bud (radius $1/C$) in order to reduce the perimeter L of the neck. If tension is very low, bud detachment can be observed. The scale of the images is of the order of 10 µm. Central image from [184], with permission from Macmillan Publishers Ltd; left and right images reproduced with permission from [265]. Copyright 2004 by American Physical Society. (B) Line tension due to phase separation induces nanotube scission at the limit of the two domains. Left: theoretical prediction (courtesy of K. Storm); right: experiments (from [266]).

these tubular structures, it changes the lipid composition of the membrane in these structures that becomes enriched in sphingolipids [275] and more prone to phase separation. Line tension due to Shiga toxin clusters can be sufficient to produce tubule scission [273]. But, in addition, actin binds *in vivo* to these tubules, which also induces lipid reorganization [276] and thus tubule scission, as shown by *in vitro* reconstitution [273]. A similar process has been proposed to facilitate scission of clathrin-mediated endocytic buds in budding yeast [277].

10.4.3 Bending membranes with proteins

10.4.3.1 Coupling between membrane curvature and protein density

As already mentioned in Section 10.1, cell membranes can be extremely curved and are continuously deformed due to intra- and extracellular traffic, the action of proteins and molecular machines or interaction with pathogens. A convenient way to describe membrane bending by proteins is to introduce a coarse-grained parameter, the membrane spontaneous curvature C_0, which represents the local mechanical effect of these proteins binding to the membrane. The asymmetric insertion or binding of proteins on a bilayer *in vivo* can produce its spontaneous bending in the absence of external constraint. Following the Helfrich model detailed in Section 10.2, the bending free energy per unit area is then:

$$F_{\text{bending}} = \frac{\kappa}{2} \times (C - C_0(\phi))^2 \qquad (10.17)$$

where C (denoted by J in Section 10.2) is the membrane curvature and ϕ the protein surface fraction on the membrane. We will detail in Section 10.4.3.2 the molecular mechanisms that produce spontaneous curvature and how it has been quantified in some cases. It was proposed initially by Leibler [278] that the protein surface fraction ϕ and membrane curvature can be simply coupled by a linear relationship:

$$C_0(\phi) = \bar{C}_{\text{p}} \phi \qquad (10.18)$$

where $\bar{C}_{\text{p}}$ is the intrinsic spontaneous curvature, a molecular parameter characteristic of the protein. This coupling together with equation (10.17) implies that proteins with a non-zero intrinsic spontaneous curvature can reduce the bending energy by concentrating in curved areas of membranes; thus proteins deforming membranes could be enriched, and thus sorted between flat and bent regions, due to membrane curvature. This gain in mechanical energy is opposed by the mixing entropy penalty due to protein sorting and potentially by the increase in the bending modulus when stiff inclusions are included in the membrane. This also shows that curvature-sensing and membrane-tubulating are two aspects of the same protein characteristics corresponding to different density ranges on a membrane.

Membrane nanotubes pulled from GUVs represent a very convenient assay for quantifying both curvature-induced sorting and the mechanical action of proteins on membranes. One method which has been successful in recent years is to couple confocal microscopy with optical tweezers to measure simultaneously protein sorting between the flat GUV and the highly curved tube as a function of its diameter and the force as a function of protein density on the membrane (see, e.g., [169,216,250,279]) (Fig. 10.14). The sorting parameter S that represents the protein enrichment in the tube can be measured from $S = (I^{\text{t}}_{\text{prot}}/I^{\text{v}}_{\text{prot}})/(I^{\text{t}}_{\text{lip}}/I^{\text{v}}_{\text{lip}})$, where the I coefficients represent the fluorescence intensity of the protein (prot) or of the lipid (lip) in the tube (t) and in the vesicle (v), respectively. We discuss some applications of this method later.

Some of the molecular mechanisms by which proteins can induce spontaneous curvature are summarized in Fig. 10.15 (for reviews see also [48,169,280–284]).

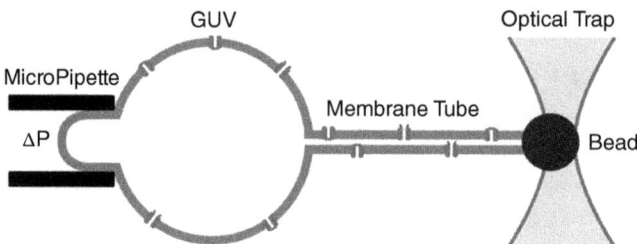

Fig. 10.14 Principle of the protein-sorting measurement using the nanotube assay. Proteins with a non-zero intrinsic positive curvature are enriched in the nanotube pulled from a GUV. Protein sorting can be detected with confocal microscopy, as well as by the mechanical effect of proteins on the tube radius. The change on the tube force due to proteins can be measured with the optical tweezers that trap the bead holding the tube.

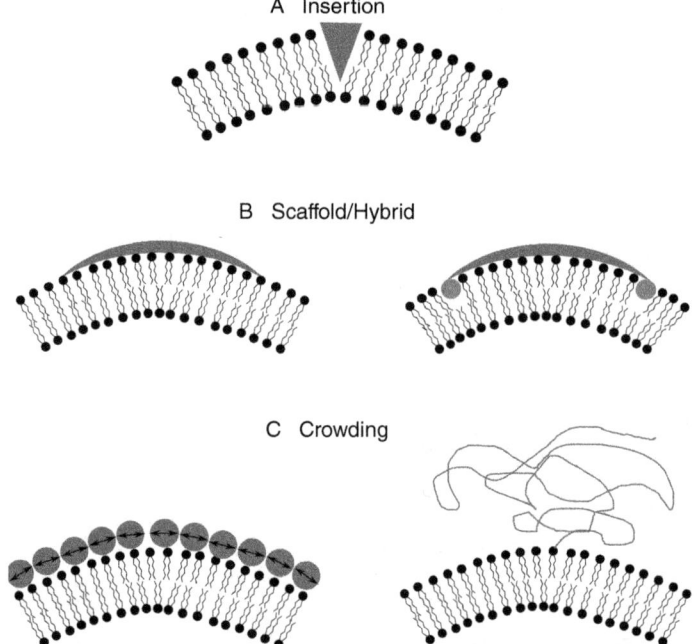

Fig. 10.15 Mechanisms by which proteins can induce spontaneous curvature. Proteins are represented in dark gray and lipids in black (A) Curvature generation by protein insertion, for instance by amphipathic helices or conical transmembrane proteins. (B) Membrane "molding" by a protein scaffold and the hybrid action of scaffold and helix insertion. (C) Steric repulsion between bound proteins causes crowding and membrane bending (left). The entropic pressure of a grafted flexible chain deforms the membrane away from graft site (right). Modified from [169], with permission from Elsevier.

10.4.3.2 Spontaneous curvature generation by proteins

Proteins with amphipathic helices Proteins that insert amphipathic or hydrophobic structures in one leaflet of a membrane create stresses across a flat membrane bilayer, thus producing a spontaneous curvature in the relaxed membrane [285] (Fig. 10.15A). This is the case with proteins such as epsin, which is involved in endocytosis and contains an unfolded N-terminal group when in solution. Upon binding to a PiP2-containing membrane, its undergoes a conformational change in which these groups fold into helices, such that the hydrophobic face is inserted into the bilayer [286]. When epsin is incubated with small liposomes containing PiP2 lipids, they are transformed into tubulated structures [280,286] which then convert into small vesicles with diameters of few tens of nanometers [287]. The spontaneous curvature created by this insertion mechanism is related to three parameters: the typical size of the insertion, the depth of insertion into the bilayer and protein surface fraction ϕ. It has been calculated by Campello et al. [285], showing that the linear relation between $\bar{C}_p$ and ϕ is valid up to 10% protein coverage on the membrane. Experimentally, a protein enrichment in nanotube has been measured [288], as expected from theoretical modeling.

BAR-domain proteins We reviewed the structural information on BAR-domain proteins in Section 10.3.3. As also mentioned, when incubated at micromolar concentrations with liposomes containing negatively charged lipids, these proteins induce membrane tubulations [78,91], or invaginations [289] for BAR or N-BAR, and I-BAR, respectively. Experiments using membrane nanotubes have been published that measure curvature-induced sorting and scaffolding capability for N-BAR proteins [250,290] and F-BAR [291]. Here, we will detail the results on amphiphysin-1 from [250]. This protein binds to the neck of clathrin-coated vesicles at a late stage just before dynamin recruitment and bud scission [292,293], when the bud neck is highly curved. The tube assay (Fig. 10.14), combined with a calibration of protein density and tube radius from fluorescence analysis, has shown a significant enrichment of amphiphysin-1 on nanotubes that increases with curvature; protein sorting on the tube was observed for the entire density range. A theoretical model using the described coupling and including mixing entropy predicted that, at low surface fraction on the GUV (less than about 3%), sorting S varies linearly with ϕ_v:

$$S = \frac{\phi_t}{\phi_v} = 1 + \frac{1}{\bar{C}_p \phi_v} \frac{1}{R_t}$$

with R_t being the tube radius. The intrinsic spontaneous curvature in the case of BAR domains is generated by membrane "molding" due to their close contact; in the case of N-BAR, the origin is hybrid with additional amphipathic helices that can insert into the bilayer (Fig. 10.15B). The relation between sorting and curvature was confirmed experimentally and the corresponding intrinsic spontaneous radius of curvature of the protein was deduced with an average $\langle \bar{C}_p \rangle^{-1} = 3 \pm 1.5$ nm. At high surface fraction on the GUV (experiments were nevertheless limited to $\phi_v < 15\%$), the tube radius is independent of tension and vesicle protein density, resulting from the formation of a scaffold with a constant radius $\langle R_a \rangle = 7 \pm 2$ nm around the tube. Scaffolding was confirmed independently by the scaling of the force with tension that is modified compared with a bare tube and depends linearly on tension instead of a square-root dependence (equation 10.14). Scaffolding occurring at densities much below full coverage of the tube (here less than 50%) was confirmed by structural experiments on a similar N-BAR protein and was initially attributed to transverse interactions between amphipathic helices [294], although this now seems not to be the case [295,296] (also Simunovic et al., unpublished). Globally, this study demonstrated the validity of using a model based on spontaneous curvature to describe coupling between protein density and membrane bending. In addition, the fact that the protein density of amphiphysin-1 at the plasma membrane is not high *in vivo* [293] suggests that its role is to sense curvature and thus recruit dynamin at the end of the budding process [297].

Trans-membrane proteins Extensive work has been done on sorting of trans-membrane proteins to identify processes governing their targeting in cells. These proteins can span the entire bilayer, or incompletely with hairpins as in the case of caveolin or reticulon [298,299]; thus their interaction with membranes should be described by an insertion mechanism (Fig. 10.15A). Specific peptide motifs have been identified corresponding to specific cell localizations [300]; moreover hydrophobic mismatch between the bilayer and the hydrophobic part of the protein also contribute to their sorting [301,302]. It has been shown that the matching of protein shape to the underlying membrane curvature might also provide a targeting mechanism. In [303] it was shown that two proteins with similar lateral size but with different shapes were differently sorted by curvature: KvAP a voltage-gated channel [304] with a conical shape is enriched in the curved area of membranes whereas aquaporin-0 with a cylindrical shape [305], and thus $\bar{C}_p = 0$ is not. The sorting mechanism is also well described by a model based on spontaneous curvature, but including the stiffening of the membrane when the density of proteins in the membrane is increased. This leads to a leveling of KvAP sorting at high curvature. Nanotube assay experiments show that the intrinsic radius of curvature of KvAP is $\langle \bar{C}_p \rangle^{-1} = 25$ nm, corresponding to a conicity of about $5°$. This relatively weak $\bar{C}_p$ value is nevertheless sufficient to produce significant curvature-induced sorting and should thus be relevant for many non-cylindrical membrane proteins. However, even at 10% surface fraction, the mechanical effect was found to be very small, whereas at the same density we would expect that the reticulon would stabilize tubular structures (such as ER) [303] based on an estimate of its $\bar{C}_p$ value [299].

Protein crowding induces membrane curvature Spontaneous curvature can also occur through steric interactions between proteins and between proteins and membranes by two distinct mechanisms. Stachowiak et al. [306] have suggested that proteins such epsin or grafted GFP, even in the absence of amphipathic helices or the ability to make scaffolds, can deform membranes when concentrated at their surface. This is a result of protein crowding, where steric interactions between proteins create a lateral pressure and a spontaneous curvature (Fig. 10.15C, left), in agreement with the theory of polymer brushes grafted to interfaces [307,308]. Quantitative estimations showed that the effects of protein crowding on membrane curvature are expected to be substantially weaker than those of shallow insertion of amphipathic or hydrophobic proteins domains such as N-terminal alpha-helices, but still could modulate to some extent the action of the latter [28]. Steric interactions between bound proteins and the membrane can also result in spontaneous curvature. The entropic-based pressure of a single polymer grafted onto a membrane has been predicted to induce membrane deformation [309,310] (Fig. 10.15C right). It has been shown experimentally that polymers grafted onto membranes at very low density (in the "mushroom" regime, where the polymers do not overlap) can generate spontaneous curvature without lateral repulsive interactions between polymers and without insertions into the bilayer [311].

10.4.4 Forces produced by molecular motors and the cytoskeleton

10.4.4.1 Nanotubes pulled by molecular motors

In cells there are many dynamic tubular structures that are formed by molecular motors moving along the cytoskeleton and pulling on cell membranes (for recent reviews see, e.g., [312,313]) (Fig. 10.16A). In the most frequent cases, these motors move along microtubules and belong to the kinesin or dynein superfamily [314,315]. These tubular carriers have been observed on different pathways, not only the secretion pathway from the ER to the Golgi apparatus [316] and from the Golgi to the plasma membrane [317–319] but also along retrograde transport routes, from endosomes to the Golgi [320] and from the Golgi to the ER [321,322]. In order to decipher the mechanism underlying the formation of membrane tubes by molecular motors, semi-*in vitro* experiments have been performed using cell extracts and cell membrane fractions [323–325]; reconstitution of membrane tubule networks along microtubules immobilized on a solid substrate was obtained. However, with such a complex protein mixture it was difficult to identify the minimum number of components that are necessary to create membrane tube networks, and if motors alone can produce sufficient force to pull tubes. Next, systems made only of purified constituents were developed, namely kinesins, stabilized microtubules, GUVs and ATP [326–328] (Fig. 10.16Ba). It was then possible to unambiguously demonstrate that molecular motors are able to pull membrane

Fig. 10.16 Membrane tubulation by molecular motors. (A) General scheme (adapted from [280], with permission from Macmillan Publishers Ltd). (B) Tubulation by kinesins: (a) *in vitro* assay using GUV with kinesins bound with streptavidin–biotin and microtubules (from [313], with permission from Elsevier); (b) network of nanotubes formed by kinesins from a GUV (bar = 5 µm) (from [326]); (c) mechanism for tube pulling by processive motors (see text). (C) Tubulation by the non-processive myosin 1b (myo1b): (a) *in vitro* assay using GUV with myo1b bound through PiP2 incorporated in the membrane, and actin bundles on the glass substrate; (b) confocal images of tubes pulled by myo1b showing the three channels (lipids, myo1b and actin) (bars = 5 µm); (c) mechanism for tube pulling by non-processive catch–bond motors (see text) (figures in part C from [338]).

328 *Membrane remodeling*

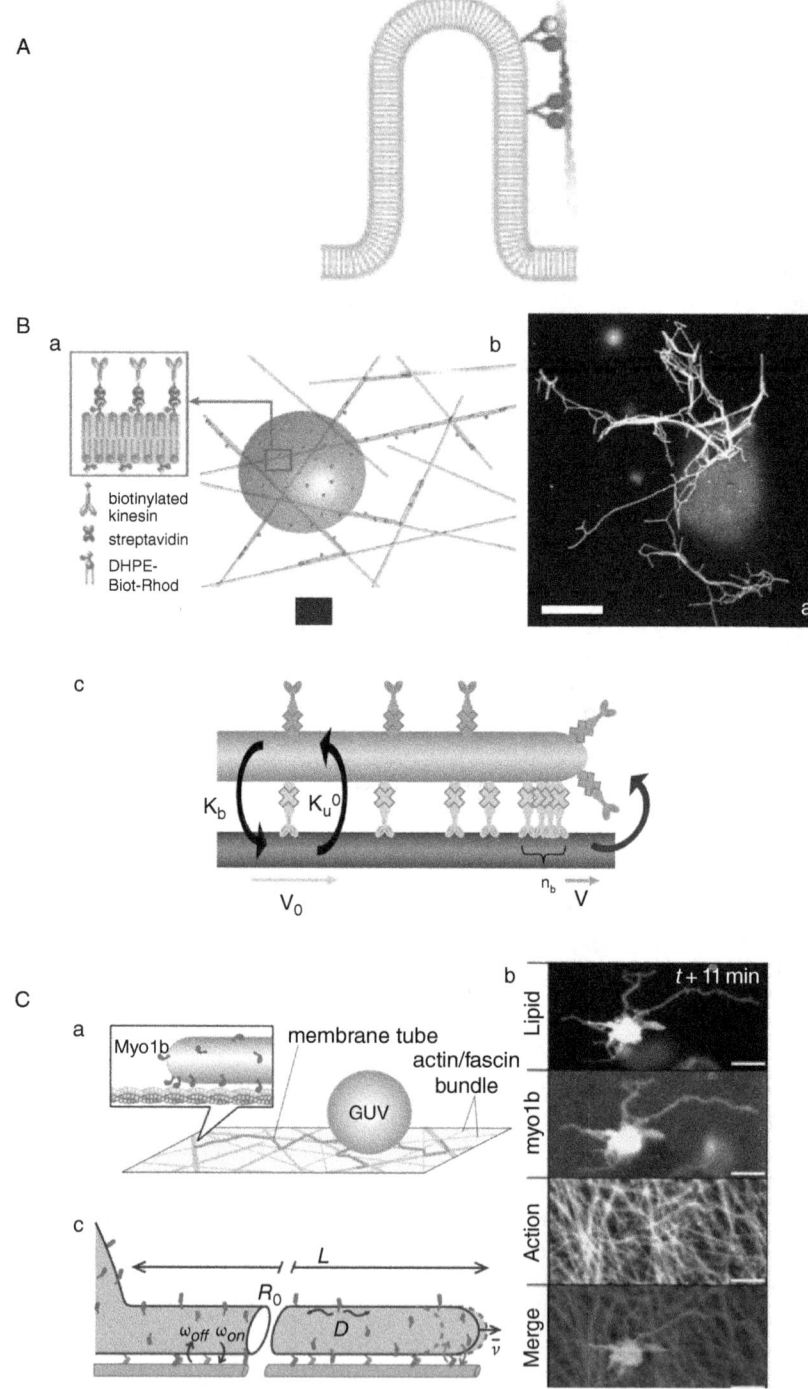

tubes, without the aid of other proteins (Fig. 10.16Bb). Since the force exerted by a single kinesin (6 pN for kinesin 1 according to [329]) is in general lower than the force required to pull a tube (between 5 pN and a few tens of pN depending on membrane tension), a mechanism leading to the clustering of the motors and the redistribution of the load on a collection of motors was proposed [327] and a general theoretical model was developed to describe how the collective behavior of processive motors results from the dynamical properties of motors and the mechanical characteristics of the membrane, in particular membrane tension (Fig. 10.16Bc) [313,328,330–332]. This model proposes that motors at the tube tip that pull the tube are slowed down (to a velocity V) by the membrane tube force compared with those behind that move freely (with velocity V_0). They create a "traffic jam" and thus accumulate at the tip with a cluster size (n_b) that is set by the tube force (and thus the membrane tension) controlling the detachment of the motors (K_d) from the microtubule and by the density of motors on the membrane. The predictions of the model, in particular the existence of a motor cluster at the tip of the tube during tube growth, and an oscillatory or stall regime, have been found to be in quantitative agreement with *in vitro* assays (see [313] for a review).

Non-processive motors—which detach from the cytoskeleton filament during each ATPase cycle and make just a single step in contrast to processive motors which make multiple steps—can also form tubes, but with different mechanisms. *In vivo*, it was shown that myosin 1b (myo1b) is indispensable for the formation of tubular carriers from the trans-Golgi network and from sorting endosomes [333,334]. Different *in vitro* experiments have confirmed tube pulling by non-processive motors. It was shown that the minus end non-processive kinesin NCD pulls tube in a rich collection of situations (growth, retraction and bidirectional movements), but requires a high density of motors and the formation of stochastic clusters on the tube and at its tip [335–337]. In contrast, myo1b was shown to pull tubes at a constant velocity always in the same direction, with a lower motor density on the membrane and without cluster formation (Fig. 10.16C) [338]. An *in vitro* assay (Figs. 10.16Ca and Cb) coupled with a theoretical model (Fig. 10.16Cc) demonstrated the following mechanism: the myo1b at the tip holds the tube until the next myo1b binds in front of it, pulls the tube further and waits for the next motor to replace it. Since the tube force strongly increases the time myo1b remains bound to actin filaments, this catch–bond property facilitates tube extension and reduces the required motor density [338].

10.4.4.2 Introduction to forces produced by an actin cytoskeleton

Membrane deformation can result from forces generated by the polymerization of cytoskeleton filaments that are bound to the membrane and pull on it (see [339] for microtubules at the ER membrane) or push against the membrane, as in the case of actin filaments that form a tubular structures a few microns long (filopodia) at the leading edge of cells (Fig. 10.17A, B) [340]. The literature on forces generated by actin on membranes and on the biochemistry of the proteins involved in the process is substantial (for review see, e.g., [341]). Here we will only briefly summarize how actin bundles can form tubular structures. Filopodia are very dynamic structures that

330 *Membrane remodeling*

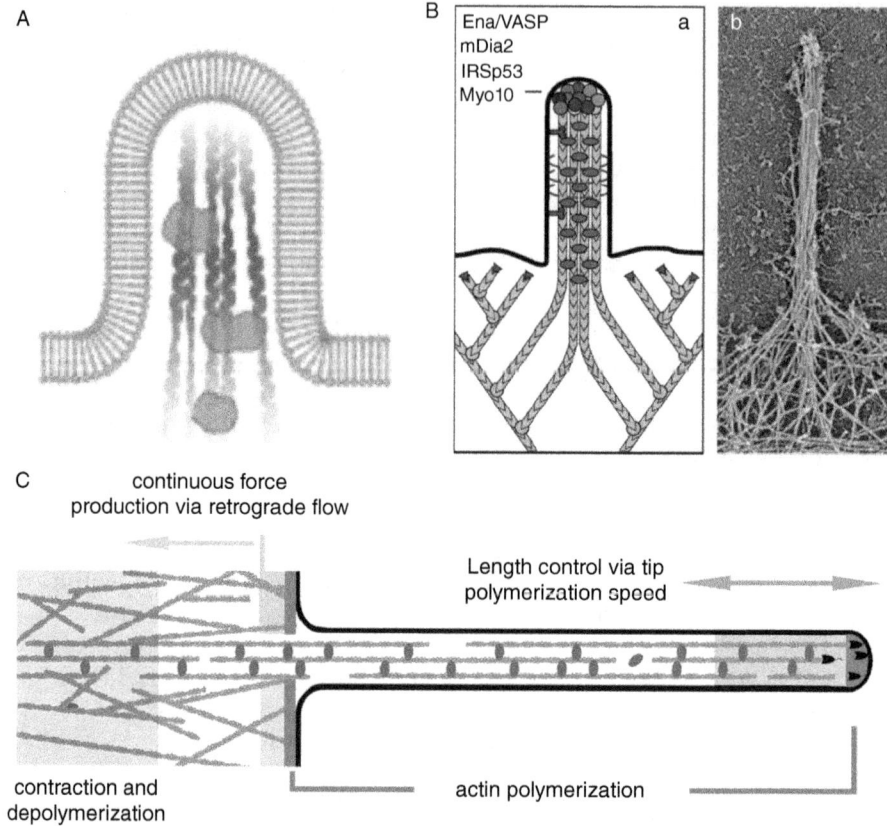

Fig. 10.17 Membrane tubulation by actin filaments. (A) General scheme (adapted from [280], with permission from Macmillan Publishers Ltd). (B) Structure of a filopodium: (a) scheme showing the plasma membrane, the actin bundle and some of the proteins interacting with actin filaments and the membrane; (b) electron microscopy image of a cell filopodium (from [347], with permission from Elsevier). (C) The dynamics and force production of filopodia is controlled to a large part by actin dynamics. In the lamellipodium, there is a continuous retrograde flow of the actin network. This is transmitted to the filopodium through high friction at the base between both actin structures. Together with membrane force, it produces a rearward force. This is opposed by extension of the filopodium due to polymerization of actin at the tip (from [342], with permission from John Wiley and Sons).

can extend and retract, depending on the activation of proteins that regulate actin dynamics (for a recent review see [342]), which compete or not with the continuous backward force due to the retrograde flow of actin in the cortex in which the filopodia are embedded (Fig. 10.17C). It order to deform the membrane and form a tubular structure, actin filaments have to form parallel bundles with adequate cross-linking proteins (e.g., fascin; green dots in Fig. 10.17C), since the force exerted by the membrane would lead to buckling of a single filament just a few microns long [343].

Membrane tubular invaginations formed that interact with actin and some associated proteins have been reconstituted *in vitro* [344]. It was shown by Liu et al. [344] that the spontaneous bundling of actin filaments onto GUV membrane occurred due to coalescence between adjacent deformations. *In vivo*, the coalescence of filopodia is prevented by embedding of the filopodial basis into the cortex and by proteins that connect actin and the membrane [345]. However, no complete reconstitution of filopodia has yet been possible due to the extreme complexity of the system. But by applying a counterforce at the tip of a filopodium, it has been demonstrated that the pulling force that a single filopodium can exert is strongly limited by the weakness of the connections between the actin bundle and membrane at the tip [346].

Acknowledgments

The work of the Kozlov group is supported by the Israel Science Foundation (grant no. 758/11) and M.M.K. holds the Joseph Klafter Chair in Biophysics. Work in the Weissenhorn lab was supported by the Deutsche Forschungsgemeinschaft SPP1175 (W.W.), the Agence Nationale de la Recherche (ANR-08-BLAN-0271-01), FINOVI, and the Labex GRAL (ANR-10-LABX-49-01). W.W. is a senior member of the "Institut Universitaire de France". We further acknowledge the use of the platforms of the Grenoble Instruct Center (ISBG; UMS 3518 CNRS-CEA-UJF-EMBL) with support from FRISBI (ANR-10-INSB-05-02) and GRAL (ANR-10-LABX-49-01). Work in the Bassereau group is supported by the Institut Curie, the Centre National pour la Recherche Scientifique, the Agence Nationale de la Recherche, the European Research Council, the Fondation pour le Recherche Médicale, la Fondation Pierre-Gilles de Gennes. The group belongs to the CNRS consortium CellTiss, to the Labex CelTisPhyBio (ANR-11-LABX0038) and to Paris Sciences et Lettres (ANR-10-IDEX-0001_02).

References

1. Voeltz GK, Prinz WA, Shibata Y, Rist JM & Rapoport TA. A class of membrane proteins shaping the tubular endoplasmic reticulum. *Cell*, 2006, **124**: 573–586.
2. Shibata Y, Voeltz GK & Rapoport TA, Rough sheets and smooth tubules. *Cell*, 2006, **126**: 435–439.
3. Terasaki M, et al. Stacked endoplasmic reticulum sheets are connected by helicoidal membrane motifs. *Cell*, 2013, **154**: 285–296.
4. Puhka M, Joensuu M, Vihinen H, Belevich I & Jokitalo E. Progressive sheet-to-tubule transformation is a general mechanism for endoplasmic reticulum partitioning in dividing mammalian cells. *Mol. Biol. Cell*, 2012, **23**: 2424–2432.
5. Palade GE. An electron microscope study of the mitochondrial structure. *J Histochem Cytochem*, 1953, **1**: 188–211.
6. Fawcett DW. Mitochondria. In: Fawcett DW (ed), *The cell*, 2nd edn. Philadelphia, WB Saunders, 1981, pp. 410–468.
7. Pearse BM. Coated vesicles from pig brain: purification and biochemical characterization. *J. Mol. Biol.*, 1975, **97**: 93–98.

8. De Camilli P & Takei K. Molecular mechanisms in synaptic vesicle endocytosis and recycling. *Neuron*, 1996, **16**: 481–486.
9. McMahon HT & Boucrot E. Molecular mechanism and physiological functions of clathrin-mediated endocytosis. *Nat. Rev. Mol. Cell Biol.*, 2011, **12**: 517–533.
10. Parton RG & Simons K. The multiple faces of caveolae. *Nat. Rev. Mol. Cell Biol.*, 2007, **8**: 185–194.
11. Stan RV. Structure of caveolae. *Biochim. Biophys. Acta*, 2005, **1746**: 334–348.
12. Cossart P & Helenius A. Endocytosis of viruses and bacteria. *Cold Spring Harbor Perspect. Biol.*, 2014, **6**: a016972.
13. Di Fiore PP & von Zastrow M. Endocytosis, signaling, and beyond. *Cold Spring Harbor Perspect. Biol.*, 2014, **6**: a016865.
14. Jahn R & Fasshauer D, Molecular machines governing exocytosis of synaptic vesicles. *Nature*, 2012, **490**: 201–207.
15. Friedman JR & Nunnari J. Mitochondrial form and function. *Nature*, 2014, **505**: 335–343.
16. Lu L, Ladinsky MS & Kirchhausen T. Cisternal organization of the endoplasmic reticulum during mitosis. *Mol. Biol. Cell*, 2009, **20**: 3471–3480.
17. Lu L, Ladinsky MS & Kirchhausen T. Formation of the postmitotic nuclear envelope from extended ER cisternae precedes nuclear pore assembly. *J. Cell Biol.*, 2011, **194**: 425–440.
18. Wang S, Romano FB, Field CM, Mitchison TJ & Rapoport TA. Multiple mechanisms determine ER network morphology during the cell cycle in *Xenopus* egg extracts. *J. Cell Biol.*, 2013, **203**: 801–814.
19. Mironov AA, et al. ER-to-Golgi carriers arise through direct en bloc protrusion and multistage maturation of specialized ER exit domains. *Dev. Cell*, 2003, **5**: 583–594.
20. Ladinsky MS, Mastronarde DN, McIntosh JR, Howell KE & Staehelin LA. Golgi structure in three dimensions: functional insights from the normal rat kidney cell. *J. Cell Biol.*, 1999, **144**: 1135–1149.
21. Sundquist WI & Krausslich HG. HIV-1 assembly, budding, and maturation. *Cold Spring Harbor Perspect. Biol.*, 2012, **2**: a006924.
22. Perez-Vargas J, et al. Structural basis of eukaryotic cell-cell fusion. *Cell*, 2014, **157**: 407–419.
23. Podbilewicz B. Virus and cell fusion mechanisms. *Annu. Rev. Cell Dev. Biol.*, 2014, **30**: 111–139.
24. Steigemann P & Gerlich DW. Cytokinetic abscission: cellular dynamics at the midbody. *Trends Cell Biol.*, 2009, **19**: 606–616.
25. Schiel JA & Prekeris R. Making the final cut—mechanisms mediating the abscission step of cytokinesis. *Sci. World J.*, 2010, **10**: 1424–1434.
26. Eggert US, Mitchison TJ & Field CM. Animal cytokinesis: from parts list to mechanisms. *Annu. Rev. Biochem.*, 2006, **75**: 543–566.
27. Tanford C. *The hydrophobic effect: formation of micelles and biological membranes*. New York, Wiley & Sons, 1973.
28. Kozlov MM, et al. Mechanisms shaping cell membranes. *Curr. Opin. Cell Biol.*, 2014, **29**: 53–60.

29. Campelo F, Arnarez C, Marrink SJ & Kozlov MM. Helfrich model of membrane bending: from Gibbs theory of liquid interfaces to membranes as thick anisotropic elastic layers. *Adv. Colloid Interfac.*, 2014, **208**: 25–33.
30. Helfrich W. Elastic properties of lipid bilayers: theory and possible experiments. *Z. Naturforsch.*, 1973, **28c**, 693–703.
31. Gibbs JW. *The scientific papers*. New York, Dover, 1961.
32. Markin VS & Kozlov MM. Is it really impermissible to shift the Gibbs dividing surface in the classical theory of capillarity. *Langmuir*, 1989, **5**: 1130–1132.
33. Kozlov MM, Leikin SL & Markin VS. Elastic properties of interfaces. Elasticity moduli and spontaneous geometrical characteristics. *J. Chem. Soc. Faraday Trans.*, 1989, **85**: 277–292.
34. Kozlov MM & Winterhalter M. Elastic moduli and neutral surface for strongly curved monolayers. Analysis of experimental results. *J. Phys. France II*, 1991, **1**: 1085–1100.
35. Kozlov MM & Winterhalter M. Elastic moduli for strongly curved monolayers. Position of the neutral surface. *J. Phys. France II*, 1991, **1**: 1077–1084.
36. Spivak MA. *Comprehensive introduction to differential geometry*. Waltham, MA, Brandeis University, 1970.
37. Singer SJ & Nicolson GL. The fluid mosaic model of the structure of cell membranes. *Science*, 1972, **175**: 720–731.
38. Landau LD & Lifshitz EM. *Statistical physics*. New York, Pergamon Press, 1969.
39. Niggemann G & Helfrich W. The bending rigidity of phosphatidylcholine bilayers: dependences on experimental method, sample cell sealing and temperature. *J. Phys. France II*, 1995, **5**: 413–425.
40. Kozlov MM, Leikin S & Rand RP. Bending, hydration and void energies quantitatively account for the hexagonal–lamellar–hexagonal reentrant phase transition in dioleoylphosphatidylethanolamine. *Biophys. J.*, 1994, **67**: 1603–1611.
41. Leikin S, Kozlov MM, Fuller NL & Rand RP. Measured effects of diacylglycerol on structural and elastic properties of phospholipid membranes. *Biophys. J.*, 1996, **71**: 2623–2632.
42. Kooijman EE, et al. Spontaneous curvature of phosphatidic acid and lysophosphatidic acid. *Biochemistry*, 2005, **44**: 2097–2102.
43. Luzzati V. X-ray diffraction studies of lipid-water systems. In: Chapman D (ed) *Biological membranes*, Vol. 1. New York, Academic Press, 1968, pp. 71–123.
44. Seddon JM & Templer RH. Polymorphism of lipid–water systems. In: Lipowsky R & Sackmann E (eds) *Structure and dynamics of membranes*, Vol. 1. Amsterdam, Elsevier, 1995, pp. 97–160.
45. Koynova R & Caffrey M. Phases and phase transitions of the hydrated phosphatidylethanolamines. *Chem. Phys. Lipids*, 1994, **69**: 1–34.
46. Gruner SM. Stability of lyotropic phases with curved interfaces. *J. Phys. Chem.*, 1989, **93**: 7562–7570.
47. Israelachvili JN. *Intermolecular and surface forces*. London, Academic Press, 1992.
48. Zimmerberg J & Kozlov MM. How proteins produce cellular membrane curvature. *Nat. Rev. Mol. Cell Biol.*, 2006, **7**: 9–19.

49. Rand RP & Fuller NL. Structural dimensions and their changes in a reentrant hexagonal–lamellar transition of phospholipids. *Biophys. J.,* 1994, **66**: 2127–2138.
50. Kozlov MM & Helfrich W. Effects of a cosurfactant on the stretching and bending elasticities—a surfactant monolayer. *Langmuir,* 1992, **8**: 2792–2797.
51. Kozlov MM. Determination of lipid spontaneous curvature from X-ray examinations of inverted hexagonal phases. *Methods Mol. Biol.,* 2007, **400**: 355–366.
52. Lorenzen S, Servuss RM & Helfrich W. Elastic torques about membrane edges—a study of pierced egg lecithin vesicles. *Biophys. J.,* 1986, **50**: 565–572.
53. Turner DC & Gruner SM. X-ray reconstitution of the inverted hexagonal (HII) phase in lipid-water system. *Biochemistry,* 1992, **31**: 1340–1355.
54. Chung H & Caffrey M. The curvature elastic-energy function of the lipid-water cubic mesophase. *Nature,* 1994, **368**: 224–226.
55. Templer RH, Khoo BJ & Seddon JM. Gaussian curvature modulus of an amphiphilic monolayer. *Langmuir,* 1998, **14**: 7427–7434.
56. Siegel DP & Kozlov MM. The Gaussian curvature elastic modulus of N-monomethylated dioleoylphosphatidylethanolamine: relevance to membrane fusion and lipid phase behavior. *Biophys. J.,* 2004, **87**: 366–374.
57. Ben-Shaul A. Molecular theory of chain packing, elasticity and lipid-protein interaction in lipid bilayers. In: Lipowsky R & Sackmann E (eds) *Structure and dynamics of membranes,* Vol. 1. Amsterdam, Elsevier, 1995, pp. 359–401.
58. Schwarz US & Gompper G. Stability of inverse bicontinuous cubic phases in lipid-water mixtures. *Phys. Rev. Lett.,* 2000, **85**: 1472–1475.
59. Kozlov MM & Helfrich W. Elastic modulus of Gaussian curvature of partially polymerized surfactant membranes. *Langmuir,* 1993, **9**: 2761–2763.
60. Schwarz US & Gompper G. Bending frustration of lipid-water mesophases based on cubic minimal surfaces. *Langmuir,* 2001, **17**: 2084–2096.
61. Helfrich W & Rennschuh H. Landau theory of the lamellar-to-cubic phase transition. *J. Phys. Coll.,* 1990, **51**: C7-189–C7-195.
62. Mitov MD. Third and fourth order curvature elasticity of lipid bilayers. *C. R. Acad. Bulg Sci.,* 1978, **31**: 513–515.
63. Helfrich W. Hats and saddles in lipid membranes. *Liq. Cryst.,* 1989, **5**: 1647–1658.
64. Kozlov MM. Some aspects of membrane elasticity. In: Poon WCK & Andelman D (eds) *Soft condensed matter physics in molecular and cell biology.* New York, CRC Press, 2006, pp. 79–93.
65. Leventis PA & Grinstein S. The distribution and function of phosphatidylserine in cellular membranes. *Annu. Rev. Biophys.,* 2010, **39**: 407–427.
66. van Meer G, Voelker DR & Feigenson GW. Membrane lipids: where they are and how they behave. *Nat. Rev. Mol. Cell Biol.,* 2008, **9**: 112–124.
67. Bevers EM, Comfurius P, van Rijn JL, Hemker HC & Zwaal RF. Generation of prothrombin-converting activity and the exposure of phosphatidylserine at the outer surface of platelets. *Eur. J. Biochem.,* 1982, **122**: 429–436.
68. Fadok VA, et al. Exposure of phosphatidylserine on the surface of apoptotic lymphocytes triggers specific recognition and removal by macrophages. *J. Immunol.,* 1992, **148**: 2207–2216.

69. Macedo-Ribeiro S, et al. Crystal structures of the membrane-binding C2 domain of human coagulation factor V. *Nature*, 1999, **402**: 434–439.
70. Corbalan-Garcia S & Gomez-Fernandez JC. Signaling through C2 domains: more than one lipid target. *Biochim. Biophys. Acta*, 2014, **1838**: 1536–1547.
71. Di Paolo G & De Camilli P. Phosphoinositides in cell regulation and membrane dynamics. *Nature*, 2006, **443**: 651–657.
72. Sutton RB & Sprang SR. Structure of the protein kinase Cbeta phospholipid-binding C2 domain complexed with Ca^{2+}. *Structure*, 1998, **6**: 1395–1405.
73. Bigay J & Antonny B. Curvature, lipid packing, and electrostatics of membrane organelles: defining cellular territories in determining specificity. *Dev. Cell*, 2012, **23**: 886–895.
74. Matsuoka K, et al. COPII-coated vesicle formation reconstituted with purified coat proteins and chemically defined liposomes. *Cell*, 1998, **93**: 263–275.
75. Mim C & Unger VM. Membrane curvature and its generation by BAR proteins. *Trends Biochem. Sci.*, 2012, **37**: 526–533.
76. Farsad K, et al. Generation of high curvature membranes mediated by direct endophilin bilayer interactions. *J. Cell Biol.*, 2001, **155**: 193–200.
77. Dawson JC, Legg JA & Machesky LM. Bar domain proteins: a role in tubulation, scission and actin assembly in clathrin-mediated endocytosis. *Trends Cell Biol.*, 2006, **16**: 493–498.
78. Itoh T & De Camilli P. BAR, F-BAR (EFC) and ENTH/ANTH domains in the regulation of membrane-cytosol interfaces and membrane curvature. *Biochim. Biophys. Acta*, 2006, **1761**: 897–912.
79. Peter BJ, et al. BAR domains as sensors of membrane curvature: the amphiphysin BAR structure. *Science*, 2004, **303**: 495–499.
80. Weissenhorn W. Crystal structure of the endophilin-A1 BAR domain. *J. Mol. Biol.*, 2005, **351**: 653–661.
81. Gallop JL, et al. Mechanism of endophilin N-BAR domain-mediated membrane curvature. *EMBO J.*, 2006, **25**: 2898–2910.
82. Masuda M, et al. Endophilin BAR domain drives membrane curvature by two newly identified structure-based mechanisms. *EMBO J.*, 2006, **25**: 2889–2897.
83. Wang Q, Kaan HY, Hooda RN, Goh SL & Sondermann H. Structure and plasticity of endophilin and sorting nexin 9. *Structure*, 2008, **16**: 1574–1587.
84. Pylypenko O, Lundmark R, Rasmuson E, Carlsson SR & Rak A. The PX-BAR membrane-remodeling unit of sorting nexin 9. *EMBO J.*, 2007, **26**: 4788–4800.
85. Li J, Mao X, Dong LQ, Liu F & Tong L. Crystal structures of the BAR-PH and PTB domains of human APPL1. *Structure*, 2007, **15**: 525–533.
86. Zhu G, et al. Structure of the APPL1 BAR-PH domain and characterization of its interaction with Rab5. *EMBO J.*, 2007, **26**: 3484–3493.
87. Itoh T, et al. Dynamin and the actin cytoskeleton cooperatively regulate plasma membrane invagination by BAR and F-BAR proteins. *Dev. Cell*, 2005, **9**: 791–804.
88. Shimada A, et al. Curved EFC/F-BAR-domain dimers are joined end to end into a filament for membrane invagination in endocytosis. *Cell*, 2007, **129**: 761–772.

89. Millard TH, et al. Structural basis of filopodia formation induced by the IRSp53/MIM homology domain of human IRSp53. *EMBO J.*, 2005, **24**: 240–250.
90. Saarikangas J, et al. Molecular mechanisms of membrane deformation by I-BAR domain proteins. *Curr. Biol.*, 2009, **19**: 95–107.
91. Cui H, et al. Understanding the role of amphipathic helices in N-BAR domain driven membrane remodeling. *Biophys. J.*, 2013, **104**: 404–411.
92. Boucrot E, et al. Membrane fission is promoted by insertion of amphipathic helices and is restricted by crescent BAR domains. *Cell*, 2012, **149**: 124–136.
93. Frost A, et al. Structural basis of membrane invagination by F-BAR domains. *Cell*, 2008, **132**: 807–817.
94. Mim C, et al. Structural basis of membrane bending by the N-BAR protein endophilin. *Cell*, 2012, **149**: 137–145.
95. Faelber K, et al. Structural insights into dynamin-mediated membrane fission. *Structure*, 2012, **20**: 1621–1628.
96. Mears JA, Ray P & Hinshaw JE. A corkscrew model for dynamin constriction. *Structure*, 2007, **15**: 1190–1202.
97. Daumke O, Roux A & Haucke V. BAR domain scaffolds in dynamin-mediated membrane fission. *Cell*, 2014, **156**: 882–892.
98. Morlot S & Roux A, Mechanics of dynamin-mediated membrane fission. *Annu. Rev. Biophys.*, 2013, **42**: 629–649.
99. Henne WM, Buchkovich NJ & Emr SD. The ESCRT pathway. *Dev. Cell*, 2011, **21**: 77–91.
100. Weissenhorn W, Poudevigne E, Effantin G & Bassereau P. How to get out: ssRNA enveloped viruses and membrane fission. *Curr Opin. Virol.*, 2013, **3**: 159–167.
101. Agromayor M & Martin-Serrano J. Knowing when to cut and run: mechanisms that control cytokinetic abscission. *Trends Cell Biol.*, 2013, **23**: 433–441.
102. McCullough J, Colf LA & Sundquist WI. Membrane fission reactions of the mammalian ESCRT pathway. *Annu. Rev. Biochem.*, 2013, **82**: 663–692.
103. Jimenez AJ, et al. ESCRT machinery is required for plasma membrane repair. *Science*, 2014, **343**: 1247136.
104. Wollert T & Hurley JH. Molecular mechanism of multivesicular body biogenesis by ESCRT complexes. *Nature*, 2010, **464**: 864–869.
105. Hurley JH & Hanson PI. Membrane budding and scission by the ESCRT machinery: it's all in the neck. *Nat. Rev. Mol. Cell Biol.*, 2010, **11**: 556–566.
106. Bajorek M, et al. Structural basis for ESCRT-III protein autoinhibition. *Nat. Struct. Mol. Biol.*, 2009, **16**: 754–762.
107. Agromayor M, et al. Essential role of hIST1 in cytokinesis. *Mol. Biol. Cell*, 2009, **20**: 1374–1387.
108. Dimaano C, Jones CB, Hanono A, Curtiss M & Babst M. Ist1 regulates vps4 localization and assembly. *Mol. Biol. Cell*, 2008, **19**: 465–474.
109. Rue SM, Mattei S, Saksena S & Emr SD. Novel Ist1–Did2 complex functions at a late step in multivesicular body sorting. *Mol. Biol. Cell*, 2008, **19**: 475–484.
110. Teis D, Saksena S & Emr SD. Ordered assembly of the ESCRT-III complex on endosomes is required to sequester cargo during MVB formation. *Dev. Cell*, 2008, **15**: 578–589.

111. Muziol T, et al. Structural basis for budding by the ESCRT-III factor CHMP3. *Dev. Cell*, 2006, **10**: 821–830.
112. Xiao J, et al. Structural basis of Ist1 function and Ist1–Did2 interaction in the multivesicular body pathway and cytokinesis. *Mol. Biol. Cell*, 2009, **20**: 3514–3524.
113. Shim S, Kimpler LA & Hanson PI. Structure/function analysis of four core ESCRT-III proteins reveals common regulatory role for extreme C-terminal domain. *Traffic*, 2007, **8**: 1068–1079.
114. Lata S, et al., Structural basis for autoinhibition of ESCRT-III CHMP3. *J. Mol. Biol.*, 2008, **378**: 818–827.
115. Ren X, et al., Hybrid structural model of the complete human ESCRT-0 complex. *Structure*, 2009, **17**: 406–416.
116. Ren X & Hurley JH. VHS domains of ESCRT-0 cooperate in high-avidity binding to polyubiquitinated cargo. *EMBO J.*, 2010, **29**: 1045–1054.
117. Komander D, Clague MJ & Urbe S. Breaking the chains: structure and function of the deubiquitinases. *Nat. Rev. Mol. Cell Biol.*, 2009, **10**: 550–563.
118. Sato Y, et al. Structural basis for specific cleavage of Lys 63-linked polyubiquitin chains. *Nature*, 2008, **455**: 358–362.
119. Kostelansky MS, et al. Molecular architecture and functional model of the complete yeast ESCRT-I heterotetramer. *Cell*, 2007, **129**: 485–498.
120. Gill DJ, et al. Structural insight into the ESCRT-I/-II link and its role in MVB trafficking. *EMBO J.*, 2007, **26**: 600–612.
121. Williams RL & Urbe S. The emerging shape of the ESCRT machinery. *Nat. Rev. Mol. Cell Biol.*, 2007, **8**: 355–368.
122. Hurley JH, et al. Piecing together the ESCRTs. *Biochem. Soc. Trans.*, 2009, **37**: 161–166.
123. Im YJ & Hurley JH. Integrated structural model and membrane targeting mechanism of the human ESCRT-II complex. *Dev. Cell*, 2008, **14**: 902–913.
124. Im YJ, Wollert T, Boura E & Hurley JH. Structure and function of the ESCRT-II-III interface in multivesicular body biogenesis. *Dev. Cell*, 2009, **17**: 234–243.
125. Teis D, Saksena S, Judson BL & Emr SD. ESCRT-II coordinates the assembly of ESCRT-III filaments for cargo sorting and multivesicular body vesicle formation. *EMBO J.*, 2010, **29**: 871–883.
126. Landsberg MJ, Vajjhala PR, Rothnagel R, Munn AL & Hankamer B. Three-dimensional structure of AAA ATPase Vps4: advancing structural insights into the mechanisms of endosomal sorting and enveloped virus budding. *Structure*, 2009, **17**: 427–437.
127. Peel S, Macheboeuf P, Martinelli N & Weissenhorn W. Divergent pathways lead to ESCRT-III catalyzed membrane fission. *Trends Biochem. Sci.*, 2011, **36**: 199–210.
128. Bodon G, et al. Charged multivesicular body protein 2B (CHMP2B) of the endosomal sorting complex required for transport-III (ESCRT-III) polymerizes into helical structures deforming the plasma membrane. *J. Biol. Chem.*, 2011, **286**: 40276–40286.

129. Buchkovich NJ, Henne WM, Tang S & Emr SD. Essential N-terminal insertion motif anchors the ESCRT-III filament during MVB vesicle formation. *Dev. Cell*, 2013, **27**: 201–214.
130. Hanson PI, Roth R, Lin Y & Heuser JE. Plasma membrane deformation by circular arrays of ESCRT-III protein filaments. *J. Cell Biol.*, 2008, **180**: 389–402.
131. Effantin G, et al. ESCRT-III CHMP2A and CHMP3 form variable helical polymers in vitro and act synergistically during HIV-1 budding. *Cell. Microbiol.*, 2013, **15**: 213–226.
132. Lata S, et al. Helical structures of ESCRT-III are disassembled by VPS4. *Science*, 2008, **321**: 1354–1357.
133. Pires R, et al. A crescent-shaped ALIX dimer targets ESCRT-III CHMP4 filaments. *Structure*, 2009, **17**: 843–856.
134. Ghazi-Tabatabai S, et al. Structure and disassembly of filaments formed by the ESCRT-III subunit Vps24. *Structure*, 2008, **16**: 1345–1356.
135. Henne WM, Buchkovich NJ, Zhao Y & Emr SD. The endosomal sorting complex ESCRT-II mediates the assembly and architecture of ESCRT-III helices. *Cell*, 2012, **151**: 356–371.
136. Guizetti J, et al.. Cortical constriction during abscission involves helices of ESCRT-III-dependent filaments. *Science*, 2011, **331**: 1616–1620.
137. Sundquist WI & Krausslich HG. HIV-1 assembly, budding, and maturation. *Cold Spring Harbor Perspect. Med.*, 2012, **2**: a006924.
138. Bieniasz PD. The cell biology of HIV-1 virion genesis. *Cell Host Microbe*, 2009, **5**: 550–558.
139. Martin-Serrano J & Neil SJ. Host factors involved in retroviral budding and release. *Nat. Rev. Microbiol.*, 2011, **9**: 519–531.
140. Usami Y, et al. The ESCRT pathway and HIV-1 budding. *Biochem. Soc. Trans.*, 2009, **37**: 181–184.
141. Hartlieb B & Weissenhorn W. Filovirus assembly and budding. *Virology*, 2006, **344**: 64–70.
142. Morita E, et al. Human ESCRT-III and VPS4 proteins are required for centrosome and spindle maintenance. *Proc. Natl. Acad. Sci. USA*, 2010, **107**: 12889–12894.
143. Morita E, et al. ESCRT-III protein requirements for HIV-1 budding. *Cell Host Microbe*, 2011, **9**: 235–242.
144. Jouvenet N, Zhadina M, Bieniasz PD & Simon SM. Dynamics of ESCRT protein recruitment during retroviral assembly. *Nat. Cell Biol.*, 2011, **13**: 394–401.
145. Carlson L-A & Hurley JH,. In vitro reconstitution of the ordered assembly of the endosomal sorting complex required for transport at membrane-bound HIV-1 Gag clusters. *Proc. Natl. Acad. Sci. USA*, 2012, **109**: 16928–16933.
146. Babst M, Katzmann DJ, Estepa-Sabal EJ, Meerloo T & Emr SD. ESCRT-III: an endosome-associated heterooligomeric protein complex required for MVB sorting. *Dev. Cell*, 2002, **3**: 271–282.
147. Saksena S, Wahlman J, Teis D, Johnson AE & Emr SD. Functional reconstitution of ESCRT-III assembly and disassembly. *Cell*, 2009, **136**: 97–109.

148. Baumgartel V, et al. Live-cell visualization of dynamics of HIV budding site interactions with an ESCRT component. *Nat. Cell Biol.*, 2011, **13**: 469–474.
149. Elia N, Sougrat R, Spurlin TA, Hurley JH & Lippincott-Schwartz J. Dynamics of endosomal sorting complex required for transport (ESCRT) machinery during cytokinesis and its role in abscission. *Proc. Natl. Acad. Sci. USA*, 2011, **108**: 4846–4851.
150. Adell MA, et al. Coordinated binding of Vps4 to ESCRT-III drives membrane neck constriction during MVB vesicle formation. *J. Cell Biol.*, 2014, **205**: 33–49.
151. Wollert T, Wunder C, Lippincott-Schwartz J & Hurley JH. Membrane scission by the ESCRT-III complex. *Nature*, 2009, **458**: 172–177.
152. Babst M, Wendland B, Estepa EJ & Emr SD. The Vps4p AAA ATPase regulates membrane association of a Vps protein complex required for normal endosome function. *EMBO J.*, 1998, **17**: 2982–2993.
153. Lenz M, Crow DJG & Joanny F-F. Membrane buckling induced by curved filaments. *Phys. Rev. Lett.*, 2009, **103**: 038101.
154. Fabrikant G, et al. Computational model of membrane fission catalyzed by ESCRT-III. *PLoS Comput. Biol.*, 2009, **5**: e1000575.
155. Cashikar AG, et al. Structure of cellular ESCRT-III spirals and their relationship to HIV budding. *eLIFE*, 2014, e02184.
156. Boura E, Ivanov V, Carlson L-A, Mizuuchi K & Hurley JH. Endosomal sorting complex required for transport (ESCRT) complexes induce phase-separated microdomains in supported lipid bilayers. *J. Biol. Chem.*, 2012, **287**: 28144–28151.
157. Allain JM, Storm C, Roux A, Ben Amar M & Joanny JF. Fission of a multiphase membrane tube. *Phys. Rev. Lett.*, 2004, **93**: 158104.
158. Boura E, et al. Solution structure of the ESCRT-I and -II supercomplex: implications for membrane budding and scission. *Structure*, 2012, **20**: 874–886.
159. Rozycki B, Boura E, Hurley JH & Hummer G. Membrane-elasticity model of coatless vesicle budding induced by ESCRT complexes. *PLoS Comput. Biol.*, 2012, **8**: e1002736.
160. Elia N, Fabrikant G, Kozlov Michael M & Lippincott-Schwartz J. Computational model of cytokinetic abscission driven by ESCRT-III polymerization and remodeling. *Biophys. J.*, 2012, **102**: 2309–2320.
161. Lafaurie-Janvore J, et al. ESCRT-III assembly and cytokinetic abscission are induced by tension release in the intercellular bridge. *Science*, 2013, **339**: 1625–1629.
162. Van Engelenburg SB, et al. Distribution of ESCRT machinery at HIV assembly sites reveals virus scaffolding of ESCRT subunits. *Science*, 2014, **343**: 653–656.
163. Langelier C, et al. Human ESCRT-II complex and its role in human immunodeficiency virus type 1 release. *J. Virol.*, 2006, **80**: 9465–9480.
164. Pineda-Molina E, et al. The crystal structure of the C-terminal domain of Vps28 reveals a conserved surface required for Vps20 recruitment. *Traffic*, 2006, **7**: 1007–1016.
165. Dordor A, Poudevigne E, Gottlinger H & Weissenhorn W. Essential and supporting host cell factors for HIV-1 budding. *Future Microbiol.*, 2011, **6**: 1159–1170.

166. Evans EA, Fleischer S & Fleischer B. Structure and deformation properties of red blood cells: concepts and quantitative methods. *Methods Enzymol.*, 1989, **173**, 3–35.
167. Chan YHM & Boxer SG. Model membrane systems and their applications. *Curr. Opin. Chem. Biol.*, 2007, **11**: 581–587.
168. Bagatolli L & Kumar PBS. Phase behavior of multicomponent membranes: experimental and computational techniques. *Soft Matter*, 2009, **5**: 3234–3248.
169. Callan-Jones A & Bassereau P. Curvature-driven membrane lipid and protein distribution. *Curr. Opin. Solid State Mater. Sci.*, 2013, **17**: 143–150.
170. Döbereiner H-G, Evans E, Kraus M, Seifert U & Wortis M. Mapping vesicles shapes into the phase diagram: a comparison of experiment and theory. *Phys. Rev. E*, 1997, **55**: 4458–4474.
171. Rodriguez N, Pincet F & Cribier S. Giant vesicles formed by gentle hydration and electroformation: a comparison by fluorescence microscopy. *Colloid Surf. B*, 2005, **42**: 125–130.
172. Hope MJ, Bally MB, Webb G & Cullis PR. Production of large unilamellar vesicles by a rapid extrusion procedure. Characterization of size distribution, trapped volume and ability to maintain a membrane potential. *Biochim. Biophys. Acta–Biomembranes*, 1985, **812**: 55–65.
173. Szoka F & Papahadjopoulos D. Procedure for preparation of liposomes with large internal aqueous space and high capture by reverse-phase evaporation. *Proc. Natl. Acad. Sci. USA*, 1978, **75**: 4194–4198.
174. Bassereau P, Sorre B & Lévy A. Bending lipid membranes: experiments after W. Helfrich's model. *Adv. Colloid Interfac.*, 2014, **208**: 47–57.
175. Reeves JP & Dowben RM. Formation and properties of thin-walled phospholipid vesicles. *J. Cell. Physiol.*, 1969, **73**: 49–60.
176. Angelova MI, Soléau S, Méléard P, Faucon JF & Bothorel P. Preparation of giant vesicles by external AC electric fields. Kinetics and applications. *Prog. Collod Polymer. Sci.*, 1992, **89**: 127–131.
177. Mathivet L, Cribier S & Devaux PF. Shape change and physical properties of giant phospholipid vesicles prepared in the presence of an AC electric field. *Biophys. J.*, 1996, **70**: 1112–1121.
178. Méléard P, Bagatolli LA & Pott T. Giant unilamellar vesicle electroformation: from lipid mixtures to native membranes under physiological conditions. *Methods Enzymol.*, 2009, **465**, 161–176.
179. Needham D & Evans E. Structure and mechanical properties of giant lipid (DMPC) vesicles bilayers from 20°C below to 10°C above the liquid crystal-crystalline phase transition at 24°C. *Biochemistry*, 1988, **27**: 8261–8269.
180. Weinberger A, et al. Gel-assisted formation of giant unilamellar vesicles. *Biophys. J.*, 2013, **105**: 154–164.
181. Korlach J, Schwille P, Webb WW & Feigenson GW. Characterization of lipid bilayer phases by confocal microscopy and fluorescence correlation spectroscopy. *Proc. Natl. Acad. Sci. USA*, 1999, **96**: 8461–8466.
182. Bagatolli LA & Gratton E. Two photon fluorescence microscopy of coexisting lipid domains in giant unilamellar vesicles of binary phospholipid mixtures. *Biophys. J.*, 2000, **78**: 290–305.

183. Veatch SL & Keller SL. Separation of liquid phases in giant vesicles of ternary mixtures of phospholipids and cholesterol. *Biophys. J.*, 2003, **85**: 3074–3083.
184. Baumgart T, Hess ST & Webb WW. Imaging coexisting fluid domains in biomembrane models coupling curvature and line tension. *Nature*, 2003, **425**: 821–824.
185. Feigenson GW. Phase behavior of lipid mixtures. *Nat. Chem. Biol.*, 2006, **2**: 560–563.
186. de Almeida RFM, Loura LMS & Prieto M. Membrane lipid domains and rafts: current applications of fluorescence lifetime spectroscopy and imaging. *Chem. Phys. Lipids*, 2009, **157**: 61–77.
187. Girard P, Pécréaux J, Falson P, Rigaud J-L & Bassereau P. A new method for the reconstitution of large concentrations of transmembrane proteins into giant unilamellar vesicles. *Biophys. J.*, 2004, **87**: 419–429.
188. Doeven MK, et al. Distribution, lateral mobility and function of membrane proteins incorporated into giant unilamellar vesicles. *Biophys. J.*, 2005, **88**: 1134–1142.
189. Kahya N. Protein-protein and protein-lipid interactions in domain-assembly: lessons from giant unilamellar vesicles. *Biochim. Biophys. Acta–Biomembranes*, 2010, **1798**: 1392–1398.
190. Aimon S, et al. Functional reconstitution of a voltage-gated potassium channel in giant unilamellar vesicles. *PLoS ONE*, 2011, **6**: e25529.
191. Dezi M, Di Cicco A, Bassereau P & Lévy D. Detergent-mediated incorporation of transmembrane proteins in giant unilamellar vesicles with controlled physiological contents. *Proc. Natl. Acad. Sci. USA*, 2013, **110**: 7276–7281.
192. Garten M, Aimon S, Bassereau P & Toombes GE. Reconstitution of the transmembrane voltage-gated potassium channel KvAP into giant unilamellar vesicles for fluorescence microscopy and patch clamp studies. *J. Visual. Exp.*, 2015, **95**: e52281.
193. Pautot S, Frisken BJ & Weitz DA. Engineering asymmetric vesicles. *Proc. Natl. Acad. Sci. USA*, 2003, **100**: 10718–10721.
194. Noireaux V, Maeda YT & Libchaber A. Development of an artificial cell, from self-organization to computation and self-reproduction. *Proc. Natl. Acad. Sci. USA*, 2011, **108**: 3473–3480.
195. Pontani L-L, et al. Reconstitution of an actin cortex inside a liposome. *Biophys. J.*, 2009, **96**: 192–198.
196. Abkarian M, Loiseau E & Massiera G. Continuous droplet interface crossing encapsulation (cDICE) for high throughput monodisperse vesicle design. *Soft Matter*, 2011, **7**: 4610–4614.
197. Campillo C, et al. Unexpected membrane dynamics unveiled by membrane nanotube extrusion. *Biophys. J.*, 2013, **104**: 1248–1256.
198. Montes L-R, Alonso A, Goni FM & Bagatolli LA. Giant unilamellar vesicles electroformed from native membranes and organic lipid mixtures under physiological conditions. *Biophys. J.*, 2007, **95**: 3548–3554.
199. Sezgin E, et al. Elucidating membrane structure and protein behavior using giant plasma membrane vesicles. *Nat. Protocols*, 2012, **7**: 1042–1051.

200. Baumgart T, et al. Large-scale fluid/fluid phase separation of proteins and lipids in giant plasma membrane vesicles. *Proc. Natl. Acad. Sci. USA*, 2007, **104**: 3165–3170.
201. Lingwood D, Ries J, Schwille P & Simons K. Plasma membranes are poised for activation of raft phase coalescence at physiological temperature. *Proc. Natl. Acad. Sci. USA*, 2008, **105**: 10005–10010.
202. Schwille P. Bottom-up synthetic biology: engineering in a tinkerer's world. *Science*, 2011, **333**: 1252–1254.
203. Nourian Z, Roelofsen W & Danelon C. Triggered gene expression in fed-vesicle microreactors with a multifunctional membrane. *Angew. Chem. Int. Edit.*, 2012, **124**: 3168–3172.
204. Tamm LK & McConnell HM. Supported phospholipid bilayers. *Biophys. J.*, 1985, **47**: 105–113.
205. Bassereau P & Pincet F. Quantitative analysis of holes in supported bilayers providing the adsorption energy of surfactants on solid substrate. *Langmuir*, 1997, **13**: 7003–7007.
206. Wong JY, Seitz M, Park CK & Israelachvili JN. Polymer-cushioned bilayers. II. An investigation of interaction forces and fusion using the surface forces apparatus. *Biophys. J.*, 1999, **77**: 1458–1468.
207. Charitat T, Bellet-Amalric E, Fragneto G & Graner F. Adsorbed and free lipid bilayers at the solid-liquid interface. *Eur. Phys. J. B*, 1999, **8**: 583–593.
208. Mennicke U & Salditt T. Preparation of solid-supported lipid bilayers by spin-coating. *Langmuir*, 2002, **18**: 8172–8177.
209. Borghi N, Rossier O & Brochard-Wyart F. Hydrodynamic extrusion of tubes from giant vesicles. *Europhys. Lett.*, 2003, **64**: 837–843.
210. Heinrich V & Waugh RE. A piconewton force transducer and its application to measurement of the bending stiffness of phospholipid membranes. *Ann. Biomed. Eng.*, 1996, **24**: 595–605.
211. Koster G, Cacciuto A, Derenyi I, Frenkel D & Dogterom M. Force barriers for membrane tube formation. *Phys. Rev. Lett.*, 2005, **94**: 068101.
212. Cuvelier D, Derényi I, Bassereau P & Nassoy P. Coalescence of membrane tethers: experiments, analysis and applications. *Biophys. J.*, 2005, **88**: 2714–2726.
213. Bo L & Waugh RE. Determination of bilayer membrane bending stiffness by tether formation from giant, thin-walled vesicles. *Biophys. J.*, 1989, **55**: 509–517.
214. Karlsson A, et al. Networks of nanotubes and containers. *Nature*, 2001, **409**: 150–152.
215. Derényi I, Jülicher F & Prost J. Formation and interaction of membrane tubes. *Phys. Rev. Lett.*, 2002, **88**: 238101.
216. Baumgart T, Capraro BR, Zhu C & Das S. Thermodynamics and mechanics of membrane curvature generation and sensing by proteins and lipids. *Annu. Rev. Phys. Chem.*, 2011, **62**: 483–506.
217. Waugh R & Evans EA. Thermoelasticity of red blood-cell membranes. *Biophys. J.*, 1979, **26**: 115–131.
218. Kwok R & Evans E. Thermoelasticity of large lecithin bilayer vesicles. *Biophys. J.*, 1981, **35**: 637–652.

219. Evans E & Rawicz W. Entropy-driven tension and bending elasticity in condensed-fluid membranes. *Phys. Rev. Lett.*, 1990, **64**: 2094–2097.
220. Rawicz W, Olbrich KC, McIntosh T, Needham D & Evans E. Effect of chain length and unsaturation on elasticity of lipid bilayers. *Biophys. J.*, 2000, **79**: 328–339.
221. Rawicz W, Smith BA, McIntosh TJ, Simon SA & Evans E. Elasticity, strength, and water permeability of bilayers that contain raft microdomain-forming lipids. *Biophys. J.*, 2008, **94**: 4725–4736.
222. Helfrich W & Servuss R-M. Undulations, steric interactions and cohesion of fluid membranes. *Nuovo Cim. D*, 1984, **3**: 137–161.
223. Fournier JB, Ajdari A & Peliti L. Effective-area elasticity and tension of micromanipulated membranes. *Phys. Rev. Lett.*, 2001, **86**: 4970–4973.
224. Schneider MB, Jenkins JT & Webb WW. Thermal fluctuations of large quasi-spherical bimolecular phospholipid vesicles. *J. Phys. France*, 1984, **45**: 1457–1472.
225. Bivas I, Hanusse P, Bothorel P, Lalanne J & Aguerre-Chariot O. An application of the optical microscopy to the determination of the curvature elastic modulus of biological and model membranes. *J. Phys. France*, 1987, **48**: 855–867.
226. Faucon JF, Mitov MD, Méléard P, Bivas I & Bothorel P. Bending elasticity and thermal fluctuations of lipid membranes. Theoretical and experimental requirements. *J. Phys. France*, 1989, **50**: 2389–2414.
227. Mutz M & Helfrich W. Bending rigidities of some biological model membranes as obtained from the Fourier-analysis of contour sections. *J. Phys. France*, 1990, **51**: 991–1002.
228. Méléard P, Faucon JF, Mitov MD & Bothorel P. Pulsed-light microscopy applied to the measurement of the bending elasticity of giant liposomes. *Europhys. Lett.*, 1992, **19**: 267–271.
229. Häckl W, Bärmann M & Sackmann E. Shape changes of self-assembled actin bilayer composite membranes. *Phys. Rev. Lett.*, 1998, **80**: 1786–1789.
230. Méléard P, et al. Mechanical properties of model membranes studied from shape transformations of giant vesicles. *Biochimie*, 1998, **80**: 401–413.
231. Döbereiner H-G, Selchow O & Lipowsky R. Spontaneous curvature of fluid vesicles induced by trans-bilayer sugar asymmetry. *Eur. Biophys J.*, 1999, **28**: 174–178.
232. Lee JB, Petrov PG & Döbereiner H-G. Curvature of zwitterionic membranes in transverse pH gradients. *Langmuir*, 1999, **15**: 8543–8546.
233. Döbereiner H-G, et al. Advanced flicker spectroscopy of fluid membranes. *Phys. Rev. Lett.*, 2003, **91**: 048301.
234. Pécréaux J, Döbereiner H-G, Prost J, Joanny J-F & Bassereau P. Refined contour analysis of giant unilamellar vesicles. *Eur. Phys. J. E*, 2004, **13**: 277–290.
235. Méléard P, Pott T, Bouvrais H & Ipsen J. Advantages of statistical analysis of giant vesicle flickering for bending elasticity measurements. *Eur. Phys. J. E*, 2011, **34**: 1–14.
236. Helfer E, et al. Microrheology of biopolymer–membrane complexes. *Phys. Rev. Lett.*, 2000, **85**: 457.

237. Betz T & Sykes C. Time resolved membrane fluctuation spectroscopy. *Soft Matter*, 2012, **8**: 5317–5326.
238. Brown AT, Kotar J & Cicuta P. Active rheology of phospholipid vesicles. *Phys. Rev. E*, 2011, **84**: 021930.
239. Seifert U & Langer S. Viscous mode of fluid bilayer membranes. *Europhys. Lett.*, 1993, **23**: 71–76.
240. Evans E & Yeung A. Hidden dynamics in rapid changes of bilayer shape. *Chem. Phys. Lipids*, 1994, **73**: 39–56.
241. Pott T & Meleard P. The dynamics of vesicle thermal fluctuations is controlled by intermonolayer friction. *Europhys. Lett.*, 2002, **59**: 87–93.
242. Rodriguez-Garcia R, et al. Bimodal spectrum for the curvature fluctuations of bilayer vesicles: pure bending plus hybrid curvature–dilation modes. *Phys. Rev. Lett.*, 2009, **102**: 128101–128104.
243. Bozic B, Svetina S & Zeks B. Theoretical analysis of the formation of membrane microtubes on axially strained vesicles. *Phys. Rev. E*, 1997, **55**: 5834–5842.
244. Svetina S, Zeks B, Waugh RE & Raphael RM. Theoretical analysis of the effect of the transbilayer movement of phospholipid molecules on the dynamic behavior of a microtube pulled out of an aspirated vesicle. *Eur. Biophys J.*, 1998, **27**: 197–209.
245. Powers TR, Huber G & Goldstein RE. Fluid-membrane tethers: minimal surfaces and elastic boundary layers. *Phys. Rev. E*, 2002, **65**: 041901.
246. Rossier O, et al. Giant vesicles under flows: extrusion and retraction of tubes. *Langmuir*, 2003, **19**: 575–584.
247. Cuvelier D, Chiaruttini N, Bassereau P & Nassoy P. Pulling long tubes from firmly adhered vesicles. *Europhys. Lett.*, 2005, **71**: 1015–1021.
248. Tian A & Baumgart T. Sorting of lipids and proteins in membrane curvature gradients. *Biophys. J.*, 2009, **96**: 2676–2688.
249. Sorre B, et al. Curvature-driven lipid sorting needs proximity to a demixing point and is aided by proteins. *Proc. Natl. Acad. Sci. USA*, 2009, **106**: 5622–5626.
250. Sorre B, et al. Nature of curvature-coupling of amphiphysin with membranes depends on its bound density. *Proc. Natl. Acad. Sci. USA*, 2012, **109**: 173–178.
251. Evans E, Ritchie K & Merkel R. Sensitive force technique to probe molecular adhesion and structural linkages at biological interfaces. *Biophys. J.*, 1995, **68**: 2580–2587.
252. Merkel R, Nassoy P, Leung A, Ritchie K & Evans E. Energy landscapes of receptor–ligand bonds explored with dynamic force microscopy. *Nature*, 1999, **397**: 50–53.
253. Evans E & Kinoshita K. Using force to probe single-molecule receptor–cytoskeletal anchoring beneath the surface of a living cell. *Methods Cell Biol.*, 2007, **83**: 373–396.
254. Rakshit S, Zhang Y, Manibog K, Shafraz O & Sivasankar S. Ideal, catch, and slip bonds in cadherin adhesion. *Proc. Natl. Acad. Sci. USA*, 2012, **109**: 18815–18820.
255. Meng F, Suchyna TM & Sachs F. A fluorescence energy transfer-based mechanical stress sensor for specific proteins in situ. *FEBS J.*, 2008, **275**: 3072–3087.

256. Borghi N, et al. E-cadherin is under constitutive actomyosin-generated tension that is increased at cell–cell contacts upon externally applied stretch. *Proc. Natl. Acad. Sci. USA*, 2012, **109**: 12568–12573.
257. Morimatsu M, Mekhdjian AH, Adhikari AS & Dunn AR. Molecular tension sensors report forces generated by single integrin molecules in living cells. *Nano Lett.*, 2013, **13**: 3985–3989.
258. Julicher F & Lipowsky R. Domain-induced budding of vesicles. *Phys. Rev. Lett.*, 1993, **70**: 2964–2967.
259. Yanagisawa M, Imai M & Taniguchi T. Shape deformation of ternary vesicles coupled with phase separation. *Phys. Rev. Lett.*, 2008, **100**: 148102.
260. Semrau S, Idema T, Schmidt T & Storm C. Membrane-mediated interactions measured using membrane domains. *Biophys. J.*, 2009, **96**: 4906–4915.
261. Ursell TS, Klug WS & Phillips R. Morphology and interaction between lipid domains. *Proc. Natl. Acad. Sci. USA*, 2009, **106**: 13301–13306.
262. Semrau S & Schmidt T. Membrane heterogeneity—from lipid domains to curvature effects. *Soft Matter*, 2009, **5**: 3174–3186.
263. Sens P & Turner MS. Budded membrane microdomains as tension regulators. *Phys. Rev. E*, 2006, **73**: 031918.
264. Sens P, Johannes L & Bassereau P. Biophysical approaches to protein-induced membrane deformations in trafficking. *Curr. Opin. Cell Biol.*, 2008, **20**: 476–482.
265. Roux A. Tubes de membrane dans le trafic intracellulaire : aspects physiques et biologiques. PhD thesis, Université Paris VII, Paris, 2004.
266. Allain J-M, Storm C, Roux A, Ben Amar M & Joanny JF. Fission of a multiphase membrane tube. *Phys. Rev. Lett.*, 2004, **93**: 158104.
267. Tian A, Johnson C, Wang W & Baumgart T. Line tension at fluid membrane domain boundaries measured by micropipette aspiration. *Phys. Rev. Lett.*, 2007, **98**: 208102–208104.
268. Garcia-Saez AJ, Chiantia S & Schwille P. Effect of line tension on the lateral organization of lipid membranes. *J. Biol. Chem.*, 2007, **282**: 33537–33544.
269. Honerkamp-Smith AR, et al. Line tensions, correlation lengths, and critical exponents in lipid membranes near critical points. *Biophys. J.*, 2008, **95**: 236–246.
270. Lipowsky R. Domain-induced budding of vesicles. *Biophys. J.*, 1993, **64**: 1133–1138.
271. Roux A, et al. Role of curvature and phase transition in lipid sorting and fission of membrane tubules. *EMBO J.*, 2005, **24**: 1537–1545.
272. Allain J & Ben Amar M. Budding and fission of a multiphase vesicle. *Eur. Phys. J. E*, 2006, **20**: 409–420.
273. Römer W, et al. Actin dynamics drive membrane reorganization and scission in clathrin independent endocytosis. *Cell*, 2010, **140**: 540–553.
274. Römer W, et al. Shiga toxin induces tubular membrane invaginations for its uptake into cells. *Nature*, 2007, **450**: 670–675.
275. Safouane M, et al. Lipid co-sorting mediated by Shiga toxin induced tubulation. *Traffic*, 2010, **11**: 1519–1529.
276. Liu AP & Fletcher DA. Actin polymerization serves as a membrane domain switch in model lipid bilayers. *Biophys. J.*, 2006, **91**: 4064–4070.

277. Weinberg J & Drubin DG. Clathrin-mediated endocytosis in budding yeast. *Trends Cell Biol.*, 2012, **22**: 1–13.
278. Leibler S. Curvature instability in membranes. *J. Phys. France*, 1986, **47**: 507–516.
279. Callan-Jones A, Sorre B & Bassereau P. Curvature-driven lipid sorting in biomembranes. *Cold Spring Harbor Perspect. Biol.*, 2011, **3**: a004648.
280. McMahon HT & Gallop JL. Membrane curvature and mechanisms of dynamic cell membrane remodelling. *Nature*, 2005, **438**: 590–596.
281. Shibata Y, Hu J, Kozlov MM & Rapoport TA. Mechanisms shaping the membranes of cellular organelles. *Annu. Rev. Cell Dev. Biol.*, 2009, **25**: 329–354.
282. Kirchhausen T. Bending membranes. *Nat. Cell Biol.*, 2012, **14**: 906–908.
283. Stachowiak JC, Brodsky FM & Miller EA. A cost-benefit analysis of the physical mechanisms of membrane curvature. *Nat. Cell Biol.*, 2013, **15**: 1019–1027.
284. Johannes L, Wunder C & Bassereau P. Bending "on the rocks"—a cocktail of biophysical modules to build endocytic pathways. *Cold Spring Harbor Perspect. Biol.*, 2014, **6**: a016741.
285. Campelo F, McMahon HT & Kozlov MM. The hydrophobic insertion mechanism of membrane curvature generation by proteins. *Biophys. J.*, 2008, **95**: 2325–2339.
286. Ford MGJ, et al. Curvature of clathrin-coated pits driven by epsin. *Nature*, 2002, **419**: 361–366.
287. Boucrot E, et al. Membrane fission is promoted by shallow hydrophobic insertions while limited by crescent BAR domains. *Cell*, 2012, **149**: 124–136.
288. Capraro BR, Yoon Y, Cho W & Baumgart T. Curvature sensing by the epsin N-terminal homology domain measured on cylindrical lipid membrane tethers. *J. Am. Chem. Soc.*, 2010, **132**: 1200–1201.
289. Mattila PK, et al. Missing-in-metastasis and IRSp53 deform PI(4,5)P2-rich membranes by an inverse BAR domain-like mechanism. *J. Cell Biol.*, 2007, **176**: 953–964.
290. Zhu C, Das S & Baumgart T. Nonlinear sorting, curvature generation, and crowding of endophilin N-BAR on tubular membranes. *Biophys. J.*, 2012, **102**: 1837–1845.
291. Ramesh P, et al. FBAR Syndapin 1 recognizes and stabilizes highly curved tubular membranes in a concentration dependent manner. *Sci. Rep.*, 2013, **3**: 1565.
292. Ferguson S, et al. Coordinated actions of actin and BAR proteins upstream of dynamin at endocytic clathrin-coated pits. *Dev. Cell*, 2009, **17**: 811–822.
293. Taylor MJ, Perrais D & Merrifield CJ. A high precision survey of the molecular dynamics of mammalian clathrin mediated endocytosis. *PLoS Biol.*, 2011, **9**: e1000604.
294. Mim C, et al. Structural basis of membrane bending by the N-BAR protein endophilin. *Cell*, 2012, **149**: 137–145.
295. Simunovic M, et al. Protein-mediated transformation of lipid vesicles into tubular networks. *Biophys. J.*, 2013, **105**: 711–719.
296. Prevost C, et al. IRSp53 senses negative membrane curvature and phase separates along membrane tubules. *Nat. Commun.*, 2015, **6**: 8529.

297. Daumke O, Roux A & Haucke V. BAR domain scaffolds in dynamin-mediated membrane fission. *Cell*, 2014, **156**: 882–892.
298. Parton RG & del Pozo MA. Caveolae as plasma membrane sensors, protectors and organizers. *Nat. Rev. Mol. Cell Biol.*, 2013, **14**: 98–112.
299. Hu J, Prinz WA & Rapoport TA. Weaving the web of ER tubules. *Cell*, 2011, **147**: 1226–1231.
300. Mellman I & Nelson WJ. Coordinated protein sorting, targeting and distribution in polarized cells. *Nat. Rev. Mol. Cell Biol.*, 2008, **9**: 833–845.
301. Killian JA. Hydrophobic mismatch between proteins and lipids in membranes. *Biochim. Biophys. Acta–Rev. Biomemb.*, 1998, **1376**: 401–416.
302. Dumas F, Lebrun MC & Tocanne J-F. Is the protein/lipid hydrophobic matching principle relevant to membrane organization and functions? *FEBS Lett.*, 1999, **458**: 271–277.
303. Aimon S, et al. Membrane shape modulates trans-membrane protein distribution. *Dev. Cell*, 2014, **28**: 212–218.
304. Lee S-Y, Lee A, Chen J & MacKinnon R. Structure of the KvAP voltage-dependent K+ channel and its dependence on the lipid membrane. *Proc. Natl. Acad. Sci. USA*, 2005, **102**: 15441–15446.
305. Qiu H, Ma S, Shen R & Guo W. Dynamic and energetic mechanisms for the distinct permeation rate in AQP1 and AQP0. *Biochim. Biophys. Acta–Biomembranes*, 2010, **1798**: 318–326.
306. Stachowiak JC, et al. Membrane bending by protein-protein crowding. *Nat. Cell Biol.*, 2012, **14**: 944–949.
307. Milner ST & Witten TA. Bending moduli of polymeric surfactant interfaces. *J. Phys. France*, 1988, **49**: 1951–1962.
308. Marsh D. Elastic constants of polymer-grafted lipid membranes. *Biophys. J.*, 2001, **81**: 2154–2162.
309. Bickel T & Marques CM. Local entropic effects of polymers grafted to soft interfaces. *Eur. Phys. J. E*, 2001, **4**: 33–43.
310. Breidenich M, Netz RR & Lipowsky R. The shape of polymer-decorated membrane. *Europhys. Lett.*, 2000, **49**: 431–437.
311. Nikolov V, Lipowsky R & Dimova R. Behavior of giant vesicles with anchored DNA molecules. *Biophys. J.*, 2007, **92**: 4356–4368.
312. Anitei M & Hoflack B. Bridging membrane and cytoskeleton dynamics in the secretory and endocytic pathways. *Nat. Cell Biol.*, 2012, **14**: 11–19.
313. Leduc C, Campas O, Joanny JF, Prost J & Bassereau P. Mechanism of membrane nanotube formation by molecular motors. *Biochim. Biophys. Acta–Biomembranes*, 2010, **1798**: 1418–1426.
314. Lippincott-Schwartz J. Cytoskeletal proteins and Golgi dynamics. *Curr. Opin. Cell Biol.*, 1998, **10**: 52–59.
315. Lane J & Allan V. Microtubule-based membrane movement. *Biochim. Biophys. Acta–Rev. Biomemb.*, 1998, **1376**: 27–55.
316. Presley JF, et al. ER-to-Golgi transport visualized in living cells. *Nature*, 1997, **389**: 81–85.

317. Toomre D, Keller P, White J, Olivo JC & Simons K. Dual-color visualization of trans-Golgi network to plasma membrane traffic along microtubules in living cells. *J. Cell Sci.*, 1999, **112**: 21–33.
318. Polishchuk RS, et al. Correlative light-electron microscopy reveals the tubular-saccular ultrastructure of carriers operating between Golgi apparatus and plasma membrane. *J. Cell Biol.*, 2000, **148**: 45–58.
319. Luini A, Ragnini-Wilson A, Polishchuck RS & Matteis MAD. Large pleiomorphic traffic intermediates in the secretory pathway. *Curr. Opin. Cell Biol.*, 2005, **17**: 353–361.
320. van Weering JRT, Verkade P & Cullen PJ. SNX–BAR proteins in phosphoinositide-mediated, tubular-based endosomal sorting. *Semin. Cell Dev. Biol.*, 2010, **21**: 371–380.
321. Sciaky N, et al. Golgi tubule traffic and the effects of brefeldin A visualized in living cells. *J. Cell Biol.*, 1997, **139**: 1137–1155.
322. White J, et al. Rab6 coordinates a novel Golgi to ER retrograde transport pathway in live cells. *J. Cell Biol.*, 1999, **147**: 743–760.
323. Vale RD & Hotani H. Formation of membrane networks in vitro by kinesin-driven microtubule movement. *J. Cell Biol.*, 1988, **107**: 2233–2241.
324. Dabora SL & Sheetz MP. The microtubule-dependent formation of a tubulovesicular network with characteristics of the ER from cultured cell extracts. *Cell*, 1988, **54**: 27–35.
325. Allan VJ & Vale RD. Movement of membrane tubules along microtubules in vitro : evidence for specialised sites of motor attachment. *J. Cell Sci.*, 1994, **107**: 1885–1897.
326. Roux A, et al. A minimal system allowing tubulation using molecular motors pulling on giant liposomes. *Proc. Natl. Acad. Sci. USA*, 2002, **99**: 5394–5399.
327. Koster G, VanDuijn M, Hofs B & Dogterom M. Membrane tube formation from giant vesicles by dynamic association of motor proteins. *Proc. Natl. Acad. Sci. USA*, 2003, **100**: 15583–15588.
328. Leduc C, et al. Cooperative extraction of membrane nanotubes by molecular motors. *Proc. Natl. Acad. Sci. USA*, 2004, **101**: 17096–17101.
329. Block SM, Asbury CL, Shaevitz JW & Lang MJ. Probing the kinesin reaction cycle with a 2D optical force clamp. *Proc. Natl. Acad. Sci. USA*, 2003, **100**: 2351–2356.
330. Campas O, Kafri Y, Zeldovich KB, Casademunt J & Joanny JF. Collective dynamics of interacting molecular motors. *Phys. Rev. Lett.*, 2006, **97**: 038101.
331. Campas O, et al. Coordination of kinesin motors pulling on fluid membranes. *Biophys. J.*, 2008, **94**: 5009–5017.
332. Campas O, Leduc C, Bassereau P, Joanny JF & Prost J. Collective oscillations of processive molecular motors. *Biophys. Rev. Lett.*, 2009, **4**: 163–178.
333. Salas-Cortes L, et al. Myosin Ib modulates the morphology and the protein transport within multi-vesicular sorting endosomes. *J. Cell Sci.*, 2005, **118**: 4823–4832.

334. Almeida CG, et al. Myosin 1b promotes the formation of post-Golgi carriers by regulating actin assembly and membrane remodelling at the trans-Golgi network. *Nat. Cell Biol.*, 2011, **13**: 779–789.
335. Leduc C, et al. Cooperative extraction of membrane nanotubes by molecular motors. *Proc. Natl. Acad. Sci. USA*, 2004, **101**: 17096–17101.
336. Shaklee PM, et al. Bidirectional membrane tube dynamics driven by nonprocessive motors. *Proc. Natl. Acad. Sci. USA*, 2008, **105**: 7993–7997.
337. Shaklee PM, Bourel-Bonnet L, Dogterom M & Schmidt T. Nonprocessive motor dynamics at the microtubule membrane tube interface. *Biophys. J.*, 2010, **98**: 93–100.
338. Yamada A, et al. Catch-bond behaviour facilitates membrane tubulation by nonprocessive myosin 1b. *Nat. Commun.*, 2014, **5**: 3624.
339. Waterman-Storer CM & Salmon ED. Endoplasmic reticulum membrane tubules are distributed by microtubules in living cells using three distinct mechanisms. *Curr. Biol.*, 1998, **8**: 798–807.
340. Mattila PK & Lappalainen P. Filopodia: molecular architecture and cellular functions. *Nat. Rev. Mol. Cell Biol.*, 2008, **9**: 446–454.
341. Ridley AJ. Life at the leading edge. *Cell*, 2011, **145**: 1012–1022.
342. Bornschlögl T. How filopodia pull: what we know about the mechanics and dynamics of filopodia. *Cytoskeleton*, 2013, **70**: 590–603.
343. Mogilner A & Rubinstein B. The physics of filopodial protrusion. *Biophys. J.*, 2005, **89**: 782–795.
344. Liu AP, et al. Membrane-induced bundling of actin filaments. *Nature Phys.*, 2008, **4**: 789–793.
345. Nambiar R, McConnell RE & Tyska MJ. Myosin motor function: the ins and outs of actin-based membrane protrusions. *Cell. Mol. Life Sci.*, 2010, **67**: 1239–1254.
346. Bornschlögl T, et al. Filopodia retraction force is generated by cortical actin dynamics and controlled by reversible tethering at the tip. *Proc. Natl. Acad. Sci. USA*, 2013, **110**: 18928–18933.
347. Mellor H. The role of formins in filopodia formation. *Biochim. Biophys. Acta–Mol. Cell Res.*, 2010, **1803**: 191–200.

Part 5

Conformational changes and their implications in diseases

11
Protein conformational changes

Yves GAUDIN

Institute for Integrative Biology of the Cell (I2BC), Université Paris-Saclay, CEA, CNRS, Université Paris-Sud, Gif-sur-Yvette cedex, France

Abstract

Proteins are major components of living cells and perform a vast array of tasks. To perform their functions most proteins undergo conformational changes that can be either local or global, involving secondary structural changes and/or domain repositioning. These structural transitions need to be finely regulated in both time and space. This chapter first presents general features about protein conformational changes. The case of conformational changes in fusion glycoproteins of enveloped viruses will then be described in detail as an impressive example of huge structural transitions. As a second example, protein misfolding leading to amyloidosis and prion diseases will be also presented. Recent results indicating that functional amyloids exist and may act in support of information or cell memory will be also discussed.

Keywords

Protein, structure, conformational changes, allostery, intrinsically disordered proteins, membrane fusion, viral glycoprotein, amyloidosis, prion

From Molecules to Living Organisms: An Interplay Between Biology and Physics. First Edition.
Eva Pebay-Peyroula et al. © Oxford University Press 2016.
Published in 2016 by Oxford University Press.

Chapter Contents

11 Protein conformational changes   353
Yves GAUDIN

- 11.1 Protein conformational changes   355
  - 11.1.1 Introduction   355
  - 11.1.2 Overview of conformational changes   355
  - 11.1.3 How to investigate protein conformational changes   358
  - 11.1.4 Triggering conformational changes   360
  - 11.1.5 Allosteric transitions   361
  - 11.1.6 The case of intrinsically disordered proteins   362
  - 11.1.7 Final remarks   364
- 11.2 Conformational changes in viral fusion proteins   364
  - 11.2.1 Enveloped viruses enter their host cell by membrane fusion   364
  - 11.2.2 Several classes of viral fusion machineries   365
  - 11.2.3 pH-sensitive molecular switches   374
  - 11.2.4 Intermediates during the structural transition   375
  - 11.2.5 Final remarks   376
- 11.3 From conformational diseases to cell memory   377
  - 11.3.1 Introduction   377
  - 11.3.2 Prion proteins   377
  - 11.3.3 Amyloidoses   382
  - 11.3.4 Functional amyloids   384
  - 11.3.5 Concluding remarks   386

References   388

11.1 Protein conformational changes

11.1.1 Introduction

Proteins are major components of living cells and perform a vast array of tasks. It is clear that to perform their functions most proteins cannot be entirely rigid molecules. This was acknowledged very early on, almost from the beginning of structural biology. Indeed, a comparison between the crystalline structure of horse oxyhemoglobin and the oxygen-free form of human hemoglobin (often referred to as reduced hemoglobin) reveals an important structural rearrangement. In 1962, in his speech at the Nobel banquet, Max Perutz mentioned that "the discovery of a marked structural change accompanying the reaction of hemoglobin with oxygens suggests that there may be other enzymes which alter their structure on combination with their substrate and that this may perhaps be an important factor in certain mechanisms of enzymatic catalysis" [1].

In 1963, Jacques Monod, Jean-Pierre Changeux and François Jacob wrote in an article published in the *Journal of Molecular Biology* that the action of an allosteric effector appears to result exclusively from a conformational alteration (allosteric transition) induced in the protein when it binds the agent [2]. However, Christian Anfinsen, following his pioneering 1961 work on the folding of ribonuclease A [3], postulated that the native structure of a protein is determined by the protein's amino acid sequence [4]. This meant that, under given environmental conditions, the native structure is unique and stable and, therefore, corresponds to a kinetically accessible minimum of free energy. The emphasis put on the uniqueness of the protein structure in Anfinsen's dogma (which was enunciated for small globular proteins), together with the first crystal structures that revealed the compactness and global rigidity of proteins, suggested that conformational changes in proteins were rather limited. However, it is now clear that the magnitude of conformational changes in proteins is extremely variable and extends from local movement (of amino acid side chains) to huge transitions leading to modifications of the secondary structure and domain repositioning (Table 11.1). These conformational changes can be either reversible or irreversible (and, in this case, the initial conformation is metastable).

11.1.2 Overview of conformational changes

11.1.2.1 Local conformational changes

In solution, proteins undergo structural variation through thermal vibration and collisions with other molecules. The amplitude of the related movements is in the sub-angstrom range and will largely depend on the location of the atom being considered in the structure. An atom buried inside a protein moves much less than one exposed to the solvent, particularly those located in flexible loops.

In general, in a structure determined by X-ray crystallography, the absence of a strong electron density is a good indicator of conformational flexibility. An indication of the mobility of individual atoms is the B-factor, which is given for each atom of a crystal structure deposited in the Protein Data Bank (Fig. 11.1). The B-factor is defined as $B = 8\pi^2 <u^2>$ where $<u^2>$ is the mean squared displacement. However,

Table 11.1 Classification of protein movements

Movement scales	Amplitude	0.01–100 Å
	Associated energy	0.1–100 kcal/mol
	Time	10^{-12}–10^3 s
Movement types	Local	Fluctuations of atom positions, oscillations of side chains
	Collective	Coupled atomic fluctuations, protein elasticity
	Fast collective moving	Movement of helices, domains or protomers
	Large amplitude	Folding, unfolding or refolding of domains or complete proteins

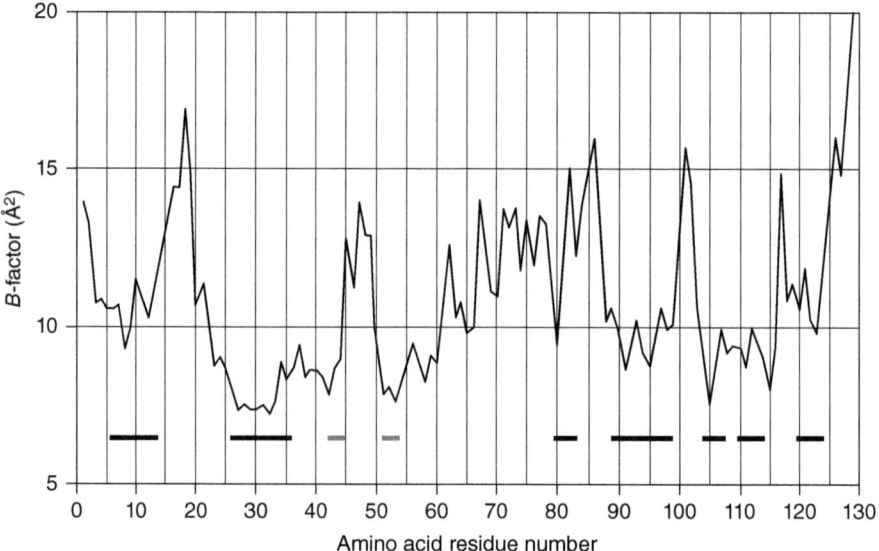

Fig. 11.1 *B*-factors are good indicators of protein flexibility. The figure shows the *B*-factor of the atoms of the main chain residues of hen egg white lysozyme. For each residue the *B*-factors of carbon Cα, carbon C′, N and O have been averaged. The position of α-helices (respectively β-strands) is indicated by black (respectively grey) bars. Note that *B*-factors are in general more elevated in loops and regions with no canonical secondary structure.

a direct interpretation of the *B*-factor in terms of atomic mobility has to be considered cautiously (particularly for a low-resolution structure) as other factors, including experimental error, influence the distribution of *B*-factors within a given structure. This explains the often observed lack of correlation between experimental *B*-factors and calculated mean square displacements using molecular dynamics simulations.

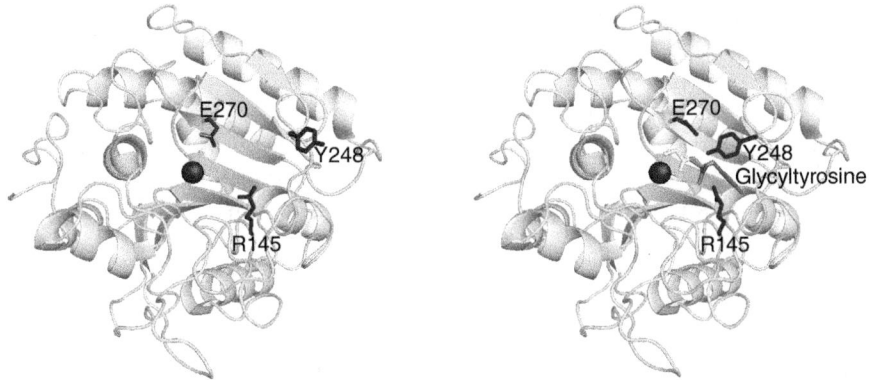

Fig. 11.2 Induced fit of carboxypeptidase A upon formation of its complex with glycyltyrosine (left, free enzyme; right, complexed enzyme). Upon binding to the substrate, the guanidium group of arginine 145 (R145) moves 2 Å to interact with the carboxyl group of the glycyltyrosine dipeptide. The carboxylate group of glutamate 270 (E270) performs a similar movement. The binding of the zinc ion (represented as a dark sphere) to the carbonyl of the substrate displaces a water molecule. At least four other water molecules are excluded from the pocket when the apolar side chain of the tyrosine substrate engages therein. The largest movement is made by the hydroxyl group of tyrosine 248 (Y248) which moves over a distance of 12 Å. This movement is due to a simple rotation around a C–C bond. R145, Y248 and D270 are represented by dark sticks.

Although local and of small amplitude, these movements can have very important functions. A 1-Å movement of a residue can be sufficient to close or open a membrane channel. Such a movement may allow a small ligand to find a pathway to its binding site. Furthermore, a local reorganization of a residue side chain can be key to the catalytic activity of an enzyme.

As a well-known example, an oxygen molecule is unable to find its way to the heme pocket in a rigid myoglobin molecule in a reasonable time scale. However, rotation of the side chain of several residues located in the vicinity of the heme allows the ligand to enter or leave the myoglobin [5].

Finally, it is noteworthy that a local movement (such as a simple rotation around a C–C bond) may have huge structural implications, as seen with the extended movement of the side chain of a tyrosine in carboxypeptidase A upon binding to its substrate (Fig. 11.2).

11.1.2.2 Global conformational changes

In a protein domain, several residues are close neighbors. These may be successive residues of the primary structure and they may also interact through hydrogen bonds, salt bridges or van der Waals interactions. As a consequence, such coupled residues will have the tendency to move collectively. This may result in the repositioning of a loop or an α-helix or in an increase curvature of a β-sheet.

Protein flexibility is obviously increased for multidomain proteins. In general, during conformational change the domains keep their tertiary structures. The repositioning of domains is often mediated by conformational changes in the hinge regions. These hinge regions are extremely diverse. During transition they may adopt another secondary structure (e.g., loop to helix transition, separation of two strands in a small β-sheet, etc.). Some α-helices may also be trapped in a bent conformation, storing elastic energy and acting as a spring during the conformational change [6].

It is also important to mention shear motions. During such a movement, two domains slide on each other at their interfaces. This movement is often controlled by a small number of amino acids located at the interface (e.g., by an aliphatic residue which jumps from one hydrophobic pocket to another).

As an illustration, several examples of huge conformational changes will be studied in detail in Section 11.2.2.4 on viral fusion glycoproteins.

11.1.3 How to investigate protein conformational changes

Virtually, any signal which varies as a function of the structure of the protein can be used to follow the conformational change. Global conformational changes are the easiest to identify and characterize and several techniques can be used.

If the conformational transition of the protein results in an important change to its shape, the hydrodynamic properties of its different states are in general quite different. It is thus expected that the sedimentation coefficient (s) and the hydrodynamic radius (r_0) (both related by the Svedberg equation, $s = m/6\pi\eta r_0$, in which m is the mass of the protein and η the viscosity of the medium) vary upon the transition. The sedimentation coefficient can be easily measured by sedimentation velocity experiments in an analytical ultracentrifuge. The hydrodynamic radius can be determined on a gel filtration column or by dynamic light scattering. Gel filtration is probably the technique of choice as it is not sensitive to supramolecular aggregates and as it is possible to work on unpurified proteins provided they can be detected at the column outlet.

Small angle X-ray scattering (SAXS) can also be used to demonstrate the existence of several conformations of a protein [7,8]. Furthermore, SAXS can reveal structural information such as a low-resolution envelope of a given conformation [9]. Finally, the SAXS profile can be used to describe the conformational ensemble in a population-weighted manner [8].

Electron microscopy of negatively stained samples can also be used to characterize protein conformational changes. However, the high local salt concentration together with the fact that the sample is dried may alter protein structure, and images have to be cautiously interpreted. For this reason, cryo-electron microscopy (cryo-EM), which preserves the sample in amorphous ice, is being increasingly used. This technique is less versatile than negative staining and the lack of contrast renders it difficult to observe small proteins directly. However, cryo-EM technology is improving steadily and its combination with high-resolution structures (obtained by X-ray crystallography and/or NMR) is extremely promising.

If the conformational change is accompanied by secondary structural modifications circular dichroism can be used, particularly when the α-helix content varies significantly during the transition. When the β-sheet content is significantly modified instead, infrared spectroscopy is the technique of choice.

Local conformational changes can be investigated using fluorescence spectroscopy, provided that the transition has an influence on a tryptophan environment or a grafted fluorophore. These spectroscopic techniques can also be used to demonstrate the existence of more than two conformations. This can be done by monitoring the kinetics of the conformational change, because the existence of only two states implies that the experimental curve can be fitted by a single exponential. If the conformational change is induced upon ligand binding, titration experiments may also reveal the existence of intermediate species. Finally, if several spectra of a protein intersect not at one or more isosbestic points but over a progressively changing wavelength, this is evidence for the involvement of at least a third species in the equilibrium (Fig. 11.3).

The structural biology techniques that allow the determination of the atomic structure of proteins can also be used to investigate changes in protein conformation. NMR is particularly valuable as it gives access to protein dynamics. The distinct conformation of a protein can be trapped in particular crystal organization, and this has been exemplified by the work performed on hemoglobin and will be further illustrated in Section 11.2.2.4. Therefore, X-ray crystallography, although giving a static picture of the protein structure, is also a tool for characterizing conformational changes in proteins.

Finally, domain motions can also be analyzed by molecular dynamics, principal component analysis and normal mode analysis. Of course, these *in silico* approaches have to be complemented by direct experiments in order to give the best results.

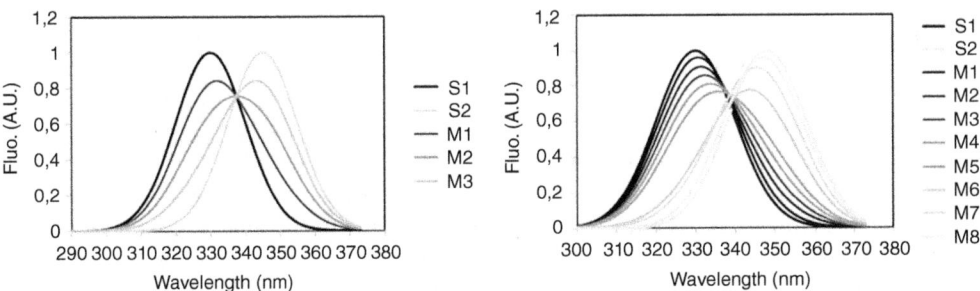

Fig. 11.3 Fluorescent spectra of a protein for which ligand binding induces a change from the S1 conformation (ligand-free form, fluorescent spectrum in black) toward the S2 conformation (ligand-bound form, fluorescent spectrum in light grey). The spectra named M are obtained upon addition of an increasing concentration of ligands. The existence of an isosbestic point on the left plot strongly suggests (but does not prove) that only two conformations of the protein exist. The absence of an isosbestic point on the right plot indicates that at least a third of the conformation is in solution.

11.1.4 Triggering conformational changes

As with any biological processes, conformational changes have to be finely regulated in both time and space. The conformation of a protein can be sensitive to its environment, such as the concentration of specific ions (e.g., Ca^{2+}). Protons, as ions, deserve particular attention. First, the pH varies from one cellular compartment to another (e.g., the endocytic pathway and the Golgi apparatus contain acidic compartments). Second, acidic residues (aspartic or glutamic acids) and histidine have a pK_a that can be finely tuned depending on their environment (either buried or exposed to the solvent, either isolated or in clusters) which makes them ideal pH-sensitive switches. Some well characterized examples will be described in Section 11.2.2.4.

The structure of a protein can also change upon ligand binding. Two mechanisms have been proposed to explain this process, namely conformational selection and induced fit. Conformational selection involves the binding of a ligand to a pre-existing rare conformation of the free protein. In an induced fit the ligand binds to the free protein which, only after binding, undergoes a conformational change to its ligand-bound conformation. Note that these mechanisms are not mutually exclusive as conformational selection by a ligand may be followed by an induced fit mechanism.

Post-translational modifications can also have an influence on the structure of a protein and can thus trigger conformational changes. An obvious example is proteolytic cleavage which is always accompanied by at least a local rearrangement of the polypeptidic chain. In this case, the associated structural transition is irreversible. Furthermore, proteolytic cleavage may also result in a metastable structure. The protein is then primed for a subsequent conformational change when it encounters an adequate trigger.

Another example of post-translational modification is covalent modification, which also constitutes an exquisite solution to the problem of triggering protein conformational change. The covalent ligands are extremely diverse: lipids, sugars, amino acids, other proteins (such as ubiquitin or the small ubiquitin-like modifier), etc.

The case of phosphorylation deserves a mention as it is certainly one of the most commonly used methods of covalent modification in eukaryotes. Phosphorylation of intracellular proteins by protein kinases serves as a molecular switch for various cellular processes including metabolism, gene expression, cell division, motility and differentiation. The replacement of neutral hydroxyl groups on serine, threonine or tyrosine with negatively charged phosphates with pK_a around 1.5 and 6.5 results in the addition of approximately two negative charges under physiological pH conditions ($\sim$7.5). This local increase in charge has, of course, an effect on the protein conformation. Importantly, phosphorylation is reversible and the protein can recover its initial state after dephosphorylation.

Finally, in situations where the conformational change of a protomer is accompanied by a change in the oligomeric status of the protein, a change in the local concentration of proteins can also trigger a structural transition.

11.1.5 Allosteric transitions

Allostery is used to qualify the mechanism of regulation of enzyme activity by regulatory ligands which bind the enzyme at an effector site that is distinct from the substrate site. The effector site can be many angstroms away from the catalytic site.

Two phenomenological models were initially proposed to explain protein allostery. The first is the concerted model of Monod, Wyman and Changeux (MWC) [10], the second is the sequential model of Koshland, Nemethy and Filmer (KNF) [11]. These models share common principles. First, both models propose that the allosteric protein can exist under two different conformations having different affinities for the ligand (a low-affinity and a high-affinity state). Second, in both models the structural transition is favored upon ligand binding. However, in the concerted model, the low-affinity (tensed, T) and the high-affinity (relaxed, R) states pre-exist in solution, whereas the sequential model postulates induced fit. As a consequence, the species that are encountered in solution are not the same (Fig. 11.4).

It is worth noting that in its original formulation the concerted model required symmetric oligomeric proteins. This was a little restrictive as, clearly, allosteric effects can be observed in non-oligomeric proteins. A well-characterized example is the Hsp70 proteins for which there is allosteric coupling between ATP binding in one domain and substrate binding in another [12–14]. Another example is the two-state

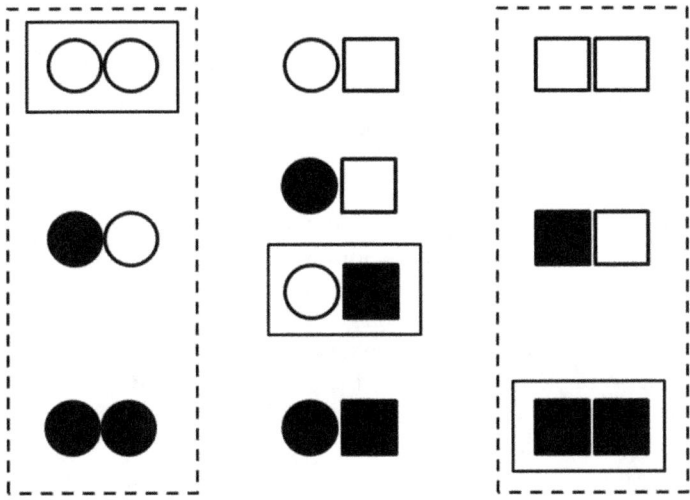

Fig. 11.4 Schematic models of protein allostery. The figure depicts all the possible states of a dimeric allosteric protein. Each subunit of the protein can adopt two distinct conformations, tensed (squares) or relaxed (circles), and two distinct modes of ligand binding, unbound (white shapes) or bound (black shapes). Dashed rectangles indicate the states allowed under the postulates of the MWC model. Rectangles in continuous lines indicate the states allowed under the postulates of the KNF model.

allosteric behavior of some single-domain signaling proteins, for which the equilibrium toward the active species is shifted by phosphorylation [15].

The crystal structure of hemoglobin revealed a symmetric tetrameric organization in agreement with the predictions of the concerted MWC model for allostery. The existence of two distinct structures having different affinities for O_2 was also revealed (reviewed in [16]). Finally, it was found that a network made of a small number of salt bridges mediates the stability of the two different conformations and their affinity for the ligand [17].

Such networks of non-covalent bonds have been observed for several other allosteric proteins [18–20]. They form a pathway between residues to connect the effector site to the binding site and, being formed and broken sequentially, they transmit information between both sites. The direct implication of those allosteric networks has been established through point mutation experiments.

However, there are an increasing number of allosteric processes that cannot be explained by structural changes alone. For example, allostery can be observed without a conformational change or in absence of a structural pathway [21,22] and more complex models are now used to account for these observations [23].

11.1.6 The case of intrinsically disordered proteins

Large segments of the primary structure in a significant proportion of proteins, especially in eukaryotes, do not have a defined tertiary structure and have a low propensity to adopt canonical secondary structures. It is predicted that about 35% (respectively 22%) of human proteins contain sequences from more than 30 (respectively 50) contiguous amino acids that do not adopt a defined structure. In extreme cases this structural disorder may extend to the whole polypeptide chain. Such proteins are called intrinsically disordered (ID) proteins, or sometimes natively unfolded proteins.

Generally, the primary structure of ID segments is enriched in polar and charged amino acids and depleted in hydrophobic amino acids such as long-chain aliphatic and aromatic amino acids. It is generally assumed that those segments sample a manifold of conformations which are in equilibrium.

One of the challenges in this field is the description of this conformational ensemble and understanding the molecular details of the disorder. NMR spectroscopy appeared to be well suited to the study of ID proteins. NMR can be used for very large ID proteins because their NMR signals exhibit the spectroscopic features of small molecules. The use of NMR, the difficulties encountered and the solutions adopted are described in detail in the review by Jensen et al. [24].

Analysis of protein–protein interaction networks reveals that a small subset of proteins have a very large number of interaction partners. This subset is enriched in proteins having large ID segments [25–28]. More remarkably, this subset can be broken into two distinct classes: (1) proteins that form stable complexes and interact simultaneously with many partners and (2) proteins that interact transiently with several partners at separate times. Only the transiently interacting subset has been found to be enriched in ID segments [26,28,29].

Therefore, disordered segments have a greater ability to interact with numerous partners [30]. Indeed, ID segments are widespread in many eukaryotic regulatory proteins involved in signal transduction or transcription. They are also present in several viral proteins known to have a huge interactome.

A question then arises: what is the role played by conformational disorder in this plasticity of interaction? Clearly, ID regions can play the role of flexible linkers between well-structured domains that are directly mediating the interaction. However, although such a role is observed, it seems that ID segments are often directly involved in the interaction.

Often, interaction with a partner protein is accompanied by a disorder to order transition, and particularly the acquisition of local secondary structure by the interacting segment. The structural element involved in the interaction is called a MoRE (for molecular recognition element) (Fig. 11.5A). Two other structural elements found in ID proteins and involved in protein–protein interaction have been described in the literature. They are the short linear motifs (SLiMs; Fig. 11.5B), which are conserved sequences of fewer that ten residues that directly bind to target proteins, and the low complexity regions (LCR) (Fig. 11.5C), which are either repetitive sequences or made of only a few types of amino acid residues. These LCR sequences (among which the best studied are the polyglutamine repeats) do not adopt any defined fold in their monomeric form. However, they are prone to oligomerize or to form aggregates in which they adopt defined secondary structures (see Section 11.3.4).

As already mentioned, recognition requires protein flexibility because it facilitates conformational reorganization and induced-fit mechanisms upon partner binding. However, the role of intrinsic disorder in recognition may seem counterintuitive. In fact, flexibility could be a way to obtain high specificity without an overly strong association constant (i.e., without extreme stabilization of the complex). Flexible proteins

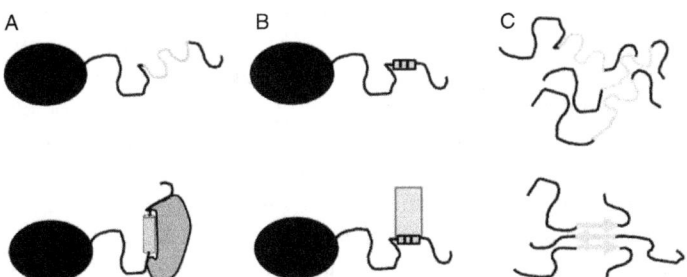

Fig. 11.5 ID segments and protein–protein interactions. (A) MoREs are short interaction segments that undergo a disorder-to-order transition upon binding to their partner. In the example, a MoRE (in grey) located in an extended ID segment adopts an α-helical structure (grey cylinder) upon binding to its partner protein. (B) SLiMs bind to several protein targets. A SLiM in an ID region is shown binding to a partner protein. (C) LCRs are regions which contain repetitive sequences or low levels of sequence variety. Shown is an example of an LCR-containing ID protein which interacts with itself to form an interchain β-sheet.

can potentially form more extended interaction surfaces than structured proteins. This allows for a precise fit to the target and, as a consequence, high specificity. However, the rigidification (associated with disorder-to-order transition) of the protein upon binding its target results in a loss of entropy and costs free energy. Thus, globally, the binding affinity is not excessive.

It is worth noting that many ID proteins remain largely disordered even in their bound state. This has led to the concept of "fuzzy complexes" [31]. The disorder may remain predominantly outside the binding zone but it seems that some ID proteins can remain entirely disordered in their bound state.

11.1.7 Final remarks

As more experimental work is performed to characterize the structure and dynamics of proteins, it is becoming evident that they can adopt a manifold of conformational states. It is worth noting that characterization of relevant conformations is essential if one wants to identify new therapeutic targets and design adequate inhibitors.

Several features indicate that protein flexibility is more present in complex, multicellular organisms (more multidomain proteins, more ID proteins, larger multiprotein complexes). This suggests that the acquisition of complex functions might be at least partly related to an increase in the global flexibility of the proteome. Indeed, protein flexibility allows functional diversity and indirectly increases the coding capacity of the genome. It also allows potentially rapid adaptation to changes in growth conditions, and can facilitate the acquisition of new functions.

11.2 Conformational changes in viral fusion proteins

11.2.1 Enveloped viruses enter their host cell by membrane fusion

The entry of enveloped viruses into host cells requires fusion between the viral and endosomal membranes [32]. This step is mediated by viral fusion glycoproteins which undergo a huge conformational change upon interaction with specific triggers such as viral receptors or a low-pH environment. During the structural transition, the fusion glycoprotein exposes hydrophobic motifs (the so-called fusion peptide or fusion loops) that interact with one or both of the participating membranes, resulting in their destabilization and merger. Based on their structure and common structural motifs, viral fusion proteins have been classified into several structural classes [33,34].

Despite the diversity of structural organization found in fusion glycoproteins, experimental data have suggested that the membrane fusion pathway is very similar for all the enveloped viruses studied so far, whatever the organization of their fusion machinery [35–37]. It is generally assumed that fusion proceeds via the formation of an initial stalk that is made by a local connection between the outer leaflets of the fusing membranes. Radial expansion of the stalk induces the formation of a hemifusion diaphragm in which the opening of a pore and its enlargement would lead to complete fusion [38] (Fig. 11.6).

Although membrane fusion is thermodynamically favorable, there are energetic barriers that have to be overcome at several stages of the process. The most

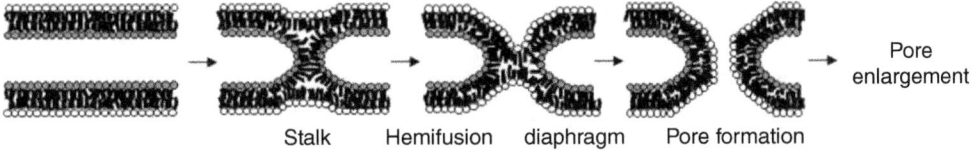

Fig. 11.6 A membrane fusion pathway according to the stalk–pore model.

energetically expensive step is apparently the expansion of the initial fusion pore [39]. It is proposed that the fusogenic transition of the viral glycoproteins is favorable and that the free energy which is released is used to overcome the different energetic barriers [34].

11.2.2 Several classes of viral fusion machineries

11.2.2.1 Class I fusion glycoprotein

Influenza HA Influenza virus hemagglutinin (HA) is the first viral fusion glycoprotein for which the ectodomain crystal structure has been determined in its pre-fusion conformation [40] (Fig. 11.7, left). It is synthesized as a precursor molecule called HA0 which is cleaved by a cellular protease to produce HA1, the sialic acid-binding domain, and HA2, the membrane-anchored fusion domain. HA0 cleavage primes the protein for the low-pH-induced fusogenic conformational change [41].

HA, in its pre-fusion conformation, is a homotrimer, each protomer consisting of a HA1 and HA2 chain connected by a single disulfide bridge. Most of HA2 forms an elongated triple coiled-coil and HA1 is located at the top of this coiled-coil. The N-terminal extremity of HA2, the result of HA0 cleavage, is particularly hydrophobic and constitutes the fusion peptide. In the pre-fusion structure it is buried in the trimer near the three-fold axis (Fig. 11.7, left) [41].

When exposed to an acidic environment (between pH 5 and 6 depending on the viral strain), HA undergoes a huge structural rearrangement. The result of this structural transition was only visualized in 1994 [42] when the post-fusion conformation of HA2 was crystallized (Fig. 11.7, right). Exposure to low pH results in the loss of HA1–HA2 interactions and the dissociation of the HA1–HA1 interface. Low-pH-induced conformational change in HA is essentially a dramatic refolding of HA2 (Fig. 11.7, center and right).

Refolding of HA2 can be described by two distinct steps. First, a long loop in native HA2 converts to a helix, resulting in an increase in the length of the central coiled-coil at its N-terminal end. This movement might translocate the fusion peptide toward the head of the molecule, i.e., toward the target membrane. At this stage, HA would be in an extended intermediate conformation. Second, the structural transition is completed by a relocation of the C-terminal membrane anchor, through a fold-back mechanism to the N-terminal end of the rod-shaped molecule. This results in a hairpin conformation in which the transmembrane segment and the fusion peptide are close together at the same end of the elongated molecule (Fig. 11.7, right) [42].

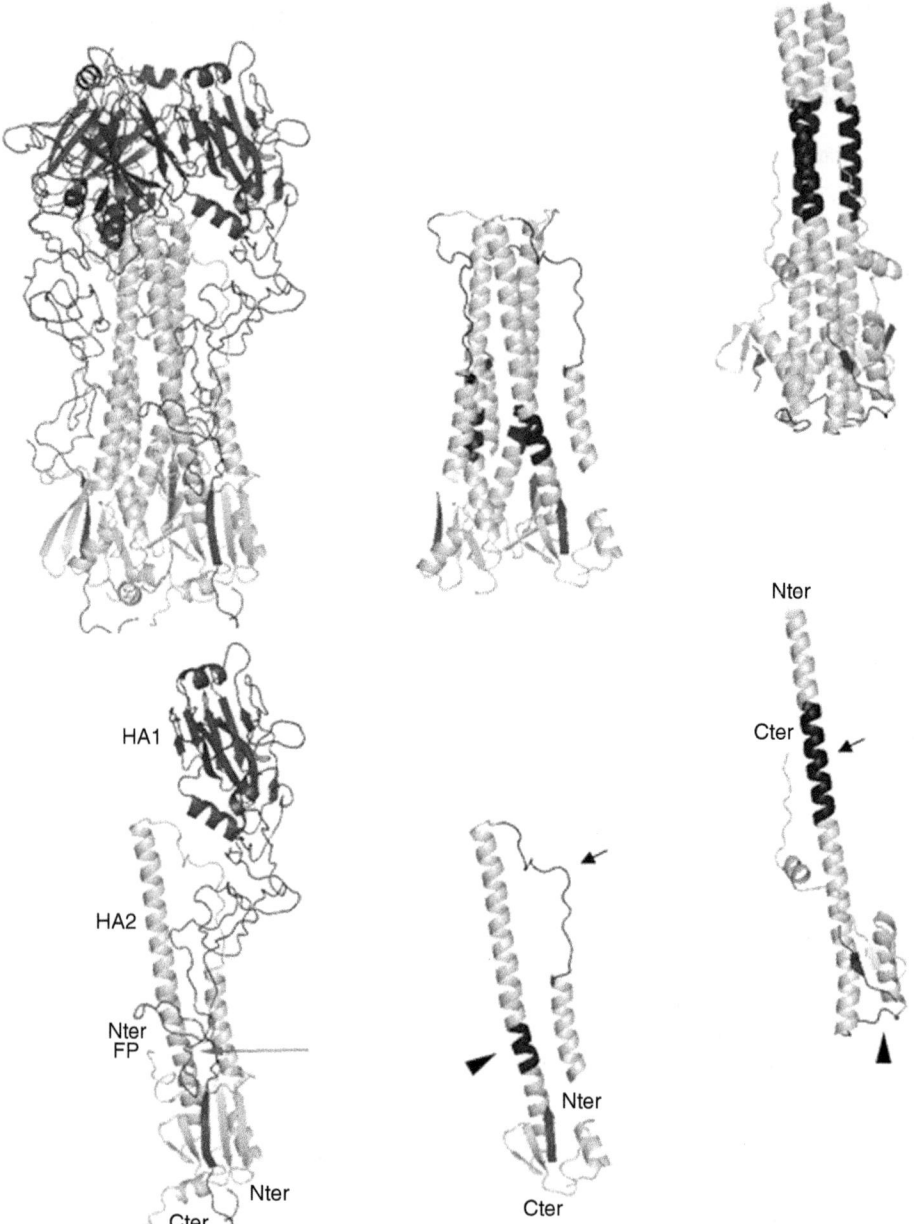

The crystal structure of the ectodomain of uncleaved HA0 has also been determined [43]. A comparison with that of cleaved HA reveals that the cleavage site is located in an exposed surface loop near a surface cavity. Cleavage of HA0 into HA1 and HA2 results in a local structural rearrangement, after which the cavity is filled by the first ten amino acids of HA2, which are non-polar. As a consequence, several ionizable

Fig. 11.7 Structures of influenza virus hemagglutinin (HA). Left: ribbon diagram of the crystal structure of BHA (the ectodomain of influenza HA cleaved by bromelain) trimer (top) or protomer (bottom). The receptor-binding domain (HA1) subunits are in black and the HA2 fusion proteins light grey. The N-terminal (Nter) and C-terminal (Cter) of each subunit and the fusion peptide (FP) of HA2 are indicated. The three fusion peptides are buried at the trimeric interface. The Cter of HA2 is anchored in the membrane. Center and right: ribbon diagram of the crystal structure of TBHA2 (the fragment resistant to proteolysis after low-pH-induced HA structural transition) trimer (top) or protomer (bottom) as in the prefusion native form (center) or as in the post-fusion form (right). The segments in black indicated by arrows or arrowheads refold during the structural transition. The Nter of HA2 is linked to the fusion peptide (which is missing in TBHA2). The Cter is associated with the transmembrane domain.

residues from the cavity are buried. Mutations of these ionizable residues have been demonstrated to affect the pH at which membrane fusion occurs [41].

Class I structural motif This structural motif, observed in the post-fusion structure of HA2, has been found in the post-fusion structure of several other viral fusion proteins including those of retroviruses [44], paramyxoviruses [45], filoviruses [46] and coronaviruses [47] (Fig. 11.8). In their post-fusion conformations, class I viral fusion glycoproteins form a central trimeric coiled-coil, at the N-termini of which are displayed the fusion peptides and against which are packed, in an antiparallel manner, the segments abutting the transmembrane domains. This motif defines class I viral fusion proteins.

Most class I fusion proteins also have an internal cleavage site located upstream of the fusion peptide. At the level of the primary structure of the protein, the class I signature is the presence of a first heptad repeat region (HRA or HR1) that is immediately downstream of the fusion peptide and forms the central trimeric coiled-coil. It is often accompanied by a second one (HRB or HR2) located in the C-terminal part of the ectodomain that constitutes the helices positioned in an antiparallel manner in the groove of the central coiled-coil (Fig. 11.8).

Apart from this common structural motif, the structures of class I fusion proteins are very different, as illustrated by the determination of the structure of the pre-fusion state of parainfluenza virus type 5 fusion protein (PIV5 F) [48] (Fig. 11.9) and the glycoprotein of Ebola virus [49].

Comparison of the pre-fusion [48] and post-fusion [50] structures of paramyxovirus F proteins reveals on the one hand the similarity of the conformational transition of class I fusion glycoproteins and on the other the different structural solutions that have been selected by evolution. During the conformational change, the HRA region, located downstream of the fusion peptide and initially collapsed in a compact set of 11 small segments, undergoes a major refolding event into a single, extended α-helix that propels the fusion peptide outward for insertion into the target membrane (Fig. 11.9). Region HRB, adjacent to the transmembrane domain, then moves to bind to HRA, forming a six-helix bundle and the post-fusion hairpin conformation.

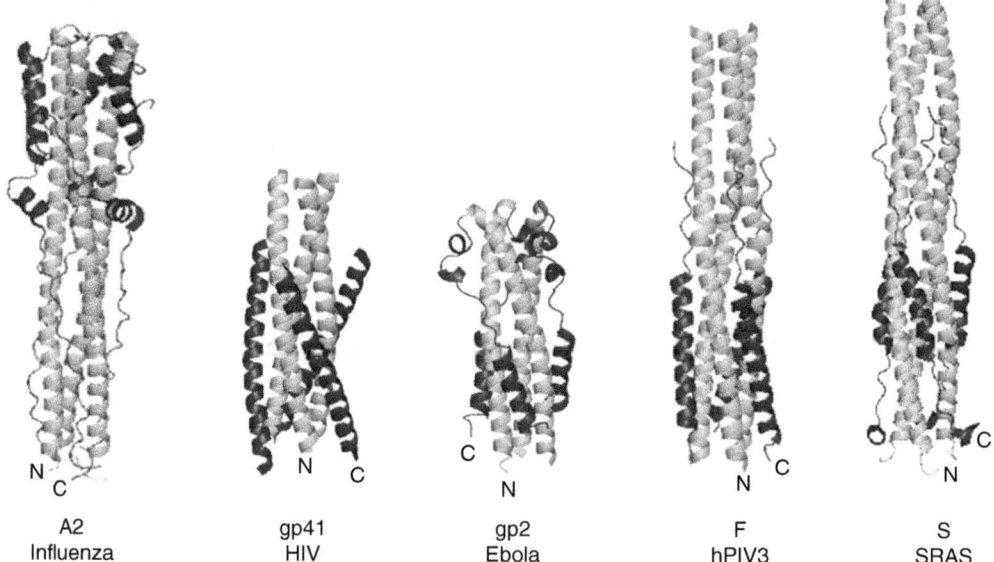

Fig. 11.8 The six-helix bundle (6HB) is the motif which is found in the post-fusion conformation of all class I viral fusion glycoproteins. Side views of the 6HB from five different viruses are shown: influenza HA2, HIV gp41, Ebola virus gp2, hPIV3 F and SRAS S2. The N-terminal coiled-coil core (located downstream of the fusion peptide, absent in these structures) is shown in light grey and the C-terminal domain (located upstream of the transmembrane domain) is in dark grey. In all structures, the trimer-of-hairpins conformation brings the N- and C-termini (corresponding to the position of the fusion peptides and transmembrane domain) into close proximity.

It is worth noting that the mechanisms for activation of different class I viral fusion proteins are distinct. For many of them, activation is not dependent on a low-pH environment. As an example, most of the retrovirus env proteins and paramyxovirus F are activated by receptor-binding events. This activation can be either by direct interaction (HIV env) [51] or indirectly through a viral attachment protein (paramyxovirus F) [52].

11.2.2.2 Class II fusion glycoproteins

The second class of fusion proteins was identified when the X-ray pre-fusion structure of the ectodomain of the fusion protein E1 of Semliki forest virus (SFV) [53], an alphavirus, was shown to have a remarkably similar secondary and tertiary structure to that of fusion protein E of tick-borne encephalitis virus (TBEV) [54], a flavivirus (Fig. 11.10). This revealed that the genes encoding these proteins were derived from a common ancestor. The alphaviruses and flaviviruses are members of the Togaviridae and Flaviviridae families, respectively, and other fusion proteins of viruses from these families, including West Nile virus E [55] and Dengue virus E [56], have been

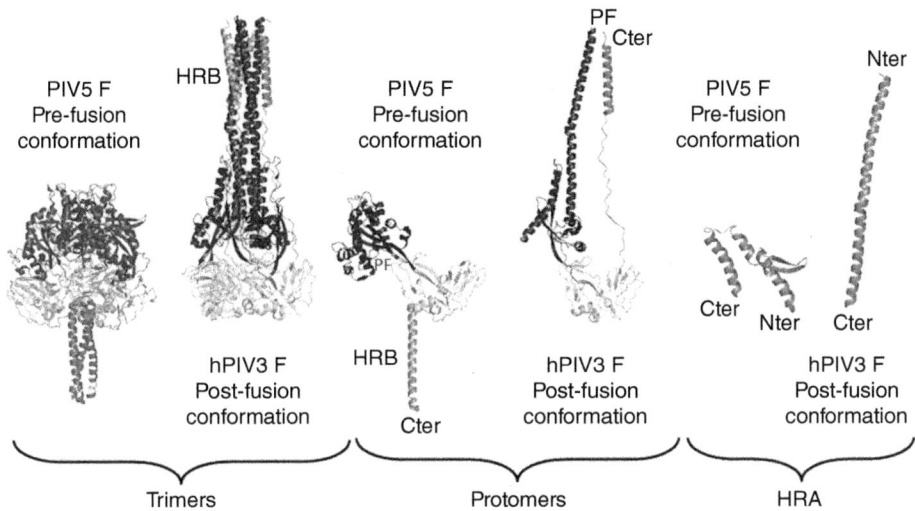

Fig. 11.9 Ribbon representation of the crystalline pre-fusion structure of parainfluenza virus 5 (PIV5) F and of the post-fusion structure of human parainfluenza virus 3 (hPIV3) F. In the trimers and protomers representation, the HRA-containing domain is in black and HRB is in dark grey. The block in light grey keeps its structure and orientation in the pre- and post-fusion structures. The positions of the C-terminal (Cter, which is linked to the transmembrane domain) and the fusion peptide (PF) are also indicated on the structure of the protomers. On the right part of the figure, the structure of HRA is drawn in both its pre- and post-fusion conformation. The fusion peptide is located at the N-terminal (Nter) end of HRA.

demonstrated to belong to class II. It has also been recently shown that the fusion glycoprotein of Rift Valley fever virus (a member of the Bunyaviridae family) also belongs to class II [57]. In their pre-fusion state anchored in the viral membrane, class II fusion proteins are organized following the icosahedral symmetry of the virus particle [53,57,58]. Remarkably, the *Caenorhabditis elegans* cell–cell fusion protein EFF-1, which is essential for nematode development, displays the same fold and quaternary structure as class II viral fusion proteins. Therefore the cellular gene which encodes EFF-1 is homologous to the genes encoding viral class II fusion proteins [59].

All class II fusion proteins identified so far fold co-translationally with a viral chaperone protein with which they form a heterodimer. This heterodimeric interaction is important for the correct folding and transport of the fusion protein [60]. The chaperone protein is cleaved by the cellular protease furin late in the secretory pathway [61,62]. This cleavage primes the fusion protein, which is then in a metastable conformation, for subsequent low-pH-induced conformational change [60].

The polypeptide chain of the class II proteins is folded in three domains essentially comprising a β-sheet [53,54,56] (Fig. 11.10). The C-terminus part of the ectodomain, which is linked to the transmembrane segment, and the internal fusion loop are found

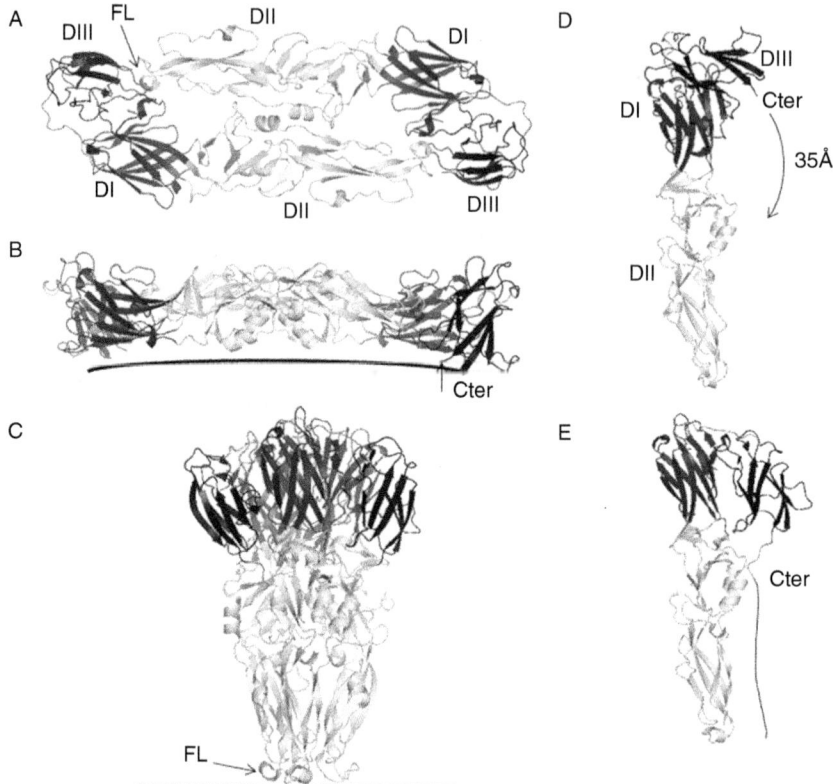

Fig. 11.10 Structures of tick-bone encephalitis (TBE) flavivirus glycoprotein E. Ribbon diagram of the pre-fusion E dimer, in top view (A) and lateral view (B), the post-fusion E trimer (C), pre-fusion E protomer (D) and post-fusion E protomer (E). Domains DI, DII and DIII of the protein correspond to, respectively, the central domain, the dimerization domain at the tip of which are located the fusion loops (FL) and the C-terminal (Cter) domain which is linked to the transmembrane domain (the segment in black drawn in E is missing in the crystal structure). In (B) and (C) the position of the membrane is indicated by a black line. In (D) the movement of DIII during the transition is also indicated.

at the two ends of an extended molecule. The fusion loop is located at the tip of an elongated β-sheet. In the pre-fusion state it is not exposed to the solvent but it is buried at a protein–protein interface (Fig. 11.10A, B).

Exposure to low pH, in the presence of a target membrane, triggers a complete rearrangement of the proteins [63–65]. First, the pre-fusion oligomers (either homodimers or heterodimers formed by the fusion glycoprotein in association with its chaperone) dissociate and then re-associate to form homotrimers (Fig. 11.10C). The conformational change is less impressive than that observed for HA or for paramyxovirus F protein as there is essentially no change in secondary structure.

Nevertheless, the intrachain interactions between domains are markedly different in the pre- and post-fusion conformation. In particular, domain III, which is connected to the C-terminal part of the molecule, moves by about 35 Å towards the fusion loop (Fig. 11.10D, E). The movement of domain III redirects the polypeptide chain, so that in the post-fusion trimeric form the transmembrane segment is located near the fusion loop region. Despite the different protein architectures, the post-fusion structure is thus reminiscent of the post-fusion hairpin structure of class I fusion proteins. A consequence of this reorganization is that the initial icosahedral symmetry of the glycoprotein organization in the viral membrane is completely disrupted [66].

11.2.2.3 Class III fusion glycoproteins

The third class of fusion proteins was identified when the structures of the ectodomains of the fusogenic protein G (G) of vesicular stomatitis virus (VSV; the prototype of the Rhabdoviridae family) [67] and glycoprotein B (gB) of herpes simplex virus 1 (HSV1) were determined [68]. This revealed an unanticipated homology between the two proteins, which belong to a negative-strand RNA virus and a DNA virus, respectively, and for which no sequence similarity had previously been identified. Epstein–Barr virus (EBV) gB [69] and baculovirus glycoprotein 64 (gp64) [70], the structures of which were further determined, have also been demonstrated to belong to this new class.

For the moment, VSV G is the only class III fusion protein for which the structures of both the pre-fusion [71] and post-fusion states [67] have been determined. The structures determined for other class III members are presumptive post-fusion conformations based on their structural similarity with the post-fusion conformation of G. Furthermore, the mechanisms for activation of HSV1 gB and VSV G are very different as conformational change in VSV G is triggered by exposure to low pH whereas HSV1 gB is indirectly activated by receptor-binding events (which involve other viral glycoproteins) [72].

The polypeptide chain of the G ectodomain folds into three distinct domains: the fusion domain (FD) is inserted in a loop of a pleckstrin homology domain (PHD) that is itself inserted in a central domain (TrD) whose long helix is involved in trimerization of the molecule in both the pre- and the post-fusion states (Fig. 11.11A). The TrD also comprises a β-sheet-rich region connected to the C-terminal segment of the ectodomain. This segment itself connects to the transmembrane (TM) domain in the intact glycoprotein. The organization of the class III fusion domain is very similar to that of class II fusion proteins with two fusion loops located at the tip of an elongated three-stranded β-sheet. However, in the pre-fusion structure, in contrast to class I and class II viral fusion proteins, the fusion loops are not buried at an oligomeric interface but point toward the viral membrane (Fig 11.11A).

VSV G structures revealed that the conformational change from the pre- to the post-fusion state involves a dramatic reorganization of the glycoprotein (Fig. 11.11A, B). During the structural transition, the three domains retain their tertiary structure but undergo large rearrangements in their relative orientation. This is due to refolding of large hinge regions (R1 to R4) connecting the domains.

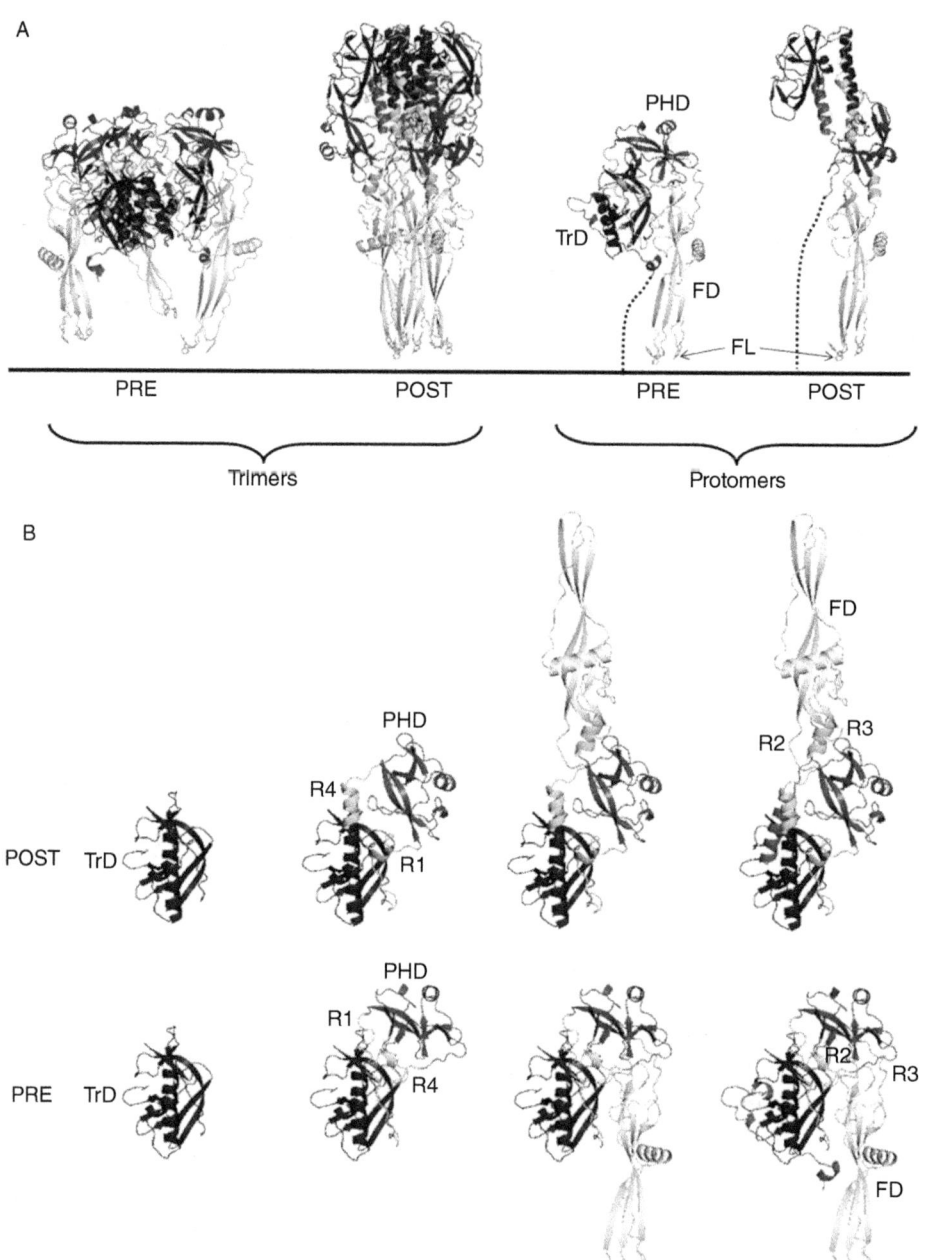

Fig. 11.11 (A) Ribbon diagram of the pre- and post-fusion crystal structures of VSV G ectodomain (trimers and protomers) shaded by domains. Hydrophobic residues of the fusion loops (FL) are as sticks. The position of the viral membrane is indicated by the horizontal black line. The dotted line in black corresponds to the C-terminal part which is not visible in the crystal. (B) Comparison between the pre- and post-fusion conformations of G protomers. Domains and segments are shaded as in (A). All the structures are superimposed on the central domain (TrD) showing that the pre- and post-fusion conformations are related by flipping both the FD and the C-terminal segment relative to the central domain through refolding of segments R1 to R4 (together with the refolding of the C-terminal).

Global refolding of G exhibits striking similarities to that of class I proteins. The pre- and post-fusion states of a protomer are related by flipping both the FD and a C-terminal segment relative to the central domain. During this movement, both the fusion loops and the TM domain move about 160 Å from one end of the molecule to the other. The FD is projected toward the membrane through the combination of two movements: a rotation around the hinge between the PHD and FD and a lengthening of the central helix (located in one of the hinge regions connecting TrD to PHD and involved in the formation of the trimeric central core of the post-fusion conformation). The movement of the TM domain is due to the refolding of the C-terminal segment into an α-helix (Fig. 11.11B). In the trimeric post-fusion state, the three C-terminal helices position themselves in the groove of the trimeric central core in an antiparallel manner to form a six-helix bundle.

One remarkable feature is that, for VSV G, the low-pH-induced structural transition is reversible and there is a pH-dependent equilibrium between the different conformations of the protein [73,74]. This is a characteristic of the fusion glycoprotein of rhabdoviruses. For the other viral families studied so far, the pre-fusion conformation is metastable and, as a consequence, the fusion-associated structural transition is irreversible [75].

11.2.2.4 Other classes of viral fusogens: viral fusion glycoproteins from pesti- and hepaciviruses

Until 2013, envelope protein structures of viruses from the Flaviviridae family were only available for members of the genus flavivirus, the fusion machinery of which belongs to class II. Therefore, it was commonly supposed that the viral fusion glycoproteins of the other genera of the family (pesti-, pegi- and hepaciviruses) would also belong to class II. Indeed, E2 envelope glycoprotein from hepatitis C virus (HCV, a hepacivirus of the family Flaviviridae) had been predicted to be a class II viral fusion glycoprotein. This was based on secondary structure predictions (combined with infrared spectroscopy and circular dichroism) and biochemical characterization of the disulfide bonding pattern [76].

However, determination of the crystal structure of the ectodomain of E2, the larger envelope protein of bovine viral diarrhea virus (BVDV) (a pestivirus), revealed an unexpected architecture consisting of two immunoglobulin (Ig) domains and an elongated β-stranded domain which does not bear any significant homology to other known protein domains [77,78]. Two structures of the HCV E2 core bound to a neutralizing antibody were subsequently determined and revealed a compact globular architecture based on an Ig-fold β-sandwich, which was different from the multidomain organization of both BVDV E2 and class II viral fusion glycoproteins [79,80].

The structures of BVDV and HCV E2 do not exhibit any of the classical features found in other viral fusion glycoproteins, such as hydrophobic motifs, a helical core or flexible hinge regions between domains. Therefore, it is now thought that it is glycoprotein E1 that is tightly associated with E2, which is the fusion glycoprotein [33].

11.2.3 pH-sensitive molecular switches

The structure of viral fusion glycoproteins with a conformational change that is triggered at low pH has allowed identification of the amino acid residues that play the role of pH-sensitive molecular switches.

For TBEV, the initiation of fusion is crucially dependent on the protonation of a single conserved histidine located at the interface between domains I and III of E, which has been identified as the critical pH sensor [81]. Its protonation results in the dissolution of domain interactions and the exposure of the fusion peptide. A similar study performed on SFV has also identified a histidine residue that acts to regulate the low-pH-dependent refolding of E1 [82].

In the pre-fusion state of VSV, three conserved histidines (H60 and H162, both located in the fusion domain, and H407, located in the C-terminal segment of the protein) cluster together [71] (Fig. 11.12A). Protonation of these residues at low pH is likely to destabilize the interaction between the C-terminal segment of G and the fusion domain in the pre-fusion conformation. This might trigger the movement of the fusion domain toward the target membrane.

Conversely, in the post-fusion form, a large number of acidic amino acids are brought close together in the six-helix bundle [67] (Fig. 11.12B). In the post-fusion state, these residues are protonated and form hydrogen bonds. The deprotonation of these residues at higher pH induces strong repulsive forces that destabilize the trimer and trigger the conformational change back to the pre-fusion state [83]. These residues are buried in the post-fusion state but exposed to solvent in the pre-fusion state [67,71]. As a consequence, they have a pK_a that is much higher in the post-fusion than in the pre-fusion conformation. This stronger affinity for protons in the post-fusion state explains the cooperativity of rhabdoviral structural transition as a function of pH [74].

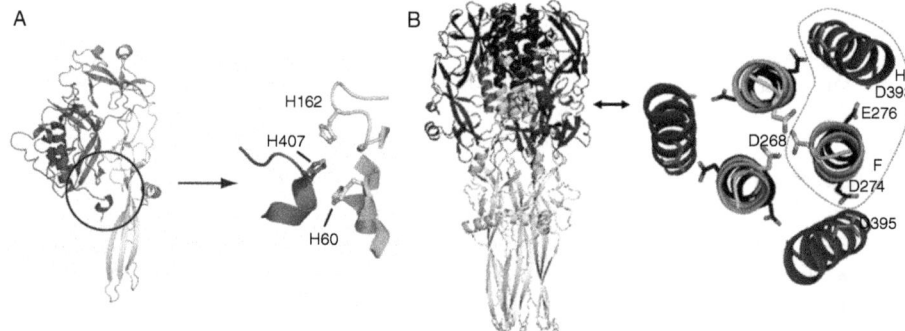

Fig. 11.12 pH-sensitive molecular switches of VSV G. (A) Close-up view of the histidine cluster located at the fusion domain/C-terminal domain interface that has to be disrupted during the structural transition from the pre- to post-fusion structure. (B) Close-up view of the VSV G trimerization domain showing the clusters of acidic residues in the post-fusion state (viewed from the bottom of the post-fusion state).

In the case of influenza HA, mutations at distinct positions affect the stability of trimer interfaces that break during the rearrangement, and therefore alter the threshold pH for fusion [84]. However, no specific amino acid residue has been identified for which protonation would trigger conformational change of HA.

11.2.4 Intermediates during the structural transition

One of the earliest stages in the membrane fusion process is the insertion of the fusion peptides (for class I fusion glycoproteins) or the fusion loops (for class II and III fusion glycoproteins) into the target membrane. This insertion has been demonstrated using hydrophobic photolabeling for both influenza virus and rhabdoviruses [85,86]. It can be concluded that an intermediate stage exposes the fusion peptides or loops so that they can interact with the target membrane.

It is proposed that when class I fusion glycoproteins interact with the target membrane they are in an extended conformation in which the central trimeric coiled-coil (made of HRA/HR1 segments) is already formed but the HR2/HRB segments against this central structure have not relocated (Fig. 11.13). The existence of such an intermediate (often called the "trimeric pre-hairpin" or "trimeric extended intermediate") is supported by two types of experimental result. First, it has been demonstrated that peptides corresponding to the HR2 segments can inhibit the fusion process (Fig. 11.13) [87,88]. Second, EM experiments have revealed some structures that have dimensions compatible with such intermediates [89,90].

Class II fusion proteins transit from a (homo- or hetero-) pre-fusion dimer to a post-fusion trimer via monomeric intermediates. However, the oligomeric status of the glycoprotein when it interacts with the target membrane and the stage of the transition at which trimerization occurs are still not known. For TBEV, the initial interaction with the target membrane seems to involve the insertion of multiple copies of E monomers and trimerization is suggested to be a late event concomitant with

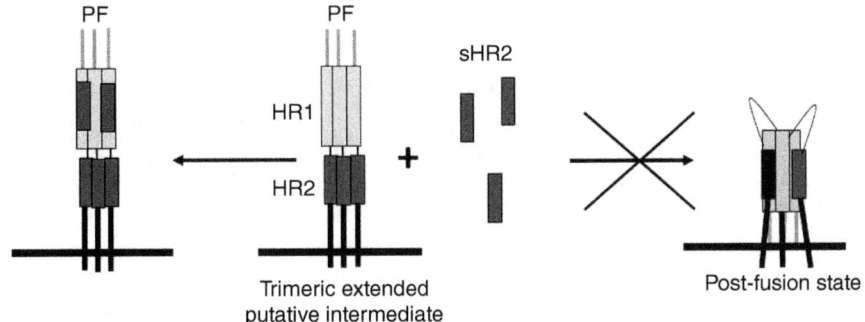

Fig. 11.13 The putative trimeric extended intermediate. Refolding of the three HR1 segments (in light grey) into a coiled-coil would propel the fusion peptides (PF) toward the target membrane. Soluble peptides (sHR2) corresponding to the HR2 region (in dark grey) inhibit the fusion process, most probably by inhibiting the relocation of HR2 in the groove of the HR1 coiled-coil.

folding back of domain III [91]. However, an exogenous domain III can function as a dominant-negative inhibitor of alphavirus and flavivirus membrane fusion [92]. This is consistent with the existence of a relatively long-lived core trimer intermediate with which domain III finally interacts (i.e., a trimeric extended intermediate). Since, for SFV, trimers made of DI and DII domains are stable in the absence of domain III, it has been proposed that DI/II trimer formation is rapid and that the folding-back of domain III is a late and relatively slow step in the structural transition pathway [93,94].

For class I and III fusion glycoproteins, the transition from a putative trimeric extended intermediate to the final trimeric post-fusion state cannot retain strict three-fold symmetry. This symmetry has to be disrupted during the folding back of the C-terminal part of the glycoprotein. Furthermore, it is worth noting that if the central trimeric core does not dissociate during the structural transition from this putative intermediate to the final post-fusion state, then it is absolutely required that the TM segments are at least temporarily separated so that they can turn on themselves to achieve the transition [95]. The complexity of such a protein movement is often underestimated in the literature.

A solution to this problem would be that the final trimerization event would be a late event occurring only when the refolding of the protomer has been achieved (or almost achieved). As already mentioned, this was initially suggested for TBEV E glycoprotein. This also seems to be the case for VSV G glycoprotein for which it has been demonstrated that the structural transition between the pre- and post-fusion trimer proceeds through monomeric intermediates [73].

11.2.5 Final remarks

The determination of an increasing number of viral fusion protein structures and recent electron microscopy studies have shed light on the molecular mechanisms of fusion machineries. This has revealed similar principles of action for machineries made up of proteins having very different shapes. Nevertheless, many questions remain unanswered.

As a first example, it is known that TM domains of viral fusion proteins play an important role at the late stages of the fusion process [96–98], but the structure of a full-length viral fusion glycoprotein (i.e., with its TM domain) has not yet been determined. Thus, further structural studies of complete pre- and post-fusion conformations of viral glycoproteins containing both membrane anchors (TM and fusion peptides) are needed. This should indicate how both regions cooperate to regulate the transition from hemifusion to fusion pore opening and/or pore enlargement.

Finally, the structures of intermediate conformations of fusion proteins, how they cooperate and how they interact with and deform the fusing membranes are still very elusive. Electron microscopy and tomography have revealed some clues [73,89,90] but the resolution of the reconstructions is too low to conclude the nature of these intermediates. Gaining high-resolution structural information on intermediate conformations of fusion proteins (ideally complete with their TM domain and in association with bilayers) is now the new frontier for scientists working in the field.

11.3 From conformational diseases to cell memory

11.3.1 Introduction

The biological functions of proteins are linked to their three-dimensional structures. Mutations which affect the structure of a protein often result in a loss of function. This may occur through a local effect, for example mutations can affect the affinity of a protein for its natural substrate or decrease its enzymatic activity. However, some other mutations have a more global effect by impeding the correct folding of the protein, resulting in a decrease in concentration of the functional protein. Generally, misfolded proteins trigger the cellular unfolded protein response (UPR) which degrades incorrectly folded proteins and activates the synthesis of chaperone proteins which will assist protein folding. However, saturation of this system can result in an excess of misfolded proteins which form aggregates. This aggregation often results in a gain of toxic function that finally cause severe damage to cells and tissues. Among the conformational diseases, the most studied are the amyloidoses and the prion diseases which will be the subject of this section.

11.3.2 Prion proteins

11.3.2.1 The prion hypothesis

Transmissible spongiform encephalopathies (TSEs) are infectious disorders characterized by motor and cognitive impairments, extensive brain damage and neuronal dysfunction. The clinical symptoms are detected after a long incubation period and the patient's condition deteriorates very rapidly.

The first TSE was initially described in sheep; the associated disease was called scrapie. In humans, there are several TSEs: the most frequent is Creutzfeld–Jakob disease (CJD) which is sporadic (about one case per million humans per year). Kuru (or laughing sickness) is another TSE which was transmitted among members of the Fore tribe of Papua New Guinea via cannibalism. There are also two autosomal dominant inherited diseases: Gerstmann–Straussler–Scheinker (GSS) syndrome, which is a rare familial form of progressive dementia, and fatal familial insomnia (FFI).

The infectious nature of TSEs was acknowledged very early on (even when CJD, GSS and FFI were known as inherited diseases). However, isolation and characterization of the infectious agent have been a long quest. It appeared to have an unusually small size and to be resistant to procedures that inactivate nucleic acids. In 1967, John Griffiths [99] was the first to propose that scrapie was caused by a protein only. He proposed three plausible models to explain how a protein could be infectious. Retrospectively, his paper (and particularly, the second model he proposed) was really visionary.

The hard work of Prusiner and colleagues demonstrated that infectivity was associated with a glycophosphatidylinositol-anchored membrane protein, called a prion (for PRotein Infectious ONly) protein (PrP). It was shown that a PrP exists in two distinct conformations, the "normal" protein, found in healthy individuals, called PrP^c (for cellular PrP) and the protein found in infected human or animal tissues, called PrP^{Sc} (for scrapie-associated PrP). The only differences between PrP^c and PrP^{Sc} reside at the level of their structure and aggregation properties. Today, the largely accepted hypothesis about the nature of the infectious agent is that it is a self-propagating aberrant conformation of the prion protein.

Experimental confirmations of this model will not be detailed here, but the most important results that support the prion hypothesis will be mentioned. First, PrP knockout mice are completely resistant to scrapie. Second, transgenic mice expressing a PrP variant with mutations associated with FFI spontaneously develop the disease and the disease can be transmitted to wild-type mice. Finally, infectious material can be generated *in vitro* in a test tube using PrP^c of recombinant origin (see Section 11.3.2.5).

11.3.2.2 Structure of PrP^c and PrP^{Sc}

The NMR structure of the mouse prion protein domain (residues 121–231) was worked out in 1996 [100] and revealed to be a globular domain. This was confirmed by the crystal structure of a complex of the same domain of ovine PrP and an antibody [101]. NMR structures of longer fragments of the protein revealed that the N-terminal part (up to about residue 120) is a flexible disordered tail [102]. The globular domain contains three α-helices and a short antiparallel β-sheet (Fig. 11.14A). Both crystal and NMR structures revealed structural disorder in several segments of the protein.

The structure of PrP^{Sc} is still not known. The available data indicate that PrP^{Sc} is a polymer composed of PrP protomers which form intermolecular β-sheet structures. Indeed, both circular dichroism and Fourier transform infrared experiments have shown that PrP^{Sc} contains predominantly β-sheet structures, in marked contrast to the essentially α-helical nature of PrP. The PrP^{Sc} polymers are substantially resistant to proteolysis. Digestion of PrP^{Sc} by proteinase K removes the flexible N-terminal part of PrP but the remaining part of the molecule retains infectivity.

The polymers of PrP^{Sc} are of different sizes and there is a dynamic equilibrium between the different species present in solution. The molecular organization of those polymers is not known, although it has been investigated by several techniques including electron microscopy, atomic force microscopy, X-ray fiber diffraction, electron crystallography and electron paramagnetic resonance spectroscopy [103,104]. Using the information obtained from those techniques, several structural models of PrP^{Sc} polymers have been proposed [105]. They differ in the importance of the secondary structural changes.

In the β-helix model based on electron microscopy data from 2D crystals [103], the two C-terminal helices of PrP^c maintain their α-helical organization while residues from about 90 to 175 form a left-handed β-helix stacked horizontally in a trimeric arrangement. PsP^{Sc} aggregates would be formed by vertical stacking of trimers along the axis of the β-helix.

Another model is the β-spiral, which has been deduced from molecular dynamics simulations [106]. It consists of a spiraling core of extended sheets made by short β-strands. In this model, the three α-helices keep their secondary structure but are relocated inside the protomer.

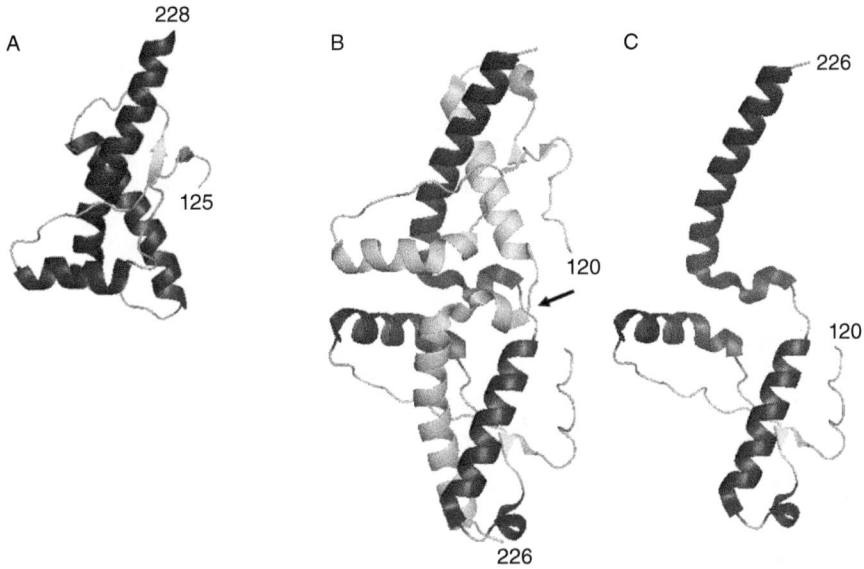

Fig. 11.14 Structure of human PrP^c. (A) Ribbon representation of the NMR structure of the globular domain of human PrP^c (residues 125–228). The protein is shaded by secondary structures. The first 120 amino acid residues are disordered and are not represented. (B) Ribbon representation the first crystal structure of PrP (111). The structure revealed a dimeric arrangement resulting from the three-dimensional swapping of the C-terminal helix. One protomer is shaded by secondary structures as in (A), the second is in light grey. The arrow indicates the position of the small interchain two-stranded antiparallel β-sheet which is formed at the dimer interface. (C) Structure of a protomer in dimeric PrP.

Finally, a radically different model proposed that PrP^{Sc} contains exclusively β-strands and short turns and forms cross β-sheet assemblies, which are characteristics of amyloids [107].

11.3.2.3 A model for the self-propagation of PrP^{Sc}

The most widely accepted hypothesis on prion propagation is the seeding–nucleation model. In this model, polymers of PrP^{Sc} act as a template for the transconformation of PrP, by incorporating it into the polymer (Fig. 11.15A). The formation of a stable nucleus (or seed) of PrP^{Sc} would be thermodynamically unfavorable. This would explain the low frequency of spontaneous disease except when a mutation (like those encountered in GSS and FFI) results in a PrP protein which is more prone to form a stable seed.

Although fibrillar polymers formed by PrP^{Sc} have a high molecular weight, attempts to characterize infectious PrP^{Sc} species have revealed a larger specific infectivity (infectious particles per PrP molecule) for non-fibrillar particles formed from about 14–28 PrP protomers. Smaller aggregates were virtually uninfectious, and the

380 *Protein conformational changes*

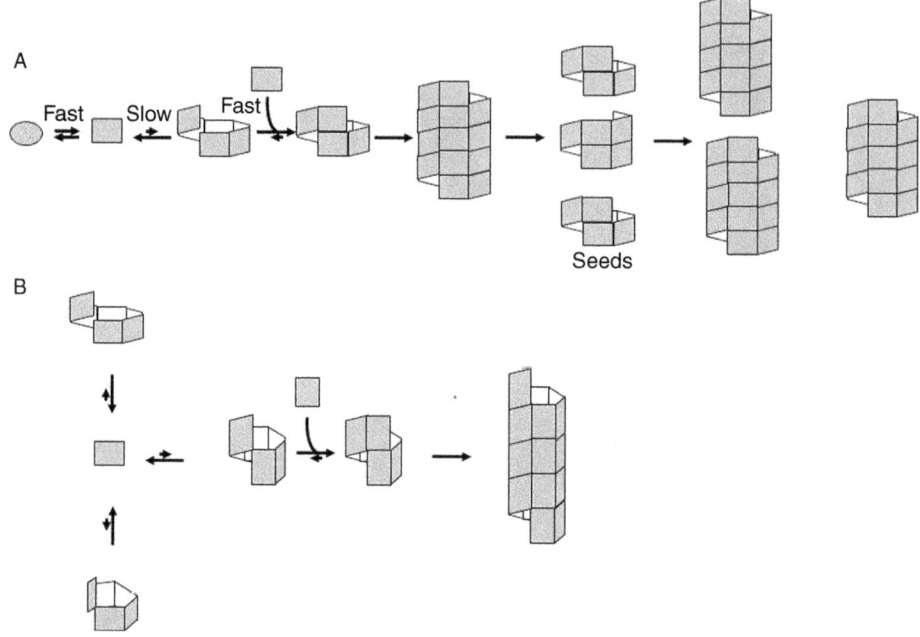

Fig. 11.15 Seeding-nucleation model of prion propagation (adapted from [138]). (A) The kinetic barrier for sporadic TSE is the formation of nucleus. Breaking the polymers generates new seeds that amplify the conversion. Therefore, infection is the result of seeding PrP polymerization by pre-formed PrP^{Sc} nuclei. (B) Strains originate from an alternative mode of polymerization (different number of Prp making up the nucleus, alternative packing of slightly different conformation of monomers).

largest aggregates made of several hundred PrP molecules exhibited 70-fold less specific infectivity. However, the specific infectivity per particle did not vary by more than two-fold between different sizes of infectious polymer [108]. This latter result, together with the fact that sonication of PrP^{Sc} decreases the size of the polymers and increases their converting activity, is a very strong argument in favor of the seeding nucleation model.

The cellular site of conversion from PrP^{c} to PrP^{Sc} has been investigated. It has been demonstrated that prion infection is extremely rapid (occurring within a minute of prion exposure) and that the plasma membrane is the primary site of PrP conversion [109]. This indicates that PrP molecules that have been transconformed into PrP^{Sc} have the characteristics of folded proteins because they have been able to leave the endoplasmic reticulum [110]. This suggests that, at the cellular surface, there is an equilibrium between several PrP conformations and that a minor conformation is trapped in the polymers.

Remarkably, the first crystal structure of PrP [111] revealed a dimeric arrangement resulting from the three-dimensional swapping of the C-terminal helix and rearrangement of the disulfide bond (Fig. 11.14B). An interchain two-stranded antiparallel β-sheet is formed at the dimer interface by residues that are located in helix 2 in the monomeric NMR structures. Whether this dimer constitutes an intermediate on the pathway of the $PrP^c \rightarrow PrP^{Sc}$ conversion is not clear. However, this strongly suggests that alternative open monomeric structures distinct from the globular form initially described by NMR exist in solution.

11.3.2.4 Prion strains

One of the arguments that has long been used against the proponents of the prion hypothesis was that, in the same animal species, several prion strains causing very different disease phenotypes could be isolated and propagated. At that time, it seemed difficult to explain such a phenomenon with only two conformations of the protein, and this was an argument in favor of a PrP^{Sc}-associated nucleic acid.

However, there is now strong evidence to show that the prion strains are encoded by distinct PrP^{Sc} conformers. For example, there are now several reports that the proteinase K sensitivity of PrP^{Sc} and the cleavage sites vary from one strain to another [112,113] and that there is a strain-dependent difference in the β-sheet content of PrP^{Sc} [114].

A question then arises about the number of PrP^{Sc} conformers and the extent of the structural differences between them. The simplest explanation is that the structural differences between PrP^{Sc} conformers are probably minor and that the differences between strains reside in the associated polymers (Fig. 11.15B). The template made by the polymer that has already formed dictates the PrP^{Sc} structure, either by conformation selection or by induced fit. This is reminiscent of viral capsid proteins which, depending on their physicochemical environment, are able to form several assemblies with very minor changes in their tertiary structure [115,116].

11.3.2.5 Generating prions in a test tube

In order to demonstrate that the infectious agent was devoid of a specific nucleic acid, several groups tried to mimic prion replication in a test tube. The first assay was developed by Caughey and colleagues. In the presence of an excess of brain-isolated PrP^{Sc} they were able to convert small amounts of radiolabelled PrP^c to a form that had similar biochemical properties to PrP^{Sc} [117]. However, the infectivity of the converted species could not be tested, owing to the small quantities and the inability to distinguish newly formed PrP^{Sc} from the original inoculum.

Based on a similar approach, an efficient *in vitro* prion replication system was developed based on cyclic amplification of PrP^{Sc} [118]. In this assay, minute amounts of brain homogenates containing PrP^{Sc} were mixed with healthy brain homogenates containing PrP^c. In this assay, newly formed PrP^{Sc} polymers were fragmented during sonication cycles, multiplying the number of seeds for conversion and resulting in an exponential increase in the concentration of PrP^{Sc}. The properties of the newly formed PrP^{Sc} were identical to those of brain-derived PrP^{Sc} but it was also highly infectious in wild-type animals. More remarkably, this assay, called PMCA (for protein-misfolding

382 Protein conformational changes

cyclic amplification), allowed the replication of prion strain properties. It is now commonly used as a highly sensitive prion-detection system.

PMCA has opened new areas of research. In particular it might allow us to answer the question about an eventual role for non-protein cellular cofactors which could assist in the conversion of PrP^c into PrP^{Sc}.

11.3.3 Amyloidoses

11.3.3.1 General

Human amyloidoses are a heterogeneous group of diseases. They are characterized by amyloid deposits, which are protein aggregates made by amyloid fibrils. Those amyloid fibrils result from the self-association of an abnormal conformation of a protein. The precursor proteins are normally soluble proteins which are extremely diverse in size, structure and function. Some biochemical and/or environmental conditions allow those proteins to explore misfolded or partially folded states that have strong self-association properties.

In the amyloid fibrils, proteins adopt a cross β-structure, and form protofilaments (2–5 nm in diameter) which can coil together or interact laterally. Amyloid fibrils are insoluble under physiological conditions of pH and ionic strength. They form non-branching fibrils of about 10 nm in diameter and between 3 and 100 μm long. They show a specific X-ray diffraction pattern with distinct intensities at 4.7 and 10 Å, corresponding to the intrastrand and stacking distances of the β-sheets, respectively (Fig. 11.16). Finally, they bind Congo red and thioflavin S/T (these are fluorescent dyes that are used to detect amyloid fibrils in clinical samples).

There are 27 different human proteins that have been involved in amyloidosis [119]. The list includes immunoglobulin light chain (in light chain amyloidosis),

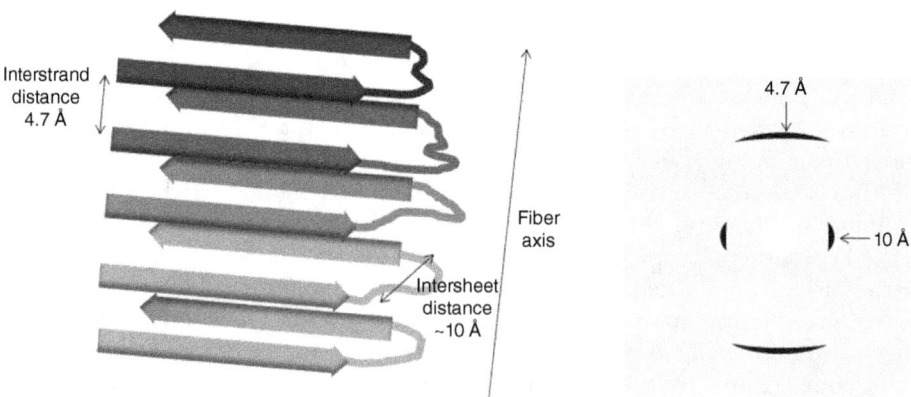

Fig. 11.16 Quaternary structure of amyloids. Left: organization of the cross β-sheet fold. Right: an X-ray fiber-diffraction pattern from aligned amyloid fibers exhibits the characteristic reflections at 4.7 and 10 Å. The meridional reflection at 4.7 Å results from the interstrand repeats, and the 10 Å equatorial reflection arises from intersheet packing.

the well-known amyloid-β peptide (in Alzheimer's disease), β2-microglobulin (in dialysis-related amyloidosis), α-synuclein (α-syn, in Parkinson's disease), huntingtin (in Huntington's disease) and tau protein (in several neurodegenerative diseases).

Amyloidoses have long been classified into two groups. The first is the localized amyloidoses, among which there are several neurodegenerative diseases such as Alzheimer's, Parkinson's and Huntington's. The associated amyloid deposits are observed just in one organ or tissue in which the amyloidogenic protein is synthesized. The second group is the systemic amyloidoses which affect several organs, and in general the kidney and liver for which the amyloid aggregates are particularly toxic.

In the localized amyloidoses, the amyloid deposits can be intracellular or extracellular. In systemic amyloidoses, the amyloid deposits are only extracellular. In this latter case, the precursor protein is secreted.

11.3.3.2 Mechanisms of amyloidogenesis

Several general mechanisms leading to the formation of amyloid fibrils have been described in the literature. They are not mutually exclusive.

The first mechanism is the seeding–nucleation model which has already been described for PrP. In this model, amyloids that have already been formed act as a template when they recruit the precursor protein. They probably preferentially recruit misfolded or partially unfolded states of the amyloidogenic protein, thus acting by conformation selection.

The second mechanism can occur when a proteolytic process leads to the accumulation of an aggregation-prone precursor. The proteases involved in this process can belong to cellular machineries in charge of degradation of misfolded proteins, but this is not necessarily the case. As an example, in the Finnish form of systemic amyloidosis related to the deposition of gelsolin fragments, a mutation that results in the loss of the calcium-binding site renders the protein susceptible to furin proteolysis, generating an abnormal fragment that undergoes a second cleavage by metalloproteinase-1 to produce the amyloidogenic 8-kDa polypeptide [120].

A third mechanism is active in familial inherited amyloidoses. The gene encoding the amyloidogenic precursor contains a mutation which destabilizes the native protein conformation and increases the population of partially unfolded, aggregation-prone states. As an example, in transthyretin (TTR) amyloidosis the amyloidogenic precursor is a monomer released from the normal TTR homotetramer. In the autosomal-dominant hereditary form, mutations render the tetramer unstable, favoring the release of monomers, which are then trapped in an energetically favorable polymerization process [121].

Finally, in the case of systemic amyloidosis, the precursor protein can be overproduced as the result of a pathological expansion of the secretory cell population. This is typically the case for light chain amyloidosis.

11.3.3.3 Transmissible amyloidosis

The similarities between the mechanisms of amyloidogenesis and those involved in the conversion of PrP^c into PrP^{Sc}, together with several recent observations, have suggested that some amyloidoses could have an infectious nature [122].

In Parkinson's disease, Lewy bodies (LB) and Lewy neurites (LN) are intracytoplasmic inclusions that develop in the cell body and neurites of affected neurons, respectively. α-syn is the primary component of these aggregates, which have been demonstrated to be amyloids. It has been recently demonstrated that α-syn is released and taken up by neurons and, as a consequence, can be transferred from one neuron to another. Remarkably, seeding activity of the propagated α-syn can be demonstrated in cell culture [123].

It has also been shown that recombinant fibrillar polyglutamine peptides (polyQ) were able to nucleate the aggregation of intracellularly expressed huntingtin protein with an extended poly-Q tract [124]. A similar seeding activity of tau has also been demonstrated in both cell culture [125] and *in vivo* [126].

Finally, as with prions, for which the existence of several strains having distinct properties has been demonstrated, the existence of two strains of α-syn has recently been demonstrated. These two strains exhibit distinct structural characteristics, different levels of cellular toxicity and different seeding and propagation properties [127].

Taken together, these results indicate that amyloidoses involved in neurodegenerative diseases are propagated through prion-like transmission mechanisms. This of course has major implications for therapeutics and prevention.

11.3.4 Functional amyloids

11.3.4.1 *A wide range of organisms have harnessed the physical properties of amyloids*

Amyloids have been associated with human diseases for more than a century. As a consequence they were considered to be toxic byproducts resulting from protein misfolding. This idea probably impeded the discovery of functional roles for amyloid assemblies.

The fact that amyloidogenic proteins are extremely diverse (see Section 11.3.3.1) indicates that amyloid assemblies can be formed by many amino acid sequences and can be considered as a peculiar type of secondary structure. In fact, almost any protein can form an amyloid *in vitro* depending on the conditions [128]. Therefore, some authors consider that amyloids are very primitive structures that have existed for as long as proteins and were a prominent feature early in the evolution of life [129]. From this point of view, the discovery of functional amyloids was not surprising.

Functional amyloids were initially identified in bacteria, fungi and insects. Amyloid fibers are often used for their unique mechanical and biochemical properties: the yield strength of amyloids is comparable to that of steel [130] and amyloids are extremely resistant to proteolysis and detergents. As an example, several bacteria use extracellular amyloid (called curli) to create a proteinaceous matrix that allows their surface adhesion. The chaplins from *Streptomyces coelicolor* are another example of amyloidogenic proteins. Chaplins are secreted proteins which assemble into an insoluble film at the air–water interface allowing *S. coelicolor* hyphae to grow into the air. Finally, it has been demonstrated that chorion proteins of insect eggs form a lamellar array of fibers protecting the eggs from harm; they are induced by physical stress, proteases and microorganisms (see review in [131]).

The first functional amyloid detected in mammals was the Pmel17 protein [132]. This protein forms amyloid-like fibrillar structures in stage II melanosomes (the organelles involved in melanin synthesis). These fibers are made by a proteolytic fragment of Pmel17 (called Mα). Several lines of evidence have shown that the role of these fibers is to act as a template for the synthesis of melanin. Indeed, it has been demonstrated that Mα fibers accelerate the polymerization of small precursor molecules such as 5,6-indolequinone (DHQ) into melanin. It is proposed that the binding of DHQ to Mα fibers increases the local concentration of DHQ and probably orientates the DHQ molecules to facilitate their polymerization. Remarkably, a similar enhancement of melanin polymerization can be obtained *in vitro* using amyloids from other proteins. Finally, by sequestering DHQ in melanosomes, Mα fibers limit their high cytotoxicity.

It is worth noting that the structure of all these aggregates might not be exactly the same as that of typical amyloids. However, they exhibit many characteristics of such structures, including Congo-red binding, fibrillar morphology and a β-sheet-rich organization with X-ray diffraction patterns consistent with amyloids.

11.3.4.2 Fungal amyloids transfer information

Several amyloid-based prions have been identified in fungi, among which the most widely studied are the proteins Sup35p and URE2p, from *Saccharomyces cerevisiae*, and HET-s, from the filamentous fungus *Podospora anserina*.

In its soluble form, Sup35p is a translation termination factor that is responsible for stopping protein synthesis at nonsense codons. Its aggregation into amyloid-like structures prevents translation termination. As a consequence, the yeast proteome is enriched in proteins having C-terminal extensions resulting from stop-codon readthrough [133].

In its soluble form, URE2p regulates nitrogen catabolism. Sequestration of Ure2p in aggregates results in constitutive expression of the Dal5 ureidosuccinate and allantoate transporter (Dal5 is normally only expressed when good nitrogen sources are unavailable) [134].

In both cases, aggregation of the proteins leads to a loss-of-function phenotype because the proteins sequestered in the aggregates cannot find their natural targets. This phenotype is transmissible upon mating and can be passed from mother to daughter cell upon division. It has also been shown that yeast populations harboring the aggregated form of the proteins have selective growth advantages under some conditions.

The *het-s* locus of *P. anserina* can encode one of two alleles: Het-S or Het-s. The *Het-s*-encoded Het-s protein generally adopts a prion structure, and this form of Het-s is denoted [Het-s], whereas the soluble form is denoted [Het-s*]. Het-s amyloid-like aggregates can be passed to progeny and the [Het-s] phenotype is heritable. The strains that do not express Het-s instead express Het-S which is normally a soluble protein. Therefore, although Het-s and Het-S share 96% sequence identity, only Het-s spontaneously exists as a prion [135].

When they come into close contact, two different strains of P. anserina can fuse and form a heterokaryon. Although heterokaryon formation is common among fungi and confers a growth advantage, P. anserina has evolved so that in most cases heterokaryon formation leads to cell death. An explanation for this is that, by limiting heterokaryon formation, P. anserina limits the spread of viral infections. This feature, called heterokaryon incompatibility, is mediated by the [Het-s] prion. It has been recently demonstrated that aggregated Het-s converts the soluble Het-S protein into an integral membrane protein which is then able to disrupt the membrane, inducing a cell death program [136] (Fig. 11.17).

11.3.4.3 Amyloids as a support for cellular memory

Normally, the exposure of yeast to pheromones induces cell-cycle arrest. However, when yeasts do not find a mating partner within a reasonable time they mount an adaptation response to resume proliferation. What is remarkable is that they keep the memory of this deceptive mating attempt and they will no longer be sensitive to pheromone exposure.

This memory is the result of the aggregation of Whi3 protein as a result of prolonged and non-productive exposure to pheromones. Normally, Whi3 is a soluble protein which binds the G1 cyclin Cln3 mRNA; as a consequence, the translation of Cln3 is inhibited. When sequestered into the aggregates, Whi3 is inactivated. Cln3 is then liberated from translational inhibition and can play its role in triggering the expression of the other G1 cyclins, thereby promoting entry into a new budding cycle [137].

The aggregation phenotype is strictly dependent on polyQ and polyasparagine (polyN) sequences of Whi3. Although not demonstrated, this strongly suggests that, in the cytoplasmic aggregates, Whi3 proteins form amyloid-like structures.

A major difference between Whi3 and yeast prions is that Whi3 aggregates are not transmitted to the progeny. They remain trapped in the mother cytoplasm and are not passed to daughters, which are therefore naive. Therefore, Whi3 aggregates are not prions—they have been called mnemons. The authors of the study mention than many of the 150 proteins containing regions predicted to promote aggregation are RNA-binding proteins. They suggest that these proteins could also be mnemons and could allow yeasts to memorize other events, so far not identified. They also mention that proteins known to be involved in memory maintenance in *Drosophila* and *Aplysia* have similar properties, revealing similarities in memory maintenance in multi- and unicellular organisms.

11.3.5 Concluding remarks

The studies performed on conformational diseases have shed light on the role of protein aggregation in non-Mendelian transmission of phenotypes.

For a long time, the current view was that there was evolutionary pressure against amyloidogenesis. However, as amyloid can be considered to be a canonical protein fold, it is probable that several organisms have harnessed the remarkable properties of this structure.

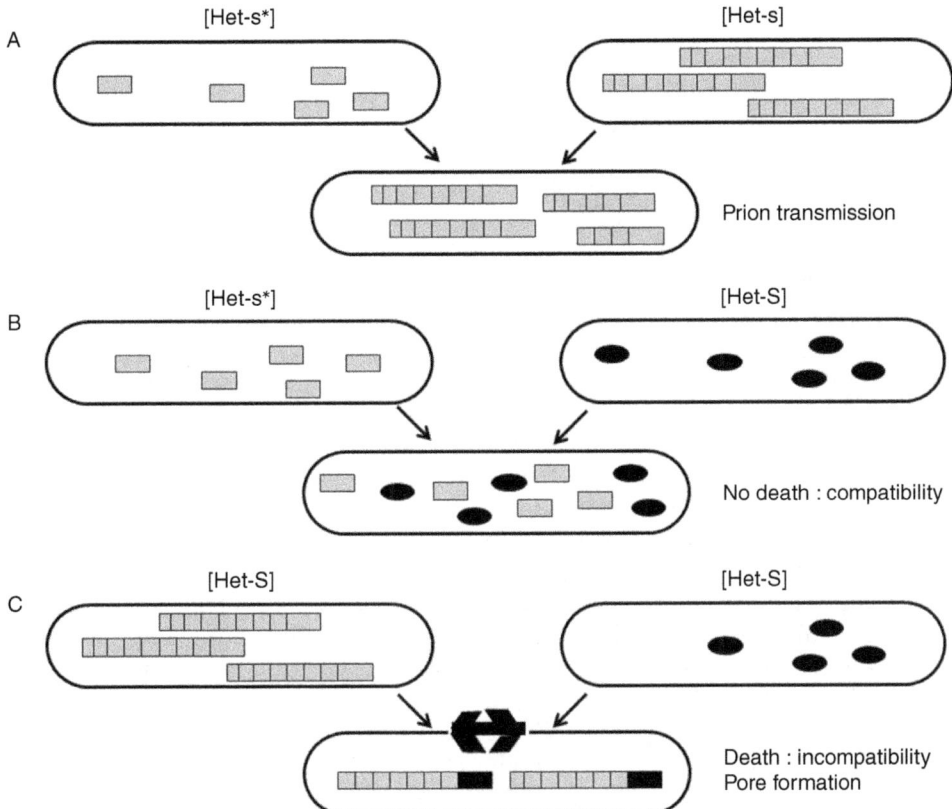

Fig. 11.17 The *P. anserina* Het-s/Het-S locus mediates heterokaryon incompatibility (adapted from [139]). When two *P. anserina* mycelia form a heterokaryon (resulting from fusion), the Het-s/Het-S heterokaryon incompatibility system dictates whether the heterokaryon will be viable (see text for details). There are three possible combinations of Het-s with Het-S. (A) A [Het-s*] strain can fuse with a [Het-s] strain resulting in a viable heterokaryon and transmission of the prion phenotype, [Het-s]. (B) A [Het-s*] strain can fuse with a [Het-S] strain resulting in a viable heterokaryon. (C) A [Het-s] strain can fuse with a [Het-S] strain resulting in a nonviable heterokaryon. Het-s amyloid-induced Het-S amyloid formation (indicated by dark rectangles) leads to a structural transition of Het-S which induces the formation of pores formed by Het-S (indicated by dark lozenges).

There is now increasing evidence that amyloids can also be functional. The results presented in this chapter strongly suggest that structural transition of proteins from a soluble form toward an aggregated state may be used to induce irreversible phenotypic switches. It would not be surprising that such a mechanism is much more widespread than currently thought. Understanding functional amyloidosis and how it is tolerated in the cell may also open up new therapeutic strategies against amyloid diseases.

References

1. Perutz MF. X-ray analysis of hemoglobin. *Science*, 1963; **140**(3569): 863–9.
2. Monod J, Changeux JP, Jacob F. Allosteric proteins and cellular control systems. *J Mol Biol*, **1963**; 6: 306–29.
3. Anfinsen CB, Haber E, Sela M, White FH, Jr. The kinetics of formation of native ribonuclease during oxidation of the reduced polypeptide chain. *Proc Natl Acad Sci USA*, 1961; **47**: 1309–14.
4. Anfinsen CB. Principles that govern the folding of protein chains. *Science*, 1973; **181**(4096): 223–30.
5. Case DA, Karplus M. Dynamics of ligand binding to heme proteins. *J Mol Biol*, 1979; **132**(3): 343–68.
6. Hayward S. Structural principles governing domain motions in proteins. *Proteins*, 1999; **36**(4): 425–35.
7. Bernado P, Mylonas E, Petoukhov MV, Blackledge M, Svergun DI. Structural characterization of flexible proteins using small-angle X-ray scattering. *J Am Chem Soc*, 2007; **129**(17): 5656–64.
8. Hammel M. Validation of macromolecular flexibility in solution by small-angle X-ray scattering (SAXS). *Eur Biophys J*, 2012; **41**(10): 789–99.
9. Mertens HD, Svergun DI. Structural characterization of proteins and complexes using small-angle X-ray solution scattering. *J Struct Biol*, 2010; **172**(1): 128–41.
10. Monod J, Wyman J, Changeux JP. On the nature of allosteric transitions: a plausible model. *J Mol Biol*, 1965; **12**: 88–118.
11. Koshland DE, Jr, Nemethy G, Filmer D. Comparison of experimental binding data and theoretical models in proteins containing subunits. *Biochemistry*, 1966; **5**(1): 365–85.
12. Mayer MP, Schroder H, Rudiger S, Paal K, Laufen T, Bukau B. Multistep mechanism of substrate binding determines chaperone activity of Hsp70. *Nat Struct Biol*, 2000; **7**(7): 586–93.
13. Bertelsen EB, Chang L, Gestwicki JE, Zuiderweg ER. Solution conformation of wild-type E. coli Hsp70 (DnaK) chaperone complexed with ADP and substrate. *Proc Natl Acad Sci USA*, 2009; **106**(21): 8471–6.
14. Kityk R, Kopp J, Sinning I, Mayer MP. Structure and dynamics of the ATP-bound open conformation of Hsp70 chaperones. *Mol Cell*, 2012; **48**(6): 863–74.
15. Volkman BF, Lipson D, Wemmer DE, Kern D. Two-state allosteric behavior in a single-domain signaling protein. *Science*, 2001; **291**(5512): 2429–33.
16. Eaton WA, Henry ER, Hofrichter J, Mozzarelli A. Is cooperative oxygen binding by hemoglobin really understood? *Nat Struct Biol*, 1999; **6**(4): 351–8.
17. Perutz MF. Stereochemistry of cooperative effects in haemoglobin. *Nature*, 1970; **228**(5273): 726–39.
18. Suel GM, Lockless SW, Wall MA, Ranganathan R. Evolutionarily conserved networks of residues mediate allosteric communication in proteins. *Nat Struct Biol*, 2003; **10**(1): 59–69.
19. Lockless SW, Ranganathan R. Evolutionarily conserved pathways of energetic connectivity in protein families. *Science*, 1999; **286**(5438): 295–9.

20. Daily MD, Gray JJ. Allosteric communication occurs via networks of tertiary and quaternary motions in proteins. *PLoS Comput Biol*, 2009; **5**(2): e1000293.
21. Popovych N, Sun S, Ebright RH, Kalodimos CG. Dynamically driven protein allostery. *Nat Struct Mol Biol*, 2006; **13**(9): 831–8.
22. Petit CM, Zhang J, Sapienza PJ, Fuentes EJ, Lee AL. Hidden dynamic allostery in a PDZ domain. *Proc Natl Acad Sci USA*, 2009; **106**(43): 18249–54.
23. Hilser VJ, Wrabl JO, Motlagh HN. Structural and energetic basis of allostery. *Annu Rev Biophys*, 2012; **41**: 585–609.
24. Jensen MR, Ruigrok RW, Blackledge M. Describing intrinsically disordered proteins at atomic resolution by NMR. *Curr Opin Struct Biol*, 2013; **23**(3): 426–35.
25. Dunker AK, Cortese MS, Romero P, Iakoucheva LM, Uversky VN. Flexible nets. The roles of intrinsic disorder in protein interaction networks. *FEBS J*, 2005; **272**(20): 5129–48.
26. Ekman D, Light S, Bjorklund AK, Elofsson A. What properties characterize the hub proteins of the protein–protein interaction network of *Saccharomyces cerevisiae*? *Genome Biol*, 2006; **7**(6): R45.
27. Haynes C, Oldfield CJ, Ji F, Klitgord N, Cusick ME, Radivojac P, et al. Intrinsic disorder is a common feature of hub proteins from four eukaryotic interactomes. *PLoS Comput Biol*, 2006; **2**(8): e100.
28. Singh GP, Ganapathi M, Dash D. Role of intrinsic disorder in transient interactions of hub proteins. *Proteins*, 2007; **66**(4): 761–5.
29. Kim PM, Sboner A, Xia Y, Gerstein M. The role of disorder in interaction networks: a structural analysis. *Mol Syst Biol*, 2008; **4**: 179.
30. Cumberworth A, Lamour G, Babu MM, Gsponer J. Promiscuity as a functional trait: intrinsically disordered regions as central players of interactomes. *Biochem J*, 2013; **454**(3): 361–9.
31. Tompa P, Fuxreiter M. Fuzzy complexes: polymorphism and structural disorder in protein-protein interactions. *Trends Biochem Sci*, 2008; **33**(1): 2–8.
32. Smith AE, Helenius A. How viruses enter animal cells. *Science*, 2004; **304**(5668): 237–42.
33. Li Y, Modis Y. A novel membrane fusion protein family in Flaviviridae? *Trends Microbiol*, 2014; **22**(4): 176–82.
34. Harrison SC. Viral membrane fusion. *Nat Struct Mol Biol*, 2008; **15**(7): 690–8.
35. Chernomordik LV, Frolov VA, Leikina E, Bronk P, Zimmerberg J. The pathway of membrane fusion catalyzed by influenza hemagglutinin: restriction of lipids, hemifusion, and lipidic fusion pore formation. *J Cell Biol*, 1998; **140**(6): 1369–82.
36. Gaudin Y. Rabies virus-induced membrane fusion pathway. *J Cell Biol*, 2000; **150**(3): 601–12.
37. Zaitseva E, Mittal A, Griffin DE, Chernomordik LV. Class II fusion protein of alphaviruses drives membrane fusion through the same pathway as class I proteins. *J Cell Biol*, 2005; **169**(1): 167–77.
38. Chernomordik LV, Kozlov MM. Mechanics of membrane fusion. *Nat Struct Mol Biol*, 2008; **15**(7): 675–83.

39. Cohen FS, Melikyan GB. The energetics of membrane fusion from binding, through hemifusion, pore formation, and pore enlargement. *J Membr Biol*, 2004; **199**(1): 1–14.
40. Wilson IA, Skehel JJ, Wiley DC. Structure of the haemagglutinin membrane glycoprotein of influenza virus at 3 A resolution. *Nature*, 1981; **289**(5796): 366–73.
41. Skehel JJ, Wiley DC. Receptor binding and membrane fusion in virus entry: the influenza hemagglutinin. *Annu Rev Biochem*, 2000; **69**: 531–69.
42. Bullough PA, Hughson FM, Skehel JJ, Wiley DC. Structure of influenza haemagglutinin at the pH of membrane fusion. *Nature*, 1994; **371**(6492): 37–43.
43. Chen J, Lee KH, Steinhauer DA, Stevens DJ, Skehel JJ, Wiley DC. Structure of the hemagglutinin precursor cleavage site, a determinant of influenza pathogenicity and the origin of the labile conformation. *Cell*, 1998; **95**(3): 409–17.
44. Weissenhorn W, Dessen A, Harrison SC, Skehel JJ, Wiley DC. Atomic structure of the ectodomain from HIV-1 gp41. *Nature*, 1997; **387**(6631): 426–30.
45. Baker KA, Dutch RE, Lamb RA, Jardetzky TS. Structural basis for paramyxovirus-mediated membrane fusion. *Mol Cell*, 1999; **3**(3): 309–19.
46. Weissenhorn W, Carfi A, Lee KH, Skehel JJ, Wiley DC. Crystal structure of the Ebola virus membrane fusion subunit, GP2, from the envelope glycoprotein ectodomain. *Mol Cell*, 1998; **2**(5): 605–16.
47. Supekar VM, Bruckmann C, Ingallinella P, Bianchi E, Pessi A, Carfi A. Structure of a proteolytically resistant core from the severe acute respiratory syndrome coronavirus S2 fusion protein. *Proc Natl Acad Sci USA*, 2004; **101**(52): 17958–63.
48. Yin HS, Wen X, Paterson RG, Lamb RA, Jardetzky TS. Structure of the parainfluenza virus 5 F protein in its metastable, prefusion conformation. *Nature*, 2006; **439**(7072): 38–44.
49. Lee JE, Fusco ML, Hessell AJ, Oswald WB, Burton DR, Saphire EO. Structure of the Ebola virus glycoprotein bound to an antibody from a human survivor. *Nature*, 2008; **454**(7201): 177–82.
50. Yin HS, Paterson RG, Wen X, Lamb RA, Jardetzky TS. Structure of the uncleaved ectodomain of the paramyxovirus (hPIV3) fusion protein. *Proc Natl Acad Sci USA*, 2005; **102**(26): 9288–93.
51. Melikyan GB. Common principles and intermediates of viral protein-mediated fusion: the HIV-1 paradigm. *Retrovirology*, 2008; **5**: 111.
52. Lamb RA, Jardetzky TS. Structural basis of viral invasion: lessons from paramyxovirus F. *Curr Opin Struct Biol*, 2007; **17**(4): 427–36.
53. Lescar J, Roussel A, Wien MW, Navaza J, Fuller SD, Wengler G, et al. The fusion glycoprotein shell of Semliki Forest virus: an icosahedral assembly primed for fusogenic activation at endosomal pH. *Cell*, 2001; **105**(1): 137–48.
54. Rey FA, Heinz FX, Mandl C, Kunz C, Harrison SC. The envelope glycoprotein from tick-borne encephalitis virus at 2 A resolution. *Nature*, 1995; **375**(6529): 291–8.
55. Kanai R, Kar K, Anthony K, Gould LH, Ledizet M, Fikrig E, et al. Crystal structure of west nile virus envelope glycoprotein reveals viral surface epitopes. *J Virol*, 2006; **80**(22): 11000–8.

56. Modis Y, Ogata S, Clements D, Harrison SC. A ligand-binding pocket in the dengue virus envelope glycoprotein. *Proc Natl Acad Sci USA*, 2003; **100**(12): 6986–91.
57. Dessau M, Modis Y. Crystal structure of glycoprotein C from Rift Valley fever virus. *Proc Natl Acad Sci USA*, 2013; **110**(5): 1696–701.
58. Ferlenghi I, Clarke M, Ruttan T, Allison SL, Schalich J, Heinz FX, et al. Molecular organization of a recombinant subviral particle from tick-borne encephalitis virus. *Mol Cell*, 2001; **7**(3): 593–602.
59. Perez-Vargas J, Krey T, Valansi C, Avinoam O, Haouz A, Jamin M, et al. Structural basis of eukaryotic cell-cell fusion. *Cell*, 2014; **157**(2): 407–19.
60. Sanchez-San Martin C, Liu CY, Kielian M. Dealing with low pH: entry and exit of alphaviruses and flaviviruses. *Trends Microbiol*, 2009; **17**(11): 514–21.
61. Stadler K, Allison SL, Schalich J, Heinz FX. Proteolytic activation of tick-borne encephalitis virus by furin. *J Virol*, 1997; **71**(11): 8475–81.
62. Zhang X, Fugere M, Day R, Kielian M. Furin processing and proteolytic activation of Semliki Forest virus. *J Virol*, 2003; **77**(5): 2981–9.
63. Bressanelli S, Stiasny K, Allison SL, Stura EA, Duquerroy S, Lescar J, et al. Structure of a flavivirus envelope glycoprotein in its low-pH-induced membrane fusion conformation. *EMBO J*, 2004; **23**(4): 728–38.
64. Gibbons DL, Vaney MC, Roussel A, Vigouroux A, Reilly B, Lepault J, et al. Conformational change and protein-protein interactions of the fusion protein of Semliki Forest virus. *Nature*, 2004; **427**(6972): 320–5.
65. Modis Y, Ogata S, Clements D, Harrison SC. Structure of the dengue virus envelope protein after membrane fusion. *Nature*, 2004; **427**(6972): 313–19.
66. Gibbons DL, Erk I, Reilly B, Navaza J, Kielian M, Rey FA, et al. Visualization of the target-membrane-inserted fusion protein of Semliki Forest virus by combined electron microscopy and crystallography. *Cell*, 2003; **114**(5): 573–83.
67. Roche S, Bressanelli S, Rey FA, Gaudin Y. Crystal structure of the low-pH form of the vesicular stomatitis virus glycoprotein G. *Science*, 2006; **313**(5784): 187–91.
68. Heldwein EE, Lou H, Bender FC, Cohen GH, Eisenberg RJ, Harrison SC. Crystal structure of glycoprotein B from herpes simplex virus 1. *Science*, 2006; **313**(5784): 217–20.
69. Backovic M, Longnecker R, Jardetzky TS. Structure of a trimeric variant of the Epstein–Barr virus glycoprotein B. *Proc Natl Acad Sci USA*, 2009; **106**(8): 2880–5.
70. Kadlec J, Loureiro S, Abrescia NG, Stuart DI, Jones IM. The postfusion structure of baculovirus gp64 supports a unified view of viral fusion machines. *Nat Struct Mol Biol*, 2008; **15**(10): 1024–30.
71. Roche S, Rey FA, Gaudin Y, Bressanelli S. Structure of the prefusion form of the vesicular stomatitis virus glycoprotein g. *Science*, 2007; **315**(5813): 843–8.
72. Eisenberg RJ, Atanasiu D, Cairns TM, Gallagher JR, Krummenacher C, Cohen GH. Herpes virus fusion and entry: a story with many characters. *Viruses*, 2012; **4**(5): 800–32.

73. Albertini AA, Merigoux C, Libersou S, Madiona K, Bressanelli S, Roche S, et al. Characterization of monomeric intermediates during VSV glycoprotein structural transition. *PLoS Pathog*, 2012; **8**(2): e1002556.
74. Roche S, Gaudin Y. Characterization of the equilibrium between the native and fusion-inactive conformation of rabies virus glycoprotein indicates that the fusion complex is made of several trimers. *Virology*, 2002; **297**(1): 128–35.
75. Gaudin Y. Reversibility in fusion protein conformational changes. The intriguing case of rhabdovirus-induced membrane fusion. *Subcell Biochem*, 2000; **34**: 379–408.
76. Krey T, d'Alayer J, Kikuti CM, Saulnier A, Damier-Piolle L, Petitpas I, et al. The disulfide bonds in glycoprotein E2 of hepatitis C virus reveal the tertiary organization of the molecule. *PLoS Pathog*, 2010; **6**(2): e1000762.
77. Li Y, Wang J, Kanai R, Modis Y. Crystal structure of glycoprotein E2 from bovine viral diarrhea virus. *Proc Natl Acad Sci USA*, 2013; **110**(17): 6805–10.
78. El Omari K, Iourin O, Harlos K, Grimes JM, Stuart DI. Structure of a pestivirus envelope glycoprotein E2 clarifies its role in cell entry. *Cell Rep*, 2013; **3**(1): 30–5.
79. Khan AG, Whidby J, Miller MT, Scarborough H, Zatorski AV, Cygan A, et al. Structure of the core ectodomain of the hepatitis C virus envelope glycoprotein 2. *Nature*, 2014; **509**: 381–4.
80. Kong L, Giang E, Nieusma T, Kadam RU, Cogburn KE, Hua Y, et al. Hepatitis C virus E2 envelope glycoprotein core structure. *Science*, 2013; **342**(6162): 1090–4.
81. Fritz R, Stiasny K, Heinz FX. Identification of specific histidines as pH sensors in flavivirus membrane fusion. *J Cell Biol*, 2008; **183**(2): 353–61.
82. Qin ZL, Zheng Y, Kielian M. Role of conserved histidine residues in the low-pH dependence of the Semliki Forest virus fusion protein. *J Virol*, 2009; **83**(9): 4670–7.
83. Ferlin A, Raux H, Baquero E, Lepault J, Gaudin Y. Characterization of pH-sensitive molecular switches that trigger the structural transition of vesicular stomatitis virus glycoprotein from the postfusion state toward the prefusion state. *J Virol*, 2014; **88**(22): 13396–409.
84. Daniels RS, Downie JC, Hay AJ, Knossow M, Skehel JJ, Wang ML, et al. Fusion mutants of the influenza virus hemagglutinin glycoprotein. *Cell*, 1985; **40**(2): 431–9.
85. Durrer P, Gaudin Y, Ruigrok RW, Graf R, Brunner J. Photolabeling identifies a putative fusion domain in the envelope glycoprotein of rabies and vesicular stomatitis viruses. *J Biol Chem*, 1995; **270**(29): 17575–81.
86. Tsurudome M, Gluck R, Graf R, Falchetto R, Schaller U, Brunner J. Lipid interactions of the hemagglutinin HA2 NH2-terminal segment during influenza virus-induced membrane fusion. *J Biol Chem*, 1992; **267**(28): 20225–32.
87. Russell CJ, Jardetzky TS, Lamb RA. Membrane fusion machines of paramyxoviruses: capture of intermediates of fusion. *EMBO J*, 2001; **20**(15): 4024–34.
88. Wild CT, Shugars DC, Greenwell TK, McDanal CB, Matthews TJ. Peptides corresponding to a predictive alpha-helical domain of human immunodeficiency

virus type 1 gp41 are potent inhibitors of virus infection. *Proc Natl Acad Sci USA*, 1994; **91**(21): 9770–4.
89. Cardone G, Brecher M, Fontana J, Winkler DC, Butan C, White JM, et al. Visualization of the two-step fusion process of the retrovirus avian sarcoma/leukosis virus by cryo-electron tomography. *J Virol*, 2012; **86**(22): 12129–37.
90. Kim YH, Donald JE, Grigoryan G, Leser GP, Fadeev AY, Lamb RA, et al. Capture and imaging of a prehairpin fusion intermediate of the paramyxovirus PIV5. *Proc Natl Acad Sci USA*, 2011; **108**(52): 20992–7.
91. Stiasny K, Kossl C, Lepault J, Rey FA, Heinz FX. Characterization of a structural intermediate of flavivirus membrane fusion. *PLoS Pathog*, 2007; **3**(2): e20.
92. Liao M, Kielian M. Domain III from class II fusion proteins functions as a dominant-negative inhibitor of virus membrane fusion. *J Cell Biol*, 2005; **171**(1): 111–20.
93. Sanchez-San Martin C, Sosa H, Kielian M. A stable prefusion intermediate of the alphavirus fusion protein reveals critical features of class II membrane fusion. *Cell Host Microbe*, 2008; **4**(6): 600–8.
94. Roman-Sosa G, Kielian M. The interaction of alphavirus E1 protein with exogenous domain III defines stages in virus-membrane fusion. *J Virol*, 2011; **85**(23): 12271–9.
95. Baquero E, Albertini AA, Vachette P, Lepault J, Bressanelli S, Gaudin Y. Intermediate conformations during viral fusion glycoprotein structural transition. *Curr Opin Virol*, 2013; **3**(2): 143–50.
96. Cleverley DZ, Lenard J. The transmembrane domain in viral fusion: essential role for a conserved glycine residue in vesicular stomatitis virus G protein. *Proc Natl Acad Sci USA*, 1998; **95**(7): 3425–30.
97. Kemble GW, Danieli T, White JM. Lipid-anchored influenza hemagglutinin promotes hemifusion, not complete fusion. *Cell*, 1994; **76**(2): 383–91.
98. Melikyan GB, Markosyan RM, Roth MG, Cohen FS. A point mutation in the transmembrane domain of the hemagglutinin of influenza virus stabilizes a hemifusion intermediate that can transit to fusion. *Mol Biol Cell*, 2000; **11**(11): 3765–75.
99. Griffith JS. Self-replication and scrapie. *Nature*, 1967; **215**(5105): 1043–4.
100. Riek R, Hornemann S, Wider G, Billeter M, Glockshuber R, Wuthrich K. NMR structure of the mouse prion protein domain PrP(121-231). *Nature*, 1996; **382**(6587): 180–2.
101. Eghiaian F, Grosclaude J, Lesceu S, Debey P, Doublet B, Treguer E, et al. Insight into the PrPC→PrPSc conversion from the structures of antibody-bound ovine prion scrapie-susceptibility variants. *Proc Natl Acad Sci USA*, 2004; **101**(28): 10254–9.
102. Zahn R, Liu A, Luhrs T, Riek R, von Schroetter C, Lopez Garcia F, et al. NMR solution structure of the human prion protein. *Proc Natl Acad Sci USA*, 2000; **97**(1): 145–50.
103. Govaerts C, Wille H, Prusiner SB, Cohen FE. Evidence for assembly of prions with left-handed beta-helices into trimers. *Proc Natl Acad Sci USA*, 2004; **101**(22): 8342–7.

104. Sim VL, Caughey B. Ultrastructures and strain comparison of under-glycosylated scrapie prion fibrils. *Neurobiol Aging*, 2009; **30**(12): 2031–42.
105. Diaz-Espinoza R, Soto C. High-resolution structure of infectious prion protein: the final frontier. *Nat Struct Mol Biol*, 2012; **19**(4): 370–7.
106. DeMarco ML, Daggett V. From conversion to aggregation: protofibril formation of the prion protein. *Proc Natl Acad Sci USA*, 2004; **101**(8): 2293–8.
107. Smirnovas V, Baron GS, Offerdahl DK, Raymond GJ, Caughey B, Surewicz WK. Structural organization of brain-derived mammalian prions examined by hydrogen-deuterium exchange. *Nat Struct Mol Biol*, 2011; **18**(4): 504–6.
108. Silveira JR, Raymond GJ, Hughson AG, Race RE, Sim VL, Hayes SF, et al. The most infectious prion protein particles. *Nature*, 2005; **437**(7056): 257–61.
109. Goold R, Rabbanian S, Sutton L, Andre R, Arora P, Moonga J, et al. Rapid cell-surface prion protein conversion revealed using a novel cell system. *Nat Commun*, 2011; **2**: 281.
110. Hammond C, Helenius A. Quality control in the secretory pathway. *Curr Opin Cell Biol*, 1995; **7**(4): 523–9.
111. Knaus KJ, Morillas M, Swietnicki W, Malone M, Surewicz WK, Yee VC. Crystal structure of the human prion protein reveals a mechanism for oligomerization. *Nat Struct Biol*, 2001; **8**(9): 770–4.
112. Bessen RA, Kocisko DA, Raymond GJ, Nandan S, Lansbury PT, Caughey B. Non-genetic propagation of strain-specific properties of scrapie prion protein. *Nature*, 1995; **375**(6533): 698–700.
113. Kim C, Haldiman T, Cohen Y, Chen W, Blevins J, Sy MS, et al. Protease-sensitive conformers in broad spectrum of distinct PrPSc structures in sporadic Creutzfeldt–Jakob disease are indicator of progression rate. *PLoS Pathog*, 2011; **7**(9): e1002242.
114. Caughey B, Raymond GJ, Bessen RA. Strain-dependent differences in beta-sheet conformations of abnormal prion protein. *J Biol Chem*, 1998; **273**(48): 32230–5.
115. Lepault J, Petitpas I, Erk I, Navaza J, Bigot D, Dona M, et al. Structural polymorphism of the major capsid protein of rotavirus. *EMBO J*, 2001; **20**(7): 1498–507.
116. Chevalier C, Lepault J, Erk I, Da Costa B, Delmas B. The maturation process of pVP2 requires assembly of infectious bursal disease virus capsids. *J Virol*, 2002; **76**(5): 2384–92.
117. Kocisko DA, Come JH, Priola SA, Chesebro B, Raymond GJ, Lansbury PT, et al. Cell-free formation of protease-resistant prion protein. *Nature*, 1994; **370**(6489): 471–4.
118. Saborio GP, Permanne B, Soto C. Sensitive detection of pathological prion protein by cyclic amplification of protein misfolding. *Nature*, 2001; **411**(6839): 810–13.
119. Sipe JD, Benson MD, Buxbaum JN, Ikeda S, Merlini G, Saraiva MJ, et al. Amyloid fibril protein nomenclature: 2010 recommendations from the nomenclature committee of the International Society of Amyloidosis. *Amyloid*, 2010; **17**(3–4): 101–4.

120. Huff ME, Page LJ, Balch WE, Kelly JW. Gelsolin domain 2 Ca2+ affinity determines susceptibility to furin proteolysis and familial amyloidosis of finnish type. *J Mol Biol*, 2003; **334**(1): 119–27.
121. Hurshman AR, White JT, Powers ET, Kelly JW. Transthyretin aggregation under partially denaturing conditions is a downhill polymerization. *Biochemistry*, 2004; **43**(23): 7365–81.
122. Brundin P, Melki R, Kopito R. Prion-like transmission of protein aggregates in neurodegenerative diseases. *Nat Rev Mol Cell Biol*, 2010; **11**(4): 301–7.
123. Hansen C, Angot E, Bergstrom AL, Steiner JA, Pieri L, Paul G, et al. alpha-Synuclein propagates from mouse brain to grafted dopaminergic neurons and seeds aggregation in cultured human cells. *J Clin Invest*, 2011; **121**(2): 715–25.
124. Ren PH, Lauckner JE, Kachirskaia I, Heuser JE, Melki R, Kopito RR. Cytoplasmic penetration and persistent infection of mammalian cells by polyglutamine aggregates. *Nat Cell Biol*, 2009; **11**(2): 219–25.
125. Frost B, Jacks RL, Diamond MI. Propagation of tau misfolding from the outside to the inside of a cell. *J Biol Chem*, 2009; **284**(19): 12845–52.
126. Clavaguera F, Bolmont T, Crowther RA, Abramowski D, Frank S, Probst A, et al. Transmission and spreading of tauopathy in transgenic mouse brain. *Nat Cell Biol*, 2009; **11**(7): 909–13.
127. Bousset L, Pieri L, Ruiz-Arlandis G, Gath J, Jensen PH, Habenstein B, et al. Structural and functional characterization of two alpha-synuclein strains. *Nat Commun*, 2013; **4**: 2575.
128. Stefani M, Dobson CM. Protein aggregation and aggregate toxicity: new insights into protein folding, misfolding diseases and biological evolution. *J Mol Med*, 2003; **81**(11): 678–99.
129. Chernoff YO. Amyloidogenic domains, prions and structural inheritance: rudiments of early life or recent acquisition? *Curr Opin Chem Biol*, 2004; **8**(6): 665–71.
130. Smith JF, Knowles TP, Dobson CM, Macphee CE, Welland ME. Characterization of the nanoscale properties of individual amyloid fibrils. *Proc Natl Acad Sci USA*, 2006; **103**(43): 15806–11.
131. Fowler DM, Koulov AV, Balch WE, Kelly JW. Functional amyloid—from bacteria to humans. *Trends Biochem Sci*, 2007; **32**(5): 217–24.
132. Fowler DM, Koulov AV, Alory-Jost C, Marks MS, Balch WE, Kelly JW. Functional amyloid formation within mammalian tissue. *PLoS Biol*, 2006; **4**(1): e6.
133. True HL, Lindquist SL. A yeast prion provides a mechanism for genetic variation and phenotypic diversity. *Nature*, 2000; **407**(6803): 477–83.
134. Lian HY, Jiang Y, Zhang H, Jones GW, Perrett S. The yeast prion protein Ure2: structure, function and folding. *Biochim Biophys Acta*, 2006; **1764**(3): 535–45.
135. Coustou V, Deleu C, Saupe S, Begueret J. The protein product of the het-s heterokaryon incompatibility gene of the fungus *Podospora anserina* behaves as a prion analog. *Proc Natl Acad Sci USA*, 1997; **94**(18): 9773–8.

136. Seuring C, Greenwald J, Wasmer C, Wepf R, Saupe SJ, Meier BH, et al. The mechanism of toxicity in HET-S/HET-s prion incompatibility. *PLoS Biol*, 2012; **10**(12): e1001451.
137. Caudron F, Barral Y. A super-assembly of Whi3 encodes memory of deceptive encounters by single cells during yeast courtship. *Cell*, 2013; **155**(6): 1244–57.
138. Lansbury PT, Jr., Caughey B. The chemistry of scrapie infection: implications of the "ice 9" metaphor. *Chem Biol*, 1995; **2**(1): 1–5.
139. Fowler DM, Kelly JW. Functional amyloidogenesis and cytotoxicity-insights into biology and pathology. *PLoS Biol*, 2012; **10**(12): e1001459.

Part 6

Membrane transporters: from structure to function

12
The dos and don'ts of handling membrane proteins for structural studies

Christine ZIEGLER

University of Regensburg, Faculty of Biology and Preclinical Studies, Department of Membrane Protein Crystallography, Regensburg, Germany

Abstract

Up to 30% of all open reading frames in cells are predicted to encode membrane proteins, and their importance in many biological pathways as well as their role in diseases is unquestionable. Intensive structure–function studies on all classes of membrane proteins have been launched. This combined effort in membrane protein biophysics and biochemistry led to tremendous progress in understanding the function of membrane proteins based on structural, biochemical, functional, bioinformatics and spectroscopic data. Despite all these efforts, the investigation of membrane proteins remains a challenge. In particular, determination of the structure of a membrane protein is still costly in terms of time and resources and has an uncertain payoff. Here, we will summarize techniques and important facts about the structural biology of membrane proteins to reveal functionally meaningful structural information.

Keywords

Crystallization, cryo-electron microscopy, detergent, lipids, membrane protein, structure

Chapter Contents

12 The dos and don'ts of handling membrane proteins for structural studies   399
Christine ZIEGLER

12.1	Introduction	401
12.2	Membrane proteins—Fragile! Handle with care!	401
	12.2.1 Why is working with membrane proteins so difficult?	401
	12.2.2 Detergents: a suitable surrogate for lipids?	402
	12.2.3 Amphiphiles and amphipols: detergents for structural biology	404
	12.2.4 Role of lipids in the purification of membrane transport proteins	405
	12.2.5 Nanodiscs: preparing transporters for biophysical investigations	406
12.3	Determining the structure of a membrane protein: difficult but not impossible	408
	12.3.1 Membrane protein crystallization	408
	12.3.2 Nanobodies: a versatile tool for transporter crystallization	408
	12.3.3 Cryo-EM: single-particle analysis and electron crystallography	410
	12.3.4 Mass spectrometry: a linkage from structure to function	410
	References	411

12.1 Introduction

Cellular membranes consist of a complex mixture of different components. A two-dimensional (2D) matrix of lipids provides the hydrophobic environment for membrane proteins, for example ion channels, transporters, receptors, pumps, respiratory and scaffolding complexes. Most of these proteins do not act independently of each other, and together they maintain the vital external and internal cellular environment, providing signaling and trafficking platforms for processes fundamental to life. Membrane proteins are ubiquitous and have diverse structures and functions. Up to 30% of all open reading frames in cells are predicted to encode membrane proteins, and their importance in many biological pathways, as well as their role in disease, is unquestionable.

Intensive structure–function studies on all classes of membrane proteins have been launched, and in the last decade combined efforts in membrane protein biophysics have led to tremendous progress in understanding the function of membrane proteins based on structural, biochemical, functional, bioinformatics and spectroscopic data.

Despite all these efforts, the investigation of membrane proteins remains a challenge. In particular the determination of the structure of a membrane protein is still costly in terms of time and resources and has an uncertain payoff. These limitations are predominantly due to the difficulties associated with obtaining sufficient quantities of the membrane protein by providing a native membrane environment *in vitro* (e.g., reconstituting the proteins in a lipid environment in functional form). However, some progress has been made with regard to overexpression of recombinant membrane proteins in different expression hosts. The development of new detergents and lipids for more efficient solubilization and crystallization is still ongoing, and has been further supported by new emerging techniques like single-particle cryo-electron microscopy (cryo-EM) or native mass spectrometry. The stability of membrane proteins can be improved systematically through mutations, deletions, engineering of fusion partners and monoclonal antibodies. On a technical level, developments in automation and in the quality of synchrotron radiation have further increased the number of atomic structures discovered.

Here, we will summarize techniques and important facts about membrane protein structural biology to reveal functionally meaningful structural information.

12.2 Membrane proteins—Fragile! Handle with care!

12.2.1 Why is working with membrane proteins so difficult?

An initial problem when studying membrane proteins is to produce functional, correctly folded and soluble material in sufficient quantities. Membranes are densely packed with a variety of different proteins and as a consequence individual membrane proteins are usually present at low levels in their native environments and have to be overexpressed, often heterologously.

The challenge after translation is to insert the membrane protein into the membrane of the expression host in sufficiently large quantities without causing its aggregation in the cytoplasm. This is a balanced interplay between the translation and insertion machinery that also involves the surrounding lipids—even bacterial membrane proteins are not trivial to produce. To make things worse, eukaryotic proteins require post-translational modifications for trafficking and therefore rely on suitable translational machinery that is not available in bacterial expression hosts like *Escherichia coli*. The next difficulty is that membrane proteins depend on the chemical and physical presence of both specific and bulk lipids for stability and function (see Section 12.2.2).

The majority of biophysical techniques that work particularly well for determining the structure and function of soluble proteins, such as NMR, X-ray crystallography, circular dichroism and isothermal titration calorimetry, appear to be more problematic for membrane proteins (and sometimes even impossible). The pre-requisite for any biophysical study is the extraction of the membrane protein from its native membrane and its reconstitution in a detergent, because these proteins are not soluble in aqueous solution *per se*. The nature and amount of detergent are crucial. Stability during purification, crystallization and functionality in *in vitro* assays depend basically on this initial solubilization step when the protein is extracted from the membrane by the action of a detergent. The ultimate question in every experiment is how closely does the solubilizing detergent mimic the natural lipid bilayer environment?

12.2.2 Detergents: a suitable surrogate for lipids?

Detergents are amphipathic molecules that, similar to lipids, consist of a polar head group and a hydrophobic chain. Due to their amphipathic nature detergents disrupt the lipid bilayer at a certain concentration. The difficulty with finding the right detergent to solubilize a membrane protein is that bulk lipids and unspecific lipids should be dissolved, while the protein structure together with functionally important lipids should remain intact.

In aqueous solutions, detergents above a certain concentration spontaneously form spherical micellar structures. This concentration is called the "critical micellar concentration" (CMC). The CMC represents a threshold below which the detergent will form monomers instead of micelles. This is important in experiments when the detergent should be removed by dialysis, as only monomers can be dialyzed. The CMC is not constant but a function of pH, ionic strength and temperature. It also depends strongly on the presence of lipids and protein. As a rule of thumb, the CMC increases with the number of double bonds and branch points and decreases with the length of the hydrophobic tail. The CMC also determines the minimum amount of free detergent required to keep membrane proteins stable, i.e., soluble. The CMC of detergents can be determined by various techniques such as surface tension measurements, isothermal calorimetry, dynamic light scattering and fluorescence-based methods [1,2].

Detergents are in a constant equilibrium between monomers, micelles and, last but not least, the fraction that is bound tightly to the transmembrane region of the membrane protein, called the surfactant layer. An example for this surfactant layer is given in Fig. 12.1. The monomeric pore-forming protein OmpG from outer membranes of *E. coli* was crystallized in complex with a total of 151 ordered LDAO (lauryldimethylamine oxide) detergent molecules mimicking the outer membrane lipid bilayer [3].

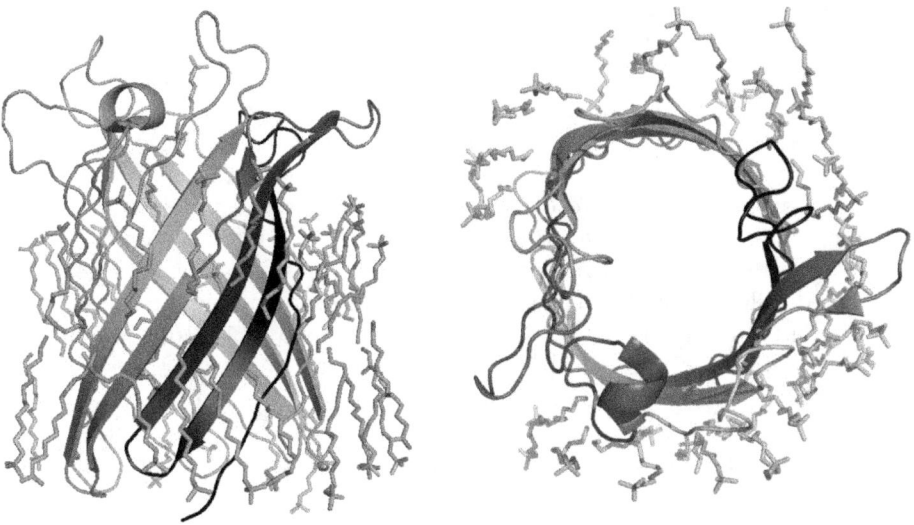

Fig. 12.1 Detergent molecules of LDAO surrounding the outer membrane porin OmpG: side view (left) and top view (right). (PDB entry code 2IWV [3].)

A membrane protein will aggregate if the surfactant concentration (not necessarily the concentration in the buffer) drops below its CMC; however, the surfactant concentration is not entirely independent of the concentration of free micelles in the buffer.

The choice of detergent for membrane protein investigations remains largely a matter of trial and error and also depends on the subsequent application—for example, will the detergent remain in complex with the membrane protein, as is the case in 3D crystallography or in some spectroscopic methods, or will it be removed after purification as in 2D crystallization and reconstitution, when the protein is mixed with lipids? Again, it is generally tricky to answer the question of how much detergent needs to be used. During 3D crystallization, minimizing free micelles in solution could reduce unproductive collisions between molecules and help the nucleation process. On the other hand, maximizing the number of free micelles might be beneficial during crystal growth to literally fill the holes in the crystal lattice of a membrane protein due to the reduced crystal contacts and the resulting high solvent content. During purification, the detergent surfactant concentration, i.e., the concentration of detergent associated with membrane protein, is altered automatically, which might also lead to stability problems. This is crucial during the concentration of proteins using centrifugal concentrators, in which detergent micelles are also enriched. Therefore, the concentration of detergent is often unknown after protein concentration, although it remains the most critical component in crystallization experiments. Colorimetric methods for determining detergent concentration are quick and simple, but limited in application to specific classes of detergents, while Fourier transform infrared spectroscopy, thin-layer chromatography and contact angle measurement are compatible with a wide range of detergents. The most sensitive techniques are gas chromatography and radio gas chromatography [1].

A large number of detergents are commercially available. Detergents are classified according to the nature of their hydrophilic groups. Ionic detergents, for example sodium dodecyl sulfate (SDS), N-laurylsarcosine and sodium cholate, have a head group with a net charge and are known for being harsh. They tend to denature membrane proteins by disrupting the hydrophobic interactions, thereby unfolding the protein core. Anionic bile acids are milder than linear ionic detergents. Non-ionic detergents, like the sugar-based maltosides and glucosides, are the most common detergents. They are considered mild as they break the lipid–lipid and protein–lipid interactions rather than the protein–protein interactions. However, short-chain (C7–C10) detergents, such as n-octyl-β-D-glucoside, can inactivate a membrane protein during the solubilization process. n-dodecyl-β-D-maltoside (DDM) and n-decyl-β-D-maltoside (DM) are the most frequently used detergents for a large group of membrane proteins. As a rule of thumb, shorter-chain detergents are suitable for crystallography while the longer-chain detergents are used in the reconstitution process. Zwitterionic detergents, for example CHAPS, CHAPSO, LDAO and Fos-12, combine properties of both ionic and non-ionic detergents and are milder than ionic detergents. They have been used in both crystallization and NMR studies. New detergents are emerging to improve the processes of protein extraction, purification and crystallization.

12.2.3 Amphiphiles and amphipols: detergents for structural biology

Amphiphiles and amphipols represent a new class of detergents and were developed to increase the resolution in structural biology studies in general (not necessarily just to obtain better diffracting crystals). The rationale behind these new detergents is rather simple: the hydrophobic surface of a solubilized membrane protein is surrounded by large numbers of flexible, disordered detergent molecules. The resulting large, dynamic detergent shell exceeds the dimension of the protein and contributes high surface entropy to the detergent–protein complex. Consequently protein–protein interactions that might serve for protein stabilization or even for crystal contacts are reduced. In amphiphiles and amphipols the size of these detergent shells is reduced, but unlike small-chain, small-micelle-forming detergents such as OG (octyl β-D-glucoside) and LDAO, which count as harsh detergents, the chemical nature of amphiphiles allows the micelle size to be reduced at elevated hydrophobicity. This is achieved by their branched-chain nature that maintains the same hydrophobicity as a longer straight-chain detergent while also forming smaller-diameter micelles. The behavior of branched-chain detergents is somewhat analogous to that of mixtures of detergents with small amphiphilic additives that partition at the polar–apolar interface of the micelle. Among the newly developed amphiphiles, glucose- and maltose-neopentyl glycol (GNG, MNG) have yielded multiple membrane proteins with enhanced structural stability that were successfully crystallized [4,5]. The high-resolution structure of a sodium-pumping pyrophosphatase (PDB entry code 4AV6) was solved in the presence of 1% octyl glucose neopentyl glycol (OGNPG) [6].

Unlike amphiphiles, the basic principle of amphipols is their high affinity for the transmembrane surfaces of a membrane protein [7]. Amphipols form very compact and stable complexes with membrane proteins due to their dense distribution of hydrophobic chains. In addition these detergents are very soluble in water. One of the most successful amphipols is A8-35, which has a molecular weight of 9–10 kDa. A8-35 comprises a short polyacrylate chain linked with octylamines and isopropylamines. Amphipols have proven to be very successful for cryo-EM single-particle studies [8]. Recently, A8-35 was used as a stabilizing detergent to determine the molecular architecture of two transient receptor potential (TRP) ion channels the nociceptor TRPA1, and led to the determination of the high-resolution structure of the thermosensitive TRPV1 channel by cryo-EM. Interestingly, A8-35 amphipols and MNG detergents share quite similar stabilizing properties and interaction mechanisms for the transmembrane region of TRP channels [8]. Another interesting application of amphipols is their use in lipidic-cubic phase crystallography. OG detergent used to solubilize and purify bacteriorhodopsin (BR) was exchanged against A8-35 and crystallized in the lipidic mesophase, yielding crystals that diffracted beyond 2 Å [9].

12.2.4 Role of lipids in the purification of membrane transport proteins

Lipids in the cell membrane provide the essential physiological environment for membrane proteins. Lipids are not only indispensable for protein function, but are also involved in the majority of regulatory processes [10] merely by changing their composition in the membrane (e.g., in a stress situation or during a disease process [11]). In general, associations of lipids seem to be necessary for correct folding and intercellular transport of nearly every integral membrane protein. The specific lipid requirements of a membrane protein (e.g., the phospholipid, glycolipid or sterol content) also affect the choice of host cell for heterologous expression because the lipid composition of bacteria, Archaea, yeasts, insects, *Xenopus* oocytes, plant and mammalian cells show huge variation [12]. Membrane proteins should always be considered as protein–lipid complexes, and crystallized membrane proteins show defined lipid–protein contacts, although these lipid–protein interactions can be transitory, making them very difficult to explore. Investigation of the cellular activities of membrane proteins that depend on lipid binding requires sensitive and quantitative tools such as computational biology, cellular lipid mapping, single-molecule imaging and lipidomics. Association of a membrane transporter with a lipid was observed for the betaine transporter BetP from the Gram-positive soil bacterium *Corynebacterium glutamicum*. This bacterium has a unique membrane composition consisting entirely of negatively charged lipids with palmitoyloleoyl (PO) fatty acid chains. When expressed heterologously in *E. coli*, BetP manages to maintain palmitoyloleoylphosphatidylglycerol (POPG) lipids tightly bound to its hydrophobic transmembrane regions during purification and crystallization (Fig. 12.2) [13].

Although detergents, artificial lipid bilayers (e.g., liposomes) or bicelles have been used to provide native-like environments for membrane proteins, *in vitro* studies

406 *Handling membrane proteins for structural studies*

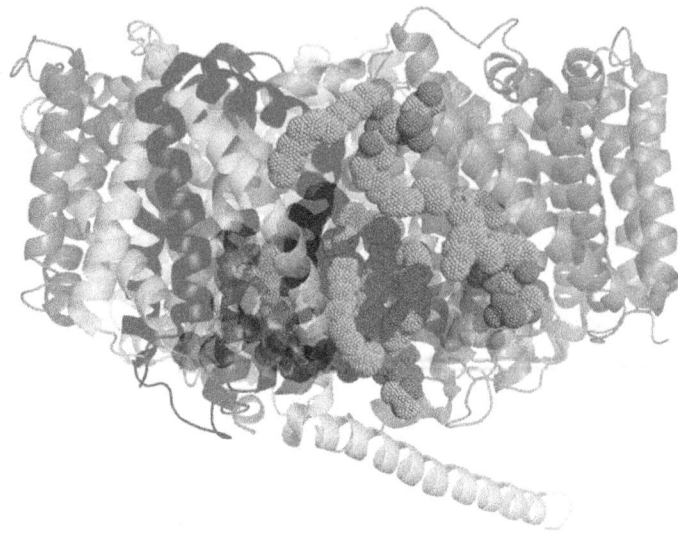

Fig. 12.2 Crystal structure of the trimeric betaine transporter BetP in complex with eight POPG lipids displayed as spheres. (PDB entry code 4C7R [13].)

should preferably be carried out with membrane proteins embedded in a lipid bilayer environment as close to the native one as possible. This means that certain parameters (the nature of the lipids, bilayer thickness and lateral pressure) have to be chosen with great care. A mismatch in the lipid environment surrounding the target membrane protein could result in distortion of membrane protein's oligomeric state and inhibition of interacting partner proteins or ligands.

The role of lipids on a membrane protein increases with the complexity of the organism [14]. Prokaryotic and eukaryotic membranes have highly diverse lipid compositions and the complexity of the lipid classes increases in higher eukaryotes. Despite all these differences, biological membranes show convergence of the surface orders, although this order is achieved by quite different means, for example in eukaryotes by sterol- and sphingolipid-based interactions and in prokaryotes, which lack sterols, by transmembrane protein domain–lipid interactions.

12.2.5 Nanodiscs: preparing transporters for biophysical investigations

Nanodiscs have become a powerful tool for studying membrane proteins by a variety of biophysical methods, for example NMR, electron paramagnetic resonance or mass spectrometry [15]. They are self-assembled phospholipid bilayers surrounded and stabilized by membrane scaffold proteins (MSP). The size of the MSPs determines in turn the size of the nanodiscs; they therefore allow selection for functionally important oligomeric states. A membrane protein reconstituted within a nanodisc gains the stability and functional activity provided by native environments and can be treated like a soluble protein. To date, a number of membrane proteins have been successfully

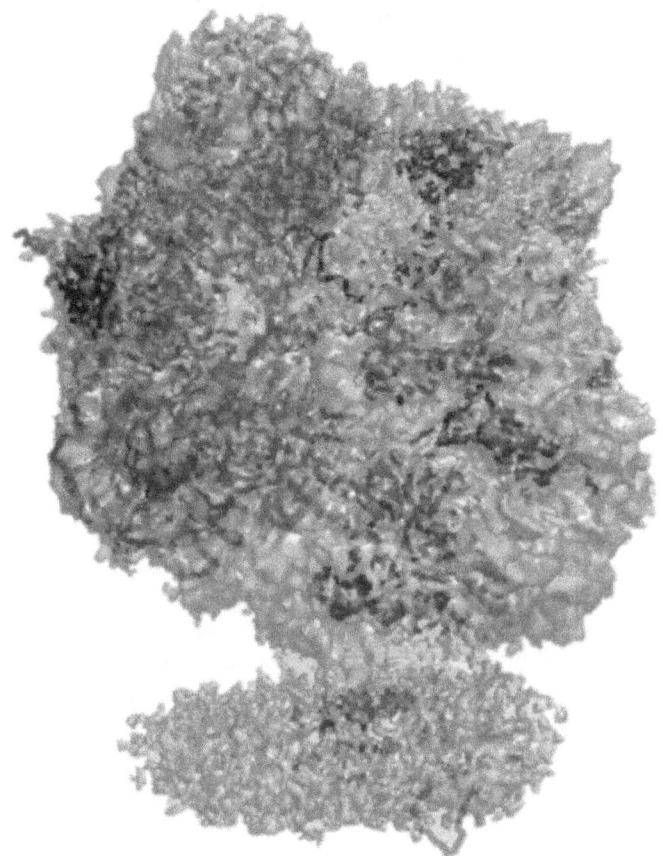

Fig. 12.3 Active *E. coli* SecYEG complex embedded in a lipid bilayer provided by nanodiscs and bound to a translating *E. coli* ribosome. (PDB entry code 4VSM [17].)

introduced into nanodiscs, reflecting a variety of protein types, numbers of the transmembrane helices and oligomeric states. In this way various functional studies, such as protein–protein interaction studies in solution, ligand-binding studies by immobilized surface plasmon resonance, characterization by dynamic light scattering (DLS) and analytical ultracentrifugation (AUC) can be applied to membrane proteins.

In combination with cell-free (CF) expression, nanodiscs could be used for the co-translational insertion of membrane proteins into defined membranes [16]. The bacterial proton pump proteorhodopsin, as well as the small multidrug resistance transporters SugE and EmrE, have been functionally reconstituted into nanodiscs. Efficient reconstitution was dependent on the size and lipid composition of the nanodiscs. Membrane proteins reconstituted into nanodiscs are also particularly suitable for cryo-EM studies. Recently, the structure of the *E. coli* SecYEG complex reconstituted into nanodiscs was solved by cryo-EM to a resolution of 7 Å [17], revealing the lipid bilayer provided by the nanodiscs (Fig. 12.3).

12.3 Determining the structure of a membrane protein: difficult but not impossible

The field of structural biology of membrane proteins has developed considerably since the first structure of a membrane protein was elucidated in 1985. At present, more than 440 unique membrane protein structures have been solved by X-ray crystallography (http://blanco.biomol.uci.edu/mpstruc, Stephen White Lab at UC Irvine) and are available in the Protein Data Bank (PDB). High-throughput technologies have enhanced the rate at which membrane protein structures can be determined. Although X-ray crystallography is still the method with the highest success rate in obtaining an atomic structure, other methods as electron microscopy, atomic force microscopy or native mass spectrometry are attracting increasing attention. The growth of well-diffracting membrane protein crystals is still a major bottleneck, and it is often necessary to apply as many biophysical and structural biology approaches as possible.

12.3.1 Membrane protein crystallization

Membrane protein crystals are classified into 2D crystals and 3D type I or type II crystals [18]. 2D crystals form from membrane proteins that have been reconstituted into an artificial membrane by simultaneous removal of detergent and the addition of lipids. As they contain only one or two bilayers, 2D crystals are too thin for X-ray crystallography studies and have to be investigated by electron microscopy, or sometimes by electron crystallography. Type I crystals are basically stacked 2D crystals in which crystal contacts are formed by protein–detergent–lipid hydrophobic interactions. Crystals grown from bicelles or lipidic cubic phase methods are known to be type I crystals. In type II crystals, proteins interact via their hydrophilic regions, while the hydrophobic regions are shielded by the presence of detergent micelles. This limits the number of possible crystal contacts, resulting in extremely fragile crystals with a large solvent content.

Setting up a crystallization experiment for membrane proteins is not very different from one for soluble proteins. Interestingly, the crystallization space, including buffers at different pHs, salts and precipitants, is narrower than for soluble proteins. For instance, polyethylene glycols with a low molecular weight between 300 and 600 have proved to be more successful than those with a higher molecular weight. A common pattern for successful crystallization of different types of membrane proteins has started to emerge. For instance, ion channels are more successfully crystallized using shorter-chain detergents such as OG, while crystals of transporters and respiratory complexes show a tendency to crystallize using dodecyl maltoside. The use of amphiphiles and mixing different detergents is an elegant way to adapt micelle size for the crystallization of individual membran proteins.

12.3.2 Nanobodies: a versatile tool for transporter crystallization

Single-domain camelid immunoglobulins, called nanobodies, are being increasingly used not only in structural biology but also in biotechnology and pharmaceutical

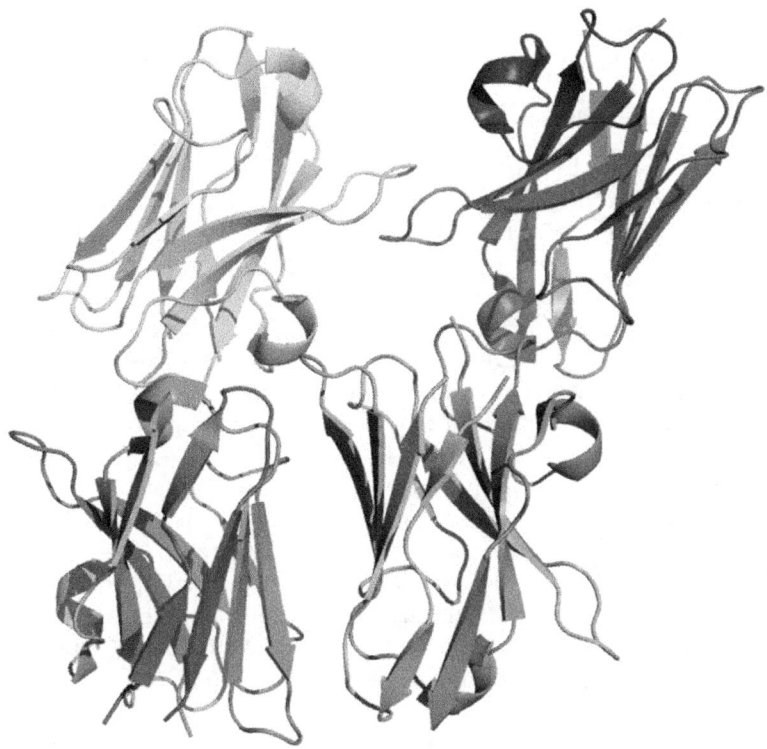

Fig. 12.4 X-ray structure of a llama-derived gelsolin nanobody (PDB entry code 2X1P) that was used as template to determine the homology model for nanobodies against the lactose permease LacY [20].

applications [19]. Due to their small size (they are single-chained; Fig. 12.4) they can access and stabilize hollow and hinge regions. In transporters these sites are often involved in substrate binding and conformational changes [20]. Therefore binding of a nanobody in this hotspot can lock the transporter into a fixed conformation. In addition, the presence of the nanobody might provide additional crystal contacts.

Associations with nanobodies have already helped to crystallize difficult proteins as they prevent domain mobility and stabilize loops by insertion into clefts or between interfaces. Moreover, in some cases they have assisted in the solubilization of proteins. Although G protein-coupled receptors are still the main target of this technique [21,22], the usefulness of nanobodies for transporters has slowly emerged too. One example is the crystallization of the solute carrier SLC11 (NRAMP) that transports transition-metal ions [23]. A complex of the transporter with a nanobody allowed researchers to obtain crystals and solve the structure at 3.1 Å resolution, while diffraction of initial crystals without the nanobody was limited to a resolution of 6.0 Å. Another example is the trapping by nanobodies of a specific transporter conformation in the lactose transporter LacY [20]. Binding of a single-domain camelid nanobody previously developed against a sugar-binding-deficient LacY mutant locked the highly conformational dynamic LacY in the outward-facing conformation.

12.3.3 Cryo-EM: single-particle analysis and electron crystallography

Cryo-EM, which was once limited to large complexes at low resolution, has, thanks to recent advances in electron detection and image processing, become a serious alternative to X-ray crystallography. The new generation of direct electron detectors allows correction for sample movements and classification of images according to different structural states [24]. Near-atomic resolution has been reached for a range of different membrane proteins. One of them is the TRPV1 channel, with a size of 300 kDa in amphipols (see Section 12.2.3). The smallest complex subjected to cryo-EM is γ-secretase, an asymmetric membrane complex of four different proteins having a total molecular weight of 170 kDa. However, for membrane proteins below 100 kDa, single-particle analysis might still not be feasible. In this special case electron crystallography could provide a powerful technique for studying structure and function in a lipid environment. In 2D crystals, the structure of both protein and lipid can be determined and lipid–protein interactions can be analyzed. The lipid bilayer of a 2D crystal closely mimics the native environment, and many important insights have been provided by electron crystallography, for example for the human water channel aquaporin-1 (AQP1). The structures of rat AQP4 and sheep AQP0 have been determined to an amazing resolution of 1.9 Å [25].

12.3.4 Mass spectrometry: a linkage from structure to function

Although mass spectrometry was developed to determine the exact molecular masses of ions, technical and biochemical improvements, particularly in the field of new detergents and nanodiscs, have allowed the method to be used to investigate the architecture and function of membrane proteins. For instance, substrate binding on the lactose transporter LacY was monitored directly in detergent solution by native mass spectrometry [26]. Here, a complex between LacY and its substrate could be ionized and complex dissociation, as the inverse process to substrate binding, could be measured, yielding the binding constants. If native mass spectrometry is to be applied to membrane proteins, labile interactions with detergent molecules must first be lost during the desolvation process by collisions with argon gas, while interactions with lipids, ligands and complex subunits must survive the flight through the mass spectrometer. Membrane proteins can also be ionized directly from lipid bilayer nanodiscs [27]. Mass spectrometry data revealed that lipid binding occurred at fixed stoichiometry and persisted even in detergent-purified complexes. In fact mass spectrometry is a suitable and elegant tool for investigating protein–lipid interactions and folding mechanisms. For LacY it could be demonstrated that specific interactions with phosphatidylethanolamine stabilize folding intermediates similarly as with conventional chaperones [26]. In light of the amazing progress achieved for membrane protein complexes and interactions, mass spectrometry appears to be one of the most promising techniques not only for investigating structure–function relationships but also the molecular physiology of membrane proteins in health and disease.

References

[1] Prince CC, Jia Z. Detergent quantification in membrane protein samples and its application to crystallization experiments. *Amino Acids*, 2013, **45**(6): 1293–302.

[2] Seddon AM, Curnow P, Booth PJ. Membrane proteins, lipids and detergents: not just a soap opera. *Biochim Biophys Acta*, 2004, **1666**(1–2): 105–17.

[3] Yildiz O, Vinothkumar KR, Goswami P, Kühlbrandt W. Structure of the monomeric outer-membrane porin OmpG in the open and closed conformation. *EMBO J*, 2006, **25**(15): 3702–13.

[4] Chae PS, Rasmussen SG, Rana RR, Gotfryd K, Chandra R, Goren MA, Kruse AC, Nurva S, Loland CJ, Pierre Y, Drew D, Popot JL, Picot D, Fox BG, Guan L, Gether U, Byrne B, Kobilka B, Gellman SH Maltose-neopentyl glycol (MNG) amphiphiles for solubilization, stabilization and crystallization of membrane proteins. *Nat Methods*, 2010, **7**(12): 1003–8.

[5] Chae PS, Rana RR, Gotfryd K, Rasmussen SG, Kruse AC, Cho KH, Capaldi S, Carlsson E, Kobilka B, Loland CJ, Gether U, Banerjee S, Byrne B, Lee JK, Gellman SH. Glucose-neopentyl glycol (GNG) amphiphiles for membrane protein study. *Chem Commun*, 2013, **49**(23): 2287–9.

[6] Kellosalo J, Kajander T, Kogan K, Pokharel K, Goldman A. The structure and catalytic cycle of a sodium-pumping pyrophosphatase. *Science*, 2012, **337**(6093): 473–6.

[7] Popot JL. Amphipols, nanodiscs, and fluorinated surfactants: three nonconventional approaches to studying membrane proteins in aqueous solutions. *Annu Rev Biochem*, 2010, **79**: 737–75.

[8] Huynh KW, Cohen MR, Moiseenkova-Bell VY. Application of amphipols for structure–functional analysis of TRP channels. *J Membr Biol*, 2014, **247**(9–10): 843–51.

[9] Polovinkin V, Gushchin I, Sintsov M, Round E, Balandin T, Chervakov P, Schevchenko V, Utrobin P, Popov A, Borshchevskiy V, Mishin A, Kuklin A, Willbold D, Chupin V, Popot JL, Gordeliy V. High-resolution structure of a membrane protein transferred from amphipol to a lipidic mesophase. *J Membr Biol*, 2014, **247**(9–10): 997–1004.

[10] Phillips R, Ursell T, Wiggins P, Sens P. Emerging roles for lipids in shaping membrane-protein function. *Nature*, 2009, **459**(7245): 379–85.

[11] Fabelo N, Martín V, Santpere G, Marín R, Torrent L, Ferrer I, Díaz M. Severe alterations in lipid composition of frontal cortex lipid rafts from Parkinson's disease and incidental Parkinson's disease. *Mol Med*, 2011, **17**(9–10): 1107–18.

[12] Opekarová M, Tanner W. Specific lipid requirements of membrane proteins—a putative bottleneck in heterologous expression. *Biochim Biophys Acta*, 2003, **1610**(1): 11–22.

[13] Koshy C, Schweikhard ES, Gärtner RM, Perez C, Yildiz O, Ziegler C. Structural evidence for functional lipid interactions in the betaine transporter BetP. *EMBO J*, 2013, **32**(23): 3096–105.

[14] Ernst AM, Brügger B. Sphingolipids as modulators of membrane proteins. *Biochim Biophys Acta*, 2014, **1841**(5): 665–70.

[15] Inagaki S, Ghirlando R, Grisshammer R. Biophysical characterization of membrane proteins in nanodiscs. *Methods*, 2013, **59**(3): 287–300.

[16] Roos C, Zocher M, Müller D, Münch D, Schneider T, Sahl HG, Scholz F, Wachtveitl J, Ma Y, Proverbio D, Henrich E, Dötsch V, Bernhard F. Characterization of co-translationally formed nanodisc complexes with small multidrug transporters, proteorhodopsin and with the *E. coli* MraY translocase. *Biochim Biophys Acta*, 2012, **1818**(12): 3098–106.

[17] Frauenfeld J, Gumbart J, Sluis EO, Funes S, Gartmann M, Beatrix B, Mielke T, Berninghausen O, Becker T, Schulten K, Beckmann R. Cryo-EM structure of the ribosome–SecYE complex in the membrane environment. *Nat Struct Mol Biol*, 2011, **18**(5): 614–21.

[18] Moraes I, Evans G, Sanchez-Weatherby J, Newstead S, Stewart PD. Membrane protein structure determination—the next generation. *Biochim Biophys Acta*, 2014, **1838**(1, Pt A): 78–87.

[19] Desmyter A, Spinelli S, Roussel A, Cambillau C. Camelid nanobodies: killing two birds with one stone. *Curr Opin Struct Biol*, 2015, **32C**: 1–8.

[20] Smirnova I, Kasho V, Jiang X, Pardon E, Steyaert J, Kaback HR. Outward-facing conformers of LacY stabilized by nanobodies. *Proc Natl Acad Sci USA*, 2014, **111**(52): 18548–53.

[21] Kruse AC, Ring AM, Manglik A, Hu JX, Hu K, Eitel K, Hubner H, Pardon E, Valant C, Sexton PM, et al. Activation and allosteric modulation of a muscarinic acetylcholine receptor. *Nature*, 2013, **504**: 101–110.

[22] Hassaine G, Deluz C, Grasso L, Wyss R, Tol MB, Hovius R, Graff A, Stahlberg H, Tomizaki T, Desmyter A, et al. X-ray structure of the mouse serotonin 5-HT3 receptor. *Nature*, 2014, **512**: 276–281.

[23] Ehrnstorfer IA, Geertsma ER, Pardon E, Steyaert J, Dutzler R. Crystal structure of a SLC11 (NRAMP) transporter reveals the basis for transition-metal ion transport. *Nat Struct Mol Biol*, 2014, **21**: 990–996.

[24] Bai XC, McMullan G, Scheres SH. How cryo-EM is revolutionizing structural biology. *Trends Biochem Sci*, 2015, **40**(1): 49–57.

[25] Wisedchaisri G, Reichow SL, Gonen T. Advances in structural and functional analysis of membrane proteins by electron crystallography. *Structure*, 2011, **19**(10): 1381–93.

[26] Weinglass AB, Whitelegge JP, Hu Y, Verner GE, Faull KF, Kaback HR. Elucidation of substrate binding interactions in a membrane transport protein by mass spectrometry. *EMBO J*, 2003, **22**: 1467–77.

[27] Hopper JT, Yu YT, Li D, Raymond A, Bostock M, Liko I, Mikhailov V, Laganowsky A, Benesch JL, Caffrey M, Nietlispach D & Robinson CV. Detergent-free mass spectrometry of membrane protein complexes. *Nat Methods*, 2013, **10**: 1206–8.

13
Molecular simulation: a virtual microscope in the toolbox of integrated structural biology

François DEHEZ

Laboratoire International Associé CNRS and University of Illinois at Urbana-Champain, UMR SRSMC N°7565, CNRS-Université de Lorraine, Nancy, France

Abstract

Accessing the full atomistic details of complex biological processes in near physiological conditions remains a challenge at the experimental level. The development of molecular modeling approaches, such as molecular dynamics, together with powerful supercomputers has allowed rationalization and prediction at the molecular level of the dynamics of large biomolecular assemblies in various conditions. Although still limited in terms of system size and time-scales, molecular dynamics (MD) simulations have proved to be a valuable technique for studying integrated structural biology. This chapter presents the standard concepts underlying molecular simulations. First, the basis of molecular mechanics and the concept of force fields are detailed. Next, the MD method and some details of its implementation in the study of biomolecular systems are presented. Finally, the complementarity of experiments and modeling are illustrated using the example of mitochondrial carriers.

Keywords

Molecular modeling, molecular dynamics, force field, coarse-grained, membrane transport, mitochondrial carriers

Chapter Contents

13 Molecular simulation: a virtual microscope in the toolbox of integrated structural biology   **413**
    François DEHEZ

13.1	Introduction	415
13.2	Modeling forces in molecular simulation: the force field	417
	13.2.1 The potential energy function	417
	13.2.2 General strategy for deriving force field parameters	419
	13.2.3 Coarse-grained force fields	420
13.3	Modeling the time evolution of biological systems using molecular dynamics	421
	13.3.1 The classical equations of motion	421
	13.3.2 Integrating the classical equations of motion	422
	13.3.3 Periodic boundary conditions	423
	13.3.4 Working at constant temperature and pressure	424
	13.3.5 Sampling rare events	425
13.4	Integrating molecular simulations with structural biology	426
13.5	Conclusion	432
	References	433

13.1 Introduction

In every scientific field, models are employed to rationalize, generalize and extrapolate experimental observations. A typical example is the standard model in physics aimed at describing interactions of elementary particles. It was refined during the twentieth century from experimental and theoretical findings, and has recently been reinforced by the discovery of the Higgs boson. Despite its sophistication, many physical phenomena, such as the acceleration of the expansion of the universe, cannot be described by the standard theory. In general, models are not universal and only describe part of the "reality"; in order to draw valid conclusions it is thus particularly important to know the limits of their application. This statement definitely holds for the models employed to describe molecular systems.

Chemists, biochemists and physicists have developed a wide range of molecular models for studying simple molecules and more complex biomolecular assemblies and materials [1–4]. In the field of structural biology, simple molecular models are often used to represent three-dimensional structures of proteins and biomolecules. Molecular representations such as wireframes, ball-and-stick models, van der Waals spheres (space filling) and cartoons or ribbons for proteins (see Fig. 13.1) represent a convenient framework for appreciating the molecular complexity of these biological objects.

Such rudimentary models, while useful for visualizing molecular structures, do not allow us to predict the dynamics of complex molecular assemblies in the condensed phase, a situation that encompasses most biological problems. Chemical and biochemical systems are obviously more than just balls and sticks. Quantum mechanics (QM)

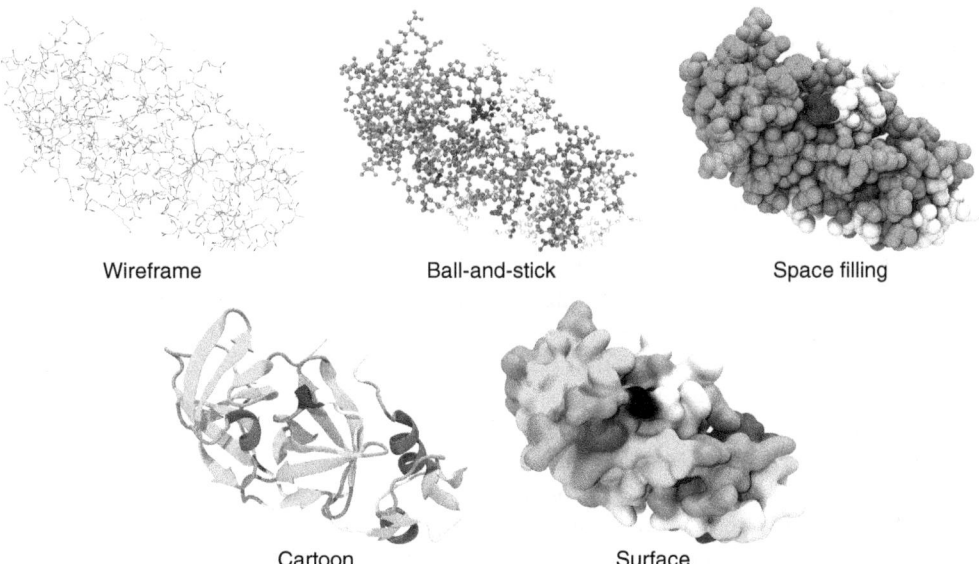

Fig. 13.1 Simple molecular models commonly used to visualize biomolecules.

offers a rigorous framework for describing molecules by representing explicitly their electrons and nuclei. Such particle assemblies are governed by the Schrödinger equation $\hat{H}|\Psi\rangle = E|\Psi\rangle$, where $\hat{H}$ is the Hamiltonian operator associated with the energy of electrons and nuclei, E, the eigenvalue of $\hat{H}$, is the energy of the system and $|\Psi\rangle$ is the wave function associated with the quantum state of the system [5]. QM approaches are commonly used to supplement structural data. They have proved to be particularly powerful for modeling biological processes involving the formation or disruption of chemical bonds or excited states (enzymatic reactions, photosensitization, etc.) and also for refining three-dimensional biological structures obtained by means of NMR or X-ray crystallography [6].

However, *ab initio* quantum calculations involve the computation of many electronic integrals, which, despite progress in computational science, limits their application to modeling of static systems restricted to a few thousand atoms. Many biological systems, such as proteins or nucleic acids, once embedded in their proper environment (water, ions, lipids, etc.) involve tens of thousands to tens of millions of atoms. Moreover, extensive sampling of the dynamics of these systems in conditions akin to *in vitro* or *in vivo* experiments is mandatory to correctly model their functioning. Extending the size and sampling limitations of QM methods is the remit of molecular mechanics (MM). In this class of models, electronic interactions are not computed explicitly but averaged and incorporated into an empirical formulation of the potential energy by means of a proper parameterization. It only takes atomic positions as variables, allowing the fast evaluation of the energy associated with a given molecular configuration. Reducing the complexity of a model is usually concomitant with lowering its precision. Since electrons are not explicitly described, MM approaches can neither model chemical reactions nor transition between ground and excited states.

Beside QM and MM models, a variety of approaches, for example hybrid QM/MM [7], docking [8] and homology [9], complete the molecular modeling arsenal. It should be kept in mind that none of these models is universal, thus the choice of one model over another is essentially dictated by the problem that needs to be solved. As molecular models rely on a number of approximations, it is mandatory to assess their validity prior to making inferences or predictions. It is thus crucial to compare them systematically with any available experimental data (structural, biochemical, thermodynamic, etc.) which may validate or refute the theoretical representation.

Modeling biological systems near physiological or experimental conditions requires the extensive sampling of their configurations at given thermodynamic conditions; hence molecular models need to be coupled to a law of evolution. Ideally, one would need to solve the time-dependent Schrödinger equation, but this method is still limited to a few degrees of freedom. Molecular dynamics (MD) simulation is the most popular approach for modeling the evolution in time of biological systems, relying on integrating the classical Newtonian equations of motion. Coupled with a molecular-mechanical description, MD is a powerful tool for describing the dynamics of biological systems at atomic resolution. It can thus be seen as a super microscope, yet a virtual one, as everything occurs in the computer. Together with the precision of the molecular model, the time and size scales represent stringent limitations on MD simulations, tightly coupled with computer architecture and the velocity of processors (central and graphics processing units). In terms of size, one of the largest systems ever studied using MD simulations is the capsid of the human immunodeficiency virus-1 (HIV-1) [10], a molecular system of about 60 million atoms comprising about 1300 proteins. Combining NMR structure analysis, electron microscopy and MD simulations, structural biologists and theoretical biophysicists have been able to produce an all-atom molecular model of HIV-1. Thanks to the computational resources of one of the most powerful computers in the world, Blue Waters, and the massively parallel MD program NAMD [11], they have been able to study the dynamics of this molecular assembly over 300 ns. The longest simulations reported so far last for 1 or 2 ms and essentially concerned the study of folding pathways of small proteins [12,13] such as BPTI or NTL9, that contain a few tens of thousands atoms. Reaching the microsecond time-scale was enabled by the development of specific computer architectures such as Anton [14] (a computer dedicated only to MD simulations) or Folding@home [15] (a network of hundreds of thousands of personal computers distributed worldwide). These examples aside, the common size of systems studied today by MD simulations is between a few tens of thousands and a few hundred thousand atoms, and using massively parallel supercomputers the time-scales commonly sampled for such systems range from a few hundred nanoseconds to a few microseconds.

In this chapter, we will first detail the basis of MM and the force field concept. Then we will present the MD method and review some implementation details commonly employed to study biomolecular systems. Finally, the complementarity of experiments and modeling will be illustrated using the example of mitochondrial carriers.

13.2 Modeling forces in molecular simulation: the force field

13.2.1 The potential energy function

In molecular mechanics, molecular assemblies are simply represented by means of interacting particles connected through springs. The associated potential energy is a function of all the coordinates of the atoms of the system. It can be written as a sum of N terms, where N is the total number of particles in the system:

$$V(\mathbf{r}^N) = \sum_i v_1(\mathbf{r}_i) + \sum_i \sum_{j>i} v_2(\mathbf{r}_i, \mathbf{r}_j) + \sum_i \sum_{j>i} \sum_{k>j>j} v_2(\mathbf{r}_i, \mathbf{r}_j, \mathbf{r}_k) + \cdots \quad (13.1)$$

where $v_1(\mathbf{r}_i)$ is the intramolecular interactions, $v_2(\mathbf{r}_i)$ the pair interaction potential and $v_N(\mathbf{r}_N)$ the N-body interaction potential. In the context of modeling biological systems, because the interaction energy is essentially governed by the interaction of pairs, the common formulation of the potential energy is usually truncated at v_2, the higher-order terms being partly incorporated in an effective pair potential $v_2^{\text{effective}}$:

$$V(\mathbf{r}^N) \approx \sum_i v_1(\mathbf{r}_i) + \sum_i \sum_{j>i} v_2^{\text{effective}}(\mathbf{r}_i, \mathbf{r}_j). \quad (13.2)$$

Various expressions for the potential energy function have been derived over the last 50 years. Here we will only detail the CHARMM formulation [16], a potential energy function used to describe biomolecular systems. Its contains intermolecular contributions for describing the change in energy associated with the deformation of chemical bonds, valence angles and dihedrals, and terms accounting for interactions between non-bonded atoms (see Fig. 13.2):

$$V(\mathbf{r}) = \sum_{\text{bonds}} k_r(r - r_0)^2$$
$$+ \sum_{\text{angles}} k_\theta(\theta - \theta_0)^2$$
$$+ \sum_{\text{dihedrals}} \sum_n V_n[1 + \cos(n\Phi - \delta)] \quad (13.3)$$
$$+ \sum_{\text{non-bonded}} \varepsilon_{ij} \left[\left(\frac{R_{\min,i,j}}{r_{ij}}\right)^{12} - 2\left(\frac{R_{\min,i,j}}{r_{ij}}\right)^{6} \right]$$
$$+ \frac{1}{4\pi\varepsilon_0\varepsilon} \sum_{\text{non-bonded}} \frac{q_i q_j}{r_{ij}}$$

where k_r and k_θ are, respectively, the force constants of the springs associated with bonds and angles between three bonded atoms and r_θ and θ_0 are their equilibrium values. V_N, n and δ are the energy barrier associated to a torsion around the central bond of four bonded atoms, its periodicity and its phase, respectively. $\varepsilon_{i,j}$ and $R_{\min i,j}$

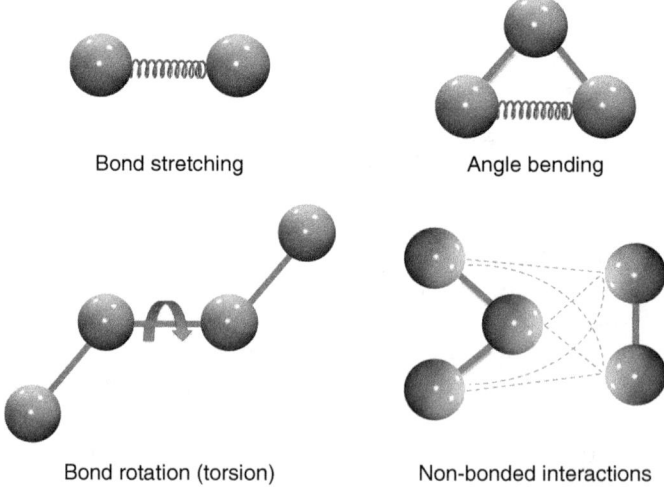

Fig. 13.2 Schematic representations of force field contributions.

are the van der Waals parameters describing repulsive and dispersive interactions between a pair of atoms. They are obtained through the following combination rules:

$$\varepsilon_{ij} = \sqrt{\varepsilon_i \varepsilon_j}$$
$$R_{\min i,j} = \frac{(R_{\min i} + R_{\min j})}{2} \tag{13.4}$$

where ε_0 and ε are the vacuum and the relative permittivity, respectively. q_i is the partial charge carried out by atom i interacting with the partial q_j by means of a Coulomb potential. Non-bonded interactions by definition refer to atoms that are not connected by one, two or three chemical bonds. Interactions atoms connected by exactly three chemical bonds, called interactions "1–4", are often taken into account into the van der Waals and Coulomb contributions. In this case a weighting factor is applied since interactions "1–4" are partially taken into account in the torsion term.

A *force field* is the ensemble of mathematical functions and parameters comprising the potential energy. In theory every molecule possesses its own force field, but deriving parameters for every single molecular system would obviously make the approach inapplicable. In practice, force fields rely on the concept of atom typing. In molecules, atoms sharing the same chemical environment, for example carbon in a methyl group or the oxygen of a carbonyl moiety, are assigned to the same atom type. Parameters are optimized once for each atom type based on a subset of model compounds. These parameters are assumed to be transferrable to molecules that are similar to those employed for the optimization procedure. Following this philosophy, a variety of pairwise additive force fields have been developed, for example CHARMM, AMBER, OPLS, GROMOS, CVFF, MM2 and MM3, targeting different classes of molecular systems (biomolecules, organic compounds, etc.) in the condensed phase.

13.2.2 General strategy for deriving force field parameters

Partitioning of the potential energy function into different contribution relies on a simplified, yet physically sound, representation of the molecular interactions. Intramolecular parameters such as equilibrium values and force constants associated with bonds or angles can be extracted from vibrational spectroscopy and small molecule X-ray crystallography of reference compounds; they can also be estimated through *ab initio* quantum chemical calculations. Energy barriers associated with rotation around dihedral angles are essentially evaluated by QM calculations. The electrostatic interaction is represented by means of fixed partial charges, which are not experimentally observable. This is a rather crude approximation, as additive models do not take polarization phenomena into account. Atomic charges are usually derived from the electrostatic potential of isolated molecules computed quantum mechanically. Polarization effects are incorporated implicitly by inflating the charges artificially. In standard biomolecular force fields, the mean polarization is assumed to be that of the bulk water. The van der Waals parameters are usually determined by molecular simulations. Starting from an initial guess, they are adjusted iteratively in order to reproduce

420 *Molecular simulation*

various experimental quantities of the condensed phase, for example vaporization enthalpies, partial molar volumes, solvation free energies, density, heat capacity and so on The same iterative procedure could finally be applied to further optimize the intramolecular parameters, particularly the torsion barriers. All this results in a potential energy function that is capable of reproducing global thermodynamic properties but its individual components do not necessary have a physical meaning.

Additive biomolecular force fields have been derived for general cases, but they may fail to model systems in which specific effects not incorporated explicitly into the potential energy function are important. For example, catalytic sites of metalloproteins, where polarization effects and charge transfer are important, can hardly be modeled by standard additive force fields. Moreover force fields are essentially valid in conditions akin to those used during the calibration procedure. Force field development is an active research field in which models and parameters are being continuously devised and refined. Important advances include the release of explicitly polarizable force fields [17].

13.2.3 Coarse-grained force fields

Time scales and size scales amenable to all-atom force fields are limited owing to the complexity of the potential energy function. Challenging these limitations requires the model to be simplified. This is exactly what "coarse graining" approaches do, by grouping several atoms into a single interacting site. Various schemes have been proposed in the literature to simplify all-atom representations. Here we will briefly introduce one of the most widely used coarse-grained force field for modeling biomolecular systems, MARTINI, developed by Marrink and co-workers [18]. MARTINI was initially developed to study the dynamics of lipid systems, but has been extended to surfactants, proteins sugars and polymers. It relies on a four-to-one mapping strategy, in which, four heavy atoms are on average gathered into a single interacting bead (see Fig. 13.3). In its latest version, MARTINI relies upon 18 bead types representing different polarities and charged groups. The interaction potential involves an electrostatic and a van der

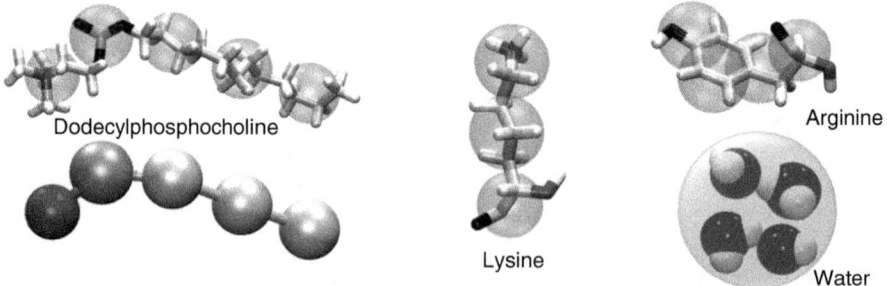

Fig. 13.3 Comparison between all-atom representations and the coarse-grained MARTINI force field.

Waals contribution, for which parameters have been optimized against partitioning free energies between polar and apolar phases of many reference compounds.

Simplifying the description of a system is necessarily concomitant with a loss of information. For instance MARTINI cannot predict the secondary structure of proteins, which has to be imposed. This limits the application of this force field to problems that do not involve conformational changes. Moreover, like many other coarse-grained force fields, parameters are hardly transferable to conditions that are different from those employed for calibration.

13.3 Modeling the time evolution of biological systems using molecular dynamics

13.3.1 The classical equations of motion

The ergodic hypothesis, one of the fundamental hypotheses of statistical mechanics [19], postulates that the mean-time average of an observable A, which can be measured experimentally, is equal to its statistical average:

$$\langle A \rangle_t = \frac{1}{N_{\text{obs}}} \sum_i^{N_{\text{obs}}} A_i(\mathbf{r}^N, \mathbf{p}^N) \qquad (13.5)$$

where A_i, the instantaneous value of A, is evaluated for N_{obs} configurations of the phase space. For a molecular system, a configuration of the phase space is defined by all the atomic positions $\mathbf{r}^N$ and momenta $\mathbf{p}^N$. The goal of MD simulations is thus to generate a trajectory in the phase space that is long enough to ensure the validity of equation (13.5).

In MD simulations, the dynamics of a molecular system is described by the classical equations of motion (Newton's second law):

$$m_i \mathbf{a}_i = m_i \frac{d^2 \mathbf{r}_i(t)}{dt^2} = \mathbf{F}_i(t)$$

$$\mathbf{F}_i(t) = -\frac{\partial V(\mathbf{r}^N)}{\partial \mathbf{r}_i(t)} \qquad (13.6)$$

where m_i, $\mathbf{a}_i$ and $\mathbf{r}_i$ are the mass, the acceleration and the position, respectively, of atom i. $\mathbf{F}_i$ is the force acting on atom i which derives from the potential energy function of the entire system. In practice, equations of motion are integrated numerically. The integration time step, δt, is about 1 femtosecond (fs) to ensure the conservation of the total energy of the system. It is usually taken to be ten times smaller than the period of the fastest molecular movement (vibration of C–H bonds). In the case of a coarse-grained description, vibrations associated with beads (groups of atoms) are much slower thus allowing for larger time steps, up to 30–40 fs. MD simulations are deterministic, providing information about each the position and velocity of each atom, and the configuration of a molecular system can theoretically be predicted at any time in the future or the past. An all-atom MD trajectory as long as 100 ns is the

13.3.2 Integrating the classical equations of motion

Many different numerical schemes have been proposed to integrate the equations of motion. The important properties of these integrators are their numerical stability, their time reversibility, their computational efficiency and finally their capacity to keep the integration errors bounded (symplectic property). The Verlet [20] algorithm is one the simplest versions. Let us write the Taylor expansions of positions $\mathbf{r}_i(t+\delta t)$ and $\mathbf{r}_i(t-\delta t)$ of atom i at time $t+\delta t$ and $t-\delta t$, respectively:

$$\mathbf{r}(t+\delta t) = \mathbf{r}(t) + \delta t \mathbf{v}(t) + \frac{1}{2}\delta t^2 \mathbf{a}(t) + \cdots$$

$$\mathbf{r}(t-\delta t) = \mathbf{r}(t) - \delta t \mathbf{v}(t) + \frac{1}{2}\delta t^2 \mathbf{a}(t) - \cdots$$
(13.7)

Combining these two expressions truncated at the third order gives:

$$\mathbf{r}(t+\delta t) = 2\mathbf{r}(t) - \mathbf{r}(t-\delta t) + \delta t^2 \mathbf{a}(t) + o(\delta t^4).$$
(13.8)

In the Verlet scheme, generating a trajectory does not require any knowledge of atomic velocities. However, their calculation is mandatory for evaluating the kinetic energy associated with the system. Atomic velocities are simply evaluated using the expression:

$$\mathbf{v}(t) = \frac{\mathbf{r}(t+\delta t) - \mathbf{r}(t-\delta t)}{2\delta t}.$$
(13.9)

The *leap-frog* algorithm is another example of a symplectic integrator employed in MD simulations. In this scheme, positions, evaluated at integer times, and velocities, evaluated at integer plus a half time step, "leap over" each other:

$$\mathbf{r}(t+\delta t) = \mathbf{r}(t) + \mathbf{v}\left(t+\frac{\delta t}{2}\right)\delta t$$

$$\mathbf{v}\left(t+\frac{\delta t}{2}\right) = \mathbf{v}\left(t-\frac{\delta t}{2}\right) + \mathbf{a}(t)\delta t$$
(13.10)

Velocities at time t ought to be computed to evaluate the kinetic contribution to the total energy:

$$\mathbf{v}(t) = \frac{\mathbf{v}[t+(\delta t/2)] - \mathbf{v}[t-(\delta t/2)]}{2\delta t}$$
(13.11)

These integration schemes essentially conserve the total energy, E, of the system. If the volume V of the system is kept constant, the resulting trajectory will be representative of the micro-canonical thermodynamics ensemble (N, E, V).

The initiation of a MD simulation requires knowledge of all the positions and velocities of the atoms comprising the system. For proteins, starting geometries are essentially experimental 3D structures obtained either by X-ray crystallography or NMR, which are further embedded in a proper environment (water molecules, ions, lipids, etc.). Initial velocities are chosen randomly from a Maxwell–Boltzmann distribution representative of the target temperature at which the simulation is to be run. All together, the initial configuration is far from being representative of thermodynamic equilibrium. Thus, any MD simulation will start with a thermalization period prior to starting data collection.

13.3.3 Periodic boundary conditions

Relating microscopic to macroscopic properties requires the limitation of edge effects due to the finite size of microscopic samples. A common strategy in MD simulations is to apply periodic boundary conditions (PBC). When using PBC, all the atoms of the system are enclosed in a central box which is replicated periodically in the three directions of space (see Fig. 13.4). Any particle in the replicated box is the image of a particle in the central cell. Boxes are generally parallelepipeds, but they can have other geometrical shapes, for example truncated octahedron or rhombic dodecahedron.

Such a system is pseudo-infinite, and can be used to formulate a series of approximations to handle the computation of non-bonded interactions. First the minimum image convention is applied. Any particle i of the central box primarily interacts with

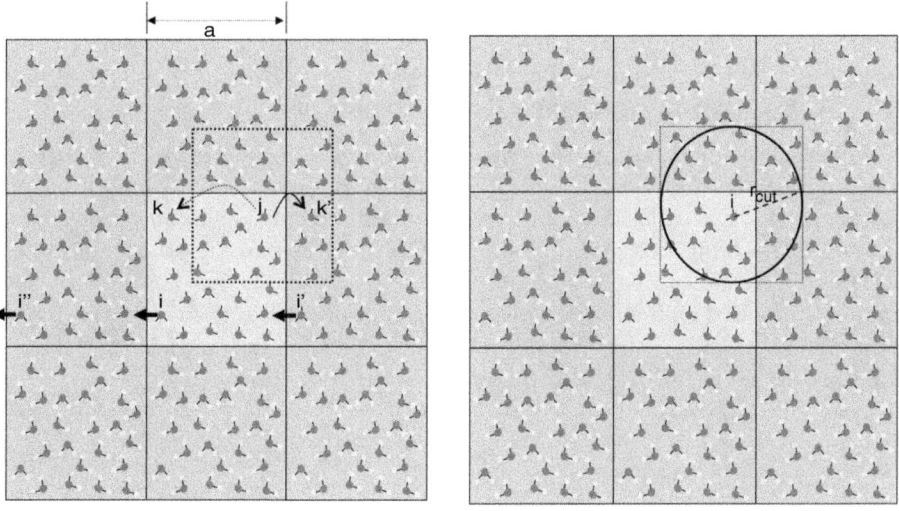

Fig. 13.4 Illustration of the PBC for a two-dimensional system. (Left) Minimum image convention: particle j primarily interacts with k' the closest image of particle k. (Right) Representation of the cut-off sphere, the energy of interaction of a particle i and any other particle located at a distance larger than r_{cut} is set to zero.

the closest image j of all the other particles. The number of non-bonded interactions increases as the square of the number of atoms. This calculation is clearly the most computing-intensive step in evaluating atomic forces. A common strategy employed to reduce the number of non-bonded interactions is to define a sphere of truncation, also called the *cut-off sphere*.

The interaction of a particle i and the nearest image of a particle j is set to zero when their relative distance apart is greater than the radius of the cut-off sphere (r_{cut}). In practice all interacting pairs of atoms are stored in a *pair list* that is regularly updated during the simulation. The value of r_{cut} cannot exceed half of the size of the central box. It should be large enough to minimize errors in the evaluation of non-bonded forces but small enough to keep the computational cost tractable. In MD simulation of biomolecular systems, r_{cut} is commonly taken to be around 12 Å. The cut-off approximation is usually valid for van der Waals interactions, which decay rapidly with interatomic distance. The situation is clearly different for electrostatic interactions which may not be negligible at r_{cut}, especially in the case of charged species. The cut-off approximation will not only underestimate the overall electrostatic interactions but also introduce a singularity in the derivatives of the potential energy function. Switching functions, which smoothly turn off the electrostatic interaction energy near r_{cut}, have been designed to prevent these singularity issues. A more rigorous approach for handling long-range electrostatic interactions in periodic systems was introduced by Ewald. The so-called *Ewald summation* replaces the summation over all pairs of atoms of the pseudo-infinite system, which hardly converges, with a short-range contribution computed in real space and a long-range contribution evaluated in reciprocal Fourier space. In direct space every charge is surrounded by a Gaussian charge distribution having the same amplitude but an opposite sign. The opposite Gaussian distribution compensates for this contribution in Fourier space.

13.3.4 Working at constant temperature and pressure

As explained above, integrating the equations of motion generates a trajectory in the micro-canonical ensemble (N, E, V). Controlling temperature and pressure appears crucial for producing trajectories in thermodynamic conditions akin to physiological or experimental conditions. Various approaches have been derived to generate MD trajectories in either the canonical ensemble (N, V, T) or the isothermal–isobaric ensemble (N, P, T). One simple strategy is to couple the system to a thermal bath by periodically rescaling the atomic velocities using an appropriate factor. In the formulation of Berendsen [21], this factor takes the form:

$$\lambda = \left[1 + \frac{\delta t}{\tau_T}\left(\frac{T_0}{T(t)} - 1\right)\right]^{1/2} \qquad (13.12)$$

where T_0 and $T(t)$ are the target and the instantaneous temperatures, respectively, and τ_T is a relaxation time associated with temperature fluctuations. It has been shown that this approach does not produce an exact canonical distribution. Similarly, coupling the system to an external barostat allows the pressure to be controlled. The target pressure can be simply maintained by periodically rescaling the atomic positions and, hence, the size of the simulation box. Nosé [22] and Hoover [23] have proposed a rigorous framework to ensure canonical control of the temperature and pressure. In their approach, the thermal bath or/and the barostat are introduced as an extra degree of freedom in the classical equations of motion.

13.3.5 Sampling rare events

We have seen that brute-force MD simulations are commonly limited to microsecond time-scales. However, many biologically relevant processes generally occur on longer time-scales such as milliseconds or seconds (see Fig. 13.5) owing to the associated free-energy barriers. Many biasing schemes have been developed to overcome this time limit by allowing efficient sampling of rare events.

Steered-MD (SMD) [24], or force probe MD [25], are techniques in which an external force is applied to pull the system along the chosen degrees of freedom (distance, dihedral, etc.), just as an atomic force microscope (AFM) would do. MD pulling velocities are much larger than values commonly employed in AFM experiments. SMD samples non-equilibrium configurations that, thanks to the Jarzynski equality [26], can be employed to evaluate equilibrium free-energy profiles associated with SMD conformational transitions. SMD has been mainly employed to model mechanical stretching and folding of biomolecules, membrane transport and ligand binding.

Adaptive biasing schemes, such as metadynamics [27] or the adaptive biasing force (ABF) [28,29], are now also widely employed in the investigation of processes involving high free-energy barriers. In the ABF approach, the force acting along collective variables ξ (distance, dihedral, angle, root-mean-square deviation, radius of gyration, eigenvector, etc.) is averaged on-the-fly and continuously subtracted from the system. When convergence is reached, and provided that the slow degrees of freedom of the system are included in the reaction coordinates, the biasing force eliminates the roughness of the free-energy landscape leading the system to freely diffuse

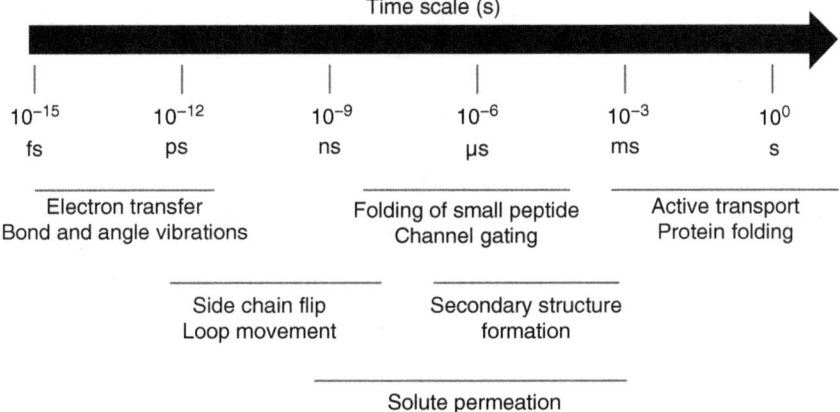

Fig. 13.5 Typical time-scales of biomolecular processes.

along ξ. The free-energy profile associated with the reaction coordinate can be further evaluated as:

$$\nabla_\xi A(\xi) = \left\langle \frac{\partial V(\mathbf{r})}{\partial \xi} \right\rangle_\xi = \langle -F_\xi \rangle_\xi \tag{13.13}$$

where A is the Helmholtz free energy and F_ξ is the average force acting along ξ.

Adaptive schemes have proved successful in the study of a broad range of biological processes such a ligand–protein binding, protein–protein binding, membrane transport and insertion, permeation and large conformational transition in proteins. However, their application remains cumbersome because it requires extensive sampling and also because we often don't know what the appropriate reaction coordinate should be for modeling the transition between two states.

13.4 Integrating molecular simulations with structural biology

In this section we will give a few examples of how to combine experimental and theoretical data in structural biology using the example of mitochondrial carriers.

Proteins from the mitochondrial carrier family (MCF) [30,31] are transmembrane proteins located in the inner mitochondrial membrane. They are involved in the selective transport of a wide range of substrates between the matrix and the cytoplasm. This family suffers from a paucity of structural data. So far only two high-resolution structures have been released for the ADP/ATP carrier (AAC), a protein that imports ADP^{3-} in the matrix and exports ATP^{4-} out of the mitochondrion. AAC is supposed to follow an alternative access mechanism in which the carrier successively navigates between conformations open toward the intermembrane space (c-state) and the matrix side (m-state). The first AAC structure, obtained from bovine heart mitochondria, was published in 2003 [32,33]. The second structure was obtained more than 10 years later, from yeast [34]. In both cases, AAC was extracted with detergents and the crystallization assays were achieved in the presence of a potent inhibitor, carboxyatractyloside (CATR). CATR is known to hamper the import of ADP^{3-}, but is the structure obtained really representative of the c-state? In other words: is the crystal structure capable of binding ADP^{3-} and how will the crystal conformation of the protein behave in physiologically relevant conditions? Starting from the crystal structure pdb:1OKC, many all-atom MD simulations have been carried out in a membrane environment to try and answer these questions [35–39]. Initial setups consist of embedding AAC in a fully hydrated environment (see Fig. 13.6), with or without CATR bound to the structure. Analysis of trajectories spanning the 100-ns time-scale have shown that, irrespective of the presence of CATR, conformations sampled in a model bilayer are comparable with the X-ray structure, indicating that 1OKC is representative of a membrane environment. Moreover, in the absence of CATR, the internal cavity of the carrier is only slightly reshaped, suggesting that CATR, albeit significantly larger than the natural nucleotide substrate, does not alter the structure of the carrier. Equilibrium simulations can also help in rationalizing the role of specific residues, such as the two patches of basic residues lining the carrier cavity of AAC.

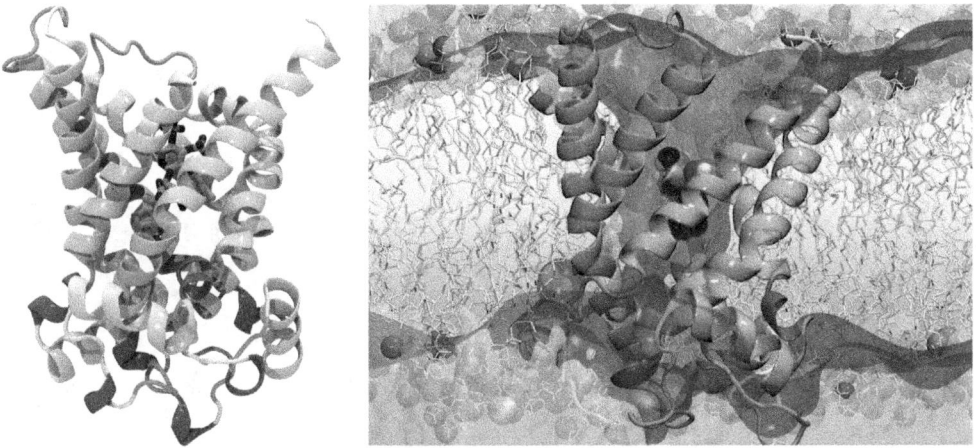

Fig. 13.6 (Left) 1OKC crystal structure in the presence of CATR. (Right) Initial MD setup, 1OKC binding ADP^{3-} is embedded in a fully hydrated POPC bilayer. The system contains about 50,000 atoms. The transparent surface represents an isocontour of the 3D electrostatic potential.

The electrostatic potential can be mapped numerically in the three dimensions of Cartesian space and averaged over configurations selected along the MD trajectory. For the ligand-free AAC, this analysis reveals a privileged electrostatic pathway, shaped like a funnel, capable of driving negatively charged species down toward the carrier cavity (see Fig. 13.6).

MD simulations can be used to investigate the binding of a substrate to a given target. To test the capacity of AAC (1OKC conformation) to recognize and stably bind ADP^{3-}, MD association assays, in which the nucleotide is positioned at the mouth of the carrier, have been repeated for several initial conditions [36]. All the simulations demonstrate that the carrier can spontaneously bind ADP^{3-}. Irrespective of the starting position, the nucleotide is rapidly attracted down the cavity (over a time-scale spanning a few tens of nanoseconds). Furthermore, the nucleotide forms a stable interaction motif with specific residues located at the bottom of the cavity (see Fig. 13.7). This prediction is validated by early biochemical studies that have demonstrated the implications of these residues for the transport activity of the protein. Determination of the mean force potential associated with the binding process ultimately shows that the observed binding motif indeed corresponds to a true minimum of the free-energy surface.

In MD simulations it is often straightforward to modify the simulation conditions. For instance, the ADP^{3-} binding process can be studied at different salt concentrations [40]. At high chloride concentrations the behavior is strikingly different from that observed at low concentrations, namely ADP^{-3} never associates with the bottom of the cavity. Activity experiments carried out in similar salt conditions corroborate this finding, showing that the transport activity of AAC is impaired at high

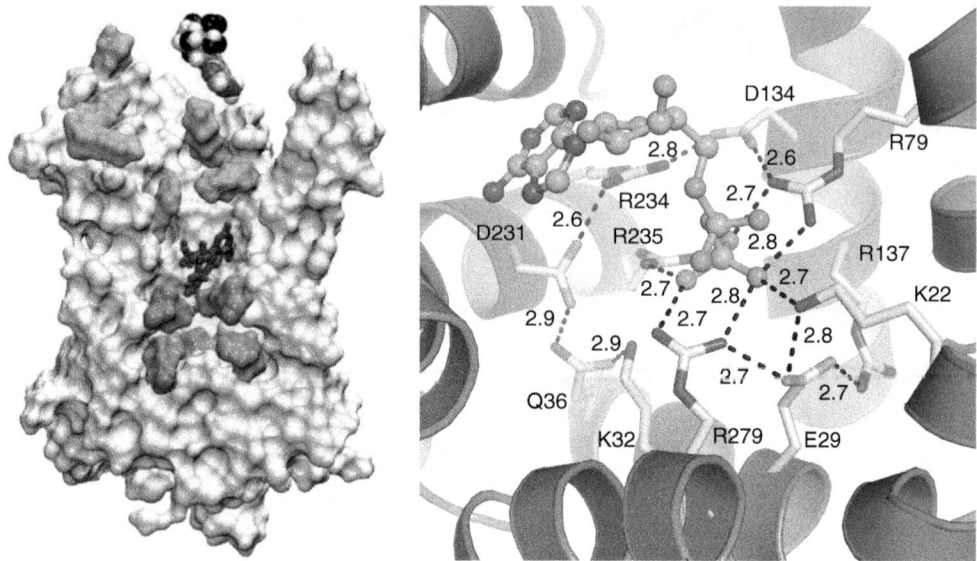

Fig. 13.7 Numerical association assays. (Left) Initial and final nucleotide positions. Dark surfaces represent the basic patches. (Right) Close-up view of the binding motif. Copyright 2013 American Chemical Society.

chloride concentrations. A critical comparison of the different trajectories revealed that inactivation is linked to a set of key residues, highly conserved in the AAC family, interacting transiently with chloride anions at low salt concentrations (0.1 M) and permanently at high chloride concentrations (0.6 M).

MD simulations can also be used to interpret and rationalize experimental findings. Early activity measurements have suggested that AAC function depends on the pH [41]. AAC selectively transports ADP^{3-} and ATP^{4-} but not their protonated forms nor other related nucleotides such as AMP, GMP, GDP or GTP^{39}. While part of the pH dependence can clearly be attributed to the protonation state of ADP^{3-}, it was suggested that some of the protein residues might also be involved. X-ray structures of proteins at standard resolution do not allow the protonation state of titratable residues to be determined. Various theoretical methods have been designed to infer the protonation state of titratable amino acids. These methods do not usually provide absolute estimates *per se*, but rather evaluate the variation of protonation probabilities between residues embedded in proteins and free in the bulk. They mostly rely on either an empirical formulation, just as a force-field would do, calibrated against measured pK_a over a wide range of proteins, or on the integration of the Poisson–Boltzmann equations. These methods are unfortunately not yet fully predictive, but they can nevertheless provide a satisfactory qualitative description, especially when the deviation from the reference bulk state is large. pK_a and protonation state probabilities have been systematically averaged over various configurations selected along the MD trajectories [42] using different approaches, namely PROPKA [43], H++ [44]

and QMBP/GMCT [45]. Results strongly indicate that only one accessible residue, namely K22, has its pK_a shifted down to values between 6 and 8, a range in which the protein activity is maximal (see Fig. 13.8). Association simulations further indicate that K22 is not capable of binding ADP^{3-} when not carrying a proton. Altogether, modeling suggests that K22 is directly involved in the pH dependence of AAC activity, but can we formally assess this prediction at the experimental level? Resorting to NMR is in principle appropriate for accessing the pK_a of amino acids in proteins. While this is true for hydrosoluble proteins, it does not apply to membrane proteins of the size of AAC. MD simulations and pK_a calculations performed for the K22R mutant revealed that the pK_a of arginine is unaffected by the cavity environment. At the experimental level, activity measurements further show that, albeit markedly diminished, nucleotide transport still occurs for this mutant (see Fig. 13.8). The topology of the pH dependence profile differs markedly from that reported for the wild-type AAC. In the case of K22R, increasing the pH from physiological to basic conditions did not alter transport—the activity rather plateaued around a maximum value (see Fig. 13.8) demonstrating the direct implication of K22 in the pH-sensing ability of AAC.

In silico mutagenesis can be used to rationalize the effects of point mutations on properties of a given biological target. Single mutations in the human AAC gene have been associated with various pathologies (AAC1 deficiency, Senger's syndrome and autosomal dominant progressive external ophthalmoplegia). Activity measurements, carried out for A123D, A90D, L98P, D104G, A114P and V289M, have unambiguously shown that the associated pathologies are the consequence of a dysfunction of the transport activity of the protein. MD simulations, which have been carried out for all the mutants, allow the gap between pathologies and their molecular origins

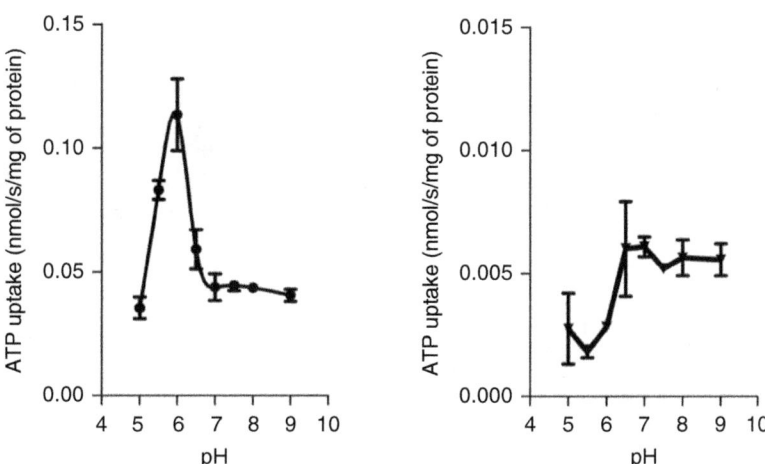

Fig. 13.8 (Left) pH dependence of [α-32P]-ATP uptake into *Escherichia coli* cells expressing hAAC1. (Right) pH dependence of [α-32P]-ATP uptake into *E. coli* cells expressing the K22R mutant of hAAC1. Copyright 2013 American Chemical Society.

to be bridged [43]. The 3D electrostatic potential, mapped for every mutant, shows that A123D leads to a topological modification of the electrostatic funnel preventing the binding of the nucleotide (see Fig. 13.9), a situation similar to K22R in basic conditions. B-factors were further computed for representative configurations selected along MD trajectories. Distributions of the differences between wild-type and mutant B-factors show a loss of flexibility upon mutation conducive to a severe impairment of AAC activity (A90D, L98P and A114P) (see Fig 13.9). The MD studies reveal two-distinct mechanisms responsible for AAC-related genetic disorders—the mutations either modulate the binding affinity of the nucleotides to the carrier or reduce the carrier's conformational plasticity.

Chou and colleagues recently published the structure of another MCF member, the uncoupling protein 2 (UCP2) [46]. It was obtained by combining a few low-resolution NMR data collected in dodecylphosphocholine (DPC) micelles, namely residual dipolar couplings (RDC) and paramagnetic relaxation enhancement (PRE) distances, with a molecular fragment replacement approach. The conclusion reached by the authors is that UCP2, in the presence of its natural inhibitor GDP^{3-}, has a structure that essentially resembles that of AAC in the presence of CATR. MD simulations of UCP2 in a POPC bilayer indeed demonstrate that the situation is clearly different compared with that of AAC [47]. First, a simple equilibrium MD simulation, with the protein backbone restrained to its NMR position, shows that contrary to AAC, the UCP2 structure is fully open, connecting the two sides of the membrane (see Fig. 13.10). Water can easily flow through the hollow cavity, the topology of which is comparable to that of porin. This situation is clearly not compatible with the alternate access model or with maintaining a proton gradient across the inner membrane of mitochondria, calling into question the physiological relevance of the proposed structure. Furthermore, when backbone restraints are released, UCP2 quickly moves away

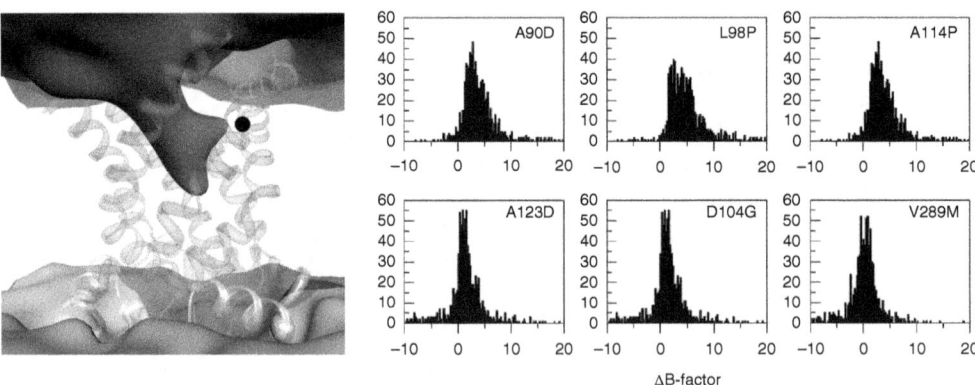

Fig. 13.9 (Left) Electrostatic funnel modified by A23D mutation. (Right) ΔB-factor $= \Delta B_{WT} - \Delta B_{mutant}$ ($Å^2$) inferred from the atomic positional fluctuations computed for the backbone atoms. A positive difference reflects a lower plasticity of the mutant with respect to the wild type. Copyright 2012 American Chemical Society.

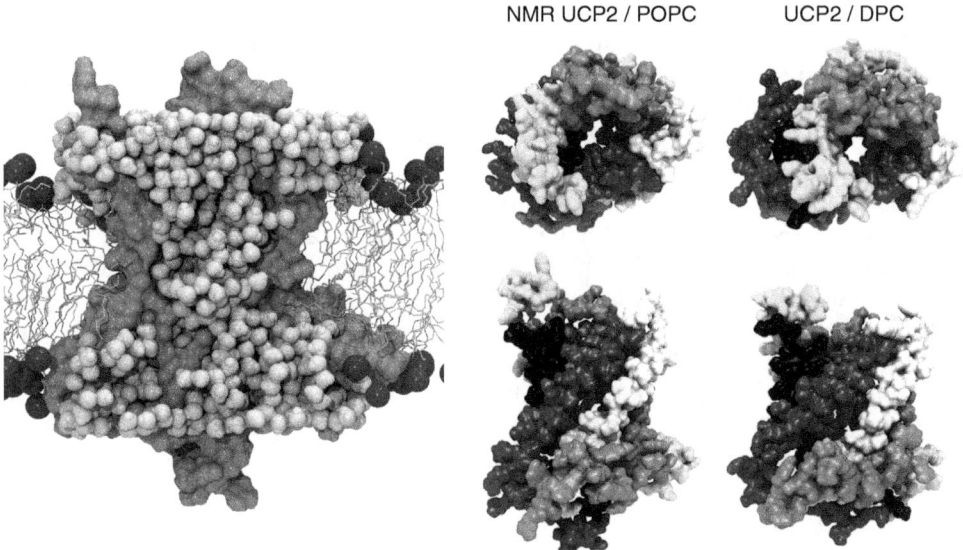

Fig. 13.10 (Left) UCP2 NMR structure (cross-section) fully opened on both sides of the bilayer, allowing for continuous water flow as revealed by MD simulations. (Right) Collapse of the UCP2 NMR structure as monitored by MD simulations in a fully hydrated POPC bilayer. Copyright 2013 American Chemical Society.

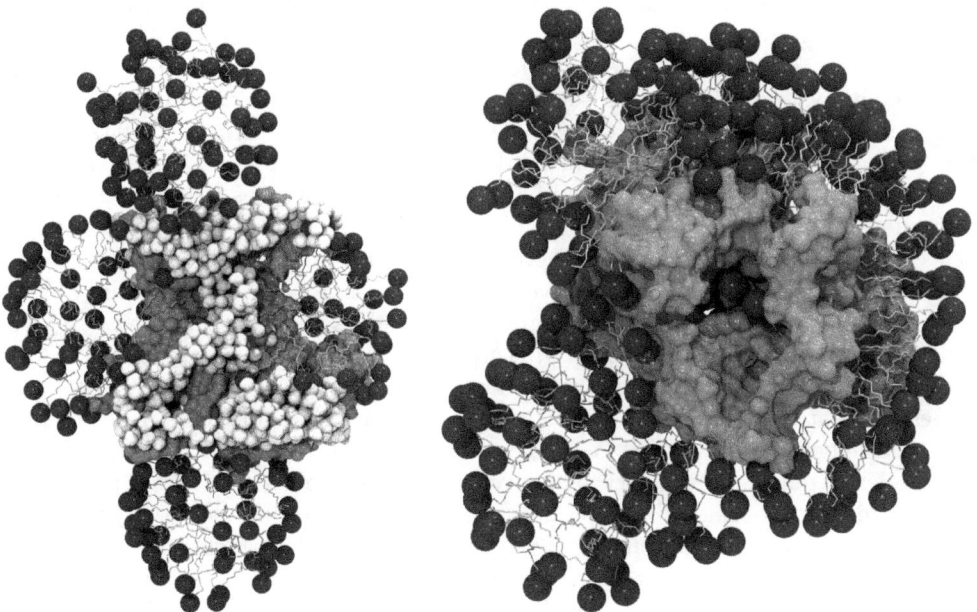

Fig. 13.11 Side and top views of a representative conformation obtained by MD simulation of UCP2 in DPC. Copyright 2013 American Chemical Society.

from its initial reference structure (see Fig. 13.10) as suggested by the evolution of quantities such as RMSD, the radius of gyration or the solvent-accessible surface area. MD thus strongly suggests that the NMR UCP2 structure is unlikely to correspond to a membrane-bound state.

UCP2 was further modeled in a solution of DPC. The self-association of detergents was studied by extensive all-atom and coarse-grained simulations in order to characterize the organization of DPC micelles around the protein (see Fig 13.11). In such an environment, unrestrained MD simulations show that small detergents such as DPC stabilize the overall folding of UCP2 in a conformation comparable to the NMR reference. MD simulations demonstrate that the structure obtained by Chou and colleagues is clearly representative of a state partly denatured by DPC. Activity measurements show that the protein is indeed inactive in this detergent, in line with the theoretical predictions. This illustrates that molecular simulations can not only be employed to rationalize and predict the properties of proteins near their equilibrium conformation, but can also address the physiological relevance of structures obtained in non-native environments.

13.5 Conclusion

Since the first MD simulation reported by Alder and Wainwright in 1957 [48], molecular modeling techniques have been continually improved and now allow for the modeling of large biological systems in relevant thermodynamic conditions. The development of this field was recently acknowledged by the 2013 Nobel Prize in chemistry awarded to Karplus, Levitt and Warshel for "the development of multiscale models for complex chemical systems".

In this chapter we have presented the fundamental concepts of MD as applied to biomolecular systems. Through a set of examples we have illustrated only a fraction of the current possibilities offered by MD in supplementing structural data. Apart from the accuracy of the force fields, MD simulations are essentially restricted by the size of the system and the sampling time. These limitations are tightly coupled to the power of computers, which, according to Moore's law, doubles every 2 years. If this assumption is true, it has been predicted that running a MD simulation on the nanosecond time-scale for an entire bacterium such as *E. coli* will be achievable in about 20 years, whereas modeling a complete mammalian cell at the atomic level will require a further 20 years [49]. Clearly, there is still a long way to go until molecular modeling of the dynamics of organisms on an appropriate time-scale is possible.

Beside all-atom simulations, a broad range of mesoscale approaches have emerged to model the behavior of entire biological systems (e.g., cells [50] and tissues [51]). Cellular models attempt to capture the life cycle of part or the whole of a cell. In a recent study, Covert and colleagues have proposed the first whole-cell model for *Mycloplasma gentitalium*, a human pathogen [52]. Their model integrates the total functionality of the cell (folding, translation, transcription, repair, metabolism), each function being represented by a mathematical model parameterized over experimental data. Using appropriate laws of evolution, the authors were able to reproduce the life cycle of this organism up to cellular division, allowing for the prediction of new biological processes.

References

1. Leach, A. R. *Molecular Modelling: Principles and Applications*, Pearson Education, 2001.
2. Frenkel, D., Smit, B. *Understanding Molecular Simulation: From Algorithms to Applications*, Academic Press, 2001.
3. Höltje, H.-D., Folkers, G. *Molecular Modeling: Basic Principles and Applications*, John Wiley and Sons, 2008.
4. Jensen, J. H. *Molecular Modeling Basics*, Taylor and Francis Group, 2010.
5. Szabo, A., Ostlund, N. S. *Modern Quantum Chemistry: Introduction to Advanced Electronic Structure Theory*, Courier Dover Publications, 2012.
6. Ryde, U., Nilsson, K. Quantum chemistry can locally improve protein crystal structures. *J. Am. Chem. Soc.* 2003, **125**, 14232–14233.
7. Merz, K. M. Using quantum mechanical approaches to study biological systems. *Acc. Chem. Res.* 2014, **47**, 2804–2811.
8. Yuriev, E., Agostino, M., Ramsland, P. A. Challenges and advances in computational docking: 2009 in review. *J. Mol. Recognit.* 2011, **24**, 149–164.
9. Cavasotto, C. N., Phatak, S. S. Homology modeling in drug discovery: current trends and applications. *Drug Discov. Today* 2009, **14**, 676–683.
10. Zhao, G., Perilla, J. R., Yufenyuy, E. L., Meng, X., Chen, B., Ning, J., Ahn, J., Gronenborn, A. M., Schulten, K., Aiken, C., Zhang, P. Mature HIV-1 capsid structure by cryo-electron microscopy and all-atom molecular dynamics. *Nature* 2013, **497**, 643–646.
11. Phillips, J. C., Braun, R., Wang, W., Gumbart, J., Tajkhorshid, E., Villa, E., Chipot, C., Skeel, R. D., Kalé, L., Schulten, K. Scalable molecular dynamics with NAMD. *J. Comput. Chem.* 2005, **26**, 1781–1802.
12. Shaw, D. E. 166 millisecond-long molecular dynamics simulations of proteins on a special-purpose machine. *J. Biomol. Struct. Dyn.* 2013, **31**, 108.
13. Lane, T. J., Shukla, D., Beauchamp, K. A., Pande, V. S. To milliseconds and beyond: challenges in the simulation of protein folding. *Curr. Opin. Struct. Biol.* 2013, **23**, 58–65.
14. Shaw, D. E., Deneroff, M. M., Dror, R. O., Kuskin, J. S., Larson, R. H., Salmon, J. K., Young, C., Batson, B., Bowers, K. J., Chao, J. C., Eastwood, M. P., Gagliardo, J., Grossman, J. P., Ho, C. R., Ierardi, D. J., Kolossváry, I., Klepeis, J. L., Layman, T., McLeavey, C., Moraes, M. A., Mueller, R., Priest, E. C., Shan, Y., Spengler, J., Theobald, M., Towles, B., Wang, S. C. Anton, a special-purpose machine for molecular dynamics simulation. In *Proceedings of the 34th Annual International Symposium on Computer Architecture, ISCA '07*, ACM, 2007, pp. 1–12.
15. Beberg, A. L., Ensign, D. L., Jayachandran, G., Khaliq, S., Pande, V. S. Folding@home: lessons from eight years of volunteer distributed computing. In *IEEE International Symposium on Parallel Distributed Processing, 2009*, IEEE, 2009, pp. 1–8.
16. MacKerell, Bashford, D., Bellott, Dunbrack, Evanseck, J. D., Field, M. J., Fischer, S., Gao, J., Guo, H., Ha, S., Joseph-McCarthy, D., Kuchnir, L., Kuczera, K., Lau,

F. T. K., Mattos, C., Michnick, S., Ngo, T., Nguyen, D. T., Prodhom, B., Reiher, W. E., Roux, B., Schlenkrich, M., Smith, J. C., Stote, R., Straub, J., Watanabe, M., Wiórkiewicz-Kuczera, J., Yin, D., Karplus, M. All-atom empirical potential for molecular modeling and dynamics studies of proteins. *J. Phys. Chem. B* 1998, **102**, 3586–3616.
17. Warshel, A., Kato, M., Pisliakov, A. V. Polarizable force fields: history, test cases, and prospects. *J. Chem. Theory Comput.* 2007, **3**, 2034–2045.
18. Marrink, S. J., Risselada, H. J., Yefimov, S., Tieleman, D. P., de Vries, A. H. The MARTINI force field: coarse grained model for biomolecular simulations. *J. Phys. Chem. B* 2007, **111**, 7812–7824.
19. Chandler, D. *Introduction to Modern Statistical Mechanics.* Oxford University Press, 1987.
20. Verlet, L. Computer "experiments" on classical fluids. I. Thermodynamical properties of Lennard-Jones molecules. *Phys. Rev.* 1967, **159**, 98–103.
21. Berendsen, H. J. C., Postma, J. P. M., Gunsteren, W. F. van, DiNola, A., Haak, J. R. Molecular dynamics with coupling to an external bath. *J. Chem. Phys.* 1984, **81**, 3684–3690.
22. Nosé, S. A molecular dynamics method for simulations in the canonical ensemble. *Mol. Phys.* 1984, **52**, 255–268.
23. Hoover, W. G. Canonical dynamics: equilibrium phase-space distributions. *Phys. Rev. A* 1985, **31**, 1695–1697.
24. Isralewitz, B., Gao, M., Schulten, K. Steered molecular dynamics and mechanical functions of proteins. *Curr. Opin. Struct. Biol.* 2001, **11**, 224–230.
25. Grubmüller, H. Force probe molecular dynamics simulations. In *Protein–Ligand Interactions*, Nienhaus, G. U., ed., Humana Press, 2005, pp. 493–515.
26. Jarzynski, C. Nonequilibrium equality for free energy differences. *Phys. Rev. Lett.* 1997, **78**, 2690–2693.
27. Laio, A., Parrinello, M. Escaping free-energy minima. *Proc. Natl. Acad. Sci. USA* 2002, **99**, 12562–12566.
28. Darve, E., Pohorille, A. Calculating free energies using average force. *J. Chem. Phys.* 2001, **115**, 9169.
29. Hénin, J., Chipot, C. Overcoming free energy barriers using unconstrained molecular dynamics simulations. *J. Chem. Phys.* 2004, **121**, 2904.
30. Walker, J. E., Runswick, M. J. The mitochondrial transport protein superfamily. *J. Bioenerg. Biomembr.* 1993, **25**, 435–446.
31. Palmieri, F. Mitochondrial carrier proteins. *FEBS Lett.* 1994, **346**, 48–54.
32. Pebay-Peyroula, E., Dahout-Gonzalez, C., Kahn, R., Trézéguet, V., Lauquin, G. J.-M., Brandolin, G. Structure of mitochondrial ADP/ATP carrier in complex with carboxyatractyloside. *Nature* 2003, **426**, 39–44.
33. Nury, H., Dahout-Gonzalez, C., Trézéguet, V., Lauquin, G., Brandolin, G., Pebay-Peyroula, E. Structural basis for lipid-mediated interactions between mitochondrial ADP/ATP carrier monomers. *FEBS Lett.* 2005, **579**, 6031–6036.
34. Ruprecht, J. J., Hellawell, A. M., Harding, M., Crichton, P. G., McCoy, A. J., Kunji, E. R. S. Structures of yeast mitochondrial ADP/ATP carriers support a

domain-based alternating-access transport mechanism. *Proc. Natl. Acad. Sci. USA* 2014, **111**, E426–E434.
35. Falconi, M., Chillemi, G., Di Marino, D., D'Annessa, I., Morozzo della Rocca, B., Palmieri, L., Desideri, A. Structural dynamics of the mitochondrial ADP/ATP carrier revealed by molecular dynamics simulation studies. *Proteins Struct. Funct. Bioinform.* 2006, **65**, 681–691.
36. Dehez, F., Pebay-Peyroula, E., Chipot, C. Binding of ADP in the mitochondrial ADP/ATP carrier is driven by an electrostatic funnel. *J. Am. Chem. Soc.* 2008, **130**, 12725–12733.
37. Wang, Y., Tajkhorshid, E. Electrostatic funneling of substrate in mitochondrial inner membrane carriers. *Proc. Natl. Acad. Sci. USA* 2008, **105**, 9598–9603.
38. Johnston, J. M., Khalid, S., Sansom, M. S. P. Conformational dynamics of the mitochondrial ADP/ATP carrier: a simulation study. *Mol. Membr. Biol.* 2008, **25**, 506–517.
39. Mifsud, J., Ravaud, S., Krammer, E.-M., Chipot, C., Kunji, E. R. S., Pebay-Peyroula, E., Dehez, F. The substrate specificity of the human ADP/ATP carrier AAC1. *Mol. Membr. Biol.* 2013, **30**, 160–168.
40. Krammer, E.-M., Ravaud, S., Dehez, F., Frelet-Barrand, A., Pebay-Peyroula, E., Chipot, C. High-chloride concentrations abolish the binding of adenine nucleotides in the mitochondrial ADP/ATP carrier family. *Biophys. J.* 2009, **97**, L25–L27.
41. Broustovetsky, N., Bamberg, E., Gropp, T., Klingenberg, M. Biochemical and physical parameters of the electrical currents measured with the ADP/ATP carrier by photolysis of caged ADP and ATP. *Biochemistry (Moscow)* 1997, **36**, 13865–13872.
42. Bidon-Chanal, A., Krammer, E.-M., Blot, D., Pebay-Peyroula, E., Chipot, C., Ravaud, S., Dehez, F. How do membrane transporters sense pH? The case of the mitochondrial ADP–ATP carrier. *J. Phys. Chem. Lett.* 2013, **4**, 3787–3791.
43. Ravaud, S., Bidon-Chanal, A., Blesneac, I., Machillot, P., Juillan-Binard, C., Dehez, F., Chipot, C., Pebay-Peyroula, E. Impaired transport of nucleotides in a mitochondrial carrier explains severe human genetic diseases. *ACS Chem. Biol.* 2012, **7**, 1164–1169.
44. Gordon, J. C., Myers, J. B., Folta, T., Shoja, V., Heath, L. S., Onufriev, A. H. A server for estimating pK_as and adding missing hydrogens to macromolecules. *Nucleic Acids Res.* 2005, **33**, W368–W371.
45. Ullmann, R. T., Ullmann, G. M. GMCT: A Monte Carlo simulation package for macromolecular receptors. *J. Comput. Chem.* 2012, **33**, 887–900.
46. Berardi, M. J., Shih, W. M., Harrison, S. C., Chou, J. J. Mitochondrial uncoupling protein 2 structure determined by NMR molecular fragment searching. *Nature* 2011, **476**, 109–113.
47. Zoonens, M., Comer, J., Masscheleyn, S., Pebay-Peyroula, E., Chipot, C., Miroux, B., Dehez, F. Dangerous liaisons between detergents and membrane proteins. The case of mitochondrial uncoupling protein 2. *J. Am. Chem. Soc.* 2013, **135**, 15174–15182.
48. Alder, B. J., Wainwright, T. E. Phase transition for a hard sphere system. *J. Chem. Phys.* 1957, **27**, 1208–1209.

49. Van Gunsteren, W. F., Bakowies, D., Baron, R., Chandrasekhar, I., Christen, M., Daura, X., Gee, P., Geerke, D. P., Glättli, A., Hünenberger, P. H., Kastenholz, M. A., Oostenbrink, C., Schenk, M., Trzesniak, D., van der Vegt, N. F. A., Yu, H. B. Biomolecular modeling: goals, problems, perspectives. *Angew. Chem. Int. Ed*. 2006, **45**, 4064–4092.
50. Tomita, M., Hashimoto, K., Takahashi, K., Shimizu, T. S., Matsuzaki, Y., Miyoshi, F., Saito, K., Tanida, S., Yugi, K., Venter, J. C., Hutchison, C. A. E-CELL: software environment for whole-cell simulation. *Bioinformatics* 1999, **15**, 72–84.
51. Clayton, R. H., Bernus, O., Cherry, E. M., Dierckx, H., Fenton, F. H., Mirabella, L., Panfilov, A. V., Sachse, F. B., Seemann, G., Zhang, H. Models of cardiac tissue electrophysiology: progress, challenges and open questions. *Prog. Biophys. Mol. Biol*. 2011, **104**, 22 48.
52. Karr, J. R., Sanghvi, J. C., Macklin, D. N., Gutschow, M. V., Jacobs, J. M., Bolival, B., Assad-Garcia, N., Glass, J. I., Covert, M. W. A whole-cell computational model predicts phenotype from genotype. *Cell* 2012, **150**, 389–401.

The manufacturer's authorised representative in the EU for product safety is Oxford University Press España S.A. of el Parque Empresarial San Fernando de Henares, Avenida de Castilla, 2 – 28830 Madrid (www.oup.es/en or product. safety@oup.com). OUP España S.A. also acts as importer into Spain of products made by the manufacturer.

www.ingramcontent.com/pod-product-compliance
Lightning Source LLC
LaVergne TN
LVHW081516060526
838200LV00005B/193